Nonlinear Numerical Methods and Rational Approximation

Mathematics and Its Applications

Nonlinear Numerical Methods and Rational Approximation

edited by

Annie Cuyt

Universitaire Instelling Antwerpen,
Universiteit Antwerpen, Belgium

D. Reidel Publishing Company

A MEMBER OF THE KLUWER ACADEMIC PUBLISHERS GROUP

Dordrecht / Boston / Lancaster / Tokyo

Library of Congress Cataloging in Publication Data

CIP

Nonlinear numerical methods and rational approximation / edited by Annie Cuyt.
p. cm. – (Mathematics and its applications)
Includes index.
ISBN 90-277-2669-8
1. Numerical analysis—Congresses. 2. Approximation theory—Congresses.
I. Cuyt, Annie, 1956– . II. Series. III. Series: Mathematics and its applications (D. Reidel Publishing Company)
QA297.N64 1987
519.4—dc 19 87-35660
CIP

Published by D. Reidel Publishing Company,
P.O. Box 17, 3300 AA Dordrecht, Holland.

Sold and distributed in the U.S.A. and Canada
by Kluwer Academic Publishers,
101 Philip Drive, Norwell, MA 02061, U.S.A.

In all other countries, sold and distributed
by Kluwer Academic Publishers Group,
P.O. Box 322, 3300 AH Dordrecht, Holland.

Printed in The Netherlands

SERIES EDITOR'S PREFACE

Approach your problems from the right end and begin with the answers. Then one day, perhaps you will find the final question.

'The Hermit Clad in Crane Feathers' in R. van Gulik's *The Chinese Maze Murders.*

It isn't that they can't see the solution. It is that they can't see the problem.

G.K. Chesterton. *The Scandal of Father Brown* 'The point of a Pin'.

Growing specialization and diversification have brought a host of monographs and textbooks on increasingly specialized topics. However, the "tree" of knowledge of mathematics and related fields does not grow only by putting forth new branches. It also happens, quite often in fact, that branches which were thought to be completely disparate are suddenly seen to be related.

Further, the kind and level of sophistication of mathematics applied in various sciences has changed drastically in recent years: measure theory is used (non-trivially) in regional and theoretical economics; algebraic geometry interacts with physics; the Minkowsky lemma, coding theory and the structure of water meet one another in packing and covering theory; quantum fields, crystal defects and mathematical programming profit from homotopy theory; Lie algebras are relevant to filtering; and prediction and electrical engineering can use Stein spaces. And in addition to this there are such new emerging subdisciplines as "experimental mathematics", "CFD", "completely integrable systems", "chaos, synergetics and large-scale order", which are almost impossible to fit into the existing classification schemes. They draw upon widely different sections of mathematics. This programme, Mathematics and Its Applications, is devoted to new emerging (sub)disciplines and to such (new) interrelations as exempla gratia:

- a central concept which plays an important role in several different mathematical and/or scientific specialized areas;
- new applications of the results and ideas from one area of scientific endeavour into another;
- influences which the results, problems and concepts of one field of enquiry have and have had on the development of another.

The Mathematics and Its Applications programme tries to make available a careful selection of books which fit the philosophy outlined above. With such books, which are stimulating rather than definitive, intriguing rather than encyclopaedic, we hope to contribute something towards better communication among the practitioners in diversified fields.

Rational or Padé approximation, the subject of this volume, is still something of a mystery to this editor. Not the basic idea itself, which is lucid enough. But why is the technique so enormously efficient, and numerically useful, in so many fields ranging from physics to electrical engineering with continued fractions, orthogonal polynomials, and completely integrable systems tossed in for good measure.

Anyway, that it is, as a topic, slowly, beginning to be appreciated, is shown, as the editor notes, by the (exponentially) increasing number of papers and conferences on the topic. Reason enough to take stock and present something of a (partial) survey of the current state-of-the-art. That is what the current volume does.

The unreasonable effectiveness of mathematics in science ...

Eugene Wigner

Well, if you know of a better 'ole, go to it.

Bruce Bairnsfather

What is now proved was once only imagined.

William Blake

As long as algebra and geometry proceeded along separate paths, their advance was slow and their applications limited.

But when these sciences joined company they drew from each other fresh vitality and thenceforward marched on at a rapid pace towards perfection.

Joseph Louis Lagrange.

Bussum, December 1987

Michiel Hazewinkel

TABLE OF CONTENTS

EDITOR'S PREFACE

These are the proceedings of a conference on "Nonlinear numerical methods and Rational approximation" organised by Annie Cuyt and Luc Wuytack at the University of Antwerp (Belgium), 20–24 April 1987. The conference focused on the use of rational functions in different fields of Numerical Analysis with sections on Padé approximation and rational interpolation, rational approximation, multidimensional and multivariate problems, orthogonal polynomials and the moment problem, continued fractions, convergence acceleration and their applications. The conference took place at the new campus of UIA, one of the three schools of the University of Antwerp. The organisation of such a conference has almost become a tradition. In the past 15 years the area of rational approximation has developed very rapidly and widely as one can tell from the existing literature. What's more, the number of participants in conferences on this subject keeps growing. This time a total of 58 participants from 16 countries took part. Previous international conferences on the subject were held at Boulder (1972) [11], Canterbury (1972) [8,9], Toulon (1974), Toulon (1975) [4], Tampa (1976) [12], Lille (1977), Lille (1978), Antwerp (1979) [15], Amsterdam (1980) [5], Leuven (1981) [3], Warsaw (1981) [6], Kõja (1982) [13], Bad Honnef (1983) [14], Tampa (1983) [10], Bar-le-Duc (1984) [2], Łańcut (1985) [7], Marseille (1985) [1], Segovia (1986) and Havana (1987).

Several one-hour lectures were held by specialists in the field and also a number of short communications were presented. All these lectures were grouped in main sections. Each chapter in the proceedings deals with such a section from the conference. For each section the speakers in that section and the papers contained in the chapter are listed. Thus the proceedings very much reflect the structure and organization of the conference. The included papers are both original research papers and survey papers. I hereby want to thank all the referees whose comments and advices were greatly appreciated. Their work contributed enormously to the speedy publication of this volume.

The organizers would also like to thank the National Fund for Scientific Research (NFWO), the Department of Education (Ministerie van Onderwijs) and the University of Antwerp (UIA) for the financial support as well as IBM, ASLK, SABENA and MIVA for the logistic support. Thanks are also due to "Antwerpen Congresstad" for their very kind assistance and the many interesting brochures, to the "Congrescentrum UIA" for their hospitality and technical know-how, to L. Janssens and F. Schoeters for their administrative help. Without all these people the conference wouldn't have been possible. Last but not least I want to thank the participants: I very much enjoyed meeting all of you !

Annie Cuyt.
Antwerp, 20 October 1987.

[1] Brezinski C. (ed.), "Proceedings of a conference on Extrapolation and Padé approximation", J. Comp. Appl. Math. 19(2), 1987.

[2] Brezinski C., Draux A., Magnus A. P., Maroni P. and Ronveaux A. (eds), "Polynômes orthogonaux et Applications", Lecture Notes in Mathematics 1171, Springer, New York, 1985.

[3] Bultheel A. and Dewilde P. (eds), "Rational approximation in Systems engineering", Birkhäuser, Boston, 1983.

[4] Cabannes H. (ed.), "Padé approximant method and its applications in mechanics", Lecture Notes in Physics 47, Springer, New York, 1976.

[5] de Bruin M. and van Rossum H. (eds), "Padé approximation and its applications", Lecture Notes in Mathematics 888, Springer, New York, 1981.

[6] Gilewicz J. (ed.), "Padé approximation", Centre de Physique Théorique, CNRS, Luminy, 1981.

[7] Gilewicz J., Pindor M. and Siemaszko W. (eds), "Rational approximation and its applications in mathematics and physics", Lecture Notes in Mathematics 1237, Springer, New York, 1987.

[8] Graves-Morris P. (ed.), "Padé approximants and their applications", Academic Press, New York, 1973.

[9] Graves-Morris P. (ed.), "Padé approximants", The Institute of Physics, London, 1973.

[10] Graves-Morris P., Saff E. B. and Varga R. S. (eds), "Rational approximation and interpolation", Lecture Notes in Mathematics 1105, Springer, New York, 1984.

[11] Jones W. B. and Thron W. J. (eds), "Proceedings of the International Conference on Padé approximants, Continued fractions and Related topics", Rocky Mountain J. Math. 4(2), 1972.

[12] Saff E. B. and Varga R. S. (eds), "Padé and Rational approximation theory and Applications", Academic Press, New York, 1977.

[13] Waadeland H. and Wallin H. (eds), "Padé approximants and continued fractions", Det Kong. Norske Vid. Selsk. 1, 1983.

[14] Werner H. and Bünger H. J. (eds), "Padé approximation and its applications", Lecture Notes in Mathematics 1071, Springer, New York, 1984.

[15] Wuytack L. (ed.), "Padé approximation and its applications", Lecture Notes in Mathematics 765, Springer, New York, 1979.

LIST OF PARTICIPANTS

G. A. Baker Jr.
T-11 MS-B262, Los Alamos National Laboratory
Los Alamos, 87545 New Mexico, U.S.A.

C. Brezinski
U.F.R.-IEEA-Informatique, Universite de Lille 1
F-59655 Villeneuve D'Ascq, France

C. Chaffy-Camus
Equipe TIM 3, IMAG Tour des Mathématiques BP 68
F-38402 St-Martin d'Heres, France

R. Charron
Dept Mathematics and Statistics, Memorial Univ. Newfoundland
St.John's, A1C 5S7 Newfoundland, Canada

A. K. Common
Mathematical Institute, University of Kent
Canterbury, Kent CT2 7NF, U.K.

S. Cooper
Department of Mathematics, Colorado State University
CO 80523 Fort Collins, U.S.A.

F. Cordellier
U.F.R.-IEEA-Informatique, Universite de Lille 1
F-59655 Villeneuve D'Ascq, France

A. Croft
Crewe and Alsager College of Higher Education
Crewe, Cheshire CW1 1DU, U.K.

A. Cuyt
Departement Wiskunde en Informatica, Universiteit Antwerpen (UIA)
Universiteitsplein 1, B-2610 Wilrijk, Belgium

M. de Bruin
Interdisciplinaire Wiskunde, Universiteit Amsterdam
Roetersstraat 15, NL-1018 WB Amsterdam, Nederland

E. De Clerck
Instituut voor Theoretische Natuurkunde
Celestijnenlaan 200D, B-3030 Heverlee, Belgium

C. Detaille
Dept. de Mathematiques, Fac. Univ. Notre-Dame de la Paix
Rempart de la Vierge 8, B-5000 Namur, Belgium

A. Draux
U.F.R.-IEEA-Informatique, Universite de Lille 1
F-59655 Villeneuve D'Ascq, France

W. B. Gragg
Department of Mathematics, University of Kentucky
Lexington, Kentucky 40506-0027, U.S.A.

P. Graves-Morris
Department of Mathematics, University of Bradford
BD7 1DP Bradford, West Yorks, U.K.

M. Gutknecht
Seminar fur Angewandte Mathematik, ETH-Zurich
Ramirstr. 101, CH–8092 Zurich, Switzerland

E. Hendriksen
Dept. of Mathematics, University of Amsterdam
Roetersstraat 15, NL–1018 WB Amsterdam, Nederland

L. Jacobsen
Institutt for Matematik og Statistikk, Universitetet i Trondheim
N–7055 Dragvoll, Norway

W. B. Jones
Dept of Mathematics, University of Colorado at Boulder
Campus Box 426, CO 80309 Boulder, U.S.A.

J. Karlsson
Department of Mathematics, Chalmers Institute of Technology
Sven Hultins gata 6, S–412 96 Göteborg, Sweden

R. Kovacheva
Institute of Mathematics, Bulgarian Acad. Science
PO Box 373, 1090 Sofia, Bulgaria

F. Lambert
D.T.E.N.A., Vrije Universiteit Brussel
Pleinlaan 2, B–1050 Brussel, Belgium

A. Lembarki
U.F.R.-IEEA-Informatique, Universite de Lille 1
F–59655 Villeneuve D'Ascq, France

P. Levrie
Dept. Computerwetenschappen, Katholieke Universiteit Leuven
Celestijnenlaan 200A, B–3030 Heverlee, Belgium

C. Lutterodt
Department of Mathematics, Howard University
DC 20059 Washington, U.S.A.

A. P. Magnus
Institut de Mathématique, Université Catholique de Louvain
Chemin du cyclotron 2, B–1348 Louvain-La-Neuve, Belgium

F. Marcellan
Departemento de Matematicas, ETS Ingenieros Industriales
Jose Gutierres Abascal 2, 28006 Madrid, Espana

D. Masson
Department of mathematics, University of Toronto
Toronto M5S 1A1, Ontario, Canada

A. Matos
U.F.R.-IEEA-Informatique, Universite de Lille 1
F–59655 Villeneuve D'Ascq, France

J. Mc Cabe
Mathematical Institute, University of St-Andrews
St-Andrews Fife, KY16 9SS Scotland, U.K.

J. Meinguet
Institut de Mathématique, Université Catholique de Louvain
Chemin du cyclotron 2, B–1348 Louvain-La-Neuve, Belgium

L. Moral
Dept. Matemática Aplicada, Universidad de Zaragoza
Ciudad Universitaria, 50009 Zaragoza, Spain

O. Njåstad
Institutt for Matematikk, Universitetet i Trondheim - NTH
N–7034 Trondheim, Norway

L. Paquet
Institut de Mathématique, Université d'Etat à Mons
Avenue Maistriau 15, B–7000 Mons, Belgium

S. Paszkowski
Strukturalnych PAN, Instytut Niskich Temperatur i Badan
Pl. Katedralny 1 Box 937, 50-950 Wroclaw, Poland

I. Perez-Grasa
Departamento de Analisis Matemático, Faculdad de Ciencias Econ. y Empr.
Calle Dr. Cerrada S/N, 50005 Zaragoza, Espana

M. Pindor
Instytut Fizyki Teoretycznej, Uniwersytet Warszawski
Ul. Hoza 69, 00-681 Warszawa, Poland

W. Reid
Dept of Mathematics, University of Wisconsin
WI 54701 Eau-Claire, U.S.A.

D. Roberts
Mathematics Department, Napier College
Colinton Road, Edinburgh, EH10 5DT Scotland, U.K.

M. J. Rodrigues
Grupo de Matematica Aplicada, Faculdade de Ciencias
Universidade do Porto, 4000 Porto, Portugal

A. Ronveaux
Facultés des Sciences, F.U.N.D.P.
Rue de Bruxelles 61, B–5000 Namur, Belgium

E. Saff
Department of Mathematics, University of South Florida
FL 33620 Tampa, U.S.A

A. Sidi
Computer Science Department, Technion Israel Inst. of Technology
Technion City, 32000 Haifa, Israel

W. Siemaszko
Politechnika Rzeszowska
Ul. Poznanska 2 P.O. Box 85, 35-084 Rzeszow, Poland

H. Stahl
FB 20/Sekr. FR 6-8
Franklinstr. 28/29, D–1000 Berlin 10, BRD

J.-P. Thiran
Dept. de Mathématiques, Fac. Univ. Notre-Dame de la Paix
Rempart de la Vierge 8, B–5000 Namur, Belgium

S. Thiry
Dept. de Mathématiques, Fac. Univ. Notre-Dame de la Paix
Rempart de la Vierge 8, B–5000 Namur, Belgium

W. Thron
Dept of Mathematics, University of Colorado at Boulder
Campus Box 426, CO 80309 Boulder, U.S.A.

L. Trefethen
MIT Dept. of Mathematics
Headquarters 2-336, MA 02139 Cambridge, U.S.A.

W. Van Assche
Departement Wiskunde, Katholieke Universiteit Leuven
Celestijnenlaan 200B, B–3030 Heverlee, Belgium

M. Van Barel
Dept. Computerwetenschappen, Katholieke Universiteit Leuven
Celestijnenlaan 200A, B–3030 Heverlee, Belgium

M. Van der Straeten
Dept. Wiskunde, KULeuven
Celestijnenlaan 200B, B–3030 Heverlee, Belgium

I. Vanherwegen
Dept. Wiskunde, KULeuven
Celestijnenlaan 200B, B–3030 Heverlee, Belgium

J. Van Iseghem
U.F.R. IEEA-Informatique, Universite de Lille 1
F–59655 Villeneuve d'Ascq, France

B. Verdonk
Departement Wiskunde en Informatica, Universiteit Antwerpen (UIA)
Universiteitsplein 1, B–2610 Wilrijk, Belgium

A. L. von Bachhaus
AMC-STITEUR, I. G. Farben Hochhaus (Zi. 740)
Bremerstrasse, D–6000 Frankfurt 1, BRD

H. Waadeland
Institutt for Matematik og Statistikk, Universitetet i Trondheim
N–7055 Dragvoll, Norway

L. Wuytack
Departement Wiskunde en Informatica, Universiteit Antwerpen (UIA)
Universiteitsplein 1, B–2610 Wilrijk, Belgium

WELCOME

It is my pleasure to welcome you, the participants of the international congress on "Nonlinear numerical methods and Rational approximation". I am glad that the organising committee has chosen this university to host the congress and particularly I want to stress the involvement of the local organisers Dr. Annie Cuyt and Prof. Lucas Wuytack. This meeting gives me the opportunity to say a few words about the University of Antwerp and then to express some views — views of an outsider — on mathematics within the research enterprise in general.

The University of Antwerp is the youngest university in Flanders and has at present about 8000 students. The university is situated on three locations in and around Antwerp. Undergraduate humanities and economics is situated in the downtown area, in the shadow of the cathedral. Undergraduate science and medicine is located in the park area just North of here, close to the open air sculpture museum Middelheim. In the campus area where you are now, most of the upper undergraduate education and also the graduate schools are located with research in mathematics, computer science, physics, chemistry, biology and biochemistry, medicine, pharmacy, law, political and social sciences, literature and linguistics. A university hospital with 610 beds is connected with the medicine department and is, of course, the biggest structure in this campus.

In the program of this congress, I could see that the general topics of interest belong to the field of applied mathematics. I have the feeling that many of the talks deal with fundamental aspects of mathematics. This is certainly not surprising and illustrates the basic relation between fundamental research and science as a means for solving the real problems in life and in society. The relation between fundamental and applied research is controversial, especially now in a period of crisis. Though everybody believes that basic fundamental science is a long term necessity for the development and the economic competitivity of modern states, emphasis is very often laid on strategic and applied research work, in this country but also in the others. In his famous book "The Mathematician's Apology" G.H. Hardy wrote in 1940 (on his research work):

> "I have never done anything useful. Not one discovery of mine has ever contributed to the livability of the world, not for the better, nor for the worse. I did my part in educating mathematicians of my type; their work, as far as I did help them, is as useless as mine."

This citation illustrates the traditional way of thinking of the university scientists, who consider scientific work as an art rather than as a practical tool within society. In short Hardy bragged that his work was of absolutely no practical use to anyone. But now 47 years later it is not sure at all that Hardy's work was as useless as he made believe back in 1940. Science and its results are totally unpredictable and mathematics — even the most fundamental mathematics — are of utmost importance for present day technology and for present day science in other disciplines. That is, for instance, why the largest multinational firms such as AT&T, Exxon and IBM, just to name a few at random, have large research centers where mathematicians play a very important role.

Think for instance of the computerized axial tomography scanner, the CAT scanner which revolutionised radiology in which classical mathematical techniques play a key-role. Think of the methodology of X-ray crystallography, which was intractable until Herbert Hauptman accomplished the major breakthrough using mathematical techniques which provided him his Nobel prize last year. Think also of the field of elementary particle physics. The string theory for instance was at one time a highly abstract mathematical theory, developed by mathematicians who had no thoughts of applications in mind. Now string theory is a hot topic in elementary particle physics and a key instrument in the interpretation of the most fundamental processes controlling the universe, especially the interactions between elementary particles.

Today there is a growing pressure to make university research more productive on a short term basis. The need of financial resources makes it necessary to find mechanisms to incorporate fundamental research in techniques for solving real problems. It is your task as mathematicians to convince the policy makers of the importance of your discipline in modern society. If mathematics and the fundamental part of it is so important within the scientific complex, let me tell you that more than the practitioners of the other sciences, you mathematicians have an image problem. Your work is more than that of the other sciences inaccessible and your works are often written in a lapidary style. Nobel prize winner Steven Weinberg termed it up this way:

> "When physicists write an article, they generally start with a paragraph saying in more or less understandable terms what the research is all about, but in mathematics I have seen not only articles but entire books in which the first sentence in the preface was "Let A be a nilpotent subgroup"."

The idea is that there should be no word that is not absolutely necessary inserted to help the reader understand what is going on. This also implies that mathematicians are inexperienced in translating their work, and its importance for the general public and hence for the policy makers. I realize however how difficult this must be: major and critical aspects of mathematics often involve the essence of the technical language, and hence by nature are difficult to explain, except to a mathematically literate audience.

We are glad that our university can give you the opportunity to interchange ideas on occasion of this congress. We hope that you will enjoy your stay on our campus. I wish you a pleasant and successful meeting !

F. Adams, rector UIA.
Antwerp, 21 April 1987.

PADE APPROXIMATION AND RATIONAL INTERPOLATION

Chairmen:

A. Cuyt

Invited communications:

G. A. Baker Jr.
Integral approximants for functions of higher monodromic dimension.

M. Gutknecht*
Werner's algorithm for rational interpolation
and the block structure of the Newton-Padé table.

Short communications:

R. Charron*
Nonlinear discrete Fourier methods.

H. Stahl
Asymptotics of Hermite-Padé Polynomials and related convergence results:
a summary of results.

---

* Lecture notes are not included.

INTEGRAL APPROXIMANTS FOR FUNCTIONS OF HIGHER MONODROMIC DIMENSION

George A. Baker, Jr.
Theoretical Division
Los Alamos National Laboratory
University of California
Los Alamos, NM 87545 USA

ABSTRACT. In addition to the description of multiform, locally analytic functions as covering a many sheeted version of the complex plane, Riemann also introduced the notion of considering them as describing a space whose "monodromic" dimension is the number of linearly independent coverings by the monogenic analytic function at each point of the complex plane. I suggest that this latter concept is natural for integral approximants (sub-class of Hermite-Padé approximants) and discuss results for both "horizontal" and "diagonal" sequences of approximants. Some theorems are now available in both cases and make clear the natural domain of convergence of the horizontal sequences is a disk centered on the origin and that of the diagonal sequences is a suitably cut complex-plane together with its identically cut pendant Riemann sheets.

1. MONODROMIC DIMENSION APPROACH

Integral approximants[1] are a special case of Hermite-Padé approximants[2] of the Latin type. The integral approximants are to a formal power series $f(z) = \Sigma_{j=0}^{\infty} f_j z^j$. One defines an integral approximant as follows: First let

$$\sum_{j=0}^{m} Q_j^{(\vec{q})}(z)\, f^{(j)}(z) + P^{(\vec{q})}(z) = O(z^{s+1}) \quad . \tag{1.1}$$

Define by the accuracy through order of principal the polynomials $Q_j^{(\vec{q})}$, $P^{(\vec{q})}$ of degrees less than or equal to q_j, p where

$$\vec{q} = (q_0, q_1, \ldots, q_m, p), \; s = m + p - 1 + \sum_{i=0}^{m} q_i \quad , \tag{1.2}$$

A. Cuyt (ed.), Nonlinear Numerical Methods and Rational Approximation, 3–22.

with the convention $q_i = -1$ implies $Q_i \equiv 0$. These $(\vec{Q},P)$ always exist and $\vec{Q} \neq 0$, by a lemma of Baker and Lubinsky.[3] If $Q_m^{(\vec{q})}(0) \neq 0$, then they are unique. In contrast to the Padé case[4], we cannot be sure that we can usefully find an infinite subsequence where $Q_m^{(\vec{q})}(0) \neq 0$ as 0 might possibly be a singular point on a different Riemann sheet from the first one where we have the expansion given for f(z) as a power series of z. From the polynomials thus defined, we may next compute y(z) from

$$\sum_{j=0}^{m} Q_j^{(\vec{q})}(z)y^{(j)}(z) + P^{(\vec{q})}(z) = 0 \quad , \tag{1.3}$$

$$y(0) = f(0),\ y'(0) = f'(0),\ \ldots\ ,\ y^{(m-1)}(0) = f^{(m-1)}(0). \tag{1.4}$$

This solution is the integral approximant to f(z) and is denoted by

$$[p/q_0;\ q_1;\ \ldots;\ q_m]_f(z) \equiv y(z) \quad . \tag{1.5}$$

If $Q_m^{(\vec{q})}(0) \neq 0$ then this solution will exist. If however $Q_m^{(\vec{q})}(0) = 0$, then there may be a restriction on the initial conditions to achieve a regular solution. In the work which I will report here this problem is overcome by the proofs of convergence in cases where the limiting value of $Q_m(0) \neq 0$.

Next let us define the concept of monodromic dimension. We will begin with a functional element,

$$f(z) = \Sigma_{j=0}^{\infty} f_j z^j \quad , \tag{1.6}$$

which converges in some neighborhood of z = 0. It defines the complete monogenic analytic function. This function consists of a finite or at most denumerably infinite number of coverings of the complex plane. At this point, for reference, I remind you of the

Monodromy theorem.[5] If f(z) is regular in a simply connected region G, then f(z) is uniform (single valued) there.

The simplest extension of the situation described in the monodromy theorem is to a multiply connected region. Here the situation is quite different. Let me give a couple of examples.

$$f(z) = z^{1/m} \quad , \qquad f_2(z) = e^{2\pi i/m} f_1(z) \quad , \tag{1.7}$$

$$f(z) = \ln z \quad , \qquad f_n(z) = f_1(z) + 2\pi i(n-1) \quad . \tag{1.8}$$

In the example of Eq. (1.7) the function has exactly m sheets. In the example of Eq. (1.8) the function has an infinite number of sheets. Riemann[6] had the idea of classifying these functions according to the number of linearly independent coverings generated by the initial functional element. The example of Eq. (1.7) has just one such linearly independent covering. The example of Eq. (1.8) has exactly two linearly independent coverings even though it has an infinite number of Riemann sheets. Obviously the number of linearly independent coverings is less than or equal to the number of Riemann sheets so it can be a more efficient description or function classification.

Definition.[7] The monodromic dimension of a functional element is the number of linearly independent coverings of the complex plane generated by the associated monogenic analytic function.

Suppose we consider a function with monodromic dimension m and exactly n singular points in the whole complex plane. At some regular point z_0 from each of the m linearly independent coverings we can define m functional elements, and therefore m monogenic analytic functions, $y_1, \ldots, y_m$. If we encircle the ith one of the n branch points then we get

$$y_j \to \sum_{k=1}^{m} M_{jk}^{(i)}\, y_k \tag{1.9}$$

This equation defines n m×m matrices M_{jk}. There is one for each of the n singularities. Form Eq. (1.9) must hold is a consequence of the supposition that there are at most m linearly independent coverings. In addition one can prove using Cauchy's theorem that for an appropriate order

$$M^{(1)} \ldots M^{(n)} = I \quad . \tag{1.10}$$

Thus the monodromy matrices, $M^{(i)}$ generate the Monodromy Group $\mathcal{M}$ of the $\vec{y}$ system. Now define the class

$$Q\begin{pmatrix} a_1 & \ldots & a_n & : z \\ M^{(1)} & \ldots & M^{(n)} & \end{pmatrix} \tag{1.11}$$

of all $\vec{y}$-systems with these monodromy properties, plus the added property that there are no singularities of infinite order, that is to say, there exist A and r such that as z tends to a_i

$$\left|\frac{y_k'}{y_k}\right| \le \frac{A}{|z-a_i|^r} \tag{1.12}$$

holds for all i, k.

Theorem (Monodromy, Riemann[6]). For any m+1 systems $\vec{y}_j$ j = 1, ..., m+1 belonging to the same class Q there exists a linear homogeneous relation with polynomial coefficients in z, the independent variable, such that

$$\sum_{j=1}^{m+1} A_j(z)\vec{y}_j(z) = 0 \quad . \tag{1.13}$$

Corollary. If $\vec{y}$ is an element of class Q then

$$\sum_{j=0}^{n} A_j(z)\, \vec{y}^{(j)}(z) = 0 \quad . \tag{1.14}$$

Proof: If y is an element of class Q then so also are y′, y″, ... as can be seen by differentiating the monodromy group equations. Thus the corollary follows directly from the theorem.

We can conclude therefore that for functions of class Q the integral approximants for large enough q_i, (p=0) are exact! This approximation procedure will, if carried to adequate order, yield the exact answer (higher orders are degenerate and essentially are equal to the exact answer). Thus it seems reasonable to study the theory of integral approximants in the context of the idea of monodromic dimension. Later on we will prove a theorem which shows that integral approximants cannot converge outside this class, in the sense that they cannot converge on more Riemann sheets simultaneously than can be accomodated by the monodromic dimension of the solution of the differential equation defining the integral approximant as understood by the standard theory of differential equations.

For integral approximants there are a great variety of possible sequences of $(\vec{q}) \to \infty$ to consider in studying the convergence behavior of approximants defined by Eq. (1.1). I will first discuss "horizontal" sequences where the q_i i=0, ..., m are all fixed and finite and p tends to ∞. Later I will discuss the diagonal sequences where all the q_i and p tend to ∞ together.

2. THE HORIZONTAL SEQUENCE CONVERGENCE PROBLEM

Ordinary Padé approximants[4] are defined by

$$[p/q] = -P(z)/Q(z) \tag{2.1a}$$

where

$$Q(z)f(z) + P(z) = O(z^{q+p+1}) \quad , \quad Q(0) = 1 \quad . \tag{2.1b}$$

They have been proven to converge pointwise, except at poles of f(z), for an appropriately selected set of degrees q of Q(z) when p → ∞ if f(z) is meromorphic (theorem by de Montessus de Ballore)[8] in a set of nested disks about the origin. Plainly[3] this conclusion implies that sequences can be found which converge in the whole complex plane (i.e., on any compact subset) except for poles of $f(z)$ and ∞.

Are the same results also true for "horizontal" sequences of integral approximants? The answer, as we shall see, is yes, if care is used. To investigate this question we need first to define a differential multiplier.

$$\vec{Q} = (Q_0, Q_1, \ldots, Q_m) \quad , \tag{2.2}$$

is called the differential multiplier of type $\vec{m} = (q_0, \ldots, q_m)$ for $f(z)$ in $|z| < R$, if $\vec{Q}$ has degree at most $\vec{m}$, $\vec{Q} \neq 0$ and

$$\sum_{j=0}^{m} f^{(j)}(z)\, Q_j(z) \tag{2.3}$$

is analytic in $|z| < R$. We say that $\vec{Q}$ is a unique differential multiplier type $\vec{m}$ if it is essentially unique, i.e., if any $\vec{Q}$ with these properties can be related to any other $\vec{Q}'$ by $\vec{Q} = c\vec{Q}'$ where $c \neq 0$ is a constant. $\vec{Q}$ is called pole matching if $Q_m(z) = 0$ only at poles of $f(z)$.

I now give what I call Baker-Lubinsky conditions.[3] Let f be analytic at 0 and meromorphic in $|z| < R$ $(0 < R \leq \infty)$ with ℓ distinct poles $z_1, \ldots, z_\ell$ of multiplicities $p_1, \ldots, p_\ell$ respectively. Next define

$$p = \sum_{j=1}^{\ell} p_j \quad , \; S_1(z) = \prod_{j=1}^{\ell} (z-z_j), \; S(z) = \prod_{j=1}^{\ell} (z-z_j)^{p_j} \; . \tag{2.4}$$

Let $m \geq 0$, be an integer and $q_1, \ldots, q_m \geq -1$, also be integers. Define

$$\vec{m} = (q_0, \ldots, q_m) \quad , \quad M = \sum_{j=0}^{m} (q_j+1) - 1 = p + m\ell \quad . \tag{2.5}$$

By use of these conditions Baker and Lubinsky have been able to prove a number of theorems. I review here some of them.

Theorem (Existence of the approximants, Baker-Lubinsky[3]). Assume the Baker-Lubinsky conditions hold, then there exists a differential multiplier $\vec{Q}$ of type $\vec{m}$. If in addition this multiplier is unique, then for L large enough there exist essentially unique integral approximant polynomials P_L, $\vec{Q}^{(\vec{q})}$ of type $\vec{q}$. With suitable normalization,

$$\lim_{L\to\infty} Q_j^{(\vec{q})}(z) = Q_j(z), \; j = 0,1, \ldots, m \quad , \tag{2.6}$$

$$\lim_{L\to\infty} P_L^{(\vec{q})}(z) = P(z) \quad , \tag{2.7}$$

uniformly on compact sets of $|z| < R$. P is analytic in $|z| < R$ and satisfies

$$\sum_{j=0}^{m} f^{(j)}(z)\, Q_j(z) + P(z) = 0 \quad , \ |z| < R \tag{2.8}$$

Theorem (Convergence, Baker-Lubinsky[3]). Assume the Baker-Lubinsky conditions hold and that the differential multiplier is unique and pole matching for f in $|z| < R$, then for L large enough $[L/q_0; \ldots; q_m]$ exists and is uniquely defined in a neighborhood of $z = 0$. It may be analytically continued to a single valued analytic function in any open simply connected set in $|z| < R$ whose closure does not contain any of the set of point $z_1, \ldots, z_\ell$. Also, uniformly on compact subsets of $|z| < R\backslash\{z_i\}$,

$$\lim_{L\to\infty} [L/q_0; \ldots; q_m] = f(z) \quad . \tag{2.9}$$

Theorem (Existence of a Unique, Differential Multiplier, Baker-Lubinsky[3]). Assume the Baker-Lubinsky conditions hold. Then there exists a unique differential multiplier of type $\vec{m}$ for f in $|z| < R$, if

$$\vec{m} = (p, \ell-1, \ldots, \ell-1) \quad , \tag{2.10}$$

or if for some $1 \leq t \leq n \leq m$,

$$\vec{m} = (p-1, \ell-1, \ldots, \ell-1, -1, \ldots, -1, t\ell, \ell-1, \ldots, \ell-1) \quad . \tag{2.11}$$

In Eq. (2.11) the terms $\ell-1$ are repeated n-t times, the terms -1 are repeated t-1 times and the terms $\ell-1$ are repeated m-n times.

Theorem (Existence of a Unique, Pole Matching, Differential Multiplier and the Rate of Convergence, Baker-Lubinsky[3]). Assume the Baker-Lubinsky conditions hold, then there is a unique differential multiplier $\vec{Q}$ of type $\vec{m}$ for f in $|z| < R$ that is pole matching if for some $1 \leq t \leq m$

$$\vec{m} = (p-1, \ell-1, \ldots, \ell-1, -1, \ldots, -1, t\ell) \tag{2.12}$$

In Eq. (2.12) the term $\ell-1$ is repeated m-t times and the term 1 is repeated t-1 times [Eq. (2.12) is a special case of the previous theorem where n=m]. The differential multiplier has the form

$$\vec{Q} = (Q_0, Q_1, \ldots, Q_{m-t}, 0, \ldots, 0, S_1{}^t) \tag{2.13}$$

If we normalize so $Q_m^{(\vec{q})}$ is a monic polynomial, then in any compact subset K of C

$$\limsup_{L\to\infty} \| Q_j^{(\vec{q})} - Q_j \|_K^{1/L} \leq (\max_k |z_k|)/R, \quad j = 0,\ldots,m, \tag{2.14}$$

while for the case $K \subset \{|z| < R\}$,

$$\limsup_{L\to\infty} \| P_L^{(\vec{q})}(z) - P(z) \|_K^{1/L} < \|z\|_K/R \quad , \tag{2.15}$$

and if K contains no poles of f(z) ,

$$\limsup_{L\to\infty} \| [L/q_0; \ldots; q_m] - f(z) \|_K^{1/L} < \|z\|_K/R \tag{2.16}$$

The results of these theorems give us the same type of pointwise convergence for horizontal sequences of integral approximants (properly selected) as the de Montessus theorem gave for Padé approximants to functions meromorphic in a disk.

It is an easy corollary[3] to construct, just as it was from the horizontal convergent sequences of Padé approximants, an appropriate convergent sequence which is made up of integral approximants of $m^{\underline{th}}$ order which converges uniformly on any given compact set in the complex plane not containing a pole of f(z) when f(z) is a meromorphic function with no limit point of poles for any finite point. In this respect, we have not lost any ground going from Padé approximants to integral appoximants. Though, of course, integral approximants are more general than Padé approximants, it is not a foregone conclusion without proof that the integral approximants would also work in a case where the Padé approximants are known by previous results to work.

Of course, it is not just to approximate meromorphic functions that one is interested in integral approximants, but for the approximation of functions which have branch points. We turn now to more general function classes. I next give a theorem, whose proof is only a minor variant of Riemann's original proof of his monodromy theorem.

Definition[9]. Given a convergent Taylor series f(z) about z = 0 and a disk $D = \{z \mid |z| \leq R\}$ we say f(z) has local monodromic dimension m if analytic continuation along all paths in D generates exactly m linearly independent coverings of D.

Disk Monodromy Theorem (Baker, Oitmaa and Velgakis[9]). Let f(z) be a convergent Taylor series about z = 0 and of local monodromic dimension m in a disk $D = \{z| \ |z| \le R\}$. Further let there be exactly $n < \infty$ singular points a_k of finite order in D, and $|a_k| < R$, k=1, ..., n. Then

$$\sum_{j=0}^{m} \rho_j(z)\, f^{(j)}(z) = 0 \quad , \tag{2.17}$$

where $\rho_m(z)$ is a polynomial of finite degree and $\rho_j(z)$, j=0, ..., m-1 are analytic in D.

I now sketch the proof because the ideas, although old, appear recurrently in this work.

Proof (Sketch). Suppose $y_j(z)$ are the m linearly independent coverings of D generated by the functional element f(z) in the neighborhood of z = 0.

At the singular point a_k we can introduce a change of basis

$$u_\ell(z) = \sum_{j=1}^{m} U_{k;\ell j}\, y_j(z) \quad , \tag{2.18}$$

where U_k is a constant matrix. This matrix is chosen such that

$$\Lambda_k = U_k\, M^{(k)}\, U_k^{-1} \quad , \tag{2.19}$$

where the matrix Λ_k is a diagonal matrix and the matrix $M^{(k)}$ is the monodromy matrix at the point $z = a_k$. With this change of basis we have,

$$\Lambda_k\, u_\ell = \lambda_{k;\ell}\, u_\ell \quad , \tag{2.20}$$

where the $\lambda_{k;\ell}$ are the eigenvalues of the matrix $M^{(k)}$. If the eigenvalues $\lambda_{k;\ell}$ are not degenerate, then

$$u_\ell = (z-a_k)^{\nu_{k;\ell}}\, h_{k;\ell}(z) \quad , \tag{2.21}$$

where the $h_{k;\ell}(z)$ is uniform (single valued) and

$$\nu_{k;\ell} - \frac{1}{2\pi i} \log \lambda_{k;\ell} = \text{whole number} \quad . \tag{2.22}$$

If the singularity is of the first order [r=1 in Eq. (1.12)], then $h_{k;\ell}(z) \ne 0$ or ∞ and is analytic at $z = a_k$. If $r > 1$ or the eigenvalues are degenerate the proof is more complex but follows classical lines. [It is easy to get an idea of the results which are to be expected here by thinking of a nearby case where the λ's are not degenerate and $(z-a_k)^{-r} \cong \Sigma_{j=1}^{r} \beta_j/(z-b_j)$, where b_j and β_j are chosen so that the $\Sigma \to (z-a_k)^{-r}$ as the b_j tend to a_k. The continuity of the solution of a differential equation as a function of its coefficients away from any singularity gives then the idea of the results. Of course the proof runs in the opposite direction.]

Now since $\vec{u}$ and its derivatives all belong to the same monodromy group it is elementary to show that

$$u_\ell^{(j)}(z) = (z-a_k)^{\nu_{k;\ell}-j} h_{j;k;\ell}(z) \quad , \tag{2.23}$$

where $h_{j;k;\ell}(z)$ is uniform in the neighborhood of a_k.

Next let us consider,

$$\sum_{j=0}^{m} c_j y_i^{(j)}(z) = 0 \ , \ i = 1, \ldots, m \ . \tag{2.24}$$

Cramer's rule gives the result,

$$c_{j_0} = \Delta_{j_0}(z) = \det\left|y_i^{(i)}(z)\right| \quad , \quad \begin{matrix} i = 1, \ldots, m \ , \\ j = 0, \ldots, m, \ j \neq j_0 \ , \end{matrix} \tag{2.25}$$

and that $\Delta_{j_0}(z) \not\equiv 0$ because, by hypothesis, the y_i are linearly independent. Using our change of basis, we find,

$$\Delta_{j_0}(z) = \det U_k \ \det|u_\ell^{(j)}(z)| \qquad \begin{matrix} \ell = 1, \ldots, m \ , \\ j = 0, \ldots, m, \ j \neq j_0 \ . \end{matrix} \tag{2.26}$$

From our representation of $u_\ell^{(j)}(z)$ we see that

$$\prod_{\ell=1}^{m} (z-a_k)^{-\nu_{k;\ell}+m} \Delta_{j_0}(z) \tag{2.27}$$

is analytic in the neighborhood of $z = a_k$. If we repeat the same argument as given above for each a_k then we may conclude

$$P_{j_0}(z) = \left[\prod_{k=1}^{n} \prod_{\ell=1}^{m} (z-a_k)^{-\nu_{k,\ell}+m}\right] \Delta_{j_0}(z) \tag{2.28}$$

is analytic at $z = a_1, \ldots, a_k$ and hence in D by construction. The P's of Eq. (2.28) exist and are not identically equal to 0 in D. By standard theorems $P_m(z)$ has only a finite number of zeros in D. We may therefore factor it as $\rho_m(z)\ \mathcal{Q}(z)$ where $\rho_m(z) \neq 0$ for $|z| > R$ and $\mathcal{Q}(z) \neq 0$ for $|z| \leq R$. The division by $\mathcal{Q}(z)$ completes the proof.

Remark: Once $\rho_m(z)$ is fixed, the linear independence of the y_i fix the other ρ_j uniquely. If fewer than m y's were independent the same argument as given above would lead to an equation of lower order which could be added to the one we have just obtained and thereby destroy uniqueness.

To make further progress with the theory of horizontal sequences of integral approximants, it is useful to note the following key property.

Separation Property (Baker, Oitmaa, and Velgakis[9]). If a function f(z), possibly multiform, can be written as $f(z) = f_i(z) + f_o(z)$, where for a disk $D = \{z|\ |z| \leq R\}$, $f_o(z)$ is analytic for all $z \in D$ and every analytic continuation of $f_i(z)$ is analytic for all z in the finite complex plane outside D, then f(z) has the separation property with respect to D.

The simplest example would be the class of meromorphic functions, with respect to an appropriate sequence of R's. We have already discussed the theory at length in this case.

Separation Property Theorem (Baker, Oitmaa and Velgakis[9]). Let f(z) have the separation property with respect to a disk D, a finite number of singular points a_i, $|a_i| > 0$ for all i, in the interior of D and none on the boundary of D. Assume further that all these singular points plus the point at ∞ for $f_i(z)$ are of finite order. Let $f_i(z)$ be of exact monodromic dimension m. Then, (i) there exists an essentially unique differential multiplier for disk D whose coefficients are polynomials of degrees $q_0, \ldots, q_m$ (q_m is chosen as the minimum possible). (ii) the integral approximants $[L/q_0; \ldots ; q_m]$ converge to f(z) as $L \to \infty$ on simply-connected, compact subsets of $D\backslash\{a_k\}$ which contain the origin.

Remarks: By Riemann's monodromy theorem[6] there exists A_j polynomials such that

$$\sum_{j=0}^{m} A_j(z)\, f_i^{(j)}(z) = 0 \quad . \tag{2.29}$$

If we add $f_o(z)$ to f_i we get

$$\sum_{j=0}^{m} A_j(z)\, f^{(j)}(z) = \Phi(z) \quad , \tag{2.30}$$

where $\phi(z)$ is analytic in D. The existence of an essentially unique $A_m(z)$ of minimum degree follows from a proof by contradiction. The convergence part follows by arguments of standard type and the path used from the origin defines which sheet of f(z) is meant.

For that class of functions with the separation property, we see by this theory that the de Montessus[8] type theorem just given assures us that the integral approximants converge in a pointwise manner much as would have been expected from the analogy with corresponding results for the Padé approximants to the meromorphic function class. In fact it is, I think, this class of functions which is the correct analogy to meromorphic functions for the theory of horizontal sequences of integral approximants. This class, as we saw, is equivalent to f such that

$$\sum_{j=0}^{m} \rho_j(z)\, f^{(j)}(z) = \phi(z) \quad , \tag{2.31}$$

where the ρ_j are polynomials and $\phi(z)$ is analytic in the disk D. The general class corresponding to f(z) of local monodromic dimension m plus a uniform analytic background would have $\rho_m(z)$ a polynomial and $\rho_j(z)$, $j = 0, \ldots, m-1$ and $\phi(z)$ analytic in D.

To illustrate the case when one does not have the separation property, I look at the simplest possible, non-trivial case. Suppose f(z) is regular in a disk $|z| \le \rho$, $\rho > 1$ except for a regular singular point at $z = 1$ and further that f(z) in $|z| \le \rho$ consists of a uniform background plus an $m = 1$ piece. Then, further simplifying to a first order singularity,

$$(1-z)\, f'(z) + G(z)\, f(z) = H(z) \quad , \tag{2.32}$$

where G and H are analytic in $|z| \le \rho$. The solution for f(z), assuming further that $G(z) = \Sigma\, G_i (1-z)^i$ and that G_0 is not an integer ($\gamma = -G_0$), is

$$f(z) = A(z)(1-z)^{-\gamma} + B(z) \quad , \tag{2.33}$$

where A and B are analytic in $|z| \le \rho$.

Let us use the following notation for the [L/M-1;1] approximant,

$$\{\lambda^{(L)}(1-z) + \alpha^{(L)}\}\, f'(z) + g^{(L)}(z)\, f(z) = h^{(L)}(z) + O(z^{L+M+1}) \quad , \tag{2.34}$$

where $g^{(L)}(z)$ is a polynomial of degree M-1 and $h^{(L)}(z)$ is a polynomial of degree L.

<u>Theorem (Baker and Graves-Morris[10])</u>. Under the above hypotheses, for L sufficiently large we have, normalizing $\lambda^{(L)} = 1$,

$$\alpha^{(L)} = O(L^{-M-1}) \quad ,$$

$$g_i^{(L)} = G_i + O(L^{M+i}) \quad i = 0,1, \ldots, M-1 \quad , \tag{2.35}$$

therefore,

$$g^{(L)}(z) \underset{L\to\infty}{\to} \Sigma_{i=0}^{M-1} \ G_i(1-z)^i \quad ,$$

$$h^{(L)}(z) \underset{L\to\infty}{\to} H(z) - \{G(z) - \Sigma_{i=0}^{M-1}(1-z)^i G_i\}f(z) \quad . \tag{2.36}$$

Remark: The estimates in this theorem show that we get a good representation of G(z) near the singular point z=1 but however the polynomial $h^{(L)}(z)$ has a limit function defined by a series which diverges for $|z| > 1$. These results suffice to establish:

Theorem (Baker and Graves-Morris[10]). Under the hypotheses of the previous theorem,

$$\lim_{L\to\infty} [L/M;1] = f(z) \quad , \ |z| < 1 \quad , \tag{2.37}$$

on all Riemann sheets accessible in the disk $|z| \leq \rho$. Further for L large enough

$$\gamma^{(L)} = \gamma + O\ (L^{-M}) \tag{2.38}$$

Note: The [L/M;1] on the second, ... Riemann sheets are defined by integrating the approximant around z = 1.

Remark: These sequences of approximants are useful in analyzing the closest singularity to the origin, even without the separability condition. For farther singularities from the origin the situation is not proven.

Further results for the case of higher monodromic dimension are known.

3. DIAGONAL SEQUENCES

For Padé approximants the "diagonal" sequences [L/M], $M \to \infty$, $L/M \to 1$ are much more powerful methods[4] of approximate analytic continuation than "horizontal" type sequences. Unfortunately the convergence theorems are harder to prove and generally for a weaker form of convergence than the pointwise convergence that we have obtained for the "horizontal" sequences. The same situation, so far at least, seems to hold for integral approximants and the problem, as we saw for horizontal sequences, is more complex.

As with the Padé approximants we need global information on the analyticity properties[11] to obtain convergence theorems (so far). The results that follow on diagonal sequences of integral approximants are generalizations of the results of Stahl[12] for Padé approximants.

Theorem (limit on the domain of convergence). A sequence of mth order integral approximants $[p/q_0; \ldots; q_m]$ can not converge simultaneously on m + 2 coverings of the complex plane to an f(z) which has a structure of a uniform function (may be zero) plus a part with a monodromic dimension greater than m.

Proof: On any Riemann sheet we may write

$$\sum_{j=0}^{m} Q_j^{(\vec{q})}(z)\, f^{(j)}(z^{(i)}) + P^{(\vec{q})}(z) = R^{(\vec{q})}(z^{(i)}) \tag{3.1}$$

where $\pi(z^{(i)}) = z$ is the projection on the complex plane from the Riemann surface and $R^{(\vec{q})}(z^{(i)})$ is the remainder. If we solve this equation using m + 2 coverings $f(z^{(i)})$, i = 1, ..., m+2 for $Q_m^{(\vec{q})}$ in terms of $R^{(\vec{q})}(z^{(i)})$ we get

$$\det \begin{vmatrix} f^{(m)}(z^{(1)}) & \ldots f(z^{(1)}) & 1 \\ \vdots & \ddots & \vdots \\ f^{(m)}(z^{(m+2)}) & \ldots f(z^{(m+2)}) & 1 \end{vmatrix} = \frac{\det \begin{vmatrix} R^{(\vec{q})}(z^{(1)}) & f^{(m-1)} & \ldots 1 \\ \vdots & \vdots & \ddots \\ R^{(\vec{q})}(z^{(m+2)}) & f^{(m-1)} & \ldots 1 \end{vmatrix}}{Q_m^{(\vec{q})}(z)}, \tag{3.2}$$

for example. Solutions for other subscripts than m are also possible and are of similar form. If we expand the determinant on the left-hand side of Eq. (3.2) along the last row we obtain that it is equal to

$$\det \begin{vmatrix} g^{(m)}(z^{(1)}) & \ldots & g(z^{(1)}) \\ \vdots & \ddots & \vdots \\ g^{(m)}(z^{(m+1)}) & \ldots & g(z^{(m+1)}) \end{vmatrix}, \tag{3.3}$$

where we define,

$$g(z^{(i)}) = f(z^{(i+1)}) - f(z^{(i)}) \quad . \tag{3.4}$$

Note that the determinant in Eq. (3.3) is independent of $(\vec{q})$ and not identically equal to 0 by our hypothesis that the monodromic dimension is greater than m. Therefore we can't have $R^{(\vec{q})}(z^{(i)})$, i = 1, ..., m + 2, all vanish simultaneously. Baker and Lubinsky[3] have proven $\vec{Q} \neq 0$. If we select that j for which $|Q_j^{(q)}(z)|$ is maximum and apply

Hadamard determinant inequality, we can derive a lower bound for the magnitude of the vector of $R^{(\vec{q})}(z^{(i)})$ on the right-hand side of eq. (3.2), divided by $|Q_j^{(\vec{q})}(z)|$. If we then take the minimum such bound over j, we have a uniform lower bound greater than zero over all $(\vec{q})$ for the whole sequence for $|\vec{R}^{(\vec{q})}|/\{\mathrm{Max}_j |Q_j^{(\vec{q})}|\}$. Hence the conclusion of the theorem follows.

In order to go beyond Riemann's monodromy theorem let us look at a class of functions defined by

$$\sum_{j=0}^{m} E_j(z)\, f^{(j)}(z) + E_{-1}(z) = 0 \quad , \tag{3.5}$$

where the E_j are entire functions. We shall explicitly assume that not all the E_j's are polynomials. We further assume that f(z) is analytic at z = 0 on all Riemann sheets for ease of exposition. In addition to Eq. (3.5) we need a further assumption in order to ensure that the f(z) so defined are of the full monodromic dimension assumed so that the equation is not reducible to lower order. In order to ensure irreducibility we assume that starting from the functional element f(z) at z = 0 we may select m + 1 connected sheets on which f(z) is a uniform function plus a function of monodromic degree exactly m. Finally, I assume, that the closure of this domain never produces more than m + 2 coverings of any point of the complex plane. By near-to-diagonal approximants $[p/q_0;\ldots;q_m]$ I mean that the $\lim_{p\to\infty} q_i/p = 1$ for $i = 0,\ldots,m$.

Theorem (diagonal convergence). Let f(z) belong to the above defined class, then at least a subsequence of the near-to-diagonal approximants $[p/q_0;\ \ldots;\ q_m]$ converge to f(z) in capacity on simply-connected compact subsets of its Riemann surface which exclude the singular points of f(z) and include the origin of the defining Riemann sheet.

Proof (sketch). To make the parallelism to the proof of Stahl easier to follow we will use the expansion about z = ∞ and take the exact diagonal sequences [n/n; ...; n]. This latter is not an essential simplification. The Hermite-Padé equations now are

$$\sum_{j=0}^{m} \hat{Q}_j^{(\vec{q})}(z)\, f^{(j)}(z) + \hat{P}^{(\vec{q})}(z) = \hat{R}^{(\vec{q})}(z) = O(z^{-(m+1)n-1}) \tag{3.6}$$

where

$$\hat{Q}_j^{(\vec{q})}(z) = z^n\, Q_j^{(\vec{q})}(z^{-1}),\quad \hat{P}^{(\vec{q})}(z) = z^n\, P^{(\vec{q})}(z^{-1}) \tag{3.7}$$

The Q and P polynomials are defined in a manner analogous to Eq. (1.1) and we have multiplied them by z^n to give the $\hat{Q}$, $\hat{P}$ as polynomials. Since these polynomials can be multiplied by a nonzero constant we can write

$$\frac{1}{n} \log \max_{j=0,\dots,m} \left(|\hat{Q}_j^{(\vec{q})}(z)|,\ |\hat{P}^{(\vec{q})}(z)| \right)$$

$$= \rho(z,\mu_{\vec{q}}),\quad ||\mu_{\vec{q}}|| \le 1 \quad , \tag{3.8}$$

where ρ is the logarithmic potential of a certain measure $\mu_{\vec{q}}$ which is positive because the maximum of subharmonic functions is again subharmonic.[13] By the weak compactness of the unit ball in the space of positive measures, there exists a subsequence $\vec{q}_1 \subseteq \vec{q}$ with the property

$$\lim_{\vec{q}_1} \mu_{\vec{q}} = \mu_0 \quad . \tag{3.9}$$

This in turn implies

$$\overline{\lim}_{\vec{q}_1}\ \rho\ (z,\ \mu_{\vec{q}_1}) = \rho(z,\mu_0) \tag{3.10}$$

quasi- everywhere[13] on $\mathcal{C}$.

Consider any domain $A \subseteq R$, the Riemann surface of f(z), which has a Green's function $g(z,w;A)$, $z,w \in A$. For a measure ν in A define the Green's potential

$$g(z,\nu) = - \int g(z,w;A)\ d\nu(w) \quad , w \in R \quad , \tag{3.11}$$

on $R\sim A$, $g(z,\nu) \equiv 0$ by the properties of the Green's function. If necessary select another subsequence $\vec{q}_2 \subseteq \vec{q}_1$ we then get,

$$\overline{\lim}_{\vec{q}_2} \frac{1}{n} \log \max_{j=0,\dots,m} (|Q_j^{(\vec{q})}|,\ |P^{(\vec{q})}|)$$

$$= g_0(z) = g_0(z,\nu_0) + h_0(z) \tag{3.12}$$

$$\overline{\lim}_{\vec{q}_2} \frac{1}{n} \log (|R^{(\vec{q})}|) = g_1(z) = g_1\ (z,\nu_1) + h_1(z) \tag{3.13}$$

where h_0 and h_1 are harmonic in $A \subseteq R$. ν_0 is just μ_0 lifted onto A. The ν's are independent of A in the sense that they agree for A_1 and A_2 in $A_1 \cap A_2$. The h's depend on A.

Lemma 1. With the sets $I_0 = \{\infty^{(1)}\}$, $I_1 = \pi^{-1}(\infty)$ and $I_2 = I_1 \sim I_0$ we have

(i) $\nu_1(I_0) \geq m + 1$

(ii) $\nu_1(\mathcal{B}) \geq 0$ for all Borel set $\mathcal{B} \subseteq \mathcal{R} \sim I_2$

(iii) If $\mathcal{A} \subseteq \mathcal{R} \sim I_1$ is a domain (connected!) in which ρ is (m+1)-valent, then

(a) $0 \leq \nu_0(\mathcal{A}) \leq (m+1)|\nu_0(I_0)| \leq m + 1$

(b) $0 \leq \nu_0(\bar{\mathcal{A}}) \leq (m+2)\ |\nu_0(I_0)| - \nu_0(\pi(\mathcal{A}))$

Result (i) follows by definition, the result (ii) follows because R is of the form of a sum of polynomials times derivatives of f and it can only $\to \infty$ outside the set I_2 at those points where f does but this effect is washed out as $n \to \infty$ by the definition of the measure. Result (iii) follows because polynomials have the same number of 0's as poles by Gauss's theorem and there are at most m + 1 copies on the surface of $\mathcal{A}$. Finally (b) follows because the boundary under closure produces by hypothesis at most one extra copy and so adds $|\nu_0(I_0)|-\nu_0(\pi(\mathcal{A}))$ at most.

Lemma 2. If the domain $\mathcal{A}$ satisties the assumptions of Lemma 1, (iii) and

$$\nu_0(\bar{\mathcal{A}}) = (m+2)\ |\nu_0(i_0)| \quad , \tag{3.14}$$

then $\overline{\pi(\mathcal{A})} = C$.

Proof. Exactly parallel to Stahl's.[12]

Now define

$$d(z) = g_1(z) - g_0(z) \quad , \ z \in R \quad . \tag{3.15}$$

Note that by definition

$$d(t) \leq \lim_{n\to\infty} \frac{1}{n} \log \{ \sum_{j=0}^{n} |f^{(j)}| + 1\} = 0 \quad , \tag{3.16}$$

except at singular points of f.

Define the region

$$\tilde{\mathcal{D}} = \{z \subset R: d(z) < 0\} \quad . \tag{3.17}$$

The significance of this definition can be seen by noting the following result

$$\frac{|R|}{\max\left\{\left|Q_j^{(\vec{q})}\right|,\ P^{(\vec{q})}\right\}} \sim e^{nd(z)} \underset{n\to\infty}{\to} 0 \quad , \tag{3.18}$$

if $d(z) < 0$. As by the Hermite-Padé equation $d(\infty^{(1)}) = -\infty$ we see at once that

$$\infty^{(1)} \in \tilde{\mathcal{D}} \ . \tag{3.19}$$

Let us define $\mathcal{D}$ to be the connected component of $\tilde{\mathcal{D}}$ which contains $\infty^{(1)}$. Since $\mathcal{D}$ may have an infinite number of Riemann sheets we need to prune it before we can complete the proof. First consider $\pi(\mathcal{D}) \subseteq D \subseteq \mathcal{C}$. We can by hypothesis construct an $m + 1$ sheeted covering of D on which $f(z)$ is represented as a uniform function plus m linearly independent functions. Call this domain $\mathcal{B} \subseteq \mathcal{D} \subseteq \mathcal{R}$. Since $d < 0$ in $\mathcal{D}$ and so also in $\mathcal{B}$ we use the potential theory notion of flux to get a key inequality

$$\nu_1(\bar{\mathcal{B}}) - \nu_0(\bar{\mathcal{B}}) \le 0 \ . \tag{3.20}$$

As for d, the potential theory sources are the zeros in $\hat{R}$ that is the positive part of ν_1 and the poles in $\mathrm{Max}(|\hat{Q}_j|, |\hat{P}|)$ the negative parts of ν_0. Likewise the potential theory sinks are the poles in R, i.e., the negative part of ν_1 and the zeros in $\mathrm{Max}(|Q_j|, |P|)$ the positive part of ν_0. We may re-express Eq. (3.20) as

$$\nu_1(I_0) + |\nu_0(I_0)| + \nu_1(\bar{\mathcal{B}} \sim I_0) < \nu_0(\bar{\mathcal{B}} \sim I_0)$$

$$\le (m+2)\ |\nu_0(I_0)| - \nu_0(\pi(\mathcal{B}) \sim I_0) \ . \tag{3.21}$$

By lemma 1 and our hypothesis on $f(z)$ we get, by a little arithmetic

(a) $\nu_1(I_0) = m + 1$,

(b) $\nu_1(\bar{\mathcal{B}} \sim I_0) = 0$,

(c) $\nu_0(I_0) = -1$,

(d) $\nu_0(\partial\mathcal{B}) = (m + 2)|\nu_0(I_0)| = m + 2$,

(e) $\nu_0(\pi(\mathcal{B}) \sim I_0) = 0$.

These results show that the zeros cluster on the boundary of the region $\pi(\mathcal{B}) = D$ and so by now standard arguments the exceptional set in D is of capacity zero. Property (a) shows that the block size in the Hermite-Padé table tends to 0 (relative to n).

By a variant of the proof of Riemann or the proof of the disk monodromy theorem, sketched herein we have the result that the $f(z^{(i)})$, $i = 1, \ldots, m + 1$ uniquely determine

$$\frac{E_i(z)}{E_m(z)} \quad , \; j = -1, 0, \ldots, m-1 \quad , \qquad (3.22)$$

as uniform functions of z. Hence adapting Stahl's proof of the extremal nature of $\pi(\mathcal{D}) = D$ we are able to conclude its uniqueness in terms of the uniquely determined E_j/E_m ratios. This uniqueness of $\pi(\mathcal{D}) = D$ allows the conclusions to be extended from a subsequence to any subsequence, again by arguments parallel to Stahl's, except we require $\hat{Q}_m(z)$ to be of full degree in order that we may solve for the approximant. The proof of this theorem depends at places on the validity of the results announced by Stahl.[12] The reader should be aware that I have not had a chance to see the full proof of his results and have simply presumed them to be correct.

The knowledge of the global behavior of f(z) has allowed the proof of convergence for diagonal sequences without the separation property, or the limitation to the nearest singularity, required for horizontal sequences. I do not think, using the knowledge we have of Padé approximants, that the assumptions that I have had to make in the results reported here by any means exhaust the range of convergence of the integral approximants.

ACKNOWLEDGMENT

The author is pleased to acknowledge helpful discussions or correspondence with P. R. Graves-Morris, D. S. Lubinsky, P. Moussa, J. Nuttall, J. Oitmaa, E. B. Saff and H. Stahl. This work was performed under the auspices of the U.S. D.O.E.

REFERENCES

1. Gammel J. L., 'Review of Two Recent Generalizations of the Padé Approximant' in P. R. Graves-Morris. (ed.), Padé Approximants and Their Applications, Academic Press, London, 1973, pp. 3-9; Joyce G. S. and Guttmann A. J., 'A New Method of Series Analysis' in P. R. Graves-Morris (ed.), Padé Approximants and Their Applications Academic Press, London, 1973, pp. 163-167; Guttman A. J., 'On the Recurrance Relation Method of Series Analysis,' J. Phys. A 8 (1975), 1081-1088; -- "Derivation of 'mimic expansions' from Regular Perturbation Expansions in Fluid Mechanics," J. Inst. Math. Appl. 15 (1975), 307-315; Hunter D.

L. and Baker G. A., Jr., 'Methods of Series Anaysis III. Intregal Approximant Methods,' Phys. Rev. B 19 (1979), 3808-3821; Fisher M. E. and Au-Yang H., 'Inhomogeneous Differential Approximants for Some Power Series,' J. Phys. A 12 (1979), 1677-1692.

2. Hermite C., 'Sur la Généralisation des fractions continues algebriques,' Ann. Math., Sér. 2, 21 (1893), 289-308; Padé H., 'Sur la généralisation des fractions continués algebriques,' J. Math, Sér.4, 10 (1894), 291-329; Dora J. D. and Di-Crescenzo C., 'Approximation de Padé-Hermite' in L. Wuytack (ed.), Padé Approximation and its Applications, Lecture Notes in Mathematics 765 Springer-Verlag, Berlin, 1979, pp. 88-115; Burley S. K., John S. O. and Nuttall J., 'Vector Orthogonal Polynomials,' SIAM J. Numer. Anal. 18 (1981), 919-924.

3. Baker G. A., Jr. and Lubinsky D. S., 'Convergence Theorems for Rows of Differential and Algebraic Hermite-Padé Approximants,' J. Comp. Appl. Math. 18 (1987), 29-51.

4. Baker G. A., Jr., Essentials of Padé Approximants, Academic Press, New York, 1975; Baker G. A., Jr. and Graves-Morris P. R., Padé Approximants Part I: Basic Theory and Part II: Extensions and Applications in G.-C. Rota (ed.), Encyclopedia of Mathematics and its Applications, Vol. 13 and 14, Cambridge University Press, London, 1981.

5. Knopp K., Theory of Functions, Parts I and II, translated by F. Bagemihl, Dover Publications, New York, 1945.

6. Riemann B., Collected Works of Bernhard Riemann, H. Weber (ed.), Dover Pub. Inc., New York, 1953, pp. 379-390.

7. Baker G. A., Jr., 'Approximate Analytic Continuation Beyond the First Riemann Sheet' in P. R. Graves-Morris, E. B. Saff and R. S. Varga (eds.), Rational Approximation and Interpolation, Lecture Notes in Mathematics 1105, Springer-Verlag, Berlin, 1984, pp. 285-294.

8. de Montessus de Ballore R., 'Sur les fractions continues algebriques,' Bull. Soc. Math, France 30 (1902), 28-36; -- "Sur les fractions continues algebriques," Rend. Circ. Mat. Palermo 19 (1905), 1-73.

9. Baker G. A., Jr., Oitmaa J. and Velgakis M. J., 'Series Analysis: Multivalued Functions' (in preparation).

10. Baker G. A., Jr. and Graves-Morris P. R., (in preparation).

11. For other results, not discussed here, on diagonal sequences, see Nuttall J., 'Asymptotics of Diagonal Hermite-Padé Polynomials,' J. Approx. Theory 42 (1984), 299-386.

12. Stahl H., 'Three Different Approaches to a Proof of Convergence for Padé Approximants,' T. U. Berlin preprint (1986).

13. Landkof N. S., Foundations of Modern Potential Theory translated by A. P. Doohovskoy, Springer-Verlag, New York, 1972.

Asymptotics of Hermite-Padé Polynomials and Related Convergence Results - A Summary of Results

Herbert Stahl
TFH-Berlin/FB 2
Luxemburger Straße 10
D-1000 Berlin 65
Fed.Rep.Germany

Abstract. In this note we present new results about n-th root asymptotics of Hermite-Padé polynomials and about the convergence of algebraic and integral Hermite-Padé approximants and simultaneous rational approximants. In the situation considered here, the m+1 simultaneous Hermite-Padé polynomials of both types have identical asymptotics. It turns out that a certain determinantal condition is necessary and sufficient for the identical asymptotics. The asymptotic distributions of the zeros of the polynomials as well as the domains of convergence for the three types of approximants are characterized by certain logarithmic potentials.

1.Introduction. Let $(f_0,\dots,f_m)$ be a vector (sometimes also called a system) of $m+1$ functions that are assumed to be analytic near $\infty\in\overline{\mathbb{C}}$. Following Mahler [Ma] we distinguish two kinds of Hermite-Padé polynomials (for a detailed treatment of the material see also [Nu1]):

DEFINITION1.1 (Hermite-Padé polynomials of type I or so-called Latin polynomials) The elements of a vector of $m+1$ polynomials $(P_0,\dots,P_m)$ of multidegree at most $n=(n_0,\dots,n_m)\in\mathbb{N}^{m+1}$, $|n|:=\sum_j n_j$, not all identically zero, satisfying

$$\sum_{j=0}^{m} P_j\left(\tfrac{1}{z}\right)f_j(z)=O(z^{-|n|-m}) \qquad \text{for} \quad z\to\infty, \tag{1.1}$$

are called Hermite-Padé polynomials of type I and multidegree n. The associated polynomials

$$P_j^*(z):=z^{max(n)}P_j\left(\tfrac{1}{z}\right),\ j=0,\dots,m;\ \ max(n):=max_j(n_j), \tag{1.2}$$

A. Cuyt (ed.), Nonlinear Numerical Methods and Rational Approximation, 23–53.

are called inverse Hermite-Padé polynomials of type I. The latter ones satisfy

(1.3) $$\sum_{j=0}^{m} P_j^*(z) f_j(z) =: R(z) = O(z^{-|n|-m+\max(n)}) \text{ for } z \to \infty.$$

DEFINITION 1.2 (Hermite-Pade polynomials of type II or so-called German polynomials) The elements of a vector of $m+1$ polynomials $(Q_0, \dots, Q_m)$ of multidegree at most $\tilde{n} = (\tilde{n}_0, \dots, \tilde{n}_m) := (|n| - n_0, \dots, |n| - n_m)$, $n \in \mathbb{N}^{m+1}$, not all identically zero, satisfying

(1.4) $$Q_i(\tfrac{1}{z}) f_j(z) - Q_j(\tfrac{1}{z}) f_i(z) = O(z^{-|n|-1}) \text{ for } z \to \infty, \ i \neq j,$$

are called Hermite-Padé polynomials of type II and multidegree $\tilde{n}$. The associated polynomials

(1.5) $$Q_j^*(z) := z^{\tilde{n}_0} Q_j(\tfrac{1}{z}), \ j = 0, \dots, m,$$

are called inverse Hermite-Padé polynomials of type II. For $i, j = 0, \dots, m$, $i \neq j$, the latter ones satisfy

(1.6) $$[Q_i^* f_j - Q_j^* f_i](z) =: R_{ij}(z) = O(z^{-n_0-1}) \quad \text{for } z \to \infty.$$

Remark 1: It is easy to see that the polynomials of both types exist for every multidegree n, but they are not unique. In any case they can be multiplied by a non-zero constant, but there may exist more essential non-uniqueness.

Remark 2: If we take $f_0 \equiv 1$, then (1.4) or (1.6) is equivalent to

(1.7) $$[Q_0^* f_j - Q_j](z) = O(z^{-n_0-1}) \qquad \text{for } z \to \infty, \ j = 1, \dots, m.$$

All $Q_1^*, \dots, Q_m^*$ are polynomials if $n_0 \leq n_j$ for $j = 1, \dots, m$. Relation (1.7) is more usual as (1.4) or (1.6) in the area of simultaneous rational approximants (cf. Definition 1.5, below).

We introduce the following normalization, which is assumed to hold true throughout the paper: By Π_n, $n \in \mathbb{N}$, we denote the collection of all complex polynomials of degree at most n, and by $\mathbb{P}_n$ the subset of polynomials P normalized by

$$P(z) = \prod_{x \in Z(P)} H(z,x), \tag{1.8}$$

where

$$H(z,x) := \begin{cases} z-x & \text{for} \quad |x| \le 1 \\ (z-x)/|x| & \text{for} \quad 1 < |x| < \infty \\ 1 & \text{for} \quad x = \infty, \end{cases} \tag{1.9}$$

is the so-called standard linear factor, and $Z(P)$ the set of all zeros of a polynomial P taking into account multiplicities. As general normalization of Hermite-Padé polynomials, we assume

$$P_m^* \in \mathbb{P}_{n_m} \quad \text{and} \quad Q_0^* \in \mathbb{P}_{|n|-n_0}. \tag{1.10}$$

We remark that the polynomials P_j, P_j^*, Q_j, Q_j^*, and the remainder terms R and R_{ij} depend on the multi-index $n \in \mathbb{N}^{m+1}$, which we may emphasize, if appropriate, by writing P_{jn}, P_{jn}^*, Q_{jn}, Q_{jn}^*, R_n, and R_{ijn}, respectively.

The polynomials of Definition 1.1 and 1.2 were introduced and studied for the first time by Hermite in 1873 ([He1] and [He2]). A detailed investigation of their algebraic properties is contained in [Ma], a paper that has been published only in 1968, but was written about 1935. The investigations in [Ma] have been continued in [Ja] and [Co]. As general reference, we shall use the survey paper of J. Nuttall [Nu1].

Besides of the two types of Hermite-Padé polynomials, we consider three types of approximants, which are closely related with the two types of polynomials.

DEFINITION 1.3 (Algebraic Hermite-Padé approximants)
Let the system $(f_0, \dots, f_m)$ be defined by

$$f_j := f^j, \quad j = 0, \dots, m, \tag{1.11}$$

where f is assumed to be a function analytic near infinity. An algebraic function $A_n(z) = w(z)$, $n \in \mathbb{N}^{m+1}$, solving the equation

$$(1.12) \qquad R(z,w) := \sum_{j=0}^{m} P_j(\tfrac{1}{z}) w^j \equiv 0$$

and having a contact to f as large as possible at infinity, is called algebraic Hermite-Padé approximant to f at infinity (of order m and multidegree n). The polynomials P_j in (1.12) are assumed to satisfy (1.1).

Remark: Starting from (1.1) we arrive at (1.12) by truncating the left-hand side of (1.1) and substituting f by w. Thus, w is an algebraic function of degree m, which should approximate f near infinity at least in some sense, and indeed, will do so in a much larger domain, as we shall see in Theorem 5 1 below

The algebraic approximants of Definition 1.3 have been introduced by Padé in 1892 in his dissertation [Pa]. But he discusses them only very shortly, and passes then immediately to the special case $m = 1$, which defines rational approximants, the now so-called Padé approximants.

Instead of the algebraic equation (1.12), we can also consider differential equations. This leads to the next definition:

DEFINITION 1.4 (Integral Hermite-Padé approximants)
Let the system $(f_0, \dots, f_m)$ be defined by

$$(1.13) \qquad f_j := f^{(j)}, \quad j = 0, \dots, m,$$

where f is assumed to be analytic near infinity, and the superscript (j) denotes the j-th formal derivative of the power series in powers of $\frac{1}{z}$ representing f. The integral (or the solution) $y_n(z) = y(z)$, $n \in \mathbb{N}^{m+1}$, of the differential equation

$$(1.14) \qquad F(y^{(m)}, \dots, y', y) := \sum_{j=0}^{m} P_j(\tfrac{1}{z}) y^{(j)}(z) \equiv 0$$

having appropriate contact with f at infinity, is called integral Hermite-Padé approximant to f at infinity (with multidegree n). As in Definition 1.3, the polynomials P_j in (1.14) have to satisfy (1.1), i.e. they are Hermite-Padé polynomials of type I.

The third and last approximant to be considered in this note is based on Hermite-Padé polynomials of type II:

<u>**DEFINITION 1.5**</u> (Simultaneous rational approximants)
Let $f_0 \equiv 1$, and let $f_1, \ldots, f_m$ be m functions, which are analytic near infinity. Let $Q_0, \ldots, Q_m$ be Hermite-Padé polynomials of type II, then the vector

$$(1.15) \qquad \left(\frac{Q_1}{Q_0}, \ldots, \frac{Q_m}{Q_0}\right)$$

of rational functions of degree at most $(\tilde{n}_1/\tilde{n}_0, \ldots, \tilde{n}_m/\tilde{n}_0)$ with common denominator polynomial Q_0 is called <u>simultaneous rational approximant</u> to the vector function $(f_0, \ldots, f_m)$.

<u>**Remark 1**</u>: If $m=1$, then Definitions 1.3 and 1.5 fall together with the ordinary Padé approximant to $-f_1/f_0$ and f_1, respectively.
<u>**Remark 2:**</u> If $m>1$, then the approximants of Definition 1.3, 1.4, and 1.5 are in general not unique for every $n \in \mathbb{N}^{m+1}$, which constitutes a remarkable contrast to the unique definition of all entries in the Padé table.
<u>**Remark 3**</u>: Besides of the integral approximant defined by the homogeneous linear differential equation (1.14) in Definition 1.4, it is possible to consider integral approximants defined analogously by an inhomogeneous linear differential equation. Such a definition has, for instance, been used in [BaLu].

In this note we first present new results about the n-th root asymptotics of Hermite-Padé polynomials of both types, and then give results about the convergence of the three types of approximants introduced in Definition 1.3, 1.4, and 1.5. These latter results are based on the asymptotics for the Hermite-Padé polynomials. We state only results and will give no proofs. The general ideas underlying these proofs are discussed in [St3, Part 1]. A more detailed description of the proofs can be found in [St2] for the case $m=1$.

A survey of known results in the area of Hermite-Padé polynomials with special emphasis to asymptotics is contained in [Nu1] The first convergence results for algebraic Hermite-Pade approximants seems to be those of Shafer [Sh]. A Montessus-de-Ballore-type theorem for algebraic, as well as for integral Hermite-Padé approximants has been proved in [BaLu]. In comparison to the situation in the area of algebraic and integral approximants, there seems to exist more results and a more extensive literature about the convergence of simultaneous rational approximants. We will here mention only [GoRa], [Ni], [GMSa], and [dBr].

The outline of the paper is the following: In the next section we formulate general assumptions, which are assumed to hold true throughout all sections. Among other things, we shall describe the class of functions for which our results are valid. In Section 3, two probability measures ν_1 and ν_2 will be introduced together with certain domains S_1 and S_2, called convergence domains. The two measures will turn out to be the asymptotic distributions of the zeros of (close-to-) diagonal sequences of Hermite-Padé polynomials of type I and type II, respectively. In Section 4, we shall state these asymptotic results. In Section 5, we then come to the convergence results for algebraic and integral Hermite-Padé approximants, and in Section 6, to the convergence results for simultaneous rational approximants. The two domains S_1 and S_2 turn out to be the convergence domains for the two groups of approximants.

2. Assumptions. First, we specify the class of functions to be considered in this note, and then we furmulate certain determinantal conditions, which turn out to be the essential preconditions for the results in Section 4, 5, and 6 to hold true.

DEFINITION 2.1 By R^* we denote the collection of all functions f that are locally analytic in a domain $\bar{\mathbb{C}} \setminus E(f)$, where $E(f)$ is a compact subset of $\bar{\mathbb{C}}$ of (logarithmic) capacity $\operatorname{cap}(E(f)) = 0$.

We note that the property of a set to be of capacity zero can be defined for subsets of the whole Riemann sphere $\bar{\mathbb{C}}$ or even for subsets of an arbitrary Riemann surface over $\bar{\mathbb{C}}$, because this property is essentially a local one.

Since in Definition 2.1 we have only demanded local analyticity, the functions of R^* may have branch points, and indeed, they have to have branch points in order that Condition A and B (given below) can be satisfied.

ASSUMPTION 2.2 We assume that $f_0, \ldots, f_m \in R^*$, $m \geq 1$, and further that the functions f_j are analytic near infinity, and $f_0(\infty) \neq 0$.

The assumption $f_0(\infty)\neq 0$ has been made only because of definiteness. It can easily be established by a simple transformation.

DEFINITION 2.3: In the same way as in the case of a single analytic function, we can define a Riemann surface $\mathcal{R}=\mathcal{R}(f_0,\dots,f_m;\infty)$ by simultaneous analytic continuation of the system $(f_0,\dots,f_m)$ starting at $\infty\in\bar{\mathbb{C}}$. By $\pi:\mathcal{R}\to\bar{\mathbb{C}}$, we denote the canonical projection of $\mathcal{R}$ on $\bar{\mathbb{C}}$.

The surface $\mathcal{R}$ consists of only one sheet if, and only if, each function f_j is single-valued in $\bar{\mathbb{C}}\setminus E(f_j)$. In this case we have $\mathcal{R}=\bar{\mathbb{C}}\setminus(E(f_0)\cup\dots\cup E(f_m))$.

The surface $\mathcal{R}$ can be broken down in a finite or a denumerably infinite number of sheets $B_0, B_1,\dots$, each of which lies 'schlicht' over $\bar{\mathbb{C}}$, and covers $\bar{\mathbb{C}}$ up to a set of capacity zero. Such a decomposition can be done in various ways, however, in every case we will assume that the point infinity, to which we lift the developments (1.1), (1.3), (1.4), and (1.6), belongs to the sheet B_0, and the point itself is denoted by $\infty^{(0)}$. Generally, by $z^{(j)}\in\mathcal{R}$, we denote points on the j-th sheet B_j, $j=0,1,\dots$

DEFINITION 2.4: We say that k points $z^{(1)},\dots,z^{(k)}\in\mathcal{R}$ lie over the same basis point $z\in\bar{\mathbb{C}}$ if

(2.1) $$\pi(z^{(j)})=z \quad \text{for } j=1,\dots,k,$$

and we say that these points lie over the same basis point on adjacent sheets if (2.1) holds true, and if in addition there exists a sheet structure $\{B_0,B_1,\dots\}$ such that the k points $z^{(1)},\dots,z^{(k)}$ are connected in the union $B_0\cup\dots\cup B_k$.

Let us now assume that $\mathcal{R}$ has at least k sheets, let $z^{(0)},\dots,z^{(k-1)}$ be k points lying over the same basis point $z\in\bar{\mathbb{C}}$, and let $\{g_1,\dots,g_k\}\subset\{f_0,\dots,f_m\}$. Then we introduce the following determinant

(2.2) $$D_k(g_1,\dots,g_k;z^{(0)},\dots,z^{(k-1)}):=\begin{vmatrix} g_1(z^{(0)}) & \dots & g_k(z^{(0)}) \\ \vdots & & \vdots \\ g_1(z^{(k-1)}) & \dots & g_k(z^{(k-1)}) \end{vmatrix},$$

which will play an important role in the subsequent investigations. It is easy to see that D_k is a locally analytic function of the basis point $z = \pi(z^{(j)})$, and it is globally determined if the functions g_j, $j = 1, \dots, k$, have been selected in neighbourhoods lying over a given basis point $z \in \bar{\mathbb{C}}$, (which will normally be $\infty \in \bar{\mathbb{C}}$).

We can now formulate the two determinantal conditions, which have already been mentioned in the introduction:

CONDITION A: Let $\mathcal{R} = \mathcal{R}(f_0, \dots, f_m; \infty)$ have at least $m+1$ sheets. For any set of $m+1$ points $z^{(0)}, \dots, z^{(m)} \in \mathcal{R}$, lying over the same basis point $z \in \mathbb{C}$ on adjacent sheets with $z^{(0)}$ lying on the 0-th sheet, we have

$$D_{m+1}(f_0, \dots, f_m; z^{(0)}, \dots, z^{(m)}) \not\equiv 0. \tag{2.3}$$

CONDITION B: Let $\mathcal{R} = \mathcal{R}(f_0, \dots, f_m; \infty)$ have at least m sheets. For every subset $\{g_1, \dots, g_m\} \subset \{f_0, \dots, f_m\}$ of m functions, and for any set of m points $z^{(0)}, \dots, z^{(m-1)} \in \mathcal{R}$, lying over the same basis point $z \in \bar{\mathbb{C}}$ on adjacent sheets with $z^{(0)}$ lying on the 0-th sheet, we have

$$D_m(g_1, \dots, g_m; z^{(0)}, \dots, z^{(m-1)}) \not\equiv 0. \tag{2.4}$$

Condition A and B can also be formulated in the following, alternative way: Let $B_0, \dots, B_m$ be $m+1$ adjacent sheets of $\mathcal{R}$, and let

$$(f_0(z^{(j)}), \dots, f_m(z^{(j)})), \quad z^{(j)} \in B_j, \quad \pi(z^{(j)}) = z, \quad j = 0, \dots, m, \tag{2.5}$$

be the $m+1$ vector functions that we get on the $m+1$ sheets B_j, $j = 0, \dots, m$ by simultaneous analytic continuation of $(f_0, \dots, f_m)$ starting at $\infty^{(0)}$. Condition A is satisfied if, and only if, the $m+1$ vector functions (2.5) are linearly independent. If this is the case, one says that the system $(f_0, \dots, f_m)$ has <u>monodromic dimension</u> at least $m+1$.

In the same way, we can say that Condition B is satisfied if, and only if, every subvector of $(f_0, \dots, f_m)$ with m components has monodromic dimension at least m.

The concept of monodromic dimension goes back to Riemann [Ri], however, its significance for the investigation of Hermite-Padé polynomials was dicovered and very fruitfully used by David and Gregory Chudnovsky [Ch1], [Ch2].

We shall consider some simple examples to demonstrate the notions introduced in this section.

Example 2.5: Let be $f(z) := [1-z^{-3}]^{1/3}$, where the branch of the root is chosen so that $f(\infty)=1$. It is easy to see that $f \in \mathcal{R}^*$. As set of singularities we have $E(f) = \{0, 1, \frac{1}{2}(1 \pm i\sqrt{3})\}$. Three point of $E(f)$ are branch points, and f is therefore only locally analytic in $\overline{\mathbb{C}} \setminus E(f)$. The natural domain of definition for f is a Riemann surface $\mathcal{R} = \mathcal{R}(f;\infty)$ with three sheets. With the function f we construct a system with three elements by

$$(2.6)\qquad (f_0, f_1, f_2) = (1, f, f^2) = (1, [1-z^{-3}]^{1/3}, [1-z^{-3}]^{2/3}).$$

Simultaneous analytic continuation of the system (2.6) leads to the same Riemann surface as the single continuation of f alone, i.e. we have $\mathcal{R}(f;\infty) = \mathcal{R}(f_0, f_1, f_2; \infty)$. Further, we have $(f_0, f_1, f_2)(\infty) = (1, 1, 1)$, and it is easy to see that on the three sheets $B_0, B_1, B_2 \subset \mathcal{R}(f_0, f_1, f_2; \infty)$ with $\omega := \frac{1}{2}(1+i\sqrt{3})$ we have

$$(2.7)\qquad f_j(z^{(\ell)}) = [\omega^\ell f(z)]^j \quad \text{for } \ell, j = 0,1,2;\ z^{(\ell)} \in B_\ell,\ \pi(z^{(\ell)}) = z.$$

Using (2.7), we immediately deduce

$$(2.8)\qquad D_3(f_0, f_1, f_2; z^{(0)}, z^{(1)}, z^{(2)}) = \begin{vmatrix} 1 & f(z) & f(z)^2 \\ 1 & \omega f(z) & \omega^2 f(z)^2 \\ 1 & \omega^2 f(z) & \omega f(z)^2 \end{vmatrix} \not\equiv 0,$$

which shows that Condition A is satisfied, and further we deduce

$$(2.9a)\qquad D_2(f_0, f_1; z^{(0)}, z^{(1)}) = \begin{vmatrix} 1 & f(z) \\ 1 & \omega f(z) \end{vmatrix} \not\equiv 0$$

$$(2.9b)\qquad D_2(f_0, f_1; z^{(0)}, z^{(2)}) = \begin{vmatrix} 1 & f(z) \\ 1 & \omega^2 f(z) \end{vmatrix} \not\equiv 0$$

(2.9c) $$D_2(f_0, f_2; z^{(0)}, z^{(1)}) = \begin{vmatrix} 1 & f(z)^2 \\ 1 & \omega^2 f(z)^2 \end{vmatrix} \neq 0$$

(2.9d) $$D_2(f_0, f_2; z^{(0)}, z^{(2)}) = \begin{vmatrix} 1 & f(z)^2 \\ 1 & \omega f(z)^2 \end{vmatrix} \neq 0$$

(2.9e) $$D_2(f_1, f_2; z^{(0)}, z^{(1)}) = \begin{vmatrix} f(z) & f(z)^2 \\ \omega f(z) & \omega^2 f(z)^2 \end{vmatrix} \neq 0$$

(2.9f) $$D_2(f_1, f_2; z^{(0)}, z^{(2)}) = \begin{vmatrix} f(z) & f(z)^2 \\ \omega^2 f(z) & \omega f(z)^2 \end{vmatrix} \neq 0,$$

which shows that also Condition B is satisfied for system (2.6).

Example 2.6: Again, we choose $m = 2$ and define $f_0 \equiv 1$,

(2.10a) $$f_1(z) := \frac{1}{\pi} \int_a^b \frac{\sqrt{(x-a)(b-x)}}{x\,(x-z)}\, dx,$$

(2.10b) $$f_2(z) := \frac{1}{\pi} \int_c^d \frac{\sqrt{(x-c)(d-x)}}{x\,(x-z)}\, dx,$$

where $a < b < c < d$ are real numbers. Of course, $f_1, f_2 \in R^*$, $E(f_1) = \{a, b\}$, and $E(f_2) = \{c, d\}$. Simultaneous analytic continuation of the system $(1, f_1, f_2)$ leads to a Riemann surface $\mathcal{R}$ with four sheets, of which the tree first sheets B_0, B_1, and B_2 are most interesting for us: By B_0 we denote the sheet on which (2.10a) and (2.10b) is defined, and which is cut along the two intervals $[a, b]$ and $[c, d]$. The sheet B_1 is connected with B_0 along the cut $[a, b]$ by sticking both sheets cross-wise together, and the sheet B_2 is connected with B_0 in the same way along the cut $[c, d]$ (cf. Figure 2.1). The two sheets B_1 and B_2 are then finally connected via the fourth sheet B_3.

We consider the two functions f_1 and f_2 on the three sheets B_0, B_1, and B_2. The function f_1 has identical values on the two sheets B_0 and B_2, while the function f_2 has identical values on B_0 and B_1, i.e.

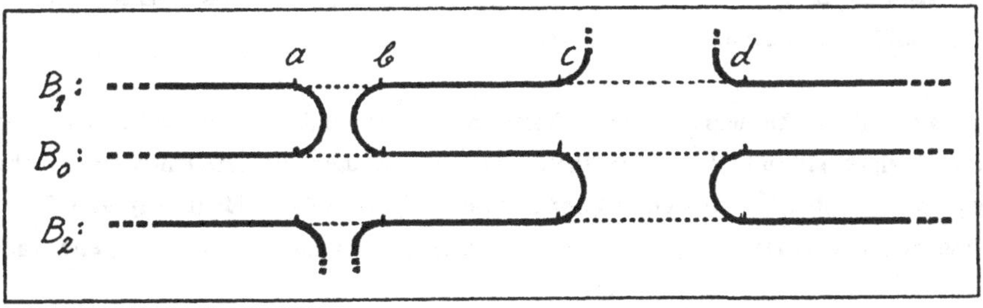

Figure 2.1

(2.11) $$f_1(z^{(0)}) \equiv f_1(z^{(2)}) \text{ and } f_2(z^{(0)}) \equiv f_2(z^{(1)}) \quad \text{for } \pi(z^{(j)}) = z.$$

From (2.11) it follows that

(2.12) $$D_3(f_0,\dots,f_2;z^{(0)},\dots,z^{(2)}) = \begin{vmatrix} 1 & f_1(z^{(0)}) & f_2(z^{(0)}) \\ 0 & f_1(z^{(1)})-f_1(z^{(0)}) & 0 \\ 0 & 0 & f_2(z^{(2)})-f_2(z^{(0)}) \end{vmatrix} \not\equiv 0,$$

which shows that Condition A is satisfied. Further, it is easy to see that the determinants analogous to (2.9a), (2.9c), (2.9e), and (2.9f) are again not identically zero, but because of the identities (2.11) we have

(2.13a) $$D_2(f_0,f_1;z^{(0)},z^{(2)}) = \begin{vmatrix} 1 & f_1(z^{(2)}) \\ 1 & f_1(z^{(2)}) \end{vmatrix} \equiv 0$$

(2.13b) $$D_2(f_0,f_2;z^{(0)},z^{(1)}) = \begin{vmatrix} 1 & f_2(z^{(1)}) \\ 1 & f_2(z^{(1)}) \end{vmatrix} \equiv 0,$$

which shows that Condition B is not satisfied in this example.

In [GoRa] Padé-Hermite polynomials of type II have been investigated for a class of functions, which includes the system considered here in Example 2.6. It turns out that the Hermite-Padé polynomials have asymptotics different from those descibed in Section 4, below. We sall come back to this aspect later on.

That the nice situation of Example 2.5, where both Conditions A and B are satisfied, has a more general background is shown by the next lemma:

LEMMA 2.7: If $f_j = f^j$, $j = 0, \ldots, m$, and $\mathcal{R} = \mathcal{R}(f; \infty)$ has at least $m+1$ sheets, then Conditions A and B hold true.

Remark: If $\mathcal{R}$ has less than $m+1$ sheets, then Condition A cannot even be formulated. If f is algebraic and $\mathcal{R}$ has less than $m+1$ sheets, then for sufficiently large multi-degrees $n \in \mathbb{N}^{m+1}$ the right-hand sides of (1.1) and (1.3) are identically zero. This phenomenon corresponds to the case of a rational function and the Padé table, where all entries are identical for large indices.

Proof: If $f_j = f^j$, $j = 0, \ldots, m$, then $\mathcal{R} = \mathcal{R}(f_0, \ldots, f_m; \infty) = \mathcal{R}(f; \infty)$, and Lemma 2.7 is an immediate consequence of the closed form of Vandermond's determinant

(2.14) $$D_{m+1}(f^0, \ldots, f^m; z^{(0)}, \ldots, z^{(m)}) = \prod_{0 \le i < j \le m} [f(z^{(i)}) - f(z^{(j)})] \neq 0.$$

⊠

Because of Lemma 2.7, we have not to worry about Condition A and B in case of algebraic Hermite-Padé appoximants introduced in Definition 1.3. Unfortunately, the situation is is not so nice for the system of functions $(f_0, \ldots, f_m)$ defined by $f_j = f^{(j)}$, $j = 0, \ldots, m$, in Definition 1.4. Here, a result analogous to Lemma 2.8 does not exist. Especially, it is is not sufficient for Condition A or B to hold true that $\mathcal{R}(f_0, \ldots, f_m; \infty)$ has more than $m+1$ sheets. To illustrate this, we give the next example:

Example 2.8: Let be

(2.15) $$f(z) := 1 + \log \frac{z+1}{z-1},$$

and let the system (f_0, f_1, f_2) be defined by $(f, f^{(1)}, f^{(2)})$. In this case $\mathcal{R} = \mathcal{R}(f_0, f_1, f_2; \infty)$ has infinitely many sheets, but we have

(2.16) $$D_3(f, f^{(1)}, f^{(2)}; z^{(0)}, z^{(1)}, z^{(2)}) \equiv 0,$$

and further

(2.17) $$D_2(f^{(1)}, f^{(2)}; z^{(0)}, z^{(1)}) \equiv 0$$

for arbitrary adjacent sheets $\mathcal{B}_0, \mathcal{B}_1, \mathcal{B}_2$. Thus, neither Condition A nor Condition B holds true in this example. The reason for this, of course, is the fact that f satisfies a second order homogeneous linear differential equation.

3. Definition of the Measures ν_1 and ν_2 and of the Convergence Domains S_1 and S_2. The two measures to be defined in the present section will turn out to be the asymptotic distribution of the zeros of near-to-diagonal sequences of Hermite-Padé polynomials of type I and II, respectively. The two domains will be fundamental for the convergence results for all three types of approximants considered in this note. We start with some notations:

For a measure μ with support $S(\mu) \subset \bar{\mathbb{C}}$, we define the <u>(logarithmic) potential</u>

$$(3.1) \qquad p(\mu; z) := \int \log |H(z,x)| \, d\mu(x),$$

where $H(z,x)$ is the standard linear factor introduced in (1.9). Let $\mathcal{D}$ denote the set of all domains $D \subset \mathcal{R}$ possessing a (generalized) <u>Green's function</u> $g_D(z,w)$, i.e. $cap(\partial D) > 0$ for all $D \in \mathcal{D}$. On $\mathcal{R}$ we consider <u>Green potentials</u>

$$(3.2) \qquad g(\mu; z) = g(\mu, D; z) := \int g_D(z,w) \, d\mu(w),$$

where μ is a measure on $\mathcal{R}$ and $D \in \mathcal{D}$. We assume throughout the paper that $g_D(z,w) = 0$ for $w \in \mathcal{R} \setminus D$. An assertion is said to hold <u>quasi everywhere</u> (<u>qu.e.</u>) on a set $S \subset \mathcal{R}$ if it holds for every $z \in S$ with possible exceptions on a set of (outer) capacity zero.

Next, we define special functions on $\mathcal{R}$, which will be needed for the definition of the measures ν_j and the domains S_j, $j = 1,2$, but will play also an autonomous role in the convergence results.

<u>DEFINITION 3.1</u>: (i) For a measure ν on $\bar{\mathbb{C}}$ we define

$$(3.3) \qquad p_\nu(z) := p(\nu; \pi(z)), \quad z \in \mathcal{R},$$

which is the lifting of the potential $p(\nu; z)$ to $\mathcal{R}$.

(ii) For a domain $D \in \mathcal{D}$ with $\infty^{(0)} \in D$, and a probability measure ν on $\bar{\mathbb{C}}$, we define two functions $r(z) = r(\nu, D; z)$ and $d(z) = d(\nu, D; z)$ by

$$(3.4) \qquad r(z) := g((m+1)\delta_{\infty^{(0)}} - \pi^{-1}(\nu), D; z)$$

$$= (m+1) g_D(z, \infty^{(0)}) - \int g_D(z, w) d\nu(\pi(z)),$$

$$(3.5) \qquad d(z) := r(z) - p_\nu(z), \quad z \in \mathcal{R},$$

where δ_z denotes Dirac's measure for the point $z \in \bar{\mathbb{C}}$.

Remark 1: The function d is a modification of the $(m+1)$-th multiple of the Green's function $g_D(z, \infty^{(0)})$. At $\infty^{(0)}$ the function d has a logarithmic pole with residuum $m+1$ and inside of D the measure ν is repeated as many times as the domain D covers $S(\nu) \subset \bar{\mathbb{C}}$. If $\pi(D) \cap S(\nu) = \phi$, then $d(z) = (m+1) g_D(z, \infty^{(0)})$.

Remark 2: The function r is characterized by the following properties:

$$(3.6) \qquad r(z) = p_\nu(z) \text{ for } z \in \mathcal{R} \setminus D,$$

$$(3.7) \qquad r(z) = m \log|\pi(z)| + O(1) \quad \text{for } z \to \infty^{(0)},$$

$r(z) - \log|\pi(z)|$ is harmonic in $D \setminus \{\infty^{(0)}\}$ and continuous qu.e. on ∂D.

THEOREM 3.2: If assumption 2.2 is satisfied and if $\mathcal{R} = \mathcal{R}(f_0, \ldots, f_m; \infty)$ has at least $m+1$ sheets, then there uniquely exist a domain $S_1 \subset \mathcal{R}$ and a probability measure ν_1, $S(\nu_1) \subset \bar{\mathbb{C}}$, such that

(i) $\infty^{(0)} \in S_1$, and

(ii) $r(\nu_1, S_1; z)$ is harmonic in a neighbourhood of ∂S_1.

Remark: Let $S \in \mathcal{D}$ be a domain with a smooth boundary ∂S, and let $\partial/\partial n$ denote the normal derivative on ∂S directed towards S. On ∂S we define a measure μ_d by

$$(3.8) \qquad d\mu_d(z) := \frac{1}{2\pi} \frac{\partial}{\partial n} d(z)\, ds_z, \quad z \in \partial S.$$

It is not difficult to see that assertion (ii) in Theorem 3.2 is equivalent to

(3.9a) $\overline{\pi(S_1)} = S(\nu_1)$

and

(3.9b) $\pi^{-1}(\nu_1)|_{S_1} = \mu_{d(\nu_1, S_1; \cdot)}$.

Thus, the domain S_1 and the measure ν_1, introduced in Theorem 3.2, are determined in such a way that the normal derivative of d at a boundary point $z \in \partial S_1$ is equal to the density of the measure ν_1 at the point $\pi(z) \in \overline{\mathbb{C}}$. In the special case $m = 1$, this implies that S_1 is contained in one sheet of $\mathcal{R}$ only, and therefore, fits in $\overline{\mathbb{C}}$, and further that d has identical normal derivatives at two boundary points lying over the same basis point. From this last property it follows that $\pi(S_1)$ can be determined by the property of minimal capacity of the set $\overline{\mathbb{C}} \setminus \pi(S_1)$.

The uniqueness part of Theorem 3.2 is a rather immediate consequence of the maximum principle for harmonic functions. The existence of the measure ν_1 and the domain S_1 has to be, or at least, can be proved simultaneously with the results formulated in Theorem 4.3, below.

DEFINITION 3.3: The domain $S_1 \subset \mathcal{R}$ and the probability measure ν_1 on $\overline{\mathbb{C}}$, the unique existence of which has been established in Theorem 3.2, are called convergence domain of type I and asymptotic distribution of type I, respectively. The two functions $r(z) = r(\nu_1, S_1; z)$ and $d(z) = d(\nu_1, S_1; z)$ are called logarithm of the n-th root asymptotics of the remainder and the error term, respectively.

Example 3.4: We consider the system introduced in Example 2.5. There we have $m = 2$ and $\mathcal{R} = \mathcal{R}(f_0, f_1, f_2; \infty)$ consists of thee sheets. The support $S(\nu_1)$ of ν_1 is shown in Figure 3.1; it consists of three symetrically positioned rays starting at the points 1, ω, ω^2; $\omega := \frac{1}{2}(1 + i\sqrt{3})$, and being connected at infinity. The domain S_1 has two sheets over $\overline{\mathbb{C}} \setminus S(\nu_1)$ and only one sheet over $S(\nu_1)$. The points $1, \omega, \omega^2$ are branch points of $\mathcal{R}$. The domain S_1 contains the sheet B_0, and besides of that, continuations around the branch points 1, ω, or ω^2. These continuations belong to S_1 as long as the basis point does not meet one of the three rays of $S(\nu_1)$. In Figure 3.1 the continuations are sketched by two sample pathes.

Example 3.3: Again, we consider the system of Example 2.5, but now only the first two functions, i.e. $m=1$ and

$$(f_0, f_1) = (1, [1-z^{-3}]^{1/3}). \tag{3.10}$$

Thus, the domain S_1 is contained in one sheet of $\mathcal{R}$. In Figure 3.2, we have sketched $S(\nu_1)$. It is an inversion of the set $S(\nu_1)$ in the last example. Since $m=1$, the domain S_1 fits in a single sheet of $\mathcal{R}$. The domain $\pi(S_1)$ is determined by the two properties: (i) $[1-z^{-3}]^{1/3}$ is single-valued in $\pi(S_1)$, and (ii) $\bar{\mathbb{C}} \setminus \pi(S_1)$ is of minimal capacity (cf. [St2]). The probability measure ν_1 is equal to the equilibrium distribution on $S(\nu_1)$.

Like in the case $m=1$, where the property of minimal capacity of a set has rather far reaching consequences for the structure of the domain $\pi(S_1)$ (cf. [St1]), similar results can be proved about the structure of the domain S_1 in the general case $m>1$. Thus, for instance, ∂S_1 is piece-wise analytic, $\pi(S_1) = S(\nu_1)$, S_1 covers $\bar{\mathbb{C}} \setminus S(\nu_1)$

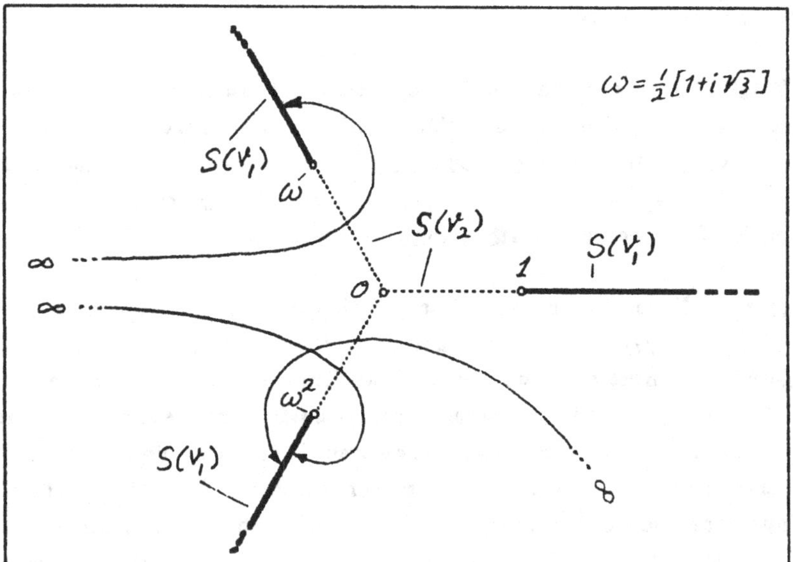

Figure 3.1

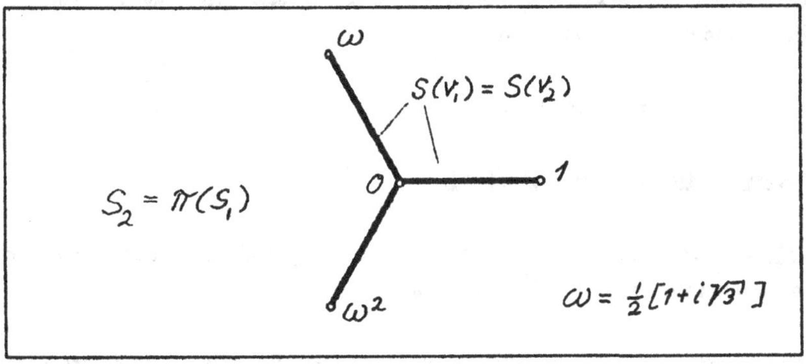

Figure 3.2

exactly m times, and $S(\nu_1)$ exactly $m-1$ times, and $\bar{S}_1$ covers $S(\nu_1)$ exactly $m+1$ times. A detailed investigation of the structure of S_1 can be found in [St3; Theorem 4.2].

We close this section with the definition of the domain S_2 and the probability measure ν_2. Starting point is the subdomain

(3.11) $$S_0 := \{z \in S_1;\ r(z) > r(w) \text{ for all } w \in S_1,\ w \neq z,\ \pi(w) = \pi(z)\}.$$

of the domain S_1. In case of $m = 1$, we have $S_0 = S_1$. Because of (3.7) $\infty^{(0)} \in S_0$, and it is easy to see that S_0 is contained in a single sheet of $\mathcal{R}$. Hence,

(3.12) $$S_2 := \pi(S_0)$$

is a domain in $\bar{\mathbb{C}}$, and $\bar{\mathbb{C}} \setminus S_2$ consists qu.e. of analytic arcs. Let $\partial/\partial n$ denote the normal derivative on ∂S_0 directed towards S_0. On ∂S_0 we define the measure μ_0 by

(3.13) $$d\mu_0(z) := \frac{1}{2\pi}\frac{\partial}{\partial n} r(\nu_1, S_1; z)\, ds_z, \quad z \in \partial S_0.$$

Since $r(\nu_1, S_1; z)$ has a logarithmic pole with residuum m at $\infty^{(0)}$, and is harmonic in $S_0 \setminus \{\infty^{(0)}\}$, we have

(3.14) $$\mu_0 \geq 0 \quad \text{and} \quad \|\mu_0\| = m.$$

DEFINITION 3.6: The domain S_2 defined in (3.12), is called convergence domain of type II, and the probability measure

$$\nu_2 := \frac{1}{m}\pi(\mu_0) \tag{3.15}$$

is called asymptotic distribution of type II.

Remark: In the special case $m=1$, the two measures ν_1 and ν_2 are equal, and they are identical to the equilibrium distribution of the set $S(\nu_1)$.

For an illustration of the definition let us turn back to Example 3.4: In Figure 3.1 the support $S(\nu_2)$ is shown by broken lines. The domain S_2 is the complement of this set. We remark that in this example ν_2 is different from the equilibrium distribution of $S(\nu_2)$. This is easy to verify: If ν_2 were the equilibrium distribution on $S(\nu_2)$, then the function $r(\nu_1, S_1; z)$ had to be constant on $S(\nu_2)$, which is not the case.

In Example 3.4 we have $m=1$, and therefore, $\nu_1 = \nu_2$, $\pi(S_1) = S_2$, and $S_1 = S_0$.

4. Asymptotics for Hermite-Padé Polynomials. In this section, n-th root asymptotics are stated for close-to-diagonal sequences of both types of Hermite-Padé polynomials.

With every polynomial P we associate a measure ν_P called zero distribution of P, which attributes to every zero of P a mass equal to the multiplicity of the zero. Hence, $\|\nu_P\| = \deg(P)$. The notation $\mu_n \xrightarrow{*} \mu$ means that the sequence $\{\mu_n\}$ of measures converges weakly to μ.

DEFINITION 4.1: We say that a sequence $\{f_n\}$ of functions, defined in a domain $D \subset \mathbb{C}$ converges in capacity in D to the function f, if for every compact set $V \subset D$, and for every $\varepsilon > 0$

$$\lim_{n\to\infty} \operatorname{cap}\{z \in V;\ |f_n(z) - f(z)| > \varepsilon\} = 0. \tag{4.1}$$

This convergence is denoted by

$$f_n \xrightarrow[D]{cap} f \quad \text{for} \quad n \to \infty. \tag{4.2}$$

Since definition (4.1) is essentially local, it can be lifted in an obvious way to domains D of a Riemann surface $\mathcal{R}$, or to subdomains of the Riemann sphere $\bar{\mathbb{C}}$. For convergence in this more general domains we also use notation (4.2).

DEFINITION 4.2: A sequence $N = \{n\} = \{n_0, \dots, n_m\}$ of multi-indices is called close-to-diagonal if

(4.3) $$\lim_{|n| \to \infty, n \in N} \frac{n_j}{|n|} = \frac{1}{m+1} \qquad \text{for} \quad j = 0, \dots, m.$$

The asymptotic results for Hermite-Padé polynomials of type I are contained in the next two theorems:

THEOREM 4.3: Let $(f_0, \dots, f_m)$ be a system satisfying Assumption 2.2, and let us assume that the Conditions A and B hold true, further let N be a close-to-diagonal sequence of multi-indices, and v_1 the asymptotic distribution of type I, introduced in Definition 3.3. Then for the sequence $\{P_{0n}^*, \dots, P_{mn}^* \, ; \, n \in N\}$ of inverse Hermite-Padé polynomials of type I, the following asymptotics hold true:

(i) For every $j = 0, \dots, m$ we have

(4.4) $$\frac{1}{n_j} v_{P_{jn}^*} \xrightarrow{\;*\;} v_1 \qquad \text{for} \quad |n| \to \infty, \; n \in N.$$

(ii) For every $j = 0, \dots, m$ we have

(4.5) $$\limsup_{|n| \to \infty, n \in N} \left| P_{jn}^*(z) \right|^{\frac{1}{n_j}} \leq \exp(\, p(v_1; z))$$

locally uniformly for $z \in \mathbb{C}$, equality holds true in (4.5) qu.e. in $\bar{\mathbb{C}}$, and

(4.6) $$\left| P_{jn}^* \right|^{1/n_j} \xrightarrow[\bar{\mathbb{C}} \setminus S(v_1)]{cap} \exp(p(v_1; \cdot)) \quad \text{for} \quad |n| \to \infty, \; n \in N.$$

Remark 1: The most remarkable feature of Theorem 4.3 can perhaps be seen in the fact that all $m+1$ polynomials $P_0^*, \dots, P_m^*$ have identical n-th root asymptotics. This is essentially a consequence of Condition B. The investigations in the second part of [St3] show that there also exist n-th root asymptotics for the polynomials $P_0^*, \dots, P_m^*$ if

Condition B is not satisfied, however, the asymptotics may then be different for different indices j.

Remark 2: In Assumption 2.2, it has been demanded that all functions f_j, $j = 0, \dots, m$, belong to the class R^*. Actually, this assumption can be weakened to the requirement that the $m+1$ functions are analytic only in the convergence domain of type I $S_1 \subset \mathcal{R}$, have sufficiently smooth boundary values qu.e. on ∂S_1, and have non-vanishing discontinuity functions associated with the arcs of ∂S_1.

Remark 3: From (4.4) it follows that

$$1 = \lim_{|n| \to \infty, n \in N} \frac{1}{n_j} \deg(P_{jn}) = \lim_{|n| \to \infty, n \in N} \frac{1}{n_j} \deg(P_{jn}^*) \tag{4.7}$$

for every $j = 0, \dots, m$.

THEOREM 4.4: Let S_1 be the convergence domain of type I, and r the function introduced in Definition 3.3. Under the same assumptions as in Theorem 4.3, the following asymptotics hold true for the remainder term $R = R_n$, $n \in N$, in (1.3):

(i) We have

$$\limsup_{|n| \to \infty, n \in N} |R_n(z)|^{\frac{1+m}{|n|}} \leq \exp(-r(v_1, S_1; z)) \tag{4.8}$$

locally uniformly in $\mathcal{R} \setminus [\pi^{-1}(\infty) \setminus \{\infty^{(0)}\}]$, equality holds true in (4.8) qu.e. in $\mathcal{R}$, and

$$|R_n|^{\frac{m+1}{|n|}} \xrightarrow[D]{cap} \exp(-r(v_1, S_1; \cdot)), \quad D = \bar{S} \setminus [\mathcal{R} \setminus \pi^{-1}(S(v_1))], \tag{4.9}$$

for $|n| \to \infty$, $n \in N$.

(ii) Let $V \subset S_1$ be a compact set, and let $Z(R_n, V)$ be the set of zeros of R_n on V, taking account of multiplicities, then

$$\lim_{|n| \to \infty, n \in N} \frac{1}{|n|} \operatorname{card} Z(R_n, V) = \begin{cases} \frac{m}{m+1} & \text{if } \infty^{(0)} \in V \\ 0 & \text{if } \infty^{(0)} \notin V. \end{cases} \tag{4.10}$$

Remark: In the special case $V = \{\infty^{(0)}\}$, (4.10) is a weak form of Mahler's notion of perfectness of a system [Ma]. Mahler's strong definition has been shown in [Ja] to hold true for exponential, binomial, and under certain additional conditions, also for logorithmic systems of functions. In Section 5, Theorem 4.4 is the basis of the convergence results for algebraic and integral Hermite-Padé approximants presented there.

We now come to asymptotics for Hermite-Padé polynomials of type II. Again, we primarily consider the inverse polynomials $Q_0^*, \dots, Q_m^*$, introduced in (1.6). Of course, the connection (1.5) with the original Hermite-Padé polynomials of type II immediately gives asymptotics for them as well. We have $deg(Q_{jn}) \leq \tilde{n}_j = |n| - n_j$, $j = 0, \dots, m$. With a multi-index $n \in \mathbb{N}^{m+1}$, we associate an index of type II defined as $\tilde{n} = (|n| - n_0, \dots, |n| - n_m)$, and we say that a sequence $\tilde{N} \subset \mathbb{N}^{m+1}$ is a type II variant of a close- to-diagonal sequence of indices $N \subset \mathbb{N}^{m+1}$ if

(4.11) $$\tilde{N} = \{ \tilde{n} = (\tilde{n}_0, \dots, \tilde{n}_m);\ \tilde{n}_j = |n| - n_j,\ j = 0, \dots, m,\ n \in N \},$$

and N satisfies (4.3) in Definition 4.2.

<u>THEOREM 4.5</u>: <u>Let $(1, f_0, \dots, f_m)$ be a system satisfying Assumption 2.2, and let us assume that the Conditions A and B hold true, further let N be a close-to-diagonal sequence of multi-indices, $\tilde{N}$ the associated sequence of type II, and ν_2 the asymptotic distribution of type II, introduced in Definition 3.6. Then for the sequence $\{Q_{0n}^*, \dots, Q_{mn}^*;\ n \in N\}$ of inverse Hermite-Padé polynomials of type II, the following asymptotics hold true</u>:

(i) <u>For every $j = 0, \dots, m$ we have</u>

(4.12) $$\frac{1}{\tilde{n}_j} \nu_{Q_{jn}^*} \xrightarrow{\ *\ } \nu_2 \quad \text{for} \quad |n| \to \infty,\ n \in N.$$

(ii) <u>For every $j = 0, \dots, m$ we have</u>

(4.13) $$\limsup_{|n| \to \infty,\ n \in N} |Q_{jn}^*(z)|^{1/\tilde{n}_j} \leq \exp(p(\nu_2; z))$$

<u>locally uniformly for $z \in \mathbb{C}$, equality holds true in (4.5) qu.e. in $\overline{\mathbb{C}}$, and</u>

(4.14) $$|Q_{jn}^*|^{1/\tilde{n}_j} \xrightarrow[\overline{\mathbb{C}} \setminus S(\nu_2)]{cap} e^{p(\nu_2; \cdot)} \quad \text{for} \quad |n| \to \infty,\ n \in N$$

Remark 1: Like in Theorem 4.3, Condition B has again to hold true in order that all polynomials $Q_0^*, \dots, Q_m^*$ have identical asymptotics.

Remark 2: Because of (3.13) and (3.15) all $Q_{0n}^*, \dots, Q_{mn}^*$ have, up to a constant factor, the same n-th root asymptotics in S_2, as the remainder term R_n of (1.3) has in $S_0 \subset \mathcal{R}$. (S_2 and S_0 have been defined in (3.11) and (3.12)).

Remark 3: Let $n_0 \leq n_j$ for $j = 1, \dots, m$. From (4.12), it follows that

$$(4.15) \quad 1 = \lim_{|n| \to \infty, n \in N} \frac{1}{\tilde{n}_j} \deg(Q_{jn}) = \lim_{|n| \to \infty, n \in N} \frac{1}{\tilde{n}_j} \deg(Q_{jn}^*), \; j = 0, \dots, m.$$

The last theorem in this section is concerned with the remainder terms R_{ijn}, $i \neq j$, in the definition of the Hermite-Padé polynomial of type II (cf. (1.6)). The theorem is the basis of the convergence result for simultaneous rational approximants, which will be stated in Section 6, below.

THEOREM 4.6: Let S_2 be the convergence domain of type II, introduced in Definition 3.6., r the function introduced in Definitition 3.3, and further let $g_2(z,w)$ be the Green's function of the domain S_2, and $\omega_0 = \omega_{S_0, \infty^{(0)}}$ the harmonic measure of $\infty^{(0)}$ on ∂S_0. Under the same assumptions as in Theorem 4.3, the following asymptotics hold true for the remainder terms R_{ijn}, $n \in N$, of (1.6):

(i) For $i, j = 0, \dots, m$, $i \neq j$, we have

$$(4.16) \quad \limsup_{|n| \to \infty, n \in N} |R_{ijn}(z)|^{\frac{m+1}{|n|}} \leq \exp(m\, p(v_2; z) - (m+1) g_2(z, \infty))$$

locally uniformly in $\mathbb{C}$, equality holds true in (4.16) qu.e. in $\mathbb{C}$, and

$$(4.17) \quad |R_{ijn}|^{\frac{m+1}{|n|}} \xrightarrow[S_2]{cap} \exp(m\, p(v_2; z) - (m+1) g_2(z, \infty))$$

for $|n| \to \infty$, $n \in N$.

(ii) Let $V \subset S_2$ be a compact set, and let $Z(R_{ijn}, V)$ be the set of zeros of R_{ijn} on V, taking account of multiplicities, then for $i, j = 1, \dots, m$, $i \neq j$, we have

$$(4.18) \quad \lim_{|n| \to \infty, n \in N} \frac{1}{|n|} \operatorname{card} Z(R_{ijn}, V) = \begin{cases} \frac{1}{m+1} & \text{if } \infty \in V \\ 0 & \text{if } \infty \notin V. \end{cases}$$

(iii) If we consider R_{ijn} as a function on S_0, where S_0 has been defined by (3.11), then for $i, j = 0, \dots, m$, $i \neq j$, we have

(4.19) $$\lim_{|n|\to\infty,\, n\in N} \frac{1}{|n|} \operatorname{card} Z(R_{ijn}, V) = \mu_0(V) - \omega_0(V), \quad V \subset S_0,$$

where μ_0 has been defined in (3.13).

The proofs of the Theorems 4.5 and 4.6 are based on one hand upon a multiple orthogonality, which holds true for the polynomials of type II (see, formula (6.11) in [St3]), and on the other hand on a generalization of Theorem 4.3 and 4.4 for the so-called Hermite-Padé polynomials of type I, defined by linear functionals instead of the interpolation property at ∞, which is underlying Definition 1.1.

Finally, we remark that in the results of this section, convergence in capacity can, in general, not be improved to a stronger type of convergence as for instance locally uniform convergence.

For more specific situations, power asymptotics (instead of n-th root asymptotics, which have been considered here) have been proved in [Nu2].

5. Convergence Results for Algebraic and Integral Hermite-Padé Approximants. In the present section we state convergence results for the first group of approximants of this note, namely for algebraic and integral Hermite-Padé approximants. They have been introduced in Definition 1.3 and 1.4, respectively. The results are consequences of the asymptotics for Hermite-Padé polynomials of type I, which have been stated in the last section in Theorem 4.3 and 4.4.

We start with algebraic Hermite-Padé approximants. For a given multidegree $n \in \mathbb{N}^{m+1}$ the approximant is defined as the root $w = w(z)$ of the m-th order algebraic equation

(5.1) $$R(w,z) := \sum_{j=0}^{m} w^j P_j\left(\tfrac{1}{z}\right) \equiv 0,$$

where the $P_j \in \Pi_{n_j}$ are Hermite-Padé polynomials of type I and multidegree $n \in \mathbb{N}^{m+1}$, defined in (1.1) for a system of functions $(1, f, f^2, \ldots, f^m)$. The function f, which generates this system, is assumed to be analytic near $\infty \in \bar{\mathbb{C}}$, and to belong to $\mathcal{R}^*$. From the m roots

(5.2) $$w_1(z), \ldots, w_m(z)$$

of equation (5.1), we have to select the root that has the highest contact to f at infinity. This selection is an unambiguous task if the root in question has no branch point at ∞, a situation that will be called the regular case. However, there may also arise a so-called irregular case, in which the roots with highest contact to f have a branch point at infinity.

Although the situation is somewhat more complicated in the irregular case, nevertheless, it is possible to select the branch of the approximants in an appropriate way. More details are given in [St3, Section 7].

We want to mention a further complication, which cannot arise in case of ordinary (rational) Padé approximants: The polynomials P_j, $j=0,\ldots,m$, are, in general, not unique, as has already been remarked in Section 1. Contrary to the situation of ordinary Padé approximants, this non-iquness can lead to different algebraic Hermite-Padé approximants for a given multidegree $n \in \mathbb{N}^{m+1}$. However, for functions f of the class $\mathcal{R}^*$ considered here, this possible non-uniqueness does not disturb the convergence of the approximants for $|n| \to \infty$.

Like the function f, we can lift also the m roots (5.2) to the Riemann surface $\mathcal{R}$. We will do this for the root that has highest contact to f at $\infty^{(0)}$. This root will be denoted by A_n, and it is called, in accordance with Definition 3.3, algebraic Hermite-Padé approximant to f and multidegree $n \in \mathbb{N}^{m+1}$.

Contrary to the function f, the approximant A_n may have branch points on $\mathcal{R}$. In order to make the approximant single-valued, we have to introduce cuts C_n, $n \in \mathbb{N}^{m+1}$, which are defined by

(5.3) $$C_n = C_n(P_0,\ldots,P_m) := \{z \in \mathcal{R};\ \exists i,j \in \{1,\ldots,m\},\ i \neq j,$$
$$|(f-w_i)(z)| = |(f-w_j)(z)| = \min_{1 \le l \le m} |(f-w_l)(z)|\}.$$

In the regular case, $\infty^{(0)} \notin C_n$, while in the irregular case, $\infty^{(0)} \in C_n$. It follows from the next theorem that for close-to-diagonal sequences N the capacity of the sets $S_1 \cap C_n$ tends to zero with $|n| \to \infty$, $n \in N$, where $S_1 \subset \mathcal{R}$ is the convergence domain of type I, introduced in Definition 3.3. This property implies that also the diameter of every component of $S_1 \cap C_n$ tends to zero with $|n| \to \infty$.

We now are prepared to state the main result about the convergence of algebraic Hermite-Padé approximants:

THEOREM 5.1: Let $f \in R^*$ be analytic near $\infty \in \bar{\mathbb{C}}$, let $\mathcal{R} = \mathcal{R}(f;\infty)$ have more than m sheets, let $N = \{n\} \subset \mathbb{N}^{m+1}$ be a close-to-diagonal sequence of multi-indices, $S_1 \subset \mathcal{R}$ the convergence domain of type I, and d the logarithm of the n-th root asymptotics of the error term, a function introduced in Definition 3.3. Then we have

$$A_n \xrightarrow[S_1]{cap} f \quad \text{for } |n| \to \infty,\ n \in N. \tag{5.4}$$

More precisely: For the error function we have

$$|f - A_n|^{\frac{m+1}{|n|}} \xrightarrow[S_1]{cap} e^{-d} \quad \text{for } |n| \to \infty,\ n \in N. \tag{5.5}$$

Remark 1: As already mentioned in Remark 2 after Theorem 4.3, the full extend of the analyticity of functions in R^* is not really necessary for Theorem 5.1 to hold true. It is sufficient if f is analytic in S_1, has sufficiently smooth boundary values on ∂S_1, and a non-vanishing discontinuity function associated with arcs of ∂S_1.

Remark 2: The domain $S_1 \subset \mathcal{R}$ extends over m sheets, and f is given on every sheet by a different branch. These m branches are approximated by the m roots (5.2) of equation (5.1). If we follow a path through the m sheets of $S_1 \setminus C_n$, then we run through the m branches of f and simultaneously, we run through the m roots of (5.1). Thus, on every sheet a different branch of f is approximated by a different element of (5.2).

Remark 3: For sufficiently large $|n|$, $S_1 \setminus C_n$ is connected, and the same is true for any fixed circular neighbourhood of $\infty^{(0)}$.

Remark 4: If $\mathcal{R}$ has not more than m sheets, and f has no essential singularities on $\mathcal{R}$, then f is algebraic, and therefore, $f \equiv A_n$ for $n \in N$ and $|n|$ sufficiently large. This phenomenon corresponds to the exact representation of a rational function in an ordinary Padé table.

Remark 5: Let $\mathcal{R}_n$, $n \in \mathbb{N}^{m+1}$, denote the Riemann surface with m sheets defined by the roots $w_j(z)$ of equation (5.1), and let $\mathcal{R}_m$ denote the Riemann surface that we get if we glue together pairs of arcs in ∂S_1 that have identical projections on $\bar{\mathbb{C}}$. Then $\mathcal{R}_n$ is a Riemann surface over $\mathcal{R}_m$. While the surface $\mathcal{R}_m$ is independent of $n \in N$, the genus of the surface $\mathcal{R}_n$ is, in general, increasing with $|n| \to \infty$. This growth is caused by the branch points that made necessary the cuts (5.3). The components of $S_1 \cap C_n$ can be considered as cross-sections of narrow pipes that connect sheets of $\mathcal{R}_n$, if we considered $\mathcal{R}_n$ as a Riemann surface over $\mathcal{R}_m$. With $|n| \to \infty$, $n \in N$, the capacity of these cross-sections tends to zero.

Remark 6: There is a rather general interpretation of Theorem 5.1: The Riemann surface $\mathcal{R}$ is defined by analytic properties of f, namely the analytic continuation of the power series defining f at infinity, while the algebraic Hermite-Padé approximants are defined by arithmetic operations, including the solution of an algebraic equation of order m. Both procedures use the same information, namely the coefficients of the development of f at infinity. However, the approximants use these cofficients only one by one, while analytic continuation uses them all at once. The results of Theorem 5.1 tell us that the arithmetic-algebraic recovery of f by the algebraic approximants A_n follows f exactly over m sheets of $\mathcal{R}$. Behind these m sheets we have divergence.

Remark 7: Taking $m = 1$, we see that ordinary Padé approximants are covered by Theorem 5.1. From this it follows that without a far more definite knowledge about the structure of the function f, as has been assumed in Theorem 5.1, we cannot expect that a type of convergence, essentially stronger than convergence in capacity, holds true. Especially, locally uniform convergence will in general not be true.

We now come to the second topic of the present section, the integral Hermite-Padé approximants. The name has been chosen, since the approximants y_n, $n \in \mathbb{N}^{m+1}$, are solutions, which are generally called integrals, of the differential equation

$$F(y^{(m)}, \ldots, y', y, z) := \sum_{j=0}^{m} y^{(j)} P_j(\tfrac{1}{z}) \equiv 0, \tag{5.6}$$

where again, as in (5.1), the $P_j \in \Pi_{n_j}$ are Hermite-Padé polynomials of type I and multidegree $n = (n_0, \ldots, n_m)$ associated with the system $(f, f', \ldots, f^{(m)})$. The function f generating this system is assumed to be analytic near $\infty \in \overline{\mathbb{C}}$, and to belong to R^*. By $f^{(j)}$, $j = 0, \ldots, m$, we denote the j-th formal derivative of the power series in $\frac{1}{z}$ representing f at infinity.

Like the algebraic Hermite-Padé approximants, also the integral Hermite-Padé approximants are not necessarily unique for a given multidegree $n \in \mathbb{N}^{m+1}$. But again, the possible non-uniquness does not disturb the convergence of the approximants for $|n| \to \infty$ if $f \in R^*$.

We will assume that for all multi-indices $n \in \mathbb{N}^{m+1}$, considered in the next theorem, the highest polynomial in (5.6) satisfies the condition

(5.7) $$P_m(0) \neq 0.$$

Thanks to this assumption, we have not to worry about the analyticity of the integral at infinity.

The integral Hermite-Padé approximant y_n, $n \in \mathbb{N}^{m+1}$, is defined, in accordance with Definition 1.4, as the solution of the differential equation (5.6). We lift the solution to the Riemann surface $\mathcal{R}$. There it has to satisfy the initial condition

(5.8) $$y^{(\ell)}(\infty^{(0)}) = f^{(\ell)}(\infty^{(0)}), \quad \ell = 0, \dots, m-1.$$

The approximant y_n has, in general, branch points on $\mathcal{R}$. By $\{y_{n\ell}, \ell \in L_n\}$ we denote the set of all branches defined by analytic continuation of y_n. The set L_n may be countably infinite or finite. In order to make y_n single-valued, we introduce cuts, which are defined by

(5.9) $$C_n = C_n(P_0, \dots, P_m) := \{z \in \mathcal{R};\ \exists ij \in L_n,\ i \neq j,\ |(f - y_{ni})(z)| = |(f - y_{nj})(z)| = \min_{\ell \in L_n} |(f - y_{n\ell})(z)|\}.$$

It is easy to see that y_n is single-valued in $\mathcal{R} \setminus C_n$.

The next theorem is the analogue of Theorem 5.1 for integral Hermite-Padé approximants:

THEOREM 5.2: Let $f \in R^*$ be analytic near $\infty \in \overline{\mathbb{C}}$, let Condition A be satisfied for the system of functions $(f, f', \dots, f^{(m)})$, let $N = \{n\} \subset \mathbb{N}^{m+1}$ be a close-to-diagonal sequence of multi-indices such that (5.7) is satisfied for every $n \in N$, let $S_1 \subset \mathcal{R}$ be the convergence domain of type I, and d the logarithm of the n-th root asymptotics of the error term, a function introduced in Definition 3.3. Then we have

(5.10) $$y_n \xrightarrow[S_1]{cap} f \quad \text{for } |n| \to \infty,\ n \in N.$$

More precisely: For the error function we have

(5.11) $$|f - y_n|^{\frac{m+1}{|n|}} \xrightarrow[S_1]{cap} e^{-d} \quad \text{for } |n| \to \infty,\ n \in N.$$

Remark: The results for the integral Hermite-Padé approximants are very similar to those for the algebraic Hermite-Padé approximants. An important difference, however, is the necessity to demand explicitly that Condition A is satisfied. The reason for that is the non-existence of an analogue to Lemma 2.7 in case of differential systems (1.13). Compare also Example 2.8.

6. Convergence Results for Simultaneous Rational Approximants. This last section is devoted to the convergence problem of simultanous rational approximants. The results are more or less a corollary to the Theorems 4.5 and 4.6, where the asymptotics of Hermite-Padé polynomials of type II and of the associated error terms have been stated.

Simultaneous rational approximants have been defined in Definition 1.5. For a multidegree $n = (n_0, \dots, n_m) \in \mathbb{N}^{m+1}$, and the associated multidegree of type II $\tilde{n} = (|n| - n_0, \dots, |n| - n_m)$ they are defined as the system

(6.1) $$\left(\frac{Q_{1n}(\frac{1}{z})}{Q_{0n}(\frac{1}{z})}, \dots, \frac{Q_{mn}(\frac{1}{z})}{Q_{0n}(\frac{1}{z})} \right)$$

of m rational functions with common denominator. We have $Q_{jn} \in \Pi_{\tilde{n}_j}$, $j = 0, \dots, m$ (cf. Definition 1.5).

The main result of the present section is the next theorem:

THEOREM 6.1: Let $f_j \in R^*$, $j = 1, \dots, m$, $m \geq 1$, be analytic near $\infty \in \bar{\mathbb{C}}$, let Condition A and B be satisfied for the system $(1, f_1, \dots, f_m)$, and let $N = \{n\} \subset \mathbb{N}^{m+1}$ be a close-to-diagonal sequence of multi-indices, $S_2 \subset \bar{\mathbb{C}}$ the convergence domain of type II, defined in Definition 3.6, and $g_2(z, \infty)$ the Green's function of the domain S_2. Then for every $j = 1, \dots, m$ we have

(6.2) $$\frac{Q_{jn}(\frac{1}{z})}{Q_{0n}(\frac{1}{z})} \xrightarrow[S_2]{cap} f_j(z) \text{ for } |n| \to \infty, \ n \in N.$$

More precisely, for every $j = 1, \dots, m$ the error functions $f_j(z) - Q_{jn}(\frac{1}{z})/Q_{0n}(\frac{1}{z})$ satisfy

(6.3) $$\left| f_j - \frac{Q_{jn}}{Q_{0n}} \right|^{\frac{m+1}{|n|}} \xrightarrow[S_2]{cap} e^{-g_2(\cdot, \infty)} \text{ for } |n| \to \infty, \ n \in N.$$

Remark 1: The convergence in capacity (6.3) of the error functions means that the n-th roots of the errors are asymptotically qu.e. circular. This implies that the estimate of the convergence rate given here is best possible. By the way: this observation holds also true for the two convergence theorems in Section 5.

Remark 2: Condition B is indispensible in Theorem 6.1. If Condition B does not hold true, simultaneous rational approximants may, nevertheless, converge in capacity, but in general, the convergence domain is not dense in $\bar{\mathbb{C}}$, and it may be different for different components of the approximant (cf. [St3, Part 2]).

Remark 3: In Theorem 6.1 it is not really necessary that the functions $f_j, j=1,\ldots,m$, have the full extend of analyticity, which is characteristic for the elements of R^*. The same situation, which has already been mentioned in Remark 2 to Theoerem 4.3 and Remark 1 to Theorem 5.1, is also valid in Theorem 6.1: It is sufficient that the m functions f_j are analytic in $S_1 \subset \mathcal{R} = \mathcal{R}(f_1,\ldots,f_m;\infty)$, the convergence domain of type I, have sufficiently smooth boundary values on ∂S_1, and non-vanishing discontinuity functions on ∂S_1. It may be interesting to note that it is in general not sufficient if the m functions f_j are analytic only in the smaller domain S_2 and have appropriate boundary values on ∂S_2.

With Theorem 6.1 we have completed the material we wanted to present in this note. It has already been said in the introduction that only results and no proofs are given here. The method of the proofs has been described in [St2] for the case $m=1$. A description of the ideas and general lines of the proofs for the results of the general case as stated here, can be found in [St3, Part 1]. (In Part 2 of [St3], the question has been investigated what happens if Condition B does not hold true).

ADDED IN PROOF: I am grateful to the referees of this paper for valuable corrections and clarifications.

References

[BaLu] Baker G.A., Jr. and Lubinsky D.S. (1986): Convergence theorems for rows of differential and algebraic Hermite-Padé approximants, Technical Report TWISK 407, NRIMS, Pretoria.

[dBr] de Bruin M.G. (1984): Some convergence results in simultaneous rational approximation to the set of hypergeometric functions $\{{}_1F_1(1;c_i;z); i=1,\ldots,n\}$, Springer Lect. Notes Math. 1071, pp. 12-33.

[Ch1] Chudnovsky D.V. (1980): Riemann monodromy problem, isomonodromy deformation equations and completely integrable systems, in "Bifurcation Phenomena in Mathematical Physics and Related Topics", Eds.: Bardos C. and Bessis D., Reidel, Dordrecht, pp. 385-447.

[Ch2] Chudnovsky G.V. (1980): Padé approximation and the Riemann monodromy problem, in "Bifurcation Phenomena in Mathematical Physics and Related Topics", Eds.: Bardos C. and Bessis D., Reidel, Dordrecht, pp. 449-510.

[Co] Coates J. (1966): On the algebraic approximation of functions I, II, III, IV, Proc. Kon. Nederl. Akad. Wet. Ser. A 69, pp. 421-461.

[GoRa] Gonchar A.A. and Rakhmanov E.A. (1981): On the convergence of simultaneous Padé approximants for systems of functions of Markov type (in Russian), Proc. Steklov Math. Inst. 157, pp. 31-48, (English transl. in Proc. Steklov Math. Inst. 3 (1983), pp. 31-50).

[GMSa] Graves-Morris P.R. and Saff E.B. (1984): A de Montessus theorem for vector valued rational interpolants, Springer Lect. Notes Math. 1105, pp. 227-242.

[He1] Hermite C. (1873): Sur la fonction exponentielle, in "Oeuvres" Vol. 3, pp. 150-181.

[He2] Hermite C. (1873): Sur quelques approximations algébriques, extrait d'une lettre à M. Borchardt, in "Oeuvres" Vol. 3, pp. 146-149.

[Ja] Jager H. (1964): A multidimensional generalization of the Padé table, Proc. Kon. Nederl. Akad. Wet. Ser. A 67, pp. 193-249.

[Ma] Mahler K. (1968): Perfect systems, Comp. Math. 19, pp. 95-166.

[Ni] Nikishin E. M. (1980): On simultaneous Padé approximants (in Russion), Mat. Sbornik 113, pp. 499-519, (English transl. in Math. USSR Sb. 41 (1982), pp. 409-425).

[Nu1] Nuttall J. (1984): Asymptotics of diagonal Hermite-Padé polynomials, J. Approx. Theory 42, pp. 299-386.

[Nu2] Nuttall J. (1986): Asymptotics of generalized Jacobi polynomials, will appear in Constr. Approx.

[Pa] Padé H. (1892): Sur la représentation approchée d'une fonction par des fractions rationelles, Thesis, Ann. École Nor. (3) 9, pp. 1-93, supplement.

[Ri] Riemann B. (1857): Zwei allgemeine Sätze über Differentialgleichungen mit algebraischen Coefficienten, in: "Collected Works of Bernhard Riemann", Ed.: Weber H., Dover, New York (1945), pp. 379-390.

[Sh] Shafer R. E. (1974): On quadratic approximation, SIAM J. Numer. Anal. 11, pp. 447-460.

[St1] Stahl H. (1985): The structure of extremal domains associated with an analytic function, Complex Variables 4, pp. 339-354.

[St2] Stahl H. (1987): Three different approches to a proof of convergence for Padé approximants, Springer Lect. Notes Math. 1237, pp. 79-124.

[St3] Stahl H. (1987): Asymptotics of Hermite-Padé polynomials and related approximants - A summary of results, Manuscript, Berlin TFH, February 1987.

RATIONAL APPROXIMATION

Chairmen:

J. Meinguet

L. Wuytack

Invited communications:

E. B. Saff
On the behavior of zeros and poles of best uniform polynomial and rational approximants.

J. Meinguet
Once again: the Adamjan–Arov–Krein approximation theory.

L. N. Trefethen*
Carathéodory–Féjer approximation and applications.

Short communications:

J. Karlsson*
Random notes on the sup-norm qualities of the best BMOA-approximants.

R. Kovacheva
Diagonal Padé approximants, rational Chebyshev approximants and poles of functions.

A. P. Magnus
On the use of the Carathéodory–Féjer method for investigating '1/9' and similar constants.

Lecture notes are not included.

ON THE BEHAVIOR OF ZEROS AND POLES OF BEST UNIFORM POLYNOMIAL AND RATIONAL APPROXIMANTS

R. Grothmann*
Mathem. Geogr. Fakultat
Katholische Universität
Eichstätt
Ostenstrasse 18
D-8078 Eichstätt
WEST GERMANY

E.B. Saff**
Institute for Constructive Mathematics
Department of Mathematics
University of South Florida
Tampa, Florida 33620
USA

ABSTRACT. We investigate the behavior of zeros of best uniform polynomial approximants to a function f, which is continuous in a compact set $E \subset \mathbb{C}$ and analytic on $\mathring{E}$, but not on E. Our results are related to a recent theorem of Blatt, Saff, and Simkani which roughly states that the zeros of a *subsequence* of best polynomial approximants distribute like the equilibrium measure for E. In contrast, we show that there might be another subsequence with zeros essentially all tending to ∞. Also, we investigate near best approximants. For rational best approximants we prove that its zeros and poles cannot all stay outside a neighborhood of E, unless f is analytic on E.

---

*The research of this author was done while visiting the Institute for Constructive Mathematics, University of South Florida, Tampa.

**The research of this author was supported, in part, by the National Science Foundation under grant DMS-8620098.

AMS Classification: 41A20
Key words and phrases: Polynomial approximation, rational approximation, best approximants, zero distributions.

A. Cuyt (ed.), Nonlinear Numerical Methods and Rational Approximation, 57–75.

1. STATEMENT OF RESULTS

We investigate the behavior of zeros and poles of best uniform approximants in the complex plane. Throughout this paper we will assume that $E \subset \mathbb{C}$ is a compact set and that $\overline{\mathbb{C}} \setminus E$ is connected. Using the Chebyshev norm on E,

$$\|g\|_E := \sup_{z \in E} |g(z)|, \tag{1.1}$$

we will approximate a function f, analytic on $\overset{o}{E}$ (the interior of E) and continuous in E, with respect to Π_n, the set of algebraic polynomials of degree at most n, or with respect to $\mathcal{R}_{m,n}$, the rational functions with numerator in Π_m and denominator in Π_n.

We denote by $p_n^*(f)$ the best uniform approximant to f on E with respect to Π_n, i.e.

$$e_n(f) := \|f - p_n^*(f)\|_E \le \|f - p_n\|_E \tag{1.2}$$

for all $p_n \in \Pi_n$. By a theorem of Mergelyan, we know $e_n(f) \to 0$. With $p_n^*(f)$ we associate a unit measure υ_n^*, called the *zero distribution* of $p_n^*(f)$, by

$$\upsilon_n^*(A) := \frac{\text{number of zeros of } p_n^*(f) \text{ in } A}{\deg p_n^*(f)} \tag{1.3}$$

for Borel sets $A \subset \mathbb{C}$, where we count the zeros according to their multiplicity.

The location of zeros of $p_n^*(f)$ or other converging

sequences of polynomials has been investigated by several authors. Jentzsch [J] proved in 1914 that the partial sums s_n of a power series with finite radius of convergence $r > 0$ have the property that every point on the circle $|z| = r$ is a limit point of zeros of the s_n. The related question in our setting is whether every point on the boundary of E is a limit point of zeros of the $p_n^*(f)$, if f is not analytic on E. The most complete and general answer to this question was obtained by Blatt, Saff, and Simkani [BSS] in 1986. Earlier investigations are due to Borwein [Bo] and Blatt and Saff [BS]. We should also mention similar results concerning sequences of so-called maximally convergent polynomials by Walsh [W2], who studied the case when f is analytic on E but not entire.

The results in [BSS] deal primarily with the limiting *distribution* of the zeros of the $p_n^*(f)$. The classical result concerning the distribution of zeros of partial sums of a power series is due to Szegö [Sz] and improves the above mentioned theorem of Jentzsch.

Before stating the known and new results we recall the definition of the *equilibrium measure* μ_E of E. Throughout the paper we will assume that $\overline{\mathbb{C}}\setminus E$ is *regular*; that is, $\overline{\mathbb{C}}\setminus E$ has a classical Green's function G with pole at ∞, which is continuous on ∂E (the boundary of E) and has the value zero on ∂E (cf. [T]). The function G is harmonic in $\mathbb{C}\setminus E$ and $G(z) - \log|z|$ is harmonic at ∞ and assumes the value $-\log[\mathrm{cap}(E)]$ at ∞, where $\mathrm{cap}(E) > 0$ is the *logarithmic capacity* of E (cf.[T]). The measure μ_E is the unique unit measure that is supported on ∂E and satisfies

$$(1.4) \qquad G(z) = \int \log|z - t|d\mu_E(t) - \log[\mathrm{cap}(E)].$$

It is also the unique unit measure that is supported on E and minimizes the energy integral

$$(1.5) \qquad I(\mu) := \iint \log|z - t|^{-1} d\mu(t)d\mu(z).$$

We can now state the main result of Blatt, Saff, and Simkani.

THEOREM 1.1 ([BSS]). *Let* $E \subset \mathbb{C}$ *be compact,* $\overline{\mathbb{C}}\backslash E$ *be connected and regular. Let* f *be continuous on* E, *analytic on* $\overset{o}{E}$ *but not on* E. *Furthermore, assume that* f *does not vanish identically on any component of* $\overset{o}{E}$. *Then the sequence* $\{\nu_n^*\}$ *in* (1.3) *possesses a subsequence* $\{\nu_{\ell(n)}^*\}$ *that converges weakly to the equilibrium measure* μ_E *of* E.

By the weak (vague) convergence of $\nu_{\ell(n)}^*$ we mean

$$\lim_{n\to\infty} \int \phi \, d\nu_{\ell(n)}^* = \int \phi \, d\mu_E$$

for all continuous ϕ on $\mathbb{C}$ having compact support.

Our first result shows that there may be another subsequence of $\{\nu_n^*\}$ that has a completely different behavior.

THEOREM 1.2. *Let* $E \subset \mathbb{C}$ *be compact with a connected and regular complement. Then there is a function* f *on* E *such that*

(a) f *is continuous on* E, *analytic on* $\overset{o}{E}$, *but not analytic on* E,

(b) f *has no zeros in* E, *and*

(c) *for a subsequence* $\{\nu^*_{k(n)}\}$ *of the measures* (1.3)

$$\lim_{n\to\infty} \nu^*_{k(n)}(S) = 0, \quad \textit{for all bounded } S \subset \mathbb{C}. \tag{1.6}$$

The proof of Theorem 1.2 will be given in Section 2.

We now discuss "near best approximants," i.e. we assume that we have a sequence $q_n \in \Pi_n$ such that

$$\|f - q_n\|_E \leq C\|f - p^*_n(f)\|_E, \quad \text{all } n \in \mathbf{N}, \tag{1.7}$$

for some constant C. (We caution the reader that our definition of "near best" is different from that in [BS].) As shown in the next example, we can no longer expect that every point on the boundary of E is a limit point of zeros of the q_n, let alone that the zero distributions ν_n associated with the q_n have a subsequence converging weakly to μ_E.

Example 1.1. Let $f(z) := \sqrt{z}$ for $\operatorname{Re} z \geq 0$ and set

$$K_\alpha := \{z \in \mathbb{C}: |z - \alpha| \leq \alpha\}. \tag{1.8}$$

Then, for $p_n \in \Pi_n$ and $\alpha > 0$,

$$\left\|\frac{1}{\sqrt{\alpha}} p_n(\alpha z) - f(z)\right\|_{K_1} = \frac{1}{\sqrt{\alpha}} \|p_n(w) - f(w)\|_{K_\alpha}, \tag{1.9}$$

where $\|\cdot\|_M$ denotes the Chebyshev norm on M. Thus, if $\{q_n\}_1^\infty$ is the sequence of best uniform approximants to f

on K_2, it satisfies (1.7) with $E := K_1$. But no point of $E\backslash\{0\}$ is a limit point of zeros of the q_n since $q_n(z) \to \sqrt{z}$ uniformly on K_2. An example similar to the above appears in [BIS].

Notice that in Example 1.1, the set E has a non-empty interior. The authors do not know of an analogous example for $E = [-1,1]$; however, the validity of the following conjecture would lead to such a result.

Conjecture. Set $E_1 := [-1,1]$,

(1.10) $$E_2 := \{z \in \mathbb{C}: |\mathrm{Re}\, z| \leq 2, \ |\mathrm{Im}\, z| \leq |\mathrm{Re}\, z|^2\},$$

and define

(1.11) $$f(z) := \begin{cases} z & \text{for } \mathrm{Re}\, z \geq 0, \\ -z & \text{for } \mathrm{Re}\, z < 0, \end{cases}$$

so that $f(x) = |x|$ for $x \in \mathbb{R}$. If we denote by $e_{i,n}$ the error of best uniform approximation of f on E_i with respect to Π_n, then we conjecture that there is a constant c such that

(1.12) $$e_{2,n} \leq c\, e_{1,n}, \quad \text{all} \quad n \in \mathbb{N}.$$

This conjecture would yield another example of a sequence $\{q_n\}$ satisfying (1.7), where $E = [-1,1]$ and q_n is the best approximant to f on E_2 in Π_n. Notice that no point $x \in [-1,1]\backslash\{0\}$ can be a limit point of zeros of such q_n.

The above discussion naturally raises the question as to whether near best approximants (in sense of (1.7)) necessarily have *at least one* limit point of zeros on ∂E when f is not analytic on E. Our next theorem lends evidence in the affirmative direction.

THEOREM 1.3. Let E *and* f *be as in Theorem 1.1 and assume that* f *does not vanish identically on any (open) component of* $\overset{\circ}{E}$ *or on any (closed) component of* E. *Suppose there exist constants* $c < 1$ *and* $K > 1$ *such that the errors (1.2) satisfy*

$$(1.13) \qquad \liminf_{n\to\infty} e_{[cn]}(f)/e_n(f) > K.$$

If $q_n \in \Pi_n$ *satisfies*

$$(1.14) \qquad \|f - q_n\|_E \le K\, e_n(f), \quad \text{all} \quad n \in \mathbf{N},$$

then at least one point of ∂E *must be a limit point of zeros of the sequence* $\{q_n\}_{n\in\mathbf{N}}$.

In (1.13) the symbol $[cn]$ denotes the integral part of cn. The proof of Theorem 1.3 will be given in Section 3.

Example 1.2. Theorem 1.3 in particular applies if $E = [-1,1]$ and

$$(1.15) \qquad \frac{D_1}{n^\alpha} \le e_n(f) \le \frac{D_2}{n^\alpha}, \quad \text{all} \quad n \in \mathbf{N},$$

for positive constants D_1, D_2, α. Indeed, (1.15) yields

$$(1.16) \qquad \liminf_{n\to\infty} \frac{e_{[cn]}(f)}{e_n(f)} \geq \frac{D_1}{c^\alpha D_2} > K,$$

for c small enough. By a famous result of Bernstein [Be], (1.15) is true with $\alpha = 1$ for the case $E = [-1,1]$ and $f(x) = |x|$. Thus for any $K \geq 1$, every sequence $\{q_n\}$ satisfying

$$(1.17) \qquad \big\| |x| - q_n(x) \big\|_{[-1,1]} \leq K\, e_n(|x|), \quad \text{all} \quad n \in \mathbf{N},$$

must have at least one limit point of zeros in $[-1,1]$. In fact, the origin must be a limit point of zeros of such q_n, since (1.17) remains true if $[-1,1]$ is replaced by $[-\varepsilon,\varepsilon]$, $\varepsilon > 0$, for the same sequence $\{q_n\}$.

The behavior of the zeros and poles of best *rational* approximants is far more delicate. An elementary result in this direction is the following.

THEOREM 1.4. *Let* $E \subset \mathbb{C}$ *be compact,* $\overline{\mathbb{C}}\backslash E$ *be connected and regular. Assume that* f *is continuous on* E, *analytic in* $\overset{\circ}{E}$, *and that* $r_n \in \mathcal{R}_{n,n}$ *satisfies*

$$(1.18) \qquad \|f - r_n\|_E \leq e_n(f), \quad \textit{all} \quad n \quad \textit{large}.$$

Suppose further that f *does not vanish identically on any (closed) component of* E. *If, for* n *large, all poles and zeros of* r_n *are outside a neighborhood* U *of* E, *then* f *is analytic on* E.

The proof of Theorem 1.4 will be given in Section 4. This theorem should be compared with results of A. Levin [L].

Remark 1. Theorem 1.4 applies, if r_n is a best approximant in $\mathcal{R}_{n,m(n)}$ for $m(n) \leq n$ to f.

Remark 2. In the proof of Theorem 1.4 we establish the stronger conclusion that if all zeros and poles of r_n lie outside the level curve $\Gamma_{\gamma_o} : G(z) = \log \gamma_o$, then f can be analytically continued throughout the interior of Γ_{γ_o}.

Remark 3. It seems likely (although the authors cannot now prove it) that Theorem 1.4 should remain true under the weaker assumption that only the poles of r_n lie outside U.

2. PROOF OF THEOREM 1.2

Denote by $T_n = z^n + \cdots$ the generalized Chebyshev polynomial of degree n on E, i.e.

$$\|T_n\|_E = \min_{p \in \Pi_{n-1}} \|z^n - p(z)\|_E , \tag{2.1}$$

and for $n \geq 1$ set

$$q_n := \frac{1}{n^2 \|T_n\|_E} T_n . \tag{2.2}$$

We construct an increasing sequence $\{k(n)\}$ of natural numbers, a sequence $\{m(n)\}$ of natural numbers and a sequence $\{\alpha_n\}$ of real numbers by induction. In each step we set

(2.3) $$S_N := 1 + \sum_{n=0}^{N} \left[q_{k(n)+m(n)} + \alpha_{n+1} q_{k(n+1)}\right].$$

The desired function f will be the limit of S_N, that is

(2.4) $$f := 1 + \sum_{n=0}^{\infty} \left[q_{k(n)+m(n)} + \alpha_{n+1} q_{k(n+1)}\right].$$

We will require that $k(n)+m(n) < k(n+1)$ and $0 < \alpha_{n+1} < 1$. This ensures that (2.4) converges uniformly on E. The truncations of the series (2.4) will reflect the behavior of $p_n^*(f)$ of respective degree.

We start with $k(0) = 0$ and set $S_{-1} = 1$. We choose $m(0)$ such that

(2.5) $$\sum_{j=m(0)}^{\infty} \frac{1}{j^2} < 1.$$

Thus f will have no zeros in E.

Assume now, that $k(n)$, $m(n)$, and α_n have been constructed for $0 \leq n \leq N$. Since the zero polynomial is the best approximant to $q_{k(N)+m(N)}$ in $\Pi_{k(N)}$, S_{N-1} is the best approximant to $S_{N-1} + q_{k(N)+m(N)}$ in $\Pi_{k(N)}$. By the continuity of the best approximation operator and Rouché's theorem, there is an $\epsilon_N > 0$ such that for all $f \in C(E)$ with

(2.6) $$\|f - \left[S_{N-1} + q_{k(N)+m(N)}\right]\|_E < \epsilon_N .$$

we have

(2.7) $p^*_{k(N)}(f)$ has not more zeros in B_N than S_{N-1} has in B_{2N}.

where $B_R := \{z \in \mathbb{C}: |z| < R\}$. Notice that S_{N-1} is a polynomial of exact degree $k(N)$. Thus, by the continuity of the best approximation operator, we can assume that $\epsilon_N > 0$ is so small that for all f satisfying (2.6) we also have

(2.8) $p^*_{k(N)}(f)$ is a polynomial of exact degree $k(N)$.

Choose $k(N+1)$ so that

(2.9) $$k(N+1) > (k(N) + m(N))^2,$$

and

(2.10) $$\sum_{j=k(N+1)}^{\infty} \frac{1}{j^2} < \epsilon_N .$$

By Rouché's theorem, we can also choose $0 < \alpha_{N+1} < 1$ such that

(2.11) $S_N = \left[S_{N-1} + q_{k(N)+m(N)}\right] + \alpha_{N+1} q_{k(N+1)}$ has at most $k(N) + m(N)$ zeros in B_{2N+2}.

Next we define $m(N+1)$. Again, $S_{N-1} + q_{k(N)+m(N)}$ is the best approximation to S_N in $\Pi_{k(N)+m(N)}$. Its leading coefficient is

(2.12) $$\beta_N := \frac{1}{(k(N)+m(N))^2 \|T_{k(N)+m(N)}\|_E}.$$

By the continuity of the best approximation operator we can choose $\delta_N > 0$ small enough such that for all $f \in C(E)$ with

(2.13) $$\|f - S_N\|_E \le \delta_N ,$$

we have

(2.14) the leading coefficient γ_N of $p^*_{k(N)+m(N)}(f)$ satisfies $|\gamma_N| > |\beta_N|/2$.

We now choose $m(N+1)$ so that

(2.15) $$\sum_{j=k(N+1)+m(N+1)}^{\infty} \frac{1}{j^2} < \delta_N.$$

For f in (2.4) we have (2.7), (2.8), (2.11) and (2.14), by (2.10) and (2.15) for $N \ge 2$. Thus $p^*_{k(N)}(f)$ is a polynomial of exact degree $k(N)$. By (2.11) and (2.7), it has at most $k(N-1) + m(N-1)$ zeros in B_N. Thus for the measure υ^*_n we have

(2.16) $$\upsilon^*_{k(N)}(B_N) \le \frac{k(N-1) + m(N-1)}{k(N)} < \frac{1}{k(N-1) + m(N-1)} .$$

For bounded $S \subset \mathbb{C}$ this implies

(2.17) $$\lim_{N\to\infty} \upsilon^*_{k(N)}(S) = 0.$$

Write $p^*_n(f) = a_n z^n + \cdots$. Then by (2.14),

$$(2.18) \qquad \limsup_{n\to\infty} |a_n|^{1/n} \geq \limsup_{n\to\infty} \frac{1}{\|T_n\|^{1/n}} = \frac{1}{\operatorname{cap}(E)}$$

(cf.[T]). Thus, by a result of Blatt and Saff [BS], f is not analytic on E. □

3. PROOF OF THEOREM 1.3.

Assume to the contrary that no point of ∂E is a limit point of zeros of $\{q_n\}_1^\infty$. Since (1.14) implies that $q_n \to f$ uniformly on E and f does not vanish identically on any component of $\mathring{E}$, then the set of zeros of f in $\mathring{E}$ is identical to the set of limit points in $\mathring{E}$ of the zeros of $\{q_n\}_1^\infty$ (recall Hurwitz's theorem). Thus f can have at most finitely many zeros in $\mathring{E}$ since, otherwise, either f vanishes identically on a component of $\mathring{E}$ or a point of ∂E is a limit point of zeros of $\{q_n\}_1^\infty$. Let $z_1, \ldots, z_m$ denote the zeros of f in $\mathring{E}$.

With G defined as in (1.4), set

$$(3.1) \qquad E_\gamma := E \cup \{z \in \mathbb{C}\backslash E: G(z) \leq \log \gamma\}, \quad \gamma > 1.$$

The assumption on the zeros of q_n implies that there exists $\gamma_o > 1$ such that for all n large, say $n \geq n_o$, the set E_{γ_o} contains precisely m zeros $z_{1,n}, \ldots, z_{m,n}$ of q_n, where $z_{j,n} \to z_j$ as $n \to \infty$. We claim that for each γ_1, with $1 < \gamma_1 < \gamma_o$,

$$(3.2)\qquad \limsup_{n\to\infty} \|q_n\|_{E_{\gamma_1}}^{1/n} = 1.$$

To establish (3.2) we first define

$$(3.3)\qquad \hat{q}_n(z) := q_n(z) / \prod_{j=1}^{m} (z - z_{j,n}), \quad n \geq n_o.$$

Then $\hat{q}_n \in \Pi_{n-m}$ and $\hat{q}_n$ is zero-free in E_{γ_o}. Furthermore, since the q_n are uniformly bounded on E, the Bernstein-Walsh lemma (cf.[W1,p.77]) implies that

$$(3.4)\qquad \limsup_{n\to\infty} \|\hat{q}_n\|_{E_{\gamma_o}}^{1/n} \leq \gamma_o.$$

Next we note that $\overset{o}{E}_{\gamma_o}$ (the interior of E_{γ_o}) consists of finitely many simply connected components which are bounded by Jordan curves (cf.[W1,p.66]). On any such component Ω of $\overset{o}{E}_{\gamma_o}$ we can define a single-valued analytic branch of $[\hat{q}_n]^{1/n}$, for $n \geq n_o$. Moreover, from (3.4), the $[\hat{q}_n]^{1/n}$, $n \geq n_o$, form a normal family in Ω. Since Ω must contain a component C of E and f does not vanish identically on C, then

$$\lim_{n\to\infty} \hat{q}_n(z) = f(z) / \prod_{j=1}^{m} (z - z_j) \neq 0$$

for infinitely many points of C (recall that $\overline{\mathbb{C}}\backslash E$ is regular, so E can contain no isolated points). Of course, for such points of C we have

$$(3.5) \qquad \lim_{n\to\infty} |\hat{q}_n(z)|^{1/n} = 1,$$

and so, from the normality property, we deduce that (3.5) holds uniformly on each closed subset of Ω. This fact yields the assertion of (3.2).

Next, let $p_{[cn]} \in \Pi_{[cn]}$ be the Lagrange interpolant to q_n in the $[cn] + 1$ Fekete points for the set E. Then it follows from (3.2), the properties of the Fekete points (cf.[W1,p.174]), and the Hermite remainder formula, that

$$(3.6) \qquad \limsup_{n\to\infty} \|q_n - p_{[cn]}\|_E^{1/n} \le 1/\gamma_1^c < 1.$$

Furthermore, by (1.14) and the definition of $e_n(f)$ we have

$$e_{[cn]}(f) \le \|f - p_{[cn]}\|_E \le \|f - q_n\|_E + \|q_n - p_{[cn]}\|_E$$

$$\le K\, e_n(f) + \|q_n - p_{[cn]}\|_E ,$$

and so

$$(3.7) \qquad \frac{e_{[cn]}(f)}{e_n(f)} \le K + \frac{1}{e_n(f)} \|q_n - p_{[cn]}\|_E .$$

But as f is not analytic on E, a theorem of Walsh [W1,p.78] asserts that

$$\limsup_{n\to\infty} [e_n(f)]^{1/n} = 1.$$

Letting Λ denote a subsequence of $\mathbb{N}$ for which $[e_n(f)]^{1/n} \to 1$ as $n \to \infty$, $n \in \Lambda$, we deduce from (3.6) that

$$\lim_{\substack{n\to\infty \\ n\in\Lambda}} \frac{1}{e_n(f)} \|q_n - p_{[cn]}\|_E = 0.$$

Hence, in view of (3.7), we have

$$\liminf_{n\to\infty} e_{[cn]}(f)/e_n(f) \le K,$$

which contradicts assumption (1.13). □

4. PROOF OF THEOREM 1.4

As in Section 3, let G be the classical Green's function on $\overline{\mathbb{C}}\backslash E$ with pole in ∞. Define E_γ, $\gamma > 1$, as in (3.1). For $z_o \in \overline{\mathbb{C}}\backslash E$ let $G(z,z_o)$ be the classical Green's function on $\overline{\mathbb{C}}\backslash E$ with pole at z_o (cf.[T]).

Write $r_n = p_n/q_n$, where $p_n \in \Pi_n$ has the zeros $z_1^{(n)},\ldots,z_{k(n)}^{(n)}$ and $q_n \in \Pi_n$ has the zeros $w_1^{(n)},\ldots,w_{\ell(n)}^{(n)}$, where each zero is listed according to its multiplicity. Then

$$(4.1) \qquad h_n(z) := \log |r_n(z)| + \sum_{\nu=1}^{k(n)} G(z,z_\nu^{(n)}) - \sum_{\nu=1}^{\ell(n)} G(z,w_\nu^{(n)})$$

$$- (k(n) - \ell(n))G(z)$$

is harmonic in $\overline{\mathbb{C}}\backslash E$ and, by the maximum principle, satisfies

$$(4.2) \qquad h_n(z) \le \log \|r_n\|_E , \qquad \text{for} \quad z \in \overline{\mathbb{C}}\setminus E.$$

Choose $\gamma_o > 1$ such that $E_{\gamma_o} \subset U$ and set $\Gamma_{\gamma_o} := \partial E_{\gamma_o}$. Then there are constants $1 < d < D$ such that for all $z_o \in \overline{\mathbb{C}}\setminus U$ and $z \in \Gamma_{\gamma_o}$

$$(4.3) \qquad \log d \le G(z,z_o) \le \log D.$$

Using (4.1), (4.2), and (4.3) we get, for $z \in \Gamma_{\gamma_o}$,

$$(4.4) \qquad |r_n(z)| \le \|r_n\|_E \frac{D^{\ell(n)}}{d^{k(n)}} \gamma_o^{k(n)-\ell(n)} \le \|r_n\|_E (D\gamma_o)^n.$$

By the maximum principle, (4.4) holds for all $z \in E_{\gamma_o}$.

From (1.18) we know that $r_n \to f$ uniformly on E. Using arguments similar to those of Section 3 we deduce that

$$(4.5) \qquad \limsup_{n\to\infty} \|r_n\|_{E_{\gamma_1}}^{1/n} = 1$$

for every $1 < \gamma_1 < \gamma_o$. Next, let $p_{n-1} \in \Pi_{n-1}$ be the Lagrange interpolant to r_n in the n Fekete points for E. As before (cf.(3.6)), equation (4.5) implies that

$$\text{(4.6)} \qquad \limsup_{n\to\infty} \|r_n - p_{n-1}\|_E^{1/n} \le 1/\gamma_1 < 1.$$

From (1.18) we deduce that

$$e_{n-1}(f) \le \|f - p_{n-1}\|_E \le \|f - r_n\|_E + \|r_n - p_{n-1}\|_E$$

$$\le e_n(f) + \|r_n - p_{n-1}\|_E ,$$

so that

$$\limsup_{n\to\infty} [e_{n-1}(f)-e_n(f)]^{1/n} \le \limsup_{n\to\infty} \|r_n-p_{n-1}\|_E^{1/n} \le 1/\gamma_1.$$

Hence

$$\text{(4.7)} \qquad \limsup_{n\to\infty} [e_n(f)]^{1/n} \le 1/\gamma_1 < 1,$$

which implies f is analytic on E. □

Finally we note that (4.7) yields the stronger conclusion that f is analytic in the interior of E_{γ_1} and, since $\gamma_1 < \gamma_o$ is arbitrary, f is analytic in the interior of E_{γ_o} as claimed in Remark 2.

REFERENCES

[Be] S.N. Bernstein, 'Sur la valeur asymptotique de la meilleure approximation de $|x|$', *Acta Mathematica* 37 (1913), 1-57.

[Bo] P.B. Borwein, 'The relationship between the zeros of best approximants and differentiability', *Proc. Amer. Math. Soc.* 92 (1984), 528-532.

[BIS] H.-P. Blatt, A. Iserles, E.B. Saff, 'Remarks on the behavior of best approximating polynomials and rational functions', In: *Algorithms for Approximation* (IMA conference series: new ser. 10) (J.C. Mason and M.G. Cox, ed), Oxford University Press (1987), pp. 437-445.

[BS] H.-P. Blatt and E.B. Saff, 'Behavior of zeros of polynomials of near best approximation', *J. Approx. Theory* 46, No. 4 (1986), 323-344.

[BSS] H.-P. Blatt, E.B. Saff, M. Simkani, 'Jentzsch-Szegö type theorems for the zeros of best approximants', (to appear).

[J] R. Jentzsch, 'Untersuchungen zur Theorie Analytischer Funktionen' Inaugural-Dissertation, Berlin (1914).

[L] A. Levin, 'The distribution of the poles of best approximating rational functions and the analytical properties of the approximated function', *Israel Journal of Mathematics*, 24, No. 2 (1976), 139-144.

[Sz] G. Szegö, 'Über die Nullstellen von Polynomen, die in einem Kreis gleichmäßig konvergieren', *Sitzungsbericht Ber. Math. Ges.* 21 (1922), 59-64.

[T] M. Tsuji, *Potential Theory in Modern Function Theory*, Chelsea Publishing Company, New York (1958).

[W1] J.L. Walsh, *Interpolation and Approximation by Rational Functions in the Complex Domain*, American Math. Soc. Coll. Publ., Vol. 20 (1935), 5th ed.
(1969).

[W2] J.L. Walsh, 'The analogue for maximally convergent polynomials of Jentzsch's theorem', *Duke Math. J.* 26 (1959), 605-616.

ONCE AGAIN : THE ADAMJAN-AROV-KREIN APPROXIMATION THEORY

Jean Meinguet
Université Catholique de Louvain
Institut de Mathématique Pure et Appliquée
Chemin du Cyclotron 2
B-1348 Louvain-la-Neuve, Belgium

ABSTRACT. This paper is largely devoted to a tutorial presentation of an extensive background material related to a Glover-like solution of the AAK problem of optimal Hankel-norm approximation for *discrete-time* multivariable systems of finite degree. A special emphasis is laid on the rationale for the Hankel-matrix approach and on the singular value decomposition of bounded infinite Hankel matrices of finite rank (with some original matrix-theoretic complements). Among the main AAK results, which are briefly summarized here in a form suitable for system-theoretic applications, one of the most remarkable states that the number of zeros (inside the unit circle) of the rational functions obtained by z-transforming the Schmidt pair belonging to any singular value of such a Hankel matrix is related simply to its serial number. Unlike the classical proofs, which are notoriously sophisticated, the pure matrix proof we outline here is both reasonably simple and transparent.

1. INTRODUCTION

By 'Adamjan-Arov-Krein (AAK) approximation theory', we mean the impressive collection of remarkable results obtained (in the early seventies) by these Russian mathematicians in their systematic investigations dealing with *infinite Hankel and block-Hankel matrices bounded in* ℓ^2 *and* (operator-theoretic or function-theoretic) *best approximation or extension problems associated with them.* The basic references here are [1] for the scalar case and [2] for the significant generalization to the case of operator-Hankel matrices in a Hilbert space setting. A simplified presentation (of the fundamental sections §0 and §1) of the rather difficult paper [1] can be found in our NATO(ASI)-lectures [8], written specially for applied mathematicians with an approximation-theoretic background.

In addition to its intrinsic mathematical interest, the AAK theory can prove very useful in linear systems theory, in particular for solving the *model-reduction problem* as regards *linear*, *time-invariant*, *BIBO* (or bounded-input bounded-output) *stable*, *causal*, *finite-degree* ℓ^2

A. Cuyt (ed.), Nonlinear Numerical Methods and Rational Approximation, 77–91.

systems. For a system-theoretic presentation of this interesting material (which indeed is relevant to important applications in engineering and science), the reader is strongly advised to refer to the long and most remarkable paper [4]. As a matter of fact, this recent comprehensive treatment of the realistic case of block-Hankel matrices of finite rank can be regarded as self-contained, essentially relies on (non-trivial !) linear algebra (rather than on abstract or sophisticated mathematics such as functional analysis, operator theory, complex function theory,...), leads to a complete characterization of all optimal Hankel-norm approximations to any given stable multivariable transfer function (of finite McMillan degree), derives these solutions (via results on balanced realizations, all-pass functions and the inertia of matrices) from the solutions to Lyapunov equations and, last but not least, yields novel uniform error bounds. A tutorial presentation (intended for mathematicians) of an extensive background material related to [4], together with some complements, was given in [9].

Our main purpose here is to contribute to an active understanding of the fine mathematics involved in the AAK theory and its system-theoretic applications, in continuation of [8,9]. Whereas the papers [4,9] were concerned with (block) *Hankel operators* bounded in $L^2(\mathbb{R})$ (that is with *continuous-time* systems), the present contribution deals with infinite (block) *Hankel matrices* bounded in ℓ^2 (that is with *discrete-time* systems), which is somewhat simpler. In Section 2, we give the rationale for the Hankel-matrix approach and emphasize in particular the significance of the underlying controllability and observability (infinite) matrices. We proceed next to a rather detailed analysis of the singular value decomposition (SVD) of ℓ^2 bounded infinite block-Hankel matrices of finite rank ; it turns out that this problem is equivalent to a Hermitian-definite generalized eigenproblem for finite matrices, which is clearly a prominent result (making the computations 'finite'). Section 3 also contains some original matrix-theoretic results, such as a complete answer to the uniqueness question for the SVD (interpreted as a polar decomposition) and the derivation (for complex symmetric matrices) of a Schur representation from any SVD (the ensuing existence, for ℓ^2 bounded infinite complex Hankel matrices, of balanced realizations that are selfdual is shown to have quite interesting applications). Section 4 is devoted to a brief summary (without proofs) of the main AAK results in a form suited to the optimal Hankel-norm solution of the model-reduction problem in the scalar case. In the last section, we outline a pure matrix proof (details shall be published elsewhere) of the rightly famous AAK result : the z-transform of the right singular vector belonging to the j-th singular value (supposed to be non-repeated) of an ℓ^2 bounded infinite Hankel matrix of finite rank has precisely j-1 zeros inside the unit circle.

As regards notations :

- the circumflex '^' over any function u defined on the integers denotes the (two-sided) *z-transform* of the sequence of values u(k), that is

$$\hat{u}(z) := \sum_{k=-\infty}^{\infty} u(k)/z^k,$$

where the placeholder z^{-1} actually plays the role of unit delay (the

independent variable k in this paper is indeed the time) ; for $z = e^{it}$ (t real), this Laurent-like series reduces to a formal *Fourier series*, which is convergent if and only if the unit circle lies in the annulus of convergence of the z-transform.
- the superscripts '+' and '−' are used systematically to distinguish between 'future' and 'past' ; more precisely, a notation like u^+ (resp. u^-) is to be interpreted as the one-sided sequence derived from u by substituting 0 for all the terms of subscripts ≤ 0 (resp. ≥ 1).
- inner products and norms (unless they are subscripted) are always those stemming from the standard Hilbert space ℓ^2.
- for any matrix A, A^T denotes the transpose, A^* the conjugate transpose and $\bar{A}$ the conjugate ; for any linear mapping, $\mathcal{N}$ denotes the kernel and $\mathcal{R}$ the range.

2. THE RATIONALE FOR THE HANKEL-MATRIX APPROACH

The *state-variable description* of the *discrete-time systems* we will consider here is given by the so-called *dynamical equation* :

(1a) $x(k+1) = Ax(k) + Bu(k)$ (state equation),

(1b) $y(k) = Cx(k) + Du(k)$ (output equation),

where the matrices $A \in \mathbb{C}^{n\times n}$, $B \in \mathbb{C}^{n\times m}$, $C \in \mathbb{C}^{p\times n}$ and $D \in \mathbb{C}^{p\times m}$ are constant, while the n-vector of *states* x(.), the m-vector of *inputs* u(.) and the p-vector of *outputs* y(.) have for components ℓ^2 sequences of complex numbers (indexed by integers, representing here the 'time' variable).

Provided the system is *relaxed at time* $-\infty$ (which is usually assumed in the engineering literature) and since (by BIBO stability) the matrix A is *asymptotically stable* (which means here that all its eigenvalues lie inside the unit circle), the solution of the state equation (1a) can be written explicitly in the form

(2a) $x(k+1) = \sum_{j=0}^{\infty} A^j Bu(k-j)$;

hence the explicit behavior of the output :

(2b) $y(k) = \sum_{j=-\infty}^{\infty} G(k-j)u(j)$,

where

(2c) $G(k) := (CA^{k-1}B)^+ + \delta(k)D$

is the *impulse-response* p-by-m matrix of the system ($\delta(k)$ denotes the discrete-time impulse function at time 0, that is 1 if k=0 and 0 otherwise), its values $G(k) = CA^{k-1}B$ for $k \geq 1$ being the so-called *Markov parameters* of the system ; notice here the superscript '+', in accordance with the causality assumption.

By the *two-sided z-transformation*, this description in terms of convolution products of the behavior (state and output) of the system *in the time domain* is changed into the following algebraic one *in the frequency domain* :

(3a) $\hat{x}(z) = (zI - A)^{-1}B\hat{u}(z),$

(3b) $\hat{y}(z) = \hat{G}(z)\hat{u}(z),$

the associated *transfer-function* matrix being the p-by-m rational function matrix (of the complex variable z)

(3c) $\hat{G}(z) := C(zI-A)^{-1}B + D$

obtained by z-transforming the impulse-response matrix (2c).

It turns out that quite remarkable ℓ^2 *bounded infinite (block) matrices* are naturally associated with the foregoing. In particular, the *input-output description of* the system (i.e., (2b,c) or (3b,c)) can be expressed in the form of a *doubly infinite* system of linear equations whose matrix is the (*lower triangular*) *block-Laurent matrix* :

$$L := (L_{jk})_{j,k=-\infty}^{\infty} \text{ with } L_{jk} := G(j-k).$$

In view of the (physically significant) perpendicular splitting

(4) $\ell^2 = \ell^{2-} \oplus \ell^{2+},$

it proves natural to rewrite (2b,c) in the operator-matrix form

(5) $$\begin{bmatrix} y^- \\ y^+ \end{bmatrix} = \begin{bmatrix} T^- & 0 \\ H & T^+ \end{bmatrix} \begin{bmatrix} u^- \\ u^+ \end{bmatrix} \quad \text{for all } u^\pm \in (\ell^{2\pm})^m \subset (\ell^2)^m.$$

Here

(6a) $(T^-u^-)(k) := [\sum_{j=0}^{-\infty} G(k-j)u^-(j)]^-,$

(6b) $(T^+u^+)(k) := [\sum_{j=1}^{\infty} G(k-j)u^+(j)]^+,$

so that T^- and T^+, which might be regarded as discrete analogs of causal Wiener-Hopf operators, have for matrix representations *semi-infinite block-Toeplitz matrices*, respectively of co-analytic type (i.e., *block upper triangular*) and of analytic type (i.e., *block lower triangular*). On the other hand,

(7a) $(Hu^-)(k) := [\sum_{j=0}^{\infty} G(k+j)u^-(-j)]^+ = (CA^{k-1})^+x$

with

(7b) $$x := x(1) = \sum_{j=0}^{\infty} A^j B u^-(-j),$$

so that H, which can be interpreted as the (bounded) mapping from the past inputs $u^- \in (\ell^{2-})^m$ to the future outputs $y^+ \in (\ell^{2+})^p$ via the state $x \in \mathbb{C}^n$ at time 1, has for matrix representation the (*semi-infinite*) *block-Hankel matrix* :

(7c) $$H := (H_{jk})_{j,k=0}^{\infty} \text{ with } H_{jk} := G(j+k+1) = CA^{j+k}B \;;$$

it should be realized indeed that the entries in the matrix H actually depend on the *sum* of the row and column indices, whereas for the matrices $T^{\pm}$ they depend on their *difference*.

We find it particularly enlightening and appropriate to our needs to substitute here, for the standard (finite) controllability and observability matrices used in modern textbooks on linear systems theory (see e.g. [3], [6]), a *controllability operator* C and an *observability operator* O with the (infinite) matrix representations :

(8a) $$C := [B, AB, A^2B, \dots],$$

(8b) $$O := [C^*, A^*C^*, A^{*2}C^*, \dots]^*.$$

Hence, as it readily follows from (7c), the remarkable factorization :

(9) $$H = OC,$$

which clearly reflects the *physical interpretation of the Hankel operator* we have recalled above, viz.,

(10a) $$x = x(1) = Cu^-,$$

(10b) $$y^+ = Hu^- = Ox.$$

Moreover, the definitions (8a,b) immediately suggest forming the composites :

(11a) $$P := CC^* = BB^* + ABB^*A^* + A^2BB^*A^{*2} + \dots,$$

(11b) $$Q := O^*O = C^*C + A^*C^*CA + A^{*2}C^*CA^2 + \dots,$$

which indeed are the classical n-by-n Hermitian matrices called *controllability Gramian* and *observability Gramian*, respectively ; it is an elementary consequence of the asymptotic stability of A (with respect to the unit circle) that P and Q are uniquely determined by the important *Stein matrix equations*

(12a) $$P - APA^* = BB^*,$$

(12b) $Q - A^*QA = C^*C,$

which are also called the *discrete-time Lyapunov equations*.

For convenience and definiteness, we will assume throughout this paper that (A,B,C) is a *minimal* state-space *realization* of the transfer-function matrix $\hat{G}(z)$. It is a basic result from realization theory that minimizing the number of state variables (i.e., the order of A) amounts to requiring that

(H) $\mathcal{C}$ is *surjective* (i.e., onto) and $\mathcal{O}$ is *injective* (i.e., one-to-one into),

which means that (A,B) is *controllable* and (A,C) is *observable* or, equivalently, that the order of A is given by

$$\text{rank}(\mathcal{H}) = \text{McMillan degree of } \hat{G}(z)$$

(i.e., the degree of the least common denominator of all minors of $\hat{G}(z)$). It must be emphasized that these results actually hold valid for any *proper* (i.e., finite at infinity) rational function matrix, whether all its poles lie inside the unit circle or not. In the present asymptotically stable case, we have the further characterization result that (A,B) is controllable (resp. (A,C) is observable) if and only if P (resp. Q) is positive definite.

3. THE SINGULAR VALUE DECOMPOSITION OF INFINITE BLOCK-HANKEL MATRICES OF FINITE RANK.

Consider the positive semi-definite self-adjoint operator $\mathcal{H}^*\mathcal{H}$ in the Hilbert space $(\ell^{2-})^m$. It is of rank n, just like the infinite block-Hankel matrix $\mathcal{H}$ defined by (7c), or by (9) with (8a,b), under assumption (H). The spectrum of $\mathcal{H}^*\mathcal{H}$, with the only exception of the point 0, accordingly consists of n *eigenvalues*, counting multiplicities, which may be interpreted as the squares of the *singular values* of $\mathcal{H}$:

$$\sigma_1 \geqslant \sigma_2 \geqslant \ldots \geqslant \sigma_n > 0,$$

and to which belong normalized $(\ell^{2-})^m$ *eigenvectors* $u^{1-}, u^{2-}, \ldots, u^{n-}$ which may always be chosen to be pairwise orthogonal, and consequently form an orthonormal basis for the space $\mathcal{N}(\mathcal{H})^{\perp} = \mathcal{R}(\mathcal{C}^*)$. It turns out that the corresponding $(\ell^{2+})^p$ vectors

(13) $y^{j+}/\sigma_j := \mathcal{H}u^{j-}/\sigma_j$ for $j=1,\ldots,n,$

form an orthonormal basis for the space $\mathcal{R}(\mathcal{H}) = \mathcal{R}(\mathcal{O})$, so that finally the operator $\mathcal{H}$ admits the so-called *Schmidt expansion*

(14a) $\mathcal{H} = \sum_{j=1}^{n} (.,u^{j-})y^{j+},$

which actually amounts to the *singular value decomposition* (SVD)

(14b) $$H = V\,\Sigma\,U^* \text{ with } V^*V = U^*U = I$$

of the block matrix H ; as a matter of fact, y^{j+}/σ_j (i.e., the j-th column of the isometry $V : \mathbb{C}^n \to (\ell^{2+})^p$) and u^{j-} (i.e., the j-th column of the isometry $U : \mathbb{C}^n \to (\ell^{2-})^m$) play here, respectively, the role of *left singular vector* and of *right singular vector* belonging to the j-th *singular value* σ_j (i.e., the (j,j) component of the diagonal n-by-n matrix Σ) of H.

It is quite remarkable that *the SVD problem for this infinite block-Hankel matrix H of rank* n *is equivalent to a Hermitian-definite generalized eigenproblem for* n-*by*-n *matrices*, viz.,

$$Qx = \lambda P^{-1}x,$$

where P and Q are the (positive definite) Gramians defined by (11a) and (11b), respectively. This essential simplification of the original problem can be achieved surprisingly simply, by setting (as suggested by (10a))

(15a) $$x^j = Cu^{j-} \text{ for } j = 1,\ldots,n,$$

or, equivalently,

(15b) $$u^{j-} = C^*P^{-1}x^j \text{ for } j = 1,\ldots,n,$$

(remember indeed that $u^{j-} \in R(C^*)$ with C^* injective). In view of these relations, the original Hermitian eigenproblem

(16) $$H^*Hu^{j-} = \sigma_j^2\, u^{j-} \text{ for } j = 1,\ldots,n,$$

strictly amounts to the *algebraic* equation

(17) $$Qx^j = \sigma_j^2 P^{-1}x^j \text{ for } j = 1,\ldots,n,$$

which shows in particular that the spectrum of H^*H (with the exception of the point 0) is $\lambda(Q,P^{-1})$ (i.e., the set of eigenvalues, counting multiplicities, of the Hermitian-definite pencil $Q-\lambda P^{-1}$) or, equivalently, the spectrum of the similar n-by-n matrices PQ and QP.

In conclusion, the j-th *Schmidt pair* $(u^{j-}, y^{j+}/\sigma_j)$ for the Hankel operator H defined by (9) with (8a,b) under assumption (H) is given explicitly by

(18a) $$u^{j-} := C^*P^{-1}x^j = [\,B,AB,A^2B,\ldots\,]^*P^{-1}x^j,$$

(18b) $$y^{j'} := \mathcal{O}x^j = [\,C^*,A^*C^*,A^{*2}C^*,\ldots]^*x^j,$$

in the time domain or, equivalently, by

(19a) $$\hat{u}^{j-}(z) := B^*(I - zA^*)^{-1}P^{-1}x^j,$$

(19b) $\hat{y}^{j+}(z) := C(zI - A)^{-1}x^j,$

in the frequency domain. As for the precise states x^j to be used in these expressions for $j = 1,\ldots,n$, they must evidently satisfy (17), together with those biorthogonality conditions which stem from the originally required biorthogonality of the u^{j-}'s, viz.,

(20a) $(x^k)^*P^{-1}x^j = (u^{j-},u^{k-}) = \delta_{jk}$ for $j,k = 1,\ldots,n,$

or, equivalently,

(20b) $(x^k)^*Qx^j = (y^{j+},y^{k+}) = \sigma_j\sigma_k\delta_{jk}$ for $j,k = 1,\ldots,n.$

As regards the (positive) singular values σ_j themselves, called in [4] *Hankel singular values of* (any system 'defined' by) *the transfer-function matrix* $\hat{G}(z)$ and denoted accordingly $\sigma_j(\hat{G}(z))$, it must be emphasized that they depend only on the Hankel operator H (irrespective of concrete realizations and state-space transformations) and are hence *fundamental input/output invariants* related to both gain and complexity. In particular, the so-called *Hankel norm* of (any system defined by) $\hat{G}(z)$, that is

(21) $\|\hat{G}(z)\|_H := \sup_{u^-} \|Hu^-\| / \|u^-\| = \|y^{1+}\| / \|u^{1-}\| = \|H\|_s = \sigma_1,$

can be interpreted physically as the ℓ^2-gain from past inputs to future outputs ($\|.\|_s$ denotes here the usual spectral norm).

Like every finite SVD, the matrix representation (14b) of H is unique up to an arbitrary change (defined by a unitary matrix M) of orthonormal basis for $R(H)$ (i.e., $V \mapsto VM$) and for $R(H^*)$ (i.e., $U \mapsto UM$) such that $M\Sigma = \Sigma M$; this clearly amounts to characterizing M as a direct sum of arbitrary unitary matrices, each of them being of an order equal to the multiplicity of the corresponding singular value σ_j (if the σ_j are pairwise distinct, then M must be a diagonal unitary matrix). This (hardly known) *uniqueness result* readily follows by interpreting the singular value decomposition (14b) rewritten in the form

(14c) $H = (VU^*)(U\Sigma U^*)$ with $V^*V = U^*U = I,$

as the (uniquely defined) *polar decomposition* of H ; indeed, $U\Sigma U^*$ is the (unique) Hermitian positive semi-definite square root of H^*H while VU^* is a partial isometry (of kernel $N(U\Sigma U^*)$, and therefore also uniquely defined, see e.g. [5], p.68) ; it is thus clear that the only changes of basis for $R(H)$ and $R(H^*)$ that preserve the matrices U^*U, V^*V, VU^* and $U\Sigma U^*$ are those characterized above.

Unlike the SVD (14b) of H, which is 'essentially' unique, the matrix representation (9) (with C surjective and O injective) depends strongly on the state-space coordinates. As a matter of fact, since the columnsof $V\Sigma^{1/2}$ and of O (resp. of $U\Sigma^{1/2}$ and of C^*) form two bases for the space $R(H)$ (resp. $R(H^*)$), it directly follows from the identification of (9) with (14b) that

(22) $\mathcal{C} = T(\Sigma^{1/2}U^*)$ and $\mathcal{O} = (V\Sigma^{1/2})T^{-1}$,

where T is an arbitrary (n-by-n, nonsingular) matrix (which can be interpreted as defining a state-space transformation $x \mapsto Tx$). In view of the matrix representations (8a,b) of $\mathcal{C}$ and $\mathcal{O}$, a minimal realization (A,B,C) can be obtained from (22) as follows : B := first m columns of $\mathcal{C}$, C := first p rows of $\mathcal{O}$, A := $\mathcal{C}_s\mathcal{C}^I$ where $\mathcal{C}_s$ is the matrix $\mathcal{C}$ shifted left m columns (i.e., the submatrix A$\mathcal{C}$ of $\mathcal{C}$) and $\mathcal{C}^I := \mathcal{C}^*(\mathcal{C}\mathcal{C}^*)^{-1}$ is the *Moore-Penrose inverse* (or even any right inverse) of the linear surjection $\mathcal{C}$. The trivial choice T = I in (22) actually leads to *balanced realizations*, whose particular significance stems from their being 'as controllable as they are observable' in the sense that

(23) $P = Q = \Sigma$,

whereas the general form of the Gramians (i.e., $P = T\Sigma T^*$, $Q = T^{*-1}\Sigma T^{-1}$) depends strongly on T. As readily verified, a balanced realization can also be obtained, from any given minimal realization, by the balancing state-space transformation : $x \mapsto Tx$ with $T := \Sigma^{1/2}X^{-1}$, X denoting the matrix $(x^1,\ldots,x^n)$ of the solutions of (17,20a).

In the particularly important *scalar case* (i.e. if m = p = 1), the ℓ^2 bounded infinite Hankel matrix (7c) is symmetric. Like every finite *complex symmetric* matrix, $\mathcal{H}$ admits a *Schur representation*, that is a symmetric SVD :

(24) $\mathcal{H} = W\Sigma W^T$ with $W^*W = I$.

We will prove here this result by proceeding constructively from the foregoing (the 'classical' proofs are quite different, see e.g. [10], p. 88). Consider the matrix representation (14b) of $\mathcal{H}$ and its transpose, rewrite them in the form (14c) corresponding to the unique (right-handed) polar decompositions of $\mathcal{H}$ and $\mathcal{H}^T$; since here $\mathcal{H}$ is assumed to be symmetric, we may equal separately the partial isometries (which shows that the n-by-n matrix U^TV is symmetric) and the Hermitian square roots (which shows that U^TV commutes with Σ) ; moreover, as readily verified, U^TV is unitary. Now the concrete problem of deriving a Schur representation (24) from the SVD (14b) amounts to finding an n-by-n matrix M that is unitary, commutes with Σ and is such that

(25) $U^TV = \overline{M}\, M^*$

(this follows from the requirement : $W = VM$ with $W^T = M^*U^*$). Such a decomposition is actually possible for any unitary symmetric matrix U^TV : indeed, $\mathrm{Re}(U^TV)$ and $\mathrm{Im}(U^TV)$ are then commuting real symmetric matrices, which can therefore be diagonalized simultaneously by some *real orthogonal* transformation N, so that

$$U^TV = N\Lambda N^T \quad \text{with } \Lambda \text{ diagonal and unitary,}$$

which shows that

$$\bar{M} := N\Lambda^{1/2} \text{ with } \Lambda^{1/2} \text{ any square root of } \Lambda$$

is a particular unitary matrix verifying (25). Finally, it should be noted that *the Schur representation* (24) *is unique up to a real orthogonal transformation* $T : W^T \mapsto TW^T$ *commuting with* Σ *(i.e., a sign matrix if the* σ_j *are pairwise distinct)*.

A most interesting consequence of (24) is the existence, for every rational function $\hat{G}(z)$ that is *strictly proper of degree* n (i.e., such that the degree of the numerator is less than the degree n of the denominator) and *stable* (i.e., such that its poles are all inside the unit circle), of a balanced realization (A,B,C) that is even *selfdual*, i.e., such that

$$O = C^T := W\Sigma^{1/2} \text{ with } W^*W = I \text{ of order n}$$

or, equivalently, such that

(26) $\qquad B = C^T$ and $A = A^T$ of order n.

Since $P = Q = \Sigma$, the matrix X of the solutions $x^1,\ldots,x^n$ of (17,20a) must be of the form $\Sigma^{1/2}N$ where N is any n-by-n unitary matrix commuting with Σ. For the natural choice $N = I$, X reduces to $\Sigma^{1/2}$ so that the j-th Schmidt pair (19a,b) can be rewritten in the form

(27a) $\qquad \hat{u}^{j-}(z) = \sigma_j^{-1/2}\bar{C}(I - z\bar{A})^{-1}e^j,$

(27b) $\qquad \hat{y}^{j+}(z) = \sigma_j^{1/2}\, C(zI - A)^{-1}e^j,$

where e^j denotes the j-th column of the n-by-n identity matrix. Hence it immediately follows (which actually completes a surprisingly simple proof !) that the rational functions

(28) $\qquad \phi^j(z) := \sigma_j^{-1}\hat{y}^{j+}(z)/\hat{u}^{j-}(z)$ for $j = 1,\ldots,n,$

which are to play a prominent role in the AAK theory (see Section 4), are *unimodular* (i.e. of modulus 1 on the unit circle) or, equivalently, are *all-pass* transfer functions.

4. ESSENTIALS OF THE AAK THEORY FOR MODEL REDUCTION

The AAK theory is concerned with the systematic study of infinite block-Hankel matrices and approximation problems associated with them. Like Adamjan et al. in [1], we will restrict ourselves for simplicity to the case of scalar Hankel matrices. Our purpose here is, however, essentially concrete : we will briefly summarize the main AAK results in a form suited to the explicit solution of the *'optimal' model reduction problem*, or *optimal Hankel-norm approximation problem*, for *scalar* (i.e., single-input : m = 1, and single-output : p = 1) *systems* of the class considered above.

A mathematical formulation of this fundamental problem of linear systems theory is the following : *given an infinite Hankel matrix* H *that is of rank* n *and bounded in* ℓ^2, *it is required to find* H^{j-1}_{opt} *(for* $1 \leq j \leq n$), *that is the unique best approximation to* H *in the spectral norm by infinite Hankel matrices of rank* $\leq j-1$. By way of motivation, it is worth remembering that controlling the rank of Hankel approximants to H amounts strictly to controlling the McMillan degree of their minimal realizations (i.e., the minimal number of state variables).

Before we proceed to the AAK characterization of the solution H^{j-1}_{opt}, we want to remind the reader of certain classical results concerning infinite Hankel matrices $H = (h_{j+k+1})^{\infty}_{j,k=0}$ and their system-theoretic applications :

- the *Kronecker theorem* essentially asserts that H is of rank n if and only if the series

$$\hat{G}_c(z) := \sum_{j=1}^{\infty} h_j/z^j$$

determines a (strictly proper) rational function (whose denominator is) of exact degree n.

- such a matrix is ℓ^2 *bounded* if and only if the poles of $\hat{G}_c(z)$ are all inside the unit circle. From a system-theoretic point of view, functions of that class may be qualified as (strictly proper) *causal* (or *stable*) *rational transfer functions* (as a matter of fact, they are not only causal, as z-transforms of 'causal' sequences $(h_1,h_2,\ldots)$, but also *bounded* on the unit circle). Even though only (proper) causal systems are physically realizable, noncausal systems such as those defined by *anticausal rational transfer functions* (i.e. analytic outside the unit circle) may arise in the course of theoretical analysis.

- for any rational function $\hat{G}(z)$ bounded on the unit circle, let $H(\hat{G})$ denote the ℓ^2 bounded Hankel matrix (of *symbol* $\hat{G}$) formed from the coefficients of the negative powers of z in the Laurent series of $\hat{G}(z)$ for an annulus of convergence containing the unit circle. Then, according to the seminal *Nehari theorem*, we have the following fundamental result :

$$\|H(\hat{G})\|_s = \min_F \|\hat{G}(e^{it}) - F(e^{it})\|_\infty, \tag{29}$$

where F ranges over the set of anticausal rational functions and $\|\cdot\|_\infty$ denotes the Chebyshev norm on the unit circle. It turns out that, in the linear variety of all equivalent symbols $\hat{G}$-F of $H(\hat{G})$, there exists one and only one *optimal symbol* (or symbol of minimal Chebyshev norm), which furthermore is given by the explicit formula

$$(\hat{G} - F)_{opt}(z) := \sigma_1(H)\phi^1(z) \tag{30}$$

(this is simply the particular case j = 1 of the AAK result (34)) and has thus constant modulus everywhere on the unit circle. For a constructive approach (at a higher mathematical level) to these remarkable results, the reader is referred to [8] (see Section 3).

A main result of Adamjan et al. is the following *characterization (in the time domain)* of the solution of the optimal model-reduction problem formulated above : *if* $1 \leq j \leq n$ *is such that*

(31) $\|H\|_s = \sigma_1 \geq \ldots \geq \sigma_{j-1} > \sigma_j = \ldots = \sigma_{j+k} \ldots (> 0),$

then there exists a unique infinite Hankel matrix H_{opt}^{j-1} *of rank* $\leq j+k-1$ *such that*

(32) $\|H - H_{opt}^{j-1}\|_s = \sigma_j.$

Moreover,

(33) $H_{opt}^{j-1} := H - \sigma_j H(\phi^j),$

where $\phi^j(z)$ *is the unimodular rational function defined by* (28), *and has rank* j-1. As a matter of fact, it is quite surprising that, for *infinite* Hankel matrices H (of finite rank and ℓ^2 bounded), the best Hankel approximant of any prescribed maximal rank actually achieves the same minimal value of the spectral norm of the error as does the unconstrained approximant (which is classically obtained by truncating appropriately the Schmidt expansion (14a) of H).

Now it can be proved in a surprisingly simple way (see e.g. [8], pp. 235-238) that the above *operator-theoretic problem* is equivalent to the following *function-theoretic problem : given a rational function* $\hat{G}(z)$ *of degree* n *that is bounded on the unit circle, it is required to find the unique best approximation to it in the Chebyshev norm* $\|\cdot\|_\infty$ *by functions of the form* $\hat{G}^{j-1} + F$, *with* $\hat{G}^{j-1}$ *strictly proper and causal of degree* $\leq$ j-1 *and* F *anticausal.* A striking result (*in the frequency domain*) of the AAK theory is the following *explicit formula for the solution* of this problem :

(34) $(\hat{G}^{j-1} + F)_{opt}(z) := \hat{G}(z) - \sigma_j \phi^j(z),$

where σ_j and ϕ^j refer to the infinite Hankel matrix $H(\hat{G})$, so that

(35) $\|\hat{G}(e^{it}) - (\hat{G}^{j-1} + F)_{opt}(e^{it})\|_\infty = \sigma_j.$

It should be clearly realized that the best Hankel approximant given by (33) is formed from the coefficients of the Laurent series of the causal part (which is precisely of degree j-1 if condition (31) is satisfied) of the best Chebyshev approximant (34).

A further essential result of the AAK theory, called here (to make it short) the '*zero count*' *property*, is the following : *let the* j*-th singular value* σ_j *of an infinite Hankel matrix* H (*of finite rank* n *and bounded in* ℓ^2) *be prime* (or non-repeated, i.e., $\sigma_{j-1} > \sigma_j > \sigma_{j+1}$) ; *then the corresponding function* $\hat{u}^{j-}(z)$, which is now uniquely determined up to a scalar factor of modulus 1 by formula (27a), *has precisely* j-1 *zeros (counting multiplicities) inside the unit circle.* The theoretical and practical importance of this result stems from the fact that $\hat{u}^{j-}$ is the denominator of the function ϕ^j that plays a leading role in the AAK theory (as exemplified above). The original proof by Adamjan et al. (see [1] and, for a somewhat simplified version,[8], pp. 242-245) covers the general case but is quite sophisticated ; insofar as we are specially interested in the finite-rank case, it is natural to search after a specific proof that should be significantly simpler ; some of the results we have recently obtained in that (hardly explored !) direction are summarized in the next section.

5. A PURE MATRIX PROOF OF THE 'ZERO COUNT' PROPERTY

The basic idea here looks quite natural : why not try to make use of the so-called *inertia theory for general matrices* (like Glover did in [4] for continuous-time systems), which indeed is concerned with various important extensions of the well known Lyapunov stability theory (to be reformulated here for discrete-time systems). One of the principal achievements is the so-called *main inertia theorem (relative to the unit circle)*, the following generalization of which is apparently suited to the research on hand : *Let an* n-*by*-n *complex matrix* A *have no eigenvalues of modulus* 1. *If there exists an* n-*by*-n *nonsingular Hermitian matrix* H *such that*

$$H - A^{*}HA \text{ is positive semi-definite,}$$

then the number of eigenvalues (counting multiplicities) of A *lying inside* (resp. outside) *the unit circle is equal to the number of positive* (resp. negative) *eigenvalues of* H (see [7], p. 453).

In order to apply this inertia theorem to the counting of the zeros of $\hat{u}^{j-}(z)$ inside the unit circle or, what is theoretically equivalent (remember indeed that $\phi^{j}(z)$ is unimodular) but notationally preferable, to the *counting of the zeros of* $\hat{y}^{j+}(z)$ *outside the unit circle*, we must first of all construct an n-by-n matrix having precisely the latter zeros as eigenvalues. Such a matrix is given by

$$(36) \qquad A_{mod} = A - Ae^{j}C/C_{j} \text{ with } C_{j} := Ce^{j},$$

at least if $C_j \neq 0$ (otherwise, $z = 0$ is a zero of $\hat{u}^{j-}(z)$ that should be divided out before proceeding, as explained hereafter, with a correspondingly 'compressed' matrix A). This can be proved very simply, by making use of the so-called *Sherman-Morrison-Woodbury formula* (or *matrix inversion lemma*), which indeed can be written in the form

$$[\,D + C(zI-A)^{-1}B\,]^{-1} = D^{-1} + D^{-1}C(A-BD^{-1}C-zI)^{-1}BD^{-1}$$

provided the matrix D is square and nonsingular (here the non-zero scalar C_j and the column vector Ae^{j} must be substituted for D and B, respectively). As a matter of fact, this identity, which is known to have many applications in system theory (in particular, for inverting realizations or transfer functions given in the Kalman form (3c), see e.g. [6], p. 76), can be interpreted most naturally as the equality of two classical expressions for the (2,2) entry in the inverse of the block matrix

$$\begin{bmatrix} A-zI & B \\ C & D \end{bmatrix}.$$

Moreover, in view of a *basic determinantal formula due to Schur*, the determinant of this matrix can be written in two complementary ways, viz.,

$$\det(A-zI)\det[D+C(zI-A)^{-1}B] = \det(D)\det(A-zI-BD^{-1}C) ;$$

hence it follows that (27b) can be rewritten in the form

$$\hat{y}^{j+}(z) = (\sigma_j^{1/2}C_j/z)\det(zI-A_{mod})/\det(zI-A), \tag{37}$$

A_{mod} being defined by (36).

Now the j-th column of A_{mod} is 0, which suggests interchanging columns j and 1 (and therefore also rows j and 1, in order to preserve the symmetry of A and the diagonal form of Σ) in all the n-by-n matrices that are involved here, before partitioning them conformably (the dividing lines are inserted between rows 1 and 2, and between columns 1 and 2). It is then clear that the eigenvalues of A_{mod}, with the only exception of the eigenvalue 0 (whose algebraic multiplicity must be reduced by 1 to take into account the fact that the rational function (37) is strictly proper), are precisely the eigenvalues of the (n-1)-by-(n-1) matrix $A_{22} - A_{21}C_2/C_1$, which is therefore a natural candidate for playing the role of A in the above inertia theorem. As for the (nonsingular) Hermitian matrix H in that theorem, the simplest natural candidate is clearly the (n-1)-by-(n-1) diagonal matrix $\sigma_j I-\Sigma_2$, which indeed has n-j positive eigenvalues and j-1 negative eigenvalues. We are thus led to the following *sufficient condition* (for the AAK 'zero count' property to hold) : *the* (n-1)*-by-*(n-1) *matrix* $A_{22}-A_{21}C_2/C_1$ *has no eigenvalues of modulus* 1 *and is such that*

(38a) $(\sigma_j I-\Sigma_2)-(A_{22}-A_{21}C_2/C_1)^*(\sigma_j I-\Sigma_2)(A_{22}-A_{21}C_2/C_1)$ is positive semi-definite.

But the underlying matrices Σ and $A = A^T$, and the row vector $C = B^T$ satisfy the Stein-Lyapunov equations (12a,b), so that finally (38a) can be rewritten in the form

(38b) $\tilde{C}^*(I-A^*A)\tilde{C}$, with $\tilde{C}^* := (C_2^*,-C_1^*I)$, is positive semi-definite

(simple verification, by eliminating Σ_2).

It turns out that *the* n*-by-*n *matrix* $I-A^*A$ is *positive definite* (so that the matrix (38a) is actually positive definite, and consequently $A_{22}-A_{21}C_2/C_1$ has no eigenvalues of modulus 1). The proof is not difficult though non-trivial (details shall appear elsewhere).

REFERENCES

[1] Adamjan, V.M., Arov, D.Z. and Krein, M.G. :'Analytic properties of Schmidt pairs for a Hankel operator and the generalized Schur-Takagi problem', *Math. USSR Sbornik* 15(1971)31-73.

[2] Adamjan, V.M., Arov, D.Z. and Krein, M.G. : 'Infinite Hankel block matrices and related extension problems, *Izv. Akad. Nauk Armjan SSR* 6(1971)87-112 (Russian) - *Amer. Math. Soc. Transl.* (2) 111 (1978)133-156.

[3] Chen, C.-T. : *Linear System Theory and Design*, CBS College Publishing, New York, 1984.

[4] Glover, K. : 'All optimal Hankel-norm approximations of linear multivariable systems and their L^{∞}-error bounds', *Int. J. Control* 39(1984)1115-1193.

[5] Halmos, P.R. : *A Hilbert Space Problem Book*, D. Van Nostrand Company, Princeton, 1967.

[6] Kailath, T. : *Linear Systems*, Prentice-Hall, Englewood Cliffs, 1980.

[7] Lancaster, P. and Tismenetsky, M. : *The Theory of Matrices*, Academic Press, Orlando, 1985.

[8] Meinguet, J. : 'A simplified presentation of the Adamjan-Arov-Krein approximation theory'. In : *Computational Aspects of Complex Analysis* (H. Werner, L. Wuytack, E.Ng and H.J. Bünger, eds.), D. Reidel Publishing Company, Dordrecht, 1983 (pp. 217-248).

[9] Meinguet, J. : 'On the Glover concretization of the Adamjan-Arov-Krein approximation theory'. In : *Modelling, Identification and Robust Control* (C.I. Byrnes and A. Lindquist, eds.), Elsevier Science Publishers, Amsterdam, 1986 (pp. 325-334).

[10] Pommerenke, C. : *Univalent Functions*, Vandenhoeck & Ruprecht, Göttingen, 1975.

Diagonal Padé Approximants, Rational Chebyshev Approximants and Poles of Functions

R. Kovacheva
Institute of Mathematics with Computer Center
Bulgarian Academy of Sciences
1090 Sofia, P.O. Box 373, Bulgaria

Abstract : In this work, converse theorems relating to diagonal Padé approximants and rational Chebyshev approximants with an unfixed number of the finite poles are proved. These theorems provide sufficient conditions that a given function has a pole at a given point. The results refer to the uniform convergence on compact sets in $\mathbb{C}$ in the spherical metric.

1. Introduction :

Let $f(z) = \sum f_\nu z^\nu$ be a formal power series. For each integer $n (n \in \mathbb{N})$ we denote by $\pi_n = \pi_n(f)$ the Padé approximant to f of order n. It is known (see [1]) that $\pi_n = p/q$, where p and q are polynomials of degree $\leq n, q \not\equiv 0$ and

$$(f.q - p)(z) = \alpha_n z^{2n+1} + \ldots \tag{1}$$

The rational function $\pi_n, n \in \mathbb{N}$ always exists and is uniquely determined by (1) (see [1]). We set

$$\pi_n = P_n/Q_n$$

where the polynomials P_n and Q_n have no common divisor and Q_n is monic. The zeros $\zeta_{n,\kappa}, \kappa = 1, \ldots, m_n$, of the polynomial Q_n are called free poles of the rational function π_n; their number m_n doesn't exceed n.

We shall denote by $l = l(\pi)$ the set of concentration points of the free poles of $\pi_n, n \in \mathbb{N}$, in $\mathbb{C}$. The following theorem was found in [2] :

Theorem 1 :
Let $f(z) = \sum f_\nu z^\nu$ be given, let D be a disk, $0 \in D$ and suppose that $F = D \cap l$ is a discrete subset of D; $0 \notin F$. Then f admits a holomorphic continuation in

A. Cuyt (ed.), Nonlinear Numerical Methods and Rational Approximation, 93–103.

$D' = D - F \quad (f \in H(D'))$ and the sequence π_n converges to f uniformly on compact subsets of D' (in the euclicean metric); for any compact set $K, K \subset D'$ the speed of the convergence is given by

$$\limsup_{n \in \mathbb{N}} \|f - \pi_n\|_K^{1/n} = \|\exp(-2g_D(z,0)\|_K \quad (< 1)$$

(here $\|X\|_K$ = sup-norm on K and $g_D(z,0)$ = Green's function for D with singularity at $z = 0$).

Before we continue, we introduce the following notation : for each $a, a \in \mathbb{C}$ we renumber the free poles of π_n, $n \in \mathbb{N}$ so that

$$|\zeta_{n,\kappa}(a) - a| \leq |\zeta_{n,\kappa+1}(a) - a|, \quad \kappa = 1, \ldots, m_n - 1$$

The next theorem is a further result in the investigations, connected with the characterization of f (see [3]) :

Theorem 2 :
In the conditions of Theorem 1, suppose that there is a point $a, a \in F$ and an integer μ such that

$$\text{(2)} \qquad \limsup_{n \in \mathbb{N}} |\zeta_{n,\kappa}(a) - a|^{1/n} < 1, \kappa = 1, \ldots, \mu$$

$$\text{(3)} \qquad \liminf_{n \in \mathbb{N}} |\zeta_{n,\mu+1}(a) - a| > 0$$

Then the function f has a pole at a of order μ.

Denote, now, by Δ the segment $[-1,1]$. Let the function f be real valued and continuous on $\Delta \quad (f \in C(\Delta))$. For each integer n we denote by R_n the rational Chebyshev approximant to f on Δ of order n. Let $l = l(R)$ be the set of concentration points of the finite poles of $R_n, n \in \mathbb{N}$, in $\mathbb{C}$ and $L = L(R)$ the set of limit points.

The following theorem was proved in [4] :

Theorem 3 :
Let $f \in C(\Delta)$. Suppose $l = L, l$ is finite and $l \cap \Delta = \emptyset$. Then

1) $f \in H(\mathbb{C} - l)$ and

2) For any compact set $K \subset \mathbb{C} - l$ the relation

$$\|f - R_n\|_K^{1/n} \longrightarrow 0, \text{as } n \in \mathbb{N}$$

holds.

As above, we renumber for each $a, a \in \mathbb{C}$ the finite poles of R_n in the way that

$$|\zeta_{n,\kappa}(a) - a| \leq |\zeta_{n,\kappa+1}(a) - a|, \kappa = 1, \ldots, m_n - 1$$

A theorem, analogous to Theorem 2, was proved in [5] :

Theorem 4.
In the conditions of Theorem 3, suppose that there is a point $a, a \in l$ and an integer μ such that the conditions (2) and (3) are valid. Then the function f has a pole at a of order μ.

2. Main results.

The basic result in the present is the following theorem, which weakens condition (2) in Theorem 2.

Theorem 5 :
Let U be a disk, let $a \in U$, $a \neq 0$ and $f \in H(\overline{U} - a)$. Suppose that a is the only concentration point of the free poles of $\{\pi_n\}_{n \in \mathbb{N}}$ in $\overline{U}$. Suppose that for any compact subset K of $\overline{U} - a$ the inequality

$$(4) \qquad \limsup_{n \in \mathbb{N}} \|f - \pi_n\|_K^{1/n} = q(K) < 1$$

holds.

If there exists an integer μ such that

$$(2') \qquad |\zeta_{n,\kappa}(a) - a| = 0(1/n), \kappa = 1, \ldots, \mu; \quad n \in \mathbb{N}$$

$$(3) \qquad \liminf_{n \in \mathbb{N}} |\zeta_{n,\mu+1}(a) - a| > 0$$

then the function f has a pole at a of order μ.

Using Theorem 1 and Theorem 5, we obtain

Corollary 1 :
Let $f(z) = \sum f_\nu z^\nu$ be given, let D be a disk, $0 \in D$ and suppose that $F = D \cap l$ is a discrete subset of D, $0 \notin F$. If for each point $a, a \in F$ there exists an integer $\mu = \mu(a)$ such that (2') and (3) are valid, then the sequence π_n converges to f uniformly on compact subsets of D in the spherical metric.

The next theorem is analogous to Theorem 5 in the case of rational Chebyshev approximants to a real valued continuous function.

Theorem 6 :
Let $f \in C(\Delta)$, let $l(R) = L(R), \Delta \cap l(R) = \emptyset$ and $l(R)$ be finite. If there is a point $a, a \in l(R)$ and an integer μ such that (2') and (3') are valid, then f has a pole at a of order μ.

This result is an improvement to Theorem 4.
Analogously as above, we obtain from Theorem 3 and Theorem 6 :

Corollary 2 :
In the conditions of Theorem 3, suppose that there is a set $L', L' \subset l(R)$ such that for each point $a, a \in L'$ there exists an integer $\mu(a)$ such that (2') and (3) are valid with $\mu = \mu(a)$. Let $L'' = l(R) - L'$. Then the sequence R_n converges to f uniformly on compact subsets of $\mathbb{C} - L''$ in the spherical metric.

3. Proofs.

Proof of Theorem 5 :
If f is a rational function, then $\pi_n \equiv f$ for all n sufficiently large and the theorem is proved. We shall consider the case, when f is not rational.

We say, that the integer n is normal, if $\deg Q_n = n$. In this case the rational function π_n interpolates $f(2n+1)$ times at zero and (1) is valid with $p = P_n$ and $q = Q_n$. Let $\Lambda = \{n_i\}_{i \in \mathbb{N}}$ be the set of the normal integers; $n_i < n_{i+1}, i \in \mathbb{N}$. It is known (see [6]) that Λ is infinite iff f is not rational. If $n \in \mathbb{N}$ and n_i is the largest normal integer with the property $n_i \leq n$, then $\pi_n \equiv \pi_{n_i}$ (see [6]). For each pair $n_i, n_{i+1} \in \Lambda$ there holds the representation

$$(\pi_{n_{i+1}} - \pi_{n_i})(z) = \mathcal{A}_{n_i} z^{n_i + n_{i+1}}.(Q_{n_i}.Q_{n_{i+1}})^{-1}(z)$$

where $\mathcal{A}_{n_i}$ is a constant, $\mathcal{A}_{n_i} \neq 0$. it is given by the formula

$$\mathcal{A}_{n_i} = (P_{n_{i+1}} Q_{n_i} - P_{n_i} Q_{n_{i+1}})(b).b^{-(n_i + n_{i+1})}, b \neq 0$$

To simplify the notation, we shall assume that $\Lambda \equiv \mathbb{N}$; in this case we obtain

$$(\pi_{n+1} - \pi_n)(z) = \mathcal{A}_n z^{2n+1} (Q_n Q_{n+1})^{-1}(z)$$

and

$$\mathcal{A}_n = (P_{n+1} Q_n - P_n Q_{n+1})(b).b^{-(2n+1)}, b \neq 0 \tag{5}$$

Since $\mathcal{A}_n \neq 0$, we get

$$\zeta_{n,\kappa}(a) \neq \zeta_{n+1,l}(a), \kappa = 1, \ldots, n; l = 1, \ldots, n+1 \tag{6}$$

We set

$$\zeta_{n,\kappa}(a) = \zeta_{n,\kappa}, \quad \kappa = 1, \ldots, n$$

$$q_n(z) = \prod_{\kappa=1}^{\mu} (z - \zeta_{n,\kappa})$$

$$Q_n^*(z) = Q_n(z).q_n^{-1}(z)$$

We shall use the following notation; for each $r, r > 0$ we shall denote by U_r the disk of radius r centered at a; $\Gamma_r = \partial U_r$.

We now select a number r such that $dist(U_{2r}, 0) > 2r$ and $U_{2r} \subset U$. It is clear that (see (3))

$$|Q_n^*(z)| > 0 \quad \text{for } z \in U_{2r}, \quad n \geq n_0 = n_0(r) \tag{7}$$

Denote now by $\mathcal{P}_n^*$ the principal part of π_n at the points $\zeta_{n,1}, \ldots, \zeta_{n,\mu}$. The calculation implies that

$$P_n(\zeta_{n,\kappa}) = \mathcal{P}_n^*(\zeta_{n,\kappa}).Q_n^*(\zeta_{n,\kappa}); \kappa = 1, \ldots, \mu \tag{8}$$

Now we are going to prove that in the conditions of the theorem

$$\prod_{\kappa=1}^{\mu} |\mathcal{P}_n^*(\zeta_{n,\kappa})|^{1/n} \longrightarrow 1, \quad \text{as} \quad n \in I\!N \tag{9}$$

Indeed, evaluating $\mathcal{A}_n$ by putting in (5) $z = \zeta_{n,\kappa}$ and $z = \zeta_{n+1,l}$, we obtain

$$\frac{P_{n+1}(\zeta_{n+1,l})Q_n(\zeta_{n+1,l})}{(\zeta_{n+1,l})^{2n+1}} = -\frac{P_n(\zeta_{n,\kappa})Q_{n+1}(\zeta_{n,\kappa})}{(\zeta_{n,\kappa})^{2n+1}}$$

Multiplying both sides of the last equality, $\kappa, l = 1, \ldots, \mu$ and cancelling the factors $|\zeta_{n,\kappa} - \zeta_{n+1,l}|$, we obtain from (6)

$$\prod_{\kappa=1}^{\mu} \left| \frac{P_{n+1}(\zeta_{n+1}, \kappa)}{P_n(\zeta_{n,\kappa})} \right| = \prod_{\kappa=1}^{\mu} \frac{|Q_{n+1}^*(\zeta_{n,\kappa})|}{|Q_n^*(\zeta_{n+1,\kappa})|} \cdot \prod_{\kappa=1}^{\mu} \left| \frac{\zeta_{n+1,\kappa}}{\zeta_{n,\kappa}} \right|^{2n+1}$$

From this and from (8), we get finally

$$\prod_{\kappa=1}^{\mu} \left| \frac{\mathcal{P}_{n+1}^*(\zeta_{n+1,\kappa})}{\mathcal{P}_n^*(\zeta_{n,\kappa})} \right| = \prod_{\kappa=1}^{\mu} \left| \frac{Q_n^*(\zeta_{n,\kappa})}{Q_n^*(\zeta_{n+1,\kappa})} \cdot \frac{Q_{n+1}^*(\zeta_{n,\kappa})}{Q_{n+1}^*(\zeta_{n+1,\kappa})} \cdot \left(\frac{\zeta_{n+1,\kappa}}{\zeta_{n,\kappa}} \right)^{2n+1} \right|$$

Let $n_1, n_1 \geq n_0$ be so large, that for all $n, n \geq n_1$

$$(10) \qquad |\zeta_{n,\kappa} - a| \leq r, \kappa = 1, \dots, \mu$$

We set $d_n = \max|\zeta_{n,\kappa} - \zeta_{n+1,\kappa}| \kappa = 1, \dots, \mu$.

From (2') it follows that $d_n = 0(1/n), n \in \mathbb{N}$.

The inequalities (7) and (10) yield

$$(1 - d_n.r^{-1})^{2n+1-\mu} \leq |Q_n^*(\zeta_{n,\kappa})|.|Q_n^*(\zeta_{n+1,\kappa})|^{-1} \leq (1 + d_n.r^{-1})^{2n+1-\mu}$$

Hence, for each $\kappa, \kappa = 1, \dots, \mu$

$$(11) \qquad |Q_n^*(\zeta_{n,\kappa})|.|Q_n^*(\zeta_{n+1,\kappa})|^{-1} \longrightarrow 1, \quad \text{as} \quad n \in \mathbb{N}$$

Similarly

$$|Q_{n+1}^*(\zeta_{n,\kappa})|.|Q_{n+1}^*(\zeta_{n+1,\kappa})|^{-1} \longrightarrow 1, \quad \text{as} \quad n \in \mathbb{N}$$

and

$$|\zeta_{n+1,\kappa}.\zeta_{n,\kappa}^{-1}|^{(2n+1)} \longrightarrow 1, \quad \text{as} \quad n \in \mathbb{N}$$

Combining (11) and the last results, we get

$$\prod_{\kappa=1}^{\mu} |\mathcal{P}_{n+1}^*(\zeta_{n+1,\kappa})|.|\mathcal{P}_n^*(\zeta_{n,\kappa})|^{-1} \longrightarrow 1, \quad \text{as} \quad n \in \mathbb{N}$$

This proves (9).

From (4) we have

$$\limsup \|f - \pi_n\|_{\Gamma_r}^{1/n} = q (= q(\Gamma_r)) < 1$$

It follows from the conditions of Theorem 5 that $\pi_n \to f$ uniformly on $\overline{U}$ in m_1-measure. If the function f has a pole at a, then its order doesn't exceed μ (see [8]). Consequently there holds in $\overline{U}$ the representation

$$f(z) = \varphi(z) + p(z).(z - a)^{-\mu}$$

where $\varphi \in H(\overline{U})$ and $\deg p \leq \mu - 1$. The function f has a pole at a of order μ iff $p(a) \neq 0$; $f \in H(\overline{U})$ iff $p \equiv 0$.

Select now a positive number θ such that $q_1 = e^{\theta} q < 1$. It follows from the results of [7] that

$$(12) \qquad \|\mathcal{P}_n^*(z).q_n^{-1}(z) - p(z).(z - a)^{-\mu}\|_{\Gamma_r} \leq C_1.q_1^n, n \geq n_2 (\geq n_1)$$

(here and everywhere afterwards C_i are positive constants which don't depend on $n, i \in I\!N$).

Suppose that $f \in H(\overline{U})$. Then $p \equiv 0$. From (12) we get

$$|\mathcal{P}_n^*(\zeta,\kappa)| \leq C_2.q_1^n, \kappa = 1,\dots,\mu, \quad n \geq n_2$$

This contradicts (9).

Thus we conclude that the function f has a pole at a; therefore $p \not\equiv 0$. We shall show that for each κ, $\kappa = 1,\dots,\mu$

$$|\mathcal{P}_n^*(\zeta_{n,\kappa})|^{1/n} \longrightarrow 1 \quad \text{as} \quad n \in I\!N \tag{13}$$

Indeed, from (2') and from (12) we have

$$\begin{aligned} \|\mathcal{P}_n^* - p\|_{\overline{U}_r} &\leq \|\mathcal{P}_n^*(z) - p(z).q_n(z).(z-a)^{-\mu}\|_{\Gamma_r} + \\ &+ \|p(z).(1 - q_n(z).(z-a)^{-\mu}\|_{\Gamma_r} \leq C_3.q_1^n + C_3, n \geq n_2 \end{aligned} \tag{14}$$

Hence

$$\|\mathcal{P}_n^*\|_{\overline{U}_r} \leq C_4, \quad n \geq n_2$$

and for each $\kappa, \kappa = 1,\dots,\mu$

$$\limsup \, |\mathcal{P}_n^*(\zeta_{n,\kappa})|^{1/n} \leq 1$$

From (9) and from the last inequality, we obtain (13).

Let now θ_1 be an arbitrary positive number such that $q_1 e^{(\mu+1)\theta_1} < 1$. We set $q_2 = (q_1 e^{\theta_1})^{1/\mu}$. Obviously $q_2 < e^{-\theta_1}$.

(13) yields that for each $\kappa, \kappa = 1,\dots,\mu$ and all n sufficiently large, $n \geq n_3$ there hold the inequalities

$$e^{-n\theta_1} \leq |\mathcal{P}_n^*(\zeta_{n,\kappa})| \leq e^{n\theta_1}, n \geq n_3 (\geq n_2) \tag{15}$$

Let U_ρ be the smallest disk, $U_\rho \subseteq U_r$ such that for all $n \geq n_3$ $\zeta_{n,\kappa} \in U_\rho, \kappa = 1,\dots,\mu$.

The maximum principle and (12) provide that

$$\|\mathcal{P}_n^*(z).(z-a)^\mu - p(z).q_n(z)\|_{\overline{U_r}} \leq C_5.q_1^n, n \geq n_3$$

From this we obtain

$$\|\mathcal{P}_n^*(z).(z-a)^\mu - p(z).q_n(z)\|_{\overline{U}_\rho} \leq C_5.q_1^n, n \geq n_3 \tag{16}$$

Putting here $z = \zeta_{n,\kappa}, \kappa = 1, \ldots, \mu$ and using (15), we get

$$|\zeta_{n,\kappa} - a| \leq C_6 . q_2^n, \qquad n \geq n_3 \tag{17}$$

Repeating the arguments in the proof of (14), from (16) and (17), we obtain

$$\|\mathcal{P}_n^* - p\|_{\overline{U}_\rho} \leq C_7 . q_2^n, \quad n \geq n_3 \tag{18}$$

Now it is easy to verify that $p(a) \neq 0$. Indeed, for each $\kappa, \kappa = 1, \ldots, \mu$ we have by Cauchy's formula

$$\mathcal{P}_n^*(\zeta_{n,\kappa}) - p(a) = \frac{1}{2\pi i} \int_{\Gamma_\rho} \frac{(\mathcal{P}_n^* - p)(t)(t-a) + p(t)(\zeta_{n,\kappa} - a)}{(t-a)(t - \zeta_{n,\kappa})} dt$$

For $|\zeta_{n,\kappa} - a| \leq \rho/2$ the inequalities (18) and (17) yield

$$|\mathcal{P}_n^*(\zeta_{n,\kappa}) - p(a)| \leq C_8 . q_2^n, n \geq n_4 (\geq n_3)$$

From (15), we get

$$|p(a)| \geq e^{-n\theta_1} - C_8 . q_2^n, \quad n \geq n_4$$

Since $e^{-\theta_1} > q_2$, we obtain that $p(a) \neq 0$. Therefore the function f has a pole at a of order μ. Theorem 5 is proved.

Proof of theorem 6 :

The proof of Theorem 6 is analogous to the proof of the preceding theorem.

We shall assume that f is not a rational function. In the opposite case $f \equiv R_n$ for all n sufficiently large and the theorem is proved.

We say that the integer n is normal, if the number of all the poles of R_n in $\mathbb{C}$ equals n. Let Λ be the set of the normal integers; $\Lambda = \{n_i\}_{i \in \mathbb{N}}; n_i < n_{i+1}$. It follows from the alternation theorem that for each $n, n \in \mathbb{N}$

$$R_n (= P_n . Q_n^{-1}) \equiv R_{n_i}$$

where $n_i, n_i \in \Lambda$ is the largest integer with the property $n_i \leq n$. For each pair n_i, n_{i+1} of normal integers we have (see [4])

$$(R_{n_{i+1}} - R_{n_i})(z) = \mathcal{A}_{n_i} . w_{n_i}(z) . (Q_{n_i} Q_{n_{i+1}})^{-1}(z) \tag{19}$$

where the polynomial w_{n_i} is monic, deg $w_{n_1} = n_i + n_{i+1}$ and all its zeros are simple and belong to Δ.

We shall assume, without losing the generality, that $\Lambda \equiv I\!N$. In this case (19)is valid for each $n, n \in I\!N$; the constant $\mathcal{A}_n$ is given by the formula

$$\mathcal{A}_n = (P_{n+1}Q_n - P_nQ_{n+1})(b).w_n^{-1}(b) \tag{20}$$

where b is an arbitrary complex number, $b \in \mathbb{C} - \Delta$.

Formula (20) is analogous to formula (5). In the other respects the proof of Theorem 6 is a repetition of the considerations, used in the proof of Theorem 5.

Proof of the corollaries :
Let us remember the definition of the uniform convergence on compact sets in the spherical metric.

Let $z_1, z_2 \in \mathbb{C}$. Then (see [9])

$$(z_1; z_2) = \text{dist}_S(z_1, z_2) = 2|z_1 - z_2|(1 + |z_1|^2)^{-1/2}(1 + |z_2|^2)^{-1/2}$$

For $z_1 = \infty$, we set

$$(z; \infty) = 2|z|(1 + |z|^2)^{-1/2}$$

Let now E be a compact set in $\overline{\mathbb{C}}$ and $\varphi_n, n \in I\!N$ be a sequence of functions, given on E and continuous in $\overline{\mathbb{C}}$. We say that $\varphi_n \longrightarrow \varphi$ uniformly on E in the spherical metric, if for each $\varepsilon, \varepsilon > 0$ and for all $n, n \in I\!N$, $n \geq \mathcal{N}_1$ the inequality $\max_{z \in E}(\varphi_n; \varphi)(z) < \varepsilon$ holds. The uniform convergence in the euclidian metric yields a uniform convergence in the spherical metric (on compact sets).

Detailed explanations about the convergence in the spherical metric in general and about the convergence (in the same metric) of sequences of Padé approximants, in particular, are to be found in [9] and in [2].

Let now U be a disk and $a \in U$. Let $\varphi \in H(U - a), \varphi(z) \neq 0$ for $z \in U$ and assume that φ has a pole at a of order p. Let φ_n be a sequence of functions, meromorphic in U, such that $\varphi_n \to \varphi$ uniformly on $\overline{U}$ in m_1-measure (for each $\varepsilon, \varepsilon > 0$, there is a set $U_\varepsilon, U_\epsilon \subset U$, such that $m_1(U - U_\epsilon) < \epsilon$ and $\varphi_n \to \varphi$ uniformly on $\overline{U}_\varepsilon$ (in the euclidian metric)). It is known (see [9]) that

a) If there is a compact set $K, K \subset U$ such that $\varphi_n \in H(K), n \in I\!N$ then φ_n converges uniformly (in the euclidian metric) on K (therefore $\varphi \in H(K)$) and

b) All the functions φ_n have at least p poles in U (for $n \geq \mathcal{N}_2$) and a attracts at least p poles. We denote by μ_n and by ν_n the number of poles and of zeros of

φ_n in $U, n \geq \mathcal{N}_2$. The argument principle provides that for all $n, n \geq \mathcal{N}(\geq \mathcal{N}_2)$ the equality

$$\nu_n - \mu_n = -p \tag{21}$$

holds.

From the definition we obtain that $\varphi_n \longrightarrow \varphi$ uniformly on $\overline{U}$ in the spherical metric iff $\nu_n = 0$ for all $n, n > \mathcal{N}$. Keeping in mind (21), we see that $\nu_n = 0$ iff $\mu_n = p$ (for $n \geq \mathcal{N}$). Thus we have shown that in the case being considered $\varphi_n \rightarrow \varphi$ on $\overline{U}$ in the sphérical metric iff $\mu_n = p$ for all $n \geq \mathcal{N}$.

Now the assertions in the corollaries follow from the proved theorems and from Theorem 1 and Theorem 3. Indeed, both theorem provide a uniform convergence (on compact subsets) of D' and of $\mathbb{C} - l$, respectively, and a uniform convergence in m_1-measure in neighbourhoods of points $a, a \in L'(R)(L'(R) = L')$ on the other hand (see Corollary 1), since $F = D \cap l$ is a discrete subset of D, we see that each $a, a \in F$ is an isolated point in D and therefore Theorem 5 and the preceding considerations are applicable.

Consider again the Padé approximants $\pi_n, n \in \mathbb{N}$ to f. Let U and a be given, $a \in U, a \neq 0$. Let $f \in H(U - a)$ and suppose that f has a pole at a of order μ. Suppose that (4) is valid. As we have seen, a attracts then at least μ free poles of π_n, as $n \in \mathbb{N}$; moreover, the attraction of at least μ of them is geometrical. Indeed, to establish that, we have to calculate the ν-th derivative of the polynomial $\mathcal{P}_n^*(z)(z-a)^\mu - (p.q_n)(z), \nu = 0, \ldots, \mu - 1$, to estimate it on $\overline{U}$ after (4) and after the results of [6] (see (12)) and to put $z = a$ (here $p(a) \neq 0$ and deg $q_n \geq \mu$). Hence we see that in this case the uniform convergence on $\overline{U}$ in the spherical metric is equivalent to the conditions (2) and (3). This result was proved in [3] by Gonchar.

On the other hand, condition (2') yields together with condition (3) a pole at a of the corresponding order; then a attracts the free poles $\zeta_{n,1}, \ldots, \zeta_{n,\mu}$ geometrically. Therefore (2') and (3) are sufficient for the uniform convergence of π_n on $\overline{U}$ in the spherical metric, but not necessary.

References.

[1.] Perron O. *Die Lehre von den Kettenbruchen*, Teubner, Stuttgart, 1957.

[2.] Gonchar A. A., *On the uniform convergence of diagonal Padé approximants*, Mat. Sbornik 118 (1982) 535-565 (in Russian).

[3.] Gonchar A. A., *On the convergence of diagonal Padé approximants in the spherical metric*, Papers dedicated to academician L. Iliev's anniversary, Publishing House of the Bulgarian Academy of Sciences (1984) 29-36 (in Russian).

[4.] Lungu K. N., *Properties of functions, resulting from the assymptotics of rational Chebyshev Approximants*, International conference on constructive function theory, Varna Proceedings (1983) 103-110 (in Russian).

[5.] Kovacheva R. K., *Best Chebyshev rational approximants and poles of functions*, Lecture Notes in Math. 1237 (1986) 68-72.

[6.] Gonchar A.A., Lungu K. N., *Poles of diagonal Padé approximants and analytic continuation of functions*, Mat. Sbornik 111 (1980) 280-292 (in Russian).

[7.] Gonchar A. A., Grigoryan, *Estimates of the norm of holomorphic functions*, Mat. Sbornik 99 (1976) 634-638 (in Russian).

[8.] Gonchar A. A., *Convergence of generalized Padé approximants of meromorphic functions.* Mat. Sbornik 98 (1975) 564-577 (in Russian).

[9.] Baker G. A. Jr., Graves-Morris P., *Padé approximants Part I*, Encyclopedia of Mathematics and its Applications Vol. 13, 14, Addison-Wesley Publ. Co., London-Amsterdam, 1981.

ON THE USE OF THE CARATHÉODORY-FEJÉR METHOD FOR INVESTIGATING '1/9' AND SIMILAR CONSTANTS

Alphonse P. Magnus

Institut Mathématique Université Catholique de Louvain
Chemin du Cyclotron 2
B-1348 Louvain-la-Neuve (Belgium)

ABSTRACT . Error norms of best rational approximations of exp(-t) on $[0,\infty)$ are known to decrease like ρ^n , where n is the degree of the approximant and ρ is the famous number '1/9' = 1/9.28902549192081891875544943595174506... Trefethen and Gutknecht have demonstrated this effect on the sequence of singular values of a Hankel matrix , as an example of their use of the Carathéodory-Fejér method . It is shown here how the rate of decrease of these singular values can be estimated from their symmetric functions . The examples of rational approximation of $\exp(-t^m)$ on $[0,\infty)$, m=2,3 are also explored . The relation with the extremal polynomials method is briefly discussed .

1. INTRODUCTION .

Best real rational approximants are characterized by equioscillation (equal ripple) properties . Although well known algorithms (as in [7] , [74]) are able to produce such approximants , it is highly useful to predict quantitatively the main properties of the result . For polynomial approximation on the standard interval [-1,1] , much can be done with Chebyshev polynomial expansions

$$F(x) = a_0/2 + \sum_{1}^{\infty} a_k T_k(x) = a_0/2 + \sum_{1}^{\infty} a_k \cos k\theta \quad (x=\cos\theta). \tag{1.1}$$

In particular , the behaviour of the uniform norm of the error of the best approximation by a polynomial of degree n can be predicted for large n from the behaviour of a_n . One would like a similar tool allowing the description of rational approximation errors in terms of the coefficients of (1.1) . Several algorithms based on Padé theory ([8] [21]) have been proposed for constructing useful rational approximations to F , and a suitably modified version was powerful enough for establishing a proof of Meinardus conjecture [4] . However , for most functions continuous on [-1,1] , Padé-Chebyshev-like error norms will not decrease with n at the same rate as the actual optimal error norms E_n (see the case of Stieltjes functions , compare [21] with [3] ; see

A. Cuyt (ed.), Nonlinear Numerical Methods and Rational Approximation, 105–132.

also [6]) . But the long-sought method has finally been found! Table 1 gives a first example of its striking efficiency . The left part of the table shows least uniform error norms E_n of approximation of exp(-t) on $t \geq 0$ by rational functions of degree n . These values come from Table 1 of [7] . Rational approximation of degree n of a function G(t) on $t \geq 0$ is the same as rational approximation of degree n of F(x) on $-1 \leq x \leq 1$ provided

$$F(x) = G(t) \quad , \quad x=(1-t)/(1+t) \tag{1.2}$$

So the numbers E_n are also relevant to the approximation of F(x) = exp((x-1)/(x+1)) on [-1,1] . On the second part of table 1 are estimates of the same numbers (with signs which will be explained) constructed with the coefficients of (1.1) with this F(x) . The method of constructing these estimates has been described in [68] , and a summary will be given in next section . A first version of table 1 is in Table 5 of [68] , and the present one had already be established by L.N.Trefethen [66]

Table 1. Remes algorithm results and CF estimates for approximation of exp(-t) . Digits of agreement of the two have been underlined in the second column .

n	E_n	λ_{n+1}
0	5.000000000000000E-01	5.601715174207940E-01
1	6.683104216185045E-02	-6.680573308019967E-02
2	7.358670169580528E-03	7.355581867871742E-03
3	7.993806363356878E-04	-7.994517064498902E-04
4	8.652240695288851E-05	8.652095258749368E-05
5	9.345713153026646E-06	-9.345740936352446E-06
6	1.008454374899671E-06	1.008453857122026E-06
7	1.087497491375248E-07	-1.087497586430036E-07
8	1.172265211633491E-08	1.172265194363526E-08
9	1.263292483322314E-09	-1.263292486435714E-09
10	1.361120523345448E-10	1.361120522787731E-10
11	1.466311194937487E-11	-1.466311195036850E-11
12	1.579456837051239E-12	1.579456837033622E-12
13	1.701187076340353E-13	-1.701187076343463E-13
14	1.832174378254041E-14	1.832174378253495E-14
15	1.973138996612803E-15	-1.973138996612899E-15
16	2.124853710495224E-16	2.124853710495207E-16
17	2.288148563247892E-17	-2.288148563247895E-17
18	2.463915737765169E-18	2.463915737765169E-18
19	2.653114658063313E-19	-2.653114658063313E-19
20	2.856777383549094E-20	2.856777383549094E-20

The exponential function was the ideal test : the study of rational approximation of this function on the positive real axis $0 \leq t < \infty$ is originally motivated by the numerical solution of differential equations [9] , [30] . When it emerged that the uniform error norm E_n decreases exponentially fast with respect to the degree n [9] , the determination of the precise rate of decrease became a problem interesting on its own right. A history of the approach of this number is given in [5], [7] and [71]. For a moment, it seemed to be 1/9, and the problem of its determination is known as the '1/9' problem [60], but Gutknecht and Trefethen [68], as well as Schönhage [61], showed numerical evidence that it is somewhat smaller, about 1/9.28903. A proof that the number is definitely smaller than 1/9 is in [53]. Much more numerical evidence is collected in [7], where it is found that E_n very likely behaves according to the asymptotic expansion (from [7] Tables 3-10 and II)

$$E_n \sim 2\rho^{n+1/2} \exp(-1/(12(n+1/2)) + O(n^{-5})),$$

where ρ = 1/9.28902549192... (see Table 5 in section 6). Finally, Goncar and Rahmanov ([18] [37] [58]) have succeeded in proving the existence of the limit ρ of $(E_n)^{1/n}$ as $n\to\infty$ by the theory of extremal polynomials, which will be described briefly in section 5. From Glover-Karlsson-Trefethen inequalities [14] [33] [67], it follows that $|\lambda_n|^{1/n}$ has the same limit.

I shall try to show here how this rate of decrease can be deduced from Gutknecht and Trefethen's CF constructions. These authors have already shown in [68] the asymptotic matching of real rational approximation and CF approximation for functions analytic in very large regions (for fixed degrees), asymptotic pole matching for meromorphic functions (for unequal degrees, with degree of numerator $\to\infty$ [24]). Braess [6] has established complete asymptotic agreement between real rational approximation and CF approximation for Stieltjes functions, another very interesting test, as the rational approximants are very accurately known [3]. CF rational approximation has been adapted to regions which are neither disks nor real intervals by S.W.Ellacott [10].

To give at least one more example of an exponential-like function, here are numerical data for $\exp(-t^2)$ on $0\le t\le\infty$. Asymptotic matching still seems to hold, but things are less regular, which makes them even more interesting.

Table 2. Remes algorithm results and CF estimates for approximation of $\exp(-t^2)$. Digits of agreement of the two have been underlined in the second column.

n	E_n	λ_{n+1}
0	5.0000000000000E-01	6.4950930361073E-01
1	1.6886948942235E-01	-1.6652643752795E-01
2	2.2842588922487E-02	2.2904765064961E-02
3	6.2967346172045E-03	-6.2950348277494E-03
4	6.6571906756928E-04	6.6578772442257E-04
5	2.6217636106147E-04	-2.6218111332487E-04
6	1.7188934554376E-05	1.7195697038148E-05
7	1.3081592285798E-05	-1.3081593719916E-05
8	7.1483585294295E-07	-7.1488121795910E-07

2. CF APPROXIMATION .

In the CF method , instead of looking for nearly equioscillating approximants in the right set of functions (for instance , the rational n/n functions) , one considers an approximant in a broader set of functions , but yielding an exactly equioscillating error function . It is then necessary to modify this approximant in order to return to the required class . Only a sketch of n/n CF approximation on the real interval [-1,1] is given here . As far as possible , the notations of [68] are kept . Let F be a continuous function on [-1,1] and let (1.1) be its Chebyshev polynomial expansion . Using $z=\exp(i\theta)$, one can write

$$F(x) = a_0/2 + f^+(z)/2 + f^+(z^{-1})/2 ,$$

where

$$f^+(z) = \sum_1^\infty a_k z^k \tag{2.1}$$

(the a_k's need not be real) . One considers the approximation of f^+ by meromorphic functions in $|z|>1$ with exactly n poles in that region . Such functions can always be written in the form

$$\tilde{r}(z) = p(z)/q(z) = \left(\sum_{-\infty}^{n} d_k z^k\right)/\left(\sum_0^n e_k z^k\right)$$

where the n zeros of q must have modulus larger than 1 . With respect to the supremum norm on the unit circle , the best approximation $\tilde{r}^*$ to f^+ in this class , the "extended CF approximant" of f^+ , is characterized by the property that except in degenerate cases , the error function $f^+(z) - \tilde{r}^*(z)$ describes an exact circle of winding number 2n+1 centered at the origin , as z describes the unit circle ([1] [14] [23] [28] [41]-[43] [65] [67]-[69]) . A consequence is that the error function can then be written

$$f^+(z)-\tilde{r}^*(z) =b(z) =b_1(z)/b_2(z) = \sigma(\bar{u}_1+\bar{u}_2 z +\ldots)/(u_1 z^{-1}+u_2 z^{-2} +\ldots), \tag{2.2}$$

where the denominator b_2 of b is holomorphic in $|z|>1$ and must have exactly n zeros in $1<|z|<\infty$, which will be precisely the zeros of the denominator q of $\tilde{r}^*$:

$$b_2(z) = u_1 z^{-1}+u_2 z^{-2}+\ldots =(e_n+e_{n-1}z^{-1}+\ldots+e_0 z^{-n})v(z) = z^{-n}q(z)v(z) ,$$

where v is still holomorphic (and without zeros) in $1<|z|<\infty$. Therefore ,multiplying the two sides of (2.2) by $b_2(z)$,

$$f^+(z)b_2(z) -z^{-n}p(z)v(z) = \sigma b_1(z) \quad , \text{or}$$

$$f^+(z)(u_1 z^{-1} + u_2 z^{-2} +\ldots) = \sigma(\bar{u}_1 + \bar{u}_2 z +\ldots) + \text{negative powers of } z,$$

i.e.

$$HU = \sigma\bar{U} ,$$

where U is the vector $[u_1,u_2,\ldots]^T$ and H is the infinite Hankel matrix

$$H = [a_{k+m-1}] \ , \ k,m=1,\ldots ; \tag{2.3}$$

σ is a singular value of H (since $H\overline{U} = \sigma U$, i.e., $\bar{H}HU = \sigma^2 U$) . The fact that the (n+1)th singular value of H ($\sigma_1 \geq \sigma_2 \geq \ldots$) is indeed related to a vector U such that $u_1 z^{-1} + u_2 z^{-2} + \ldots$ has exactly n zeros in $1<|z|<\infty$ requires a deeper understanding of Hankel matrices theory [1] [14] [41]-[43] . However , the authors of [28] succeeded in getting all the fine points with only elementary tools of complex functions theory , such as Rouché's theorem . Actually , the picture presented here supposes that σ_{n+1} is non-repeated: $\sigma_n > \sigma_{n+1} > \sigma_{n+2}$ (see [1] [14] [23] [28] [41]-[43] [65] [67]-[69] for a general treatment) .
If the a_k's happen to be real , then $\sigma_{n+1} = |\lambda_{n+1}|$, the absolute value of the (n+1)th eigenvalue of H .

The last step in CF approximation is to project $\tilde{R}(x) = (a_0 + \tilde{r}^*(z) + \tilde{r}^*(z^{-1}))/2$ onto a rational n/n function of $x = (z+z^{-1})/2$. One naturally chooses $Q(x) = q(z)q(z^{-1})$ as the denominator and determines the numerator P(x) such that the Chebyshev expansions of $R_{CF}(x) = P(x)/Q(x)$ and $\tilde{R}(x)$ agree through the $T_n(x) = (z^n+z^{-n})/2$ term. This operation destroys the exact equioscillation of the error function , but the perturbation is usually very much smaller than σ_{n+1} . This further treatment will not be studied here . The incredible accuracy of CF approximation as quasioptimal real rational approximation is best demonstrated by tables 1 to 3 showing successive values of $\sigma_{n+1} = |\lambda_{n+1}|$ compared to corresponding values of E_n , n=0,1,... It is also striking to discover a Hankel matrices generalized eigenproblem in a step of the Remes algorithm [74]... wether this is another relation between Remes and CF is not clear .

In short , CF allows one to explore rational approximation of a function F by means of quantitative estimates of the eigenvalues of Hankel matrices composed of the Chebyshev coefficients of F . For this reason , some properties of the spectra of Hankel matrices are investigated now .

3. AN INTEGRAL FORM FOR HANKEL MATRIX INVARIANTS .

From now on , we consider only functions with real Chebyshev coefficients in (1.1) , so that the eigenvalues $\lambda_1, \lambda_2, \ldots$ of the Hankel matrix (2.3) are the quantities of interest . If F is continuous (a minimal requirement !) , H represents a bounded compact operator on the Hilbert sequence space ℓ^2 (these facts are nicely summarized in [41] and [76] §8, and there is a much more detailed and advanced treatment in [54] [55]) , so that its spectrum is a countable set of eigenvalues having the origin as the only limit point . Moreover , the meaning of some equalities of the preceding section becomes more rigorous as U is now a square summable sequence : the functions discussed there are actually functions in L^2 or H^2 .

How can we get information on the behaviour of the eigenvalues of H from the knowledge of F ? The simplest identity involves the trace of H , which makes immediate reference to values of F :

$$\varsigma_1 = \text{trace } H = \sum_1^\infty \lambda_n = \sum_1^\infty a_{2n-1} = (F(1)-F(-1))/2 .$$

For example , with $F(x) = [(1+x)/2]^{1/2}$, a well-worked example ([12] [32] [49] [72] , the value E_n for F equals the value E_{2n} for $|x|$) , $\varsigma_1 = 1/2$ so that , if the decrease of the λ_n's is fast , λ_1 must be expected to be near 1/2 -which is not surprising , as $E_0 = 1/2$. For more information on the subsequent λ_n's, let us consider the trace of H^2 , which is still not too hard (it is the sum of the squares of all the elements of the matrix ,= square of Frobenius norm , or Hilbert-Schmidt norm) :

$$\varsigma_2 = \text{trace } H^2 = \sum_1^\infty (\lambda_n)^2 = \sum_1^\infty n(a_n)^2 .$$

For $F(x)=[(1+x)/2]^{1/2}$, $a_n = (-1)^{n-1}/[\pi(n^2-1/4)]$ and $\varsigma_2 = 2/\pi^2$ is easily obtained . Considering only λ_1 and λ_2 , this gives $\lambda_1 \sim 0.447$ and $\lambda_2 \sim 0.053$. By numerical means , the following table is obtained ; we have added the rational least error norms E_n , completed from [26]) :

Table 3. Remes algorithm results , CF estimates for approximation of $[(x+1)/2]^{1/2}$ and empirical formula .

n	E_n	λ_{n+1}
0	0.500000	0.448264
1	0.043689	0.040407
2	0.008501	0.008008
3	0.002282	0.002170
4	0.000737	0.000705
5	0.000269	0.000258
	$8\exp(-\pi(2n+3/4)^{1/2})$	$8\exp(-\pi(2n+5/6)^{1/2})$

The link between F and ς_2 is already more difficult than for ς_1 : ς_2 requires information on the conjugate function $F_c(x) = \sum_1^\infty a_n \sin n\theta$, as $\varsigma_2 = (2/\pi)\int_0^\pi F(\cos\theta)dF_c(\cos\theta)$, by Parseval's relation . In terms of f^+ : $\varsigma_2 = (2\pi i)^{-1}\int_{|z|=1} f^+(z^{-1})df^+(z)$.

Following this line , integral forms for the ς_p's will be produced , where

$$\zeta_p = \sum_1^\infty (\lambda_n)^p . \qquad (3.1)$$

Estimates of these values will in turn yield estimates for the λ_n's although ζ_p for large p gives only the largest λ's $(\lambda_1, \lambda_2, \dots)$.

Should there be a way to estimate ζ_p for (non integer) small values of p, then knowledge of the asymptotic behaviour of λ_n, $n\to\infty$, could be reached. The theoretical foundations are in [55]; see [73] for this Zeta notation.

In order to manipulate safely these quantities, a supplementary condition is needed: H will be supposed to be trace-class, i.e., $\sum_1^\infty |\lambda_j| < \infty$. A sufficient condition is the absolute convergence of the double series of all the elements of H, turning here into $\sum_1^\infty n|a_n| < \infty$.

A necessary condition is that H is Hilbert-Schmidt: $\sum_1^\infty n(a_n)^2 < \infty$. The first condition is much too severe and does not work for instance with $F(x)=(1+x)^{1/2}$, whereas it is known ([17] [72]) that the E_n's (and consequently the σ_n's, from the Glover-Karlsson-Trefethen inequalities[14] [33] [67]) decrease as fast as $\exp(-\pi(2n)^{1/2})$... The complete characterization has been given by Peller in terms of the function F, actually in terms of f^+:

<u>Theorem 3.1.</u> The Hankel matrix $H = [a_{k+m-1}]_1^\infty$ is trace-class if and only if $f^+(z)=\sum_1^\infty a_k z^k$ is in the Besov class A_1^1 [54],

which means: f^+ analytic in $|z|<1$, measurable on $|z|=1$, and

$$\int_{-\pi}^{\pi} h^{-2}\|\Delta^2 f^+(\exp(i\theta),h)\|dh < \infty ,$$

where

$$\|\Delta^2 f^+(\exp(i\theta),h)\| = (2\pi)^{-1}\int_{-\pi}^{\pi} |f^+(\exp(i(\theta+h))) - 2f^+(\exp(i\theta))+f^+(\exp(i(\theta-h)))|d\theta .$$

The condition is satisfied by piecewise analytic continuous functions F with a finite number of power singularities: $F(x) \sim A_k+B_k(x-x_k)^\alpha$, $\alpha>0$, although the a_n's decrease no faster than $n^{-\alpha-1}$ ([34] p.43 and 72, from which the behaviour of $f^+(\exp i\theta)$ near $\theta = \arccos x_k$ may be deduced). For functions like $\exp((x-1)/(x+1))$ the decrease of the a_n's is fast enough to have $\sum n|a_n|<\infty$ (see [77] Example 7.10).

When H is trace-class ,

$$\det(I-wH) = \prod_{1}^{\infty}(1-w\lambda_j) = \sum_{0}^{\infty}\delta_n w^n \qquad (3.2)$$

also has a meaning as an entire function of w . The coefficients δ_n's are so far the best source of information on the rate of decrease of the λ_n's , as δ_n is the series made of all the products of n λ's of distinct indexes :

$$\delta_n = (-1)^n \sum_{1\le i_1<\ldots<i_n} \lambda_{i_1}\ldots\lambda_{i_n} \qquad (3.3)$$

Therefore , if the λ_n's are assumed to decrease like a power ρ^n , the δ_n's will essentially decrease like $\rho^{n^2/2}$, and the detection of such behaviour for the δ_n's will be the subject of section 4 .
Here is an example of such a behaviour : values of δ_n for $\exp(-t^m)$ (i.e., $F(x)=\exp[-((1-x)/(1+x))^m]$) , m=1,2,3 , are given together with second differences of their exponents . A crude estimate of ρ follows from a mean value of this second difference :

Table 4 . Samples of δ_n and second differences of exponents .

	exp(-t)		$\exp(-t^2)$		$\exp(-t^3)$	
n	δ_n	Δ^2	δ_n	Δ^2	δ_n	Δ^2
0	1.000E 000		1.000E 000		1.000E 000	
5	-1.736E-011	24	-6.676E-009	16	-1.822E-007	16
10	-2.636E-046	24	-1.806E-034	19	-2.920E-030	16
15	2.590E-105	24	-1.237E-078	19	3.885E-069	15
20	1.624E-188	25	6.811E-141	18	2.969E-123	16
25	-6.474E-296	23	-8.510E-222	17	-1.718E-193	17
30	-1.636E-427	25	-3.418E-320	19	-3.555E-280	16
35	2.619E-583		1.177E-437		1.141E-383	
	$10^{-24/25} = 1/9.12$		$10^{-18/25} = 1/5.25$		$10^{-16/25} = 1/4.37$	

Remark that $\sum_{0}^{\infty}\delta_n w^n = \exp(-\sum_{1}^{\infty}\zeta_k w^k/k)$, so that the δ_n's can be obtained from the ζ_k's in a finite number of steps :

$$\delta_1 = -\zeta_1 \;,\; \delta_2 = ((\zeta_1)^2-\zeta_2)/2 \;,\; \delta_3 = (-(\zeta_1)^3+3\zeta_1\zeta_2-2\zeta_3)/6\ldots \qquad (3.4)$$

Now that the interest of spectral invariants of the Hankel matrix H seems to have been established , we come to a compact formula for them , which will allow a discussion of their behaviour .
We assume the following integral form :

$$a_k = \int_C z^{k-1}\, d\alpha(z) \qquad k\ge 1 \qquad (3.5)$$

where α is a measure whose support C is in the closed unit disk . There is always at least one way to find (3.5) for any expansion (1.1) : from

$$a_k=\frac{2}{\pi}\int_{-1}^{1} F(x)T_k(x)(1-x^2)^{-1/2}dx = \int_C z^{k-1}(\pi i)^{-1}F((z+1/z)/2)\, dz, \qquad (3.6)$$

where C is the unit circle . Of course , when F is analytic in some region containing the unit circle , the integration contour can be deformed and alternative forms will be available . This happens with Stieltjes functions , where C is a part of the real interval [-1,1] ([3], [6]) .

Now , we try to find the traces ς_n using (3.5) :

<u>Theorem 3.2.</u> Let the Hankel matrix $H = [a_{k+m-1}]_{k,m=1}^{\infty}$ be trace-class , and let (3.5) hold with C in the closed unit disk . Then ,

ς_n =traceH^n =

$$\int_C\cdots\int_C (1-z_1z_2)^{-1}(1-z_2z_3)^{-1}\ldots(1-z_{n-1}z_n)^{-1}(1-z_nz_1)^{-1}d\alpha(z_1)\ldots d\alpha(z_n) \qquad (3.7)$$

which must be considered as the limit when $\eta \to 1$, $\eta<1$, of

$$\int_C\cdots\int_C (1-\eta z_1z_2)^{-1}(1-\eta z_2z_3)^{-1}\ldots(1-\eta z_{n-1}z_n)^{-1}(1-\eta z_nz_1)^{-1}d\alpha(z_1)\ldots d\alpha(z_n).$$

This asks first for an integral form for the elements of H^n . Let us start the row and column indexes at zero . Then , $H_{m,k} = \int_C z^{m+k}\, d\alpha(z)$.

If the whole of C is inside the unit disk , one shows by induction

$(H^n)_{m,k}$ =

$$\int_C\cdots\int_C (z_1)^m(1-z_1z_2)^{-1}(1-z_2z_3)^{-1}\ldots(1-z_{n-1}z_n)^{-1}(z_n)^k d\alpha(z_1)\ldots d\alpha(z_n)$$

and (3.7) follows easily . If at least a part of C touches the unit disk boundary , one must be more careful . One defines $H(\eta) = [\eta^{(m+k)/2}a_{m+k+1}]_{m,k=0}^{\infty}$. Geometric series summations can now be performed in the integrals on $\eta^{1/2}C$, and as $\|H-H(\eta)\| \to 0$ when $\eta\to 1$ ($\|H-H(\eta)\|^2 \le \Sigma n(1-\eta^{(n-1)/2})^2(a_n)^2$) , the result follows from the continuity of the trace with respect to the operator norm (see [76] p.337) . □

An interesting change of variable is $u = (1-z)/(1+z)$ ($\longleftrightarrow z = (1-u)/(1+u)$) mapping the unit disk on the right half plane . Let

$$d\beta(u)= -\pi i(1+u)^2d\alpha(z)/2 \quad , \quad u=(1-z)/(1+z) \qquad (3.8)$$

Then ,

$$(-\pi i)^n \zeta_n = \int_\Gamma \cdots \int_\Gamma (u_1+u_2)^{-1}(u_2+u_3)^{-1}\cdots(u_{n-1}+u_n)^{-1}(u_n+u_1)^{-1} d\beta(u_1)\cdots d\beta(u_n) ,$$

where $\Gamma=(1-C)/(1+C)$ is a set in the (closed) right half plane .
With $G(t)=\exp(-t)$, $d\beta(u) = \exp(-u^2)du$ (see (4.1) below) produces $\zeta_1 = 1/2$ and $\zeta_2 = 1/\pi$.

For the δ_n's , we have :

<u>Theorem 3.3.</u> Under the conditions of theorem 3.2 , the coefficient δ_n of (3.2) is

$$\delta_n = \frac{(-1)^n}{n!} \int_C \cdots \int_C S(z_1,\dots,z_n) d\alpha(z_1)\cdots d\alpha(z_n) \qquad (3.9)$$

with $$S(z_1,\dots,z_n) = \prod_{m<j} \left(\frac{z_m-z_j}{1-z_m z_j}\right)^2 \prod_1^n (1-z_j{}^2)^{-1}$$

or , with $u=(1-z)/(1+z)$, using (3.8)

$$\delta_n = (2\pi i)^{-n}(n!)^{-1} \int_\Gamma \cdots \int_\Gamma \frac{R(u_1,\dots,u_n)}{u_1\cdots u_n} d\beta(u_1)\cdots d\beta(u_n) \qquad (3.10)$$

with $$R(u_1,\dots,u_n) = \prod_{m<j} \left(\frac{u_j-u_m}{u_j+u_m}\right)^2$$

This result was first deduced by A. Hautot [27] by symbolic programmation with low values of n . The proof which follows is by J.Meinguet [44] .

We start again from (3.5) and use the fact that $(-1)^n\delta_n$ is the series of all the determinants made with the same combinations $0\le i_1<i_2< \dots <i_n$ of n rows and columns of the matrix H . From (3.5) , each of these determinants turns into a n-uple integral of the determinant of elements $z_m^{i_m+i_j}$, $1\le m,j\le n$. As in the proof of the preceding theorem , we may suppose $|z_m|\le\eta^{1/2}<1$, and take the limit $\eta\to1$ in fine , as δ_n is an expression involving a finite number of ζ's (from (3.4)) . From Binet-Cauchy theorem , n! times the (therefore convergent) series in $i_1<i_2<\dots <i_n$ happens to be the determinant of the product

$$\begin{bmatrix} 1 & z_1 & (z_1)^2 & \cdots \\ 1 & z_2 & (z_2)^2 & \cdots \\ \cdots & \cdots & \cdots & \cdots \\ 1 & z_n & (z_n)^2 & \cdots \end{bmatrix} \begin{bmatrix} 1 & 1 & \cdots & 1 \\ z_1 & z_2 & & z_n \\ (z_1)^2 & (z_2)^2 & \cdots & (z_n)^2 \\ \cdots & \cdots & \cdots & \cdots \end{bmatrix}$$

i.e., the determinant of the matrix $[(1-z_i z_j)^{-1}]_{i,j=1}^{n}$ giving

$$S(z_1,\ldots,z_n)=(-1)^n(n!)^{-1}\det[(1-z_i z_j)^{-1}]_1^n$$

$$=(-1)^n(n!)^{-1}\prod_{m<j}\left(\frac{z_m-z_j}{1-z_m z_j}\right)^2\prod_1^n(1-z_j^2)^{-1}$$ (Cauchy determinant [56]Sect.7 §1.3) □

It is still not known if a numerical implementation of the forms (3.9) or (3.10) can be achieved in a reasonable way . Tables 1-4 have been constructed from conventional numerical methods applied to a big finite section of H (with a little trick : instead of taking the Chebyshev coefficients of F(x) , those of $F((x+\alpha)/(\alpha x+1))$ with a suitable $\alpha\in(-1,1)$ give better approximations of the same eigenvalues , this is to be compared with the use of c_n in [7] p.392) .

4. ASYMPTOTIC ESTIMATES BY NUTTALL' SADDLEPOINTS METHOD .

We proceed now with tentative asymptotic estimates of the multiple integral (3.10) . Let us consider the case when the form (3.5) holds with a contour C almost completely inside the unit disk , with $F((z+z^{-1})/2)$ piecewise analytic on C . Then (3.10) holds with

$$d\beta(u) = f(u)du$$

where f is piecewise analytic on Γ almost entirely in the right-half plane . As one can always take $d\alpha(z) = (\pi i)^{-1}F((z+z^{-1})/2)dz$,

$$f(u) = F((1+u^2)/(1-u^2)) = G(-u^2) \qquad (4.1)$$

is always a valid choice , from (1.2) , (3.6) and (3.8) .

The integral (3.10) is then essentially dominated by the configurations $(u_1,\ldots,u_n)$ maximizing $|R(u_1,\ldots,u_n)f(u_1)\ldots f(u_n)|$ on Γ . Moreover , Γ can be deformed in an admissible way (singular points of f must not be crossed) into a new contour Γ_n where this latter maximum is minimized : the expected behaviour of $|\delta_n|$ is therefore essentially

$$|\delta_n| \sim \min_{\Gamma}\max_{u_i\in\Gamma} |R(u_1,\ldots,u_n)f(u_1)\ldots f(u_n)| \qquad (4.2)$$

This leads to a search for saddlepoints , i.e., solutions of the set of nonlinear equations

$$\partial(R(u_1,\ldots,u_n)f(u_1)\ldots f(u_n))/\partial u_i = 0\ , \quad i=1,\ldots,n$$

or

$$4\sum_{\substack{j=1\\ j\neq i}}^{n} \frac{u_j}{(u_i)^2-(u_j)^2} = -f'(u_i)/f(u_i)\ , \quad i=1,\dots,n \tag{4.3}$$

Similar derivations have been made by J.Nuttall ([50] §5.3, [51]) for the asymptotic description of Padé and Hermite-Padé approximants . The discussion of the solutions of (4.3) is by no means simple . For

$$G(t) = \exp(-t^m)$$

i.e.,

$$-f'(u)/f(u) = 2m(-1)^m u^{2m-1}\ ,$$

numerical tests have been made for some values of m . For m=1 , there seems to be only one solution in the right-half complex plane . For m>1 , it seems obvious that several solutions are possible . Figure 1. shows the solutions for m=1,2 and 3 that seem relevant to our problem (with a change of scale which will be explained in section 6) .

We may find a help in solving our problem of minimizing on admissible curves Γ the maximum of $|R(u_1,\dots,u_n)f(u_1)\dots f(u_n)|$ on Γ by coming to an equivalent problem in electrostatics . To this end , let us consider n particles with a positive unit charge at $u_1,\dots,u_n$ on Γ , and n particles with a negative unit charge at $v_1,\dots,v_n$ on $-\Gamma$. The particles are repelled by the particles of the same family and attracted by the particles of the other family , according to the logarithmic potential . Moreover , they are all submitted to a supplementary field derived from the potential Re $(-f(u)/2)$. The total potential energy of a configuration $\{u_1,\dots,u_n;\ v_1,\dots,v_n\}$ is therefore

$$W_n=-\sum_{i\neq j}\log|u_i-u_j|-\sum_{i\neq j}\log|v_i-v_j|+\sum_{i,j}\log|u_i-v_j|-\sum_{i}(\log|f(u_i)|+\log|f(v_i)|)/2$$

Remark that , f being even , we may assume that $v_i = -u_i$ will hold at equilibrium . Then we find easily

$$W_n = -\log|R(u_1,\dots,u_n)f(u_1)\dots f(u_n)|$$

so that we may now concentrate on the value of W_n : from (3.10) and (4.2) , a rough asymptotic estimate of $|\delta_n|$ will be given by

$$|\delta_n| = \exp(-W_n + O(n\log n))\ ,$$

provided W_n is found to be much larger than nlogn .
For Γ fixed , the particles will take positions on Γ and $-\Gamma$ so as to minimize W_n . The result will be that the forces acting on u_i

$$-\sum_{j\neq i}\overline{(u_j-u_i)^{-1}} + \overline{\sum_{j}(v_j-u_i)^{-1}} + \frac{1}{2}\,\overline{f'(u_i)/f(u_i)} \tag{4.4a}$$

and on v_i

$$-\sum_{j\neq i}\overline{(v_j-v_i)^{-1}} + \overline{\sum_{j}(u_j-v_i)^{-1}} + \frac{1}{2}\,\overline{f'(v_i)/f(v_i)} \tag{4.4b}$$

will be directed as the normal to Γ and $-\Gamma$. If we assume that the particles tend to fill a part $\tilde{\Gamma}_n$ of Γ when n is large , the potential function

$$h_n(u) = \sum_i \log|u-v_i| - \sum_i \log|u-u_i| - \frac{1}{2}\log|f(u)|$$

will take constant values $h_n(\tilde{\Gamma}_n)$ and $h_n(-\tilde{\Gamma}_n)$ near $\tilde{\Gamma}_n$ and $-\tilde{\Gamma}_n$ (because the force at u is minus the gradient of the potential which is orthogonal to the level lines of the potential ; and the gradient of a function of the form Re H(u) , H analytic , is $\overline{H'(u)}$) . This may be appreciated if we drop the contributions $\log|u-u_i|$ or $\log|u-v_i|$ when u comes to be close to u_i or v_i , typically at a distance to be compared to some (negative) power of n , causing O(logn) perturbations which are asymptotically inocuous if $h_n(u)$ is much larger than logn for u near $\pm\tilde{\Gamma}_n$. With such a rule for estimating the potential function h_n , we find

$$nh_n(\tilde{\Gamma}_n) \sim \sum_i h_n(u_i) = W_n + \frac{1}{2}\sum_i \log|f(u_i)| \ .$$

Let us introduce the approximate distribution μ_n of the u_i's on $\tilde{\Gamma}_n$, so that any sum on the u_i's will be approximated as

$$\sum_i \varphi(u_i) \sim \int_{\tilde{\Gamma}_n} \varphi(\xi)d\mu_n(\xi) \ ,$$

then the potential function h_n takes the form (assuming $v_i = -u_i$ at equilibrium)

$$h_n(u) \sim -\mathrm{Re}\ \phi_n(u) = \text{constant on } \tilde{\Gamma}_n \text{ and } -\tilde{\Gamma}_n$$

with

$$\phi_n(u) = \int_{\tilde{\Gamma}_n} \log\frac{u-\xi}{u+\xi}d\mu_n(\xi) + \frac{1}{2}\log f(u) \ . \qquad (4.5)$$

So we replaced h_n by a harmonic function taking exactly constant values on $\tilde{\Gamma}_n$ and $-\tilde{\Gamma}_n$. This will result in μ_n to be a smooth distribution approximating the actual discrete distribution of $u_1, \ldots u_n$.

At this point , Γ being given , all the unknowns can be estimated asymptotically from the solution of the following boundary problem :

To find a function ϕ_n of the form (4.5) , with

Re $\phi_n(u)$ = a constant c_n on a part $\tilde{\Gamma}_n$ of Γ ,

Re $\phi_n(u) < c_n$ on the remaining part $\Gamma\setminus\tilde{\Gamma}_n$ of Γ (minimality of potential on $\tilde{\Gamma}_n$) ,

Increase of imaginary part of ϕ_n after a circuit of $\tilde{\Gamma}_n$ = $2n\pi$ (principle of argument , or total charge on Γ) .

Then , $$W_n \sim -nc_n - \frac{1}{2}\int_{\tilde{\Gamma}_n} \log|f(\xi)|d\mu_n(\xi) .$$

Remark that the determination of $\tilde{\Gamma}_n$ out of Γ is a part of the problem .

Finally , maximizing W_n on admissible curves Γ makes us reach a (therefore unstable) further state of equilibrium where all the forces (4.4) vanish. Remark that , with $v_i = -u_i$, to have (4.4) to vanish is indeed to find a solution of (4.3) . The final conditions are then summarized by the following

<u>Rule 4.1.</u> Let us consider the real Hankel matrix (2.3) when (3.5) holds , with

$$d\alpha(z) = -2(\pi i)^{-1}(1+u)^{-2}f(u)du \quad , \quad u=(1-z)/(1+z)$$

where f is analytic in a part of the right-half complex plane , whose boundary contains the imaginary axis .

Then we expect the eigenvalues of H to decrease like

$$\exp(2c_n)$$

provided $-c_n/\log n \to \infty$ when $n \to \infty$,

where c_n is determined by the following boundary-value problem :

To find a function ϕ_n of the form (4.5) , with

Re $\phi_n(u)$ = a constant c_n on $\tilde{\Gamma}_n$,

There exists a curve Γ_n in the right-half plane , containing $\tilde{\Gamma}_n$, such that Γ_n and the imaginary axis enclose a region where f is holomorphic , and Re $\phi_n(u) < c_n$ on $\Gamma_n \backslash \tilde{\Gamma}_n$.

Increase of imaginary part of ϕ_n after a circuit of $\tilde{\Gamma}_n = 2n\pi$ (equivalent to $\int_{\tilde{\Gamma}_n} d\mu_n(\xi) = n$) .

The limit values of the derivative ϕ'_n at the two sides of $\tilde{\Gamma}_n$ are opposite :

$$\phi'_{n,+}(u) \text{ and } \phi'_{n,-}(u) = \pm\pi i\mu'_n(u) \; , \quad u\in\tilde{\Gamma}_n \qquad (4.6)$$

such that Re $\phi_n(u) > c_n$ when one leaves $\tilde{\Gamma}_n$ along the normal direction .

The last condition is a consequence of the vanishing of the forces (4.4) and Sokhotskyi-Plemelj formulas([29])

$$\phi'_{n,+}(u) \text{ and } \phi'_{n,-}(u) = \int_{\tilde{\Gamma}_n} ((u-\xi)^{-1}-(u+\xi)^{-1})d\mu_n(\xi) + \frac{f'(u)}{2f(u)} \pm\pi i\mu'_n(u) \; ,$$

$u \in \tilde{\Gamma}_n$, giving the limit values of Φ'_n at the two sides of $\tilde{\Gamma}_n$.
We still have to show that $W_n - W_{n-1} \sim -2c_n$. This comes from an estimate of $W_{n-1}-W_n$ as a mean value of $\log|f(u_n)R(u_1 \ldots u_n)/ R(u_1, \ldots, u_{n-1})|$ which gives indeed $\log|f(u)| + \int_{\tilde{\Gamma}} \log\left|\frac{u-\xi}{u+\xi}\right|^2 d\mu_n(\xi) = 2c_n$.

5. RELATION WITH THE THEORY OF EXTREMAL POLYNOMIALS .

Much of the material of the preceding section is actually found in rational approximation theory . For instance , the electrostatics picture is much used [2] [15]-[20] [51] [59] etc. The function S of (3.9) is almost in [2] . So the connection between CF , Hankel matrices spectral theory , and rational approximation in the complex plane becomes deeper and deeper . Indeed , let us start from G(t) analytic in a region R of the extended complex plane $\bar{C}$ which we want to approximate on some closed set $K \in \bar{R}$ (the positive real axis in the preceding sections) by a rational function P_n/Ω_n of degree n .
As a first step , consider that P_n interpolates $\Omega_n G$ at $t_1, \ldots, t_{n+1} \in K$. An integral form of the remainder is

$$G(t)-P_n(t)/\Omega_n(t) = (2\pi i)^{-1} \frac{\omega_n(t)}{\Omega_n(t)} \int_\Delta \frac{\Omega_n(\tau)}{\omega_n(\tau)} (\tau-t)^{-1}G(\tau)d\tau \qquad (5.1)$$

where Δ is a contour in $\bar{R}$ enclosing K and $\omega_n(\tau) = (\tau-t_1)\ldots(\tau-t_{n+1})$.
When K is the positive real axis , Δ could be a parabola-like curve . A first bound of the error norm is essentially dominated by

$$\max_{t\in K} |\omega_n(t)/\Omega_n(t)| \max_{\tau\in\Delta}|\Omega_n(\tau)G(\tau)/\omega_n(\tau)| \qquad (5.2)$$

which we can try to make as small as possible by a suitable choice of the zeros $t_1, \ldots, t_{n+1}$ of ω_n and $p_1, \ldots, p_n$ of Ω_n (extended weighted Zolotarev problem [13] [16]) .

The contour Δ can also be deformed provided it is still in $\bar{R}$ and encloses K , in order to minimize (5.2) . However , the integral in (5.1) constructed with the best rational approximant is often still very much smaller than the best possible bound (5.2) . This is explained by the fact that P_n/Ω_n can satisfy up to 2n+1 interpolation conditions , so that the integral in (5.1) vanishes at n further points $t_{n+2}, \ldots, t_{2n+1}$. Introducing $\Omega_n(t) = (t-t_1)\ldots(t-t_{2n+1})$, the new conditions become

$$\int_\Delta \frac{\Omega_n(\tau)}{\Omega_n(\tau)} \Pi(\tau)G(\tau)d\tau = 0 \quad , \text{any polynomial } \Pi \text{ of degree } < n \qquad (5.3)$$

and the remainder can now be written

$$G(t)-P_n(t)/\varrho_n(t) = (2\pi i)^{-1} \frac{\Omega_n(t)}{\varrho_n^2(t)} \int_\Delta \frac{\varrho_n^2(\tau)}{\Omega_n(\tau)}(\tau-t)^{-1}G(\tau)d\tau \ . \qquad (5.4)$$

Now, we consider again (5.2) with Ω_n and ϱ_n^2 instead of ω_n and ϱ_n, and let $2\lambda_n$ and ν_n be smooth approximations of the distributions of $\{t_1, \ldots, t_{2n+1}\}$ and $\{p_1, \ldots, p_n\}$. Then, with

$$\Psi_n(t) = \int_{\tilde{\Delta}_n} \log(t-\eta)d\nu_n(\eta) - \int_K \log(t-\eta)d\lambda_n(\eta) + \frac{1}{2}\log G(t) \qquad (5.5)$$

where $\tilde{\Delta}_n$ is the actual support of $d\nu_n$ ($\int_{\tilde{\Delta}_n} d\nu_n(\eta) = \int_K d\lambda_n(\eta) = n$), the best bound of the remainder (5.4) can be estimated as

$$\log E_n \sim 2\ [\max_{t\in K} \operatorname{Re}\ (-\Psi_n(t) + \frac{1}{2}\log(G(t)) + \max_{\tau\in\tilde{\Delta}_n} \operatorname{Re}\Psi_n(\tau)\ ]$$

and this kind of problem is again related to a boundary problem

<u>Rule</u> <u>5.1.</u>
Find Ψ_n of the form (5.5), with

$\operatorname{Re}\ (\Psi_n - \frac{1}{2}\log G(t)\ ) =$ a constant a_n on K,

$\operatorname{Re}\Psi_n(t) =$ a constant b_n on $\tilde{\Delta}_n$,

There exists a curve Δ_n in $\bar{R}$, enclosing K and containing $\tilde{\Delta}_n$, such that $\operatorname{Re}\Psi_n < b_n$ on $\Delta_n \setminus \tilde{\Delta}_n$,

$\int_{\tilde{\Delta}_n} d\nu_n(\eta) = \int_K d\lambda_n(\eta) = n$,

The limit values of the derivative Ψ'_n at the two sides of $\tilde{\Delta}_n$ are opposite :

$\Psi'_{n,+}(t)$ and $\Psi'_{n,-}(t) = \pm\pi i\nu'_n(t)$, $\quad t\in\tilde{\Delta}_n$

such that $\operatorname{Re}\Psi_n(t) > b_n$ when one leaves $\tilde{\Delta}_n$ along the normal direction.

Then, we expect

$$G(t)-P_n(t)/\varrho_n(t) \sim \exp(2b_n - 2\Psi_n(t) + \log G(t)) \ , \quad t\notin K \ , \ t\notin\tilde{\Delta}_n$$

$$E_n \sim \exp(\ 2(b_n - a_n)\ ) \ .$$

The conditions of validity of the rule are by no means simple!! They have been worked a.o. by Goncar, Lopez, Rahmanov, Stahl [15]-[20] [35] [37] [62][63] [64]. It sems important that G must be analytic in the whole complex plane excepting a set of vanishing capacity, so that Δ_n may be deformed almost at will (as before, $\tilde{\Delta}_n$ (the place where Q_n 'lives' [45]) is an unknown of the problem and is related to the boundary of so-called extremal regions [62]). The most difficult part is to show that (5.3) is asymptotically compatible with rule 5.1, that is that if (5.3) holds, the distribution of the zeros of Q_n and Ω_n behave indeed asymptotically as given by rule 5.1. This has been established by Stahl for functions with branch points [63].

Relation (5.3) may be considered as defining Q_n to be (formally) orthogonal to polynomials of degree less than n with respect to a "weight" $G(\tau)/\Omega_n(\tau)$. Should Δ be allowed to shrink to a real set where $G(\tau)/\Omega_n(\tau)>0$, then (5.3) is equivalent to saying that Q_n is the monic polynomial of degree n minimizing $\int_\Delta (Q_n(\tau))^2 G(\tau)/\Omega_n(\tau)d\tau$, whence the name extremal polynomial given to Q_n: denominators of best rational approximations behave like extremal polynomials. The real set case has received recently extremely dramatic and impressive accelerations of knowledge [36]-[39] [45] [47] [57] [70]: see the big recent surveys by two active workers in this field, P.Nevai ([47] including added note on p.144, the subject has been declared by P.Nevai [48] as one of the hottest in approximation theory) and D.Lubinsky [39].

For the complex set case, Q_n could rightly be called extremal if it would minimize a true norm $\int_\Delta |Q_n(\tau)|^2\ |G(\tau)/\Omega_n(\tau)|\,|d\tau|$ (Szegö's complex orthogonality, see [75]), and this does not seem to be the same as (5.3) and (5.4). However, near $\tilde{\Delta}_n$ but not on $\tilde{\Delta}_n$, one has $Q_n(G/\Omega_n)^{1/2} \sim \exp(\Psi_n)$ which has a discontinuous derivative when one crosses $\tilde{\Delta}_n$, whereas the true $Q_n(G/\Omega_n)^{1/2}$ is of course regular. A better description of this function on $\tilde{\Delta}_n$ is $Q_n(G/\Omega_n)^{1/2} \sim \exp(\Psi_n) + \exp(2b_n-\Psi_n)$ (see [50] 3.2.7 for such a situation) which takes into account the actual distribution of zeros of Q_n on $\tilde{\Delta}_n$. This suggests that the phase of $(Q_n)^2G/\Omega_n$ has only slow variation on $\tilde{\Delta}_n$ and that Q_n still behaves like an extremal polynomial.

Finally, here is the compatibility of rules 4.1 and 5.1:

<u>Remark</u> <u>5.2.</u> When $K = [0,\infty]$, $f(u) = G(-u^2)$ with $f(\bar{u}) = \overline{f(u)}$, let us

take $\tilde{\Gamma}_n$ symmetrically with respect to the real axis in the right half complex plane and satisfying the conditions of rule 4.1 . Then

$$\tilde{\Delta}_n = -(\tilde{\Gamma}_n)^2 \quad , \Psi_n(t) = \Psi_n(-u^2) = \Phi_n(u)$$

satisfy rule 5.1 with $a_n=0$ and $b_n=c_n$.

Indeed , as $\Gamma=\tilde{\Gamma}_n$ and $d\mu_n$ are symmetric with respect to the real axis , (4.5) may be written $\Phi_n(u) = \int_{\tilde{\Gamma}_n} \log \frac{u-\bar{\xi}}{u+\xi} d\mu_n(\xi) + \frac{1}{2}\log(f(u))$, so that

$\Phi_n(u)-\log(f(u))/2 = -2i\int_{\tilde{\Gamma}}\text{Arctan}\,[(u/i + \text{Im}\xi)/\text{Re}\,\xi]d\mu_n(\xi) = -2i\pi\kappa_n(u)$ is pure imaginary when u is on the imaginary axis , that is when $t\in K$. Remark that $\kappa_n(i\infty)-\kappa_n(-i\infty) = n$, and that $i\kappa'_n(u)>0$ on the imaginary axis . The derivative $\Phi'_n(u)-f'(u)/(2f(u))$ is given in the right half plane (avoiding $\tilde{\Gamma}_n$ where the jump of derivative values is $2\pi i\mu'_n(u)$) by the Cauchy integral $(2\pi i)^{-1}\int_{-i\infty}^{i\infty}(\xi-u)^{-1}2i\pi\kappa'_n(\xi)d\xi -(2\pi i)^{-1}\int_{\tilde{\Gamma}_n}(\xi-u)^{-1}$ $2i\pi\mu'_n(\xi)d\xi$. Multiplying by two and subtracting the known form $\int_{\tilde{\Gamma}_n}((u-\xi)^{-1}-(u+\xi)^{-1})\mu'_n(\xi)d\xi$, we get $2\int_{-i\infty}^{i\infty}(\xi-u)^{-1}d\kappa_n(\xi)$ + $\int_{\tilde{\Gamma}_n}2u(u^2-\xi^2)^{-1}d\mu_n(\xi)$. After integrating in u , one recovers a new form

$$\Phi_n(u) \quad -(\log f(u))/2 = -2 \int_{-i\infty}^{i\infty} \log(\xi-u)d\kappa_n(\xi) \quad + \int_{\tilde{\Gamma}_n}\log(\xi^2-u^2)d\mu_n(\xi) \ .$$

Finally , as κ'_n is even , the first integral is written $-2\int_0^{i\infty}\log(\xi^2-u^2)d\kappa_n(\xi)$ and we have indeed a representation of the required form (5.5) in terms of $t = -u^2$. □

So we expect the square roots of minus the zeros of $\mathcal{Q}_n$ to be distributed approximately on $\tilde{\Gamma}_n$ with the same distribution μ_n as the saddlepoints in (4.3) . For $G(t) = \exp(-t^m)$, Figure 1 shows saddle-points (squares) and $(-\text{poles})^{1/2}$ (circles) indeed almost on the same locus. Actually , these values have been divided by $n^{1/(2m)}$ (n=50 for the saddlepoints , n=12 for the rational approximation poles)

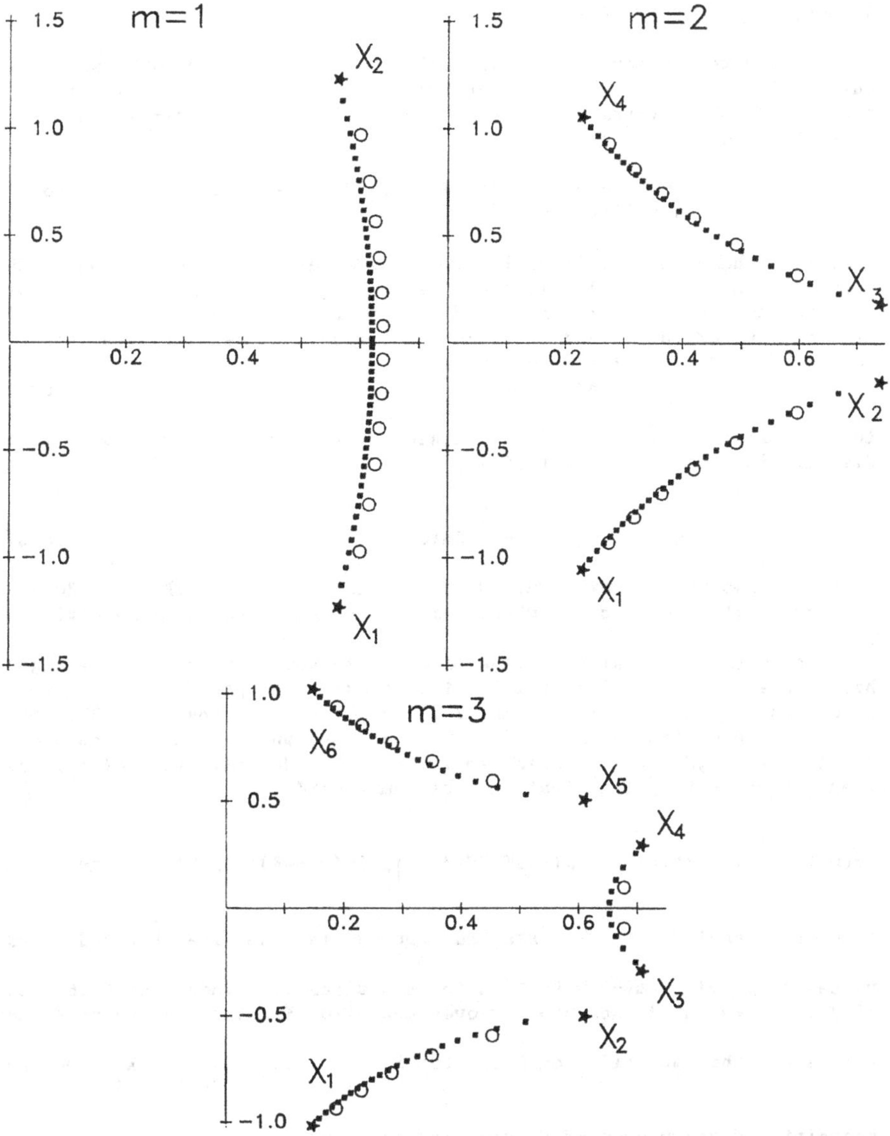

Figure 1 . Positions in the complex plane of remarkable points relevant to rational approximation of $\exp(-t^m)$ on $t \geq 0$, for m=1,2,3 . Solid squares : scaled saddlepoints from the integral form of the 50th elementary symmetric function of eigenvalues of Hankel matrix . Hollow circles : scaled square roots of minus the poles of 12th degree approximant . Stars : calculated endpoints of the common locus drawn by the preceding points .

6. SOLUTION IN TERMS OF (HYPER)ELLIPTIC INTEGRALS .

We come now to a discussion of the solution of (4.3) , or an example of rule 4.1 , when $G(t) = \exp(-t^m)$, $m=1,2,\ldots$, i.e., $f(u) = \exp((-1)^{m+1}u^{2m})$ (from 4.1) , so that $f'(u)/f(u) = 2m(-1)^{m+1}u^{2m-1}$: (4.3) becomes

$$4\sum_{\substack{j=1\\ j\neq i}}^{n} \frac{u_j}{(u_i)^2-(u_j)^2} = 2m(-1)^m(u_i)^{2m-1},\ i=1,\ldots,n \tag{6.1}$$

A simple change of scale will immediately settle the variation with respect to n : in (6.1) , if the u_j's behave like some function $\varphi(n)$ of n , each term of the sum will be of the order of $1/\varphi(n)$, the sum itself will be like $n/\varphi(n)$, and the right-hand side behaves like $(\varphi(n))^{2m-1}$, so that $\varphi(n) = n^{1/(2m)}$. So we expect

$$v_i = n^{-1/(2m)}u_i\ ,\quad i=1,\ldots,n \tag{6.2}$$

to be ultimately distributed on a fixed system of arcs $\tilde{\Gamma}$ according to a distribution $d\mu$ of total weight 1 , and (4.5) becomes $\Phi_n(u) = n\phi(v)$, where

$$\phi(v) = \int_{\tilde{\Gamma}} \log\frac{v-\xi}{v+\xi}\,d\mu(\xi) + (-1)^{m+1}v^{2m}/2 \tag{6.3}$$

Rule 4.1 becomes : eigenvalues of H will decrease like $\exp(2c_n) = \exp(2nc) = \rho^n$ ($\rho=\exp(2c)$) , where c is the constant value of Re $\phi(v)$ on $\tilde{\Gamma}$. We assume now that $\tilde{\Gamma}$ is made of a finite number p arcs in the right half plane (it will be shown later that $p\leq m$) . The proof of such a fact is difficult (see [62] for example) . Goncar and Rahmanov [18] [37] [58] were able to do that at least for the case m=1 and they have a complete proof that $(\log E_n)/n \to 2c$ when $n\to\infty$, the actual determination of c being then a mere technicality that will be summarized now .

From (6.3) , $\phi'(v) + m(-1)^m v^{2m-1} = \int_{\tilde{\Gamma}} 2\xi(v^2-\xi^2)^{-1}d\mu(\xi)$ is an even function , analytic in the extended complex plane outside $\pm\tilde{\Gamma}$, and takes values $\pm\pi i\mu'(v) + m(-1)^m v^{2m-1}$ on the two sides of $\tilde{\Gamma}$ (adapting (4.6) to (6.3)) . We build now another even analytic function taking opposite values on the two sides of $\tilde{\Gamma}$. Let X_1 , X_2 ,... X_{2p-1} , X_{2p} be the endpoints of the p arcs of $\tilde{\Gamma}$(see fig. 1) , and

$$X(v) = v^{4p}+\chi_1 v^{4p-2}+\ldots+\chi_{2p} = (v^2-X_1^2)\ldots(v^2-X_{2p}^2)\ . \tag{6.4}$$

We define then $X^{1/2}(v)$ as the square root of X behaving like v^{2p} for large v, and continuous outside $\tilde{\Gamma}$. This function takes opposite values $\pm X_+^{1/2}(v)$ on the two sides of $\tilde{\Gamma}$, so that the function $(\phi'(v) + m(-1)^m v^{2m-1})/X^{1/2}(v)$ takes the values $\pm m(-1)^m v^{2m-1}/X_+^{1/2}(v) + \pi i\mu'(v)/X_+^{1/2}(v)$ on the two sides of $\tilde{\Gamma}$. Applying again Sokhotskyi-Plemelj formulas [29], we find

$$(\phi'(v) + m(-1)^m v^{2m-1})/X^{1/2}(v) =$$

$$= (\pi i)^{-1} \int_{\tilde{\Gamma}} 2\xi(v^2-\xi^2)^{-1} m(-1)^m \xi^{2m-1} X_+^{-1/2}(\xi)d\xi$$

$$= 2m(-1)^m(\pi i)^{-1} \int_{\tilde{\Gamma}} \xi^{2m}(v^2-\xi^2)^{-1} X_+^{-1/2}(\xi)d\xi \tag{6.5}$$

So everything is known if we find the 2p complex numbers $X_1, \ldots X_{2p}$ (and p)! From (6.3), the left hand side of (6.5) behaves like $(2\int_{\tilde{\Gamma}}\xi d\mu(\xi))v^{-2p-2}$ for large v, with $\int_{\tilde{\Gamma}}\xi d\mu(\xi) > 0$ (because $\tilde{\Gamma}$ is in the right-half plane and is symmetric with respect to the real axis). So the expansion of the right-hand side of (6.5) in negative powers of v must start with a positive coefficient times v^{-2p-2}, whence the p conditions

$$M_j = \int_{\tilde{\Gamma}} \xi^{2m+2j} X_+^{-1/2}(\xi)d\xi = 0, \quad j=0,\ldots, p-1, \tag{6.6}$$

and $(-1)^m i^{-1} M_p > 0$; p-1 further real conditions will come from Re $\phi(v)$ = the same constant c on all the p arcs of $\tilde{\Gamma}$, we shall assume $\tilde{\Gamma}$ to be symmetric with respect to the real axis (so that $X_{2p-j+1} = \overline{X_j}$), and $\pi\int_{\tilde{\Gamma}} d\mu(\xi) = \sum_1^p \text{Im}[\phi(X_{2j})-\phi(X_{2j-1})] = \pi$ will be the last condition. Now we transform (6.5) using a technique of [22] (pp.266 and 283): we multiply (6.5) by the even polynomial $v^2X(v)$ which is written $v^2X(v)-\xi^2X(\xi) + \xi^2X(\xi)$ inside the integral (6.5) which turns into two integrals, the first involving the polynomial $[v^2X(v)-\xi^2X(\xi)]/(v^2-\xi^2)$. A careful calculation of this integral leaves

$$v^2X^{1/2}(v)\phi'(v) + m(-1)^m v^{2m+1}X^{1/2}(v) =$$

$$2m(-1)^m(\pi i)^{-1}\{ \sum_{j=0}^{p} X_j \sum_{k=p}^{2p-j} M_k \, v^{4p-2j-2k} + \int_{\tilde{\Gamma}} \xi^{2m+2} X_+^{1/2}(\xi)(v^2-\xi^2)^{-1}d\xi \}.$$

Differentiating (in v), and integrating by parts (in ξ) the last

integral , it appears to be $\int_{\tilde\Gamma}(v^2-\xi^2)^{-1}\,P(\xi)\,X^{-1/2}(\xi)\,\xi^{2m}d\xi$, where $P(\xi)$ is the polynomial $(2m+1)X(\xi)+\xi X'(\xi)/2$. Using the trick $P(\xi) = P(\xi)-P(v)+P(v)$ as before , we get rid of P in the integral and recover the integral of (6.5) . The result is a differential equation for ϕ
$\{X^{1/2}(v)[v^2\phi'(v)+m(-1)^m v^{2m+1}]\}' = vY(v)+vP(v)[\phi'(v)+m(-1)^m v^{2m-1}]/X^{1/2}(v)$,
where Y is the polynomial of degree 2p-2

$$Y(v) = 2m(-1)^m(\pi i)^{-1}\sum_{j=0}^{p-1} X_j \sum_{k=p}^{2p-j-1} (2p-j-2k-2m-1)M_k\, v^{4p-2j-2k-2} . \tag{6.7}$$

The differential equation for ϕ' turns as
$X^{1/2}(v)[v^{1-2m}\phi'(v)]' = v^{-2m}Y(v)$, so that

$$\phi'(v) = m(-1)^{m-1}v^{2m-1} + v^{2m-1}\int_\infty^v \xi^{-2m}Y(\xi)X^{-1/2}(\xi)d\xi$$

$$= v^{2m-1}\int_{X_1}^v \xi^{-2m}Y(\xi)X^{-1/2}(\xi)d\xi \tag{6.8}$$

with the conditions

$$\int_\infty^{X_1} \xi^{-2m}Y(\xi)X^{-1/2}(\xi)d\xi = m(-1)^m ,$$

$$\int_{X_j}^{X_{j+1}} \xi^{-2m}Y(\xi)X^{-1/2}(\xi)d\xi = 0 , \quad j=1,2,\dots,2p-1 ,$$

as ϕ' must vanish at each X_j , this is the only possible way for ϕ' to take opposite values on the two sides of each arc of $\tilde\Gamma$. More precisions on $\tilde\Gamma$ can now be given :

<u>Proposition</u> <u>6.1 .</u> One must have $p\le m$ and ϕ' has exactly m-p further zeros in the right half-plane outside $\tilde\Gamma$.

Indeed , $(\phi')^2$ is holomorphic in the right half-plane , and has already 2p zeros $X_1 , \dots , X_{2p}$ there . For large v , $(\phi')^2 \sim m^2v^{4m-2}$. According to the discussion of the remark 5.2 , $\phi'(v) = -2i\kappa'(v)-(-1)^m mv^{2m-1}$ on the imaginary axis , with $-2i\kappa'(v) > 0$, so that the total increase of argument of $(\phi')^2$ on a big contour enclosing the right half-plane is $4m\pi$: this accounts for the 2p known zeros of $(\phi')^2$ and m-p more (double) zeros . □

If p<m , it seems that the remaining critical points of ϕ do not allow the existence of a curve Γ (required by rule 4.1) where Re ϕ is everywhere $\le$ c , but this should be proved . Tests with m>1 and p=1 gave indeed wrong results . Figure 1 shows $\tilde\Gamma$ for m=1,2,3 with 2m endpoints (five point stars) calculated as below .

For ϕ itself , integration of (6.8) yields

$$\phi(v) = (-1)^m v^{2m}/2 + (2m)^{-1} \int_{\infty}^{v} (v^{2m} - \xi^{2m}) \xi^{-2m} Y(\xi) X^{-1/2}(\xi) d\xi$$

$$= v\phi'(v)/(2m) - (2m)^{-1} \int_{\infty}^{v} Y(\xi) X^{-1/2}(\xi) d\xi ,$$

so that

$$c = -(2m)^{-1} \operatorname{Re} \int_{\infty}^{X_1} Y(\xi) X^{-1/2}(\xi) d\xi$$

and where the first kind (hyper)elliptic integrals

$$\int_{X_{2j-1}}^{X_{2j}} Y(\xi) X^{-1/2}(\xi) d\xi \ , \ j=1, \dots , p$$

must be pure imaginary numbers and have a sum πi . Working all these conditions results in table 5 below ($p=m$) . For $m=1$, elliptic integrals identities yield several equivalent forms . From the conditions above , one finds that $2c = -\pi K'/K$, where K and K' are the complete elliptic integrals of first kind related to modulii k and $(1-k^2)^{1/2}$ such that $K = 2E$, where E is the corresponding elliptic integral of second kind (see [11] [31] [46] for notations ; solving equation (6.6) can be interpreted as looking for a zero of the complete elliptic integral of the second kind considered as a function of its modulus [44]) . Goncar and Rahmanov presented the equation $\sum_{n=1}^{\infty} n\rho^n/[1-(-1)^n \rho^n] = 1/8$ for $\rho = \exp(2c)$. Equivalence of the two formulations can be established from various identities for Jacobi theta and zeta functions [46] . It is very remarkable that another problem involving theta functions , posed more than 100 years ago by G.H.Halphen [25] , requires the very same number ρ in its solution! Another equation for ρ is $\sum_{n=0}^{\infty} (2n+1)^2 (-\rho)^{n(n+1)/2} = 0$ ([25] pp. 287 and 427) .

Table 5 . Results for m=1 , 2 and 3 .

m=1

X_1, X_2	0.564412701731271∓1.230228033100522i
Y	−3.664045422603946
$2c$	−2.228833648714334
ρ	1/9.289025491920819

m=2

X_1, X_4	0.230299569523605∓1.056043075724618i
X_2, X_3	0.740872935366573∓0.181255581301698i
Y	$-4.089738435714034u^2 + 3.018259576896660$

2c -1.679203056619678

ρ 1/5.361281630239104

<u>m=3</u>

X_1, X_6 0.147466522850900∓1.019380088475020i

X_2, X_5 0.610266123777351∓0.504086300206854i

X_3, X_4 0.707976575689678∓0.293205472234699i

Y -5.038105707137323u^4 + 3.413952831945491u^2 -2.465551268021017

2c -1.472035162993397

ρ 1/4.358095556608086

ACKNOWLEDGEMENTS .

It is a pleasure to thank D.Braess , T.Ganelius , M.Gutknecht , A.Hautot , D.S.Lubinsky , P.Nevai , H.U.Opitz , E.A.Rahmanov , E.B.Saff , H.Stahl and especially J.Meinguet and L.N.Trefethen for all their assistance and kind interest .

A part of the computations have used the routines C02AEF , C05NCF , F01AJF and F02AMF of the NAG library implemented on the IBM 4381 of Louvain-la-Neuve .

REFERENCES .

1. V.M.ADAMYAN , D.Z.AROV , M.G.KREIN : 'Analytic properties of Schmidt pairs for a Hankel operator and the generalized Schur-Takagi problem' , <u>Mat.Sb</u>.<u>86</u> (<u>128</u>) (1971) 34-75 = <u>Math.USSR Sb</u>. <u>15</u> (1971) 31-73 .
2. T.BAGBY : 'On interpolation by rational functions' , <u>Duke Math. J</u>. <u>36</u> (1969) 95-104 .
3. W.BARRETT : 'On the convergence of sequences of rational approximations to analytic functions of a certain class' , <u>J. Inst. Maths Applics</u> <u>7</u> (1971) 308-323 ; 'On best rational approximations to certain functions defined as integrals' , <u>ibid</u>. <u>13</u>(1974) 107-116 .
4. D.BRAESS : 'On the conjecture of Meinardus on rational approximation of e^x ' , <u>J.Approx. Th</u>. <u>36</u> (1982) 317-320 ; II : <u>ibid</u>. <u>40</u> (1984) 375-379 .
5. D.BRAESS : <u>Nonlinear Approximation Theory</u> , Springer Series in Computational Mathematics Vol.7 , Springer , Berlin 1987.
6. D.BRAESS : 'Rational Approximation of Stieltjes functions by the Carathéodory-Fejér method' . <u>Constr. Approx</u>. <u>3</u> (1987) 43-50 .
7. A.J.CARPENTER , A.RUTTAN , R.S.VARGA : 'Extended numerical computations on the " 1/9" conjecture in rational approximation theory' , pp.383-411 in <u>Rational Approximation and Interpolation</u> , Tampa 1983 , P.R.GRAVES-MORRIS , E.B.SAFF , R.S.VARGA editors , Springer Lect. Notes Math. 1105 , Berlin 1984 .
8. C.W.CLENSHAW , K.LORD : 'Rational approximation from Chebyshev series' , pp.95-113 in <u>Studies in Numerical Analysis</u> , B.K.P.SCAIFE editor , Ac. Press London 1974 .

9. W.CODY , G.MEINARDUS , R.VARGA : 'Chebyshev rational approximation to e^{-x} and applications to heat-conduction problems' , J.Approx.Th. 2 (1969) 50-65 .
10. S.W.ELLACOTT : 'On the Faber transform and efficient numerical rational approximation' , SIAM J. Num. An. 20 (1983) 989-1000 .
11. A.ERDELYI et al. , editors : Higher Transcendental Functions McGraw-Hill , N.Y. 1953
12. T.GANELIUS : 'Rational approximation in the complex plane and on the line' , Ann. Acad. Sci. Fenn. 2 (1976) 129-145 .
13. T.GANELIUS : 'Degree of rational approximation' , pp.9-78 in Lectures on Approximation and Value Distribution , Sém. Math. Sup.-Sém. Sci. OTAN , Pr.Univ. Montréal , Québec 1982 .
14. K.GLOVER : 'All optimal Hankel-norm approximations of linear multivariable systems and their L^{∞}-error bounds' , Int.J.Control 39 (1984) 1115-1193 .
15. A.A.GONCAR : 'On a generalized analytic continuation' , Math. USSR Sbornik 3 (1968) 129-140 .
16. A.A.GONČAR : 'Zolotarev problems connected with rational functions' , Math. USSR Sbornik 7 (1969) 623-635 .
17. A.A.GONCAR : 'The rate of rational approximation and the property of single-valuedness of an analytic function in the neighborhood of an isolated singular point' , Math. USSR Sbornik 23 (1974) 254-270 .
18. A.A.GONCAR : Int. Congress Math. Berkeley 1986 .
19. A.A.GONCAR , G.LOPEZ : 'On Markov's theorem for multipoint Padé approximation' , Math. USSR Sb. 34 (1978) 449-459 .
20. A.A.GONCHAR , E.A.RAKHMANOV : 'Equilibrium measure and the distribution of extremal polynomials' , Math. USSR Sb. 53 (1986) 119-130 .
21. W.B.GRAGG : 'Laurent , Fourier , and Chebyshev-Padé tables' , pp. 61-72 in Padé and Rational Approximation , E.B.SAFF , R.S.VARGA editors , Academic Press , N.Y. 1977 .
22. C.C.GROSJEAN : 'The weight functions , generating functions and miscellaneous properties of the sequences of orthogonal polynomials of the second kind associated with the Jacobi and the Gegenbauer polynomials ', J. Comp. Appl. Math. 16 (1986) 259-307 .
23. M.H.GUTKNECHT : 'Rational Carathéodory-Fejér approximation on a disk , a circle , and an interval' , J.Approx. Th. 41 (1984) 257-278 .
24. M.H.GUTKNECHT , E.B.SAFF : 'A de Montessus-type theorem for CF approximation' , J.Comp. Appl. Math. 16 (1986) 251-254 .
25. G.H. HALPHEN : Traité des fonctions elliptiques et de leurs applications I , Gauthier-Villars , Paris 1886 .
26. C.HASTINGS Jr. : Approximations for Digital Computers , Princeton U.P. , Princeton 1955 .
27. André HAUTOT , private communication 14 April 1985 .
28. E.HAYASHI , L.N.TREFETHEN and M.H.GUTKNECHT : 'The CF table' , submitted to Constr. Approx.
29. P.HENRICI : Applied and Computational Complex Analysis III , Wiley

N.Y. 1986 .
30. A.ISERLES : 'Rational interpolation to exp(-x) with application to certain stiff systems' , SIAM J. Numer. An. 18 (1981) 1-12 .
31. E.JAHNKE , F.EMDE and F.LÖSCH : Tables of Higher Functions , 6th ed. , Teubner , Stuttgart 1960 .
32. J.KARLSSON : 'Bulgarian achievements in rational approximation' , Preprint Umea 1980 .
33. J.KARLSSON : 'Rational approximation in BMOA and H^{∞}' , Preprint Göteborg 1987 .
34. M.J.LIGHTHILL : Introduction to Fourier Analysis and Generalised Functions , Cambridge U.P. , London 1958 .
35. G.L.LÓPEZ : 'Survey on multipoint approximation to Markov type meromorphic functions and asymptotic properties of the orthogonal polynomials generated by them' , pp. 309-316 in Polynômes Orthogonaux et Applications , Proceedings Bar-le-Duc 1984 (C.BREZINSKI et al. , editors) , Lect. Notes in Math. 1171 , Springer-Verlag , Berlin 1985 .
36. G.L.LÓPEZ : 'On the asymptotic of the ratio of orthogonal polynomials and convergence of multipoint Padé approximants' , Math. USSR Sb. 56(1987) 207-219 .
37. G.LÓPEZ ,E.A.RAHMANOV : 'Rational approximations , orthogonal polynomials and equilibrium distributions' , 2nd Int. Symposium on Orthogonal Polynomials and their Applications , Segovia (Spain) , Sept. 22-27 , 1986 .
38. D.S.LUBINSKY , E.B.SAFF : 'Strong asymptotics for extremal errors and extremal polynomials associated with weights on $(-\infty,\infty)$' , Preprint NRIMS CSIR , Pretoria 1987 .
39. D.S.LUBINSKY : 'A survey of general orthogonal polynomials for weights on finite and infinite intervals' , to appear in Acta Applicandae Mathamaticae .
40. A.P.MAGNUS and ? : 'Why is CF so good ?' In preparation .
41. J.MEINGUET : 'A simplified presentation of the Adamjan-Arov-Krein approximation theory' , pp.217-248 in Computational Aspects of Complex Analysis , H.WERNER , L.WUYTACK , E.NG , H.J.BÜNGER editors , Reidel , Dordrecht , 1983 .
42. J.MEINGUET : 'On the Glover concretization of the Adamjan-Arov-Krein approximation theory' , pp. 325-334 in Modelling , Identification and Robust Control , C.I.BYRNES and A.LINDQUIST editors , North-Holland 1986 .
43. J.MEINGUET : 'Once again : the Adamjan-Arov-Krein approximation theory' , these Proceedings .
44. J.MEINGUET Private communication 30-31 Dec. 1985 .
45. H.N.MHASKAR , E.B.SAFF : 'Where does the sup norm of a weighted polynomial live ? (a generalization of incomplete polynomials)' , Constr. Approx. 1 (1985) 71-91 .
46. L.M. MILNE-THOMSON : 'Jacobian elliptic functions and theta

functions ; elliptic integrals' , chapters 16 & 17 in Handbook of Mathematical Functions, M.ABRAMOWITZ , I.STEGUN editors , Dover NY 1965.
47. P.NEVAI : 'Geza Freud , orthogonal polynomials and Christoffel functions' , J.Approx. Th. 48 (1986) 3-167.
48. P.NEVAI : Private communication 6 March 1987 .
49. D.J.NEWMAN : 'Rational approximation to $|x|$' , Mich. Math.J. 11 (1964) 11-14 .
50. J.NUTTALL : 'Asymptotics of diagonal Hermite-Padé polynomials' , J.Approx. Th. 42 (1984) 299-386 .
51. J.NUTTALL : 'Location of poles of Padé approximants to entire functions' , pp.354-363 in Rational Approximation and Interpolation , Tampa 1983 , P.R.GRAVES-MORRIS , E.B.SAFF , R.S.VARGA editors , Springer Lect. Notes Math. 1105 , Berlin 1984 .
52. J.NUTTALL : 'Asymptotics of generalized Jacobi polynomials' , Constr. Approx. 2 (1986) 59-77 .
53. H.U.OPITZ , K.SCHERER : 'On the rational approximation of e^{-x} on $[0,\infty)$' , Constr. Approx. 1 (1985) 195-216 .
54. V.V.PELLER : 'Hankel operators of class S_p and their applications (rational approximation , Gaussian processes , the problem of majorizing operators)' , Mat.Sb. 113(155) (1980) = Math. USSR Sb. 41 (1982) 443-479 .
55. V.V.PELLER : 'A description of Hankel operators of class S_p for $p>0$, an investigation of the rate of rational approximation , and other applications' , Mat.Sb. 122(164) (1983) = Math. USSR Sb. 50 (1985) 465-494 .
56. G.PÓLYA , G.SZEGÖ : Problems and Theorems in Analysis II , Springer , Berlin 1976 .
57. E.A.RAHMANOV : 'Asymptotic properties of polynomials orthogonal on the real axis' , Math. USSR Sb. 47 (1984) 155-193 .
58. E.A.RAHMANOV : Private communication to G.López , P.Nevai and the author , Segovia , 23 Sept. 1986 .
59. E.B.SAFF , J.L.ULLMAN and R.S.VARGA : 'Incomplete polynomials : an electrostatics approach' , pp. 769-782 in Approximation Theory III , E.W.CHENEY editor , Ac. Press , N.Y. 1980 .
60. E.B.SAFF , R.S.VARGA : 'Some open problems concerning polynomials and rational functions' , pp. 483-488 in Padé and Rational Approximation , E.B.SAFF , R.S.VARGA editors , Academic Press , N.Y. 1977 .
61. A.SCHÖNHAGE : 'Rational approximation to e^{-x} and related L^2-problems' , SIAM J. Num. An. 19 (1982) 1067-1080 .
62. H.STAHL : 'Extremal domains associated with an analytic function' , I : Complex Variables 4 (1985) 311-324 , II : ibid. 4 (1985) 325-338 .
63. H.STAHL : 'Orthogonal polynomials with complex valued weight function' , I : Constr. Approx. 2 (1986) 225-240 ; II : ibid. 2

(1986) 241-251 .
64. H.STAHL : 'A note on three conjectures by Gonchar on rational approximation' , J. Approx. Th. 50 (1987) 3-7 .
65. L.N.TREFETHEN : 'Square blocks and equioscillation in the Padé , Walsh , and CF tables' , pp. 170-181 in Rational Approximation and Interpolation , Tampa 1983 , P.R.GRAVES-MORRIS , E.B.SAFF , R.S.VARGA editors , Springer Lect. Notes Math. 1105 , Berlin 1984 .
66. L.N.TREFETHEN : Private communication of MIT CF memos 10 (11/16/82) , 15(9/25/84) , 16(9/27/85)
67. L.N.TREFETHEN : 'CF approximation and applications , Conference on Nonlinear numerical methods and rational approximation' , Antwerpen , April 1987 , these Proceedings .
68. L.N.TREFETHEN , M.GUTKNECHT : 'The Carathéodory-Fejér method for real rational approximation' , SIAM J. Numer. Anal. 20 (1983) 420-436.
69. L.N.TREFETHEN , M.GUTKNECHT : 'Padé , stable Padé , and Chebyshev-Padé approximation' , in IMA Conf. on Algorithms for Approx. of Functions and Data , J.C.MASON et M.COX , eds , Oxford U.P.
70. W. VAN ASSCHE : Asymptotics for Orthogonal Polynomials , Lecture Notes Math. 1265 , Springer-Verlag , Berlin , 1987 .
71. R.S.VARGA : Topics in Polynomial and Rational Interpolation and Approximation , Sém. Math. Sup.-Sém. Sci. OTAN , Pr.Univ. Montréal , Québec 1982 .
72. N.S.VJACESLAVOV : 'On the approximation of x^{α} by rational functions' , Math.USSR Izv. 16 (1981) 83-101.
73. A.VOROS : 'Correspondance semi-classique et résultats exacts : cas des spectres d'opérateurs de Schrödinger homogènes' , C.R.Acad.Sc. Paris 293 (1981) ser.1 709-712 .
74. H.WERNER : 'Rationale Tschebyscheff-Approximation , Eigenwerttheorie und Differenzenrechnung' , Arch. Rat. Mech. An. 13 (1963) 330-347 .
75. H.WIDOM : 'Extremal polynomials associated with a system of curves in the complex plane' , Adv. in Math. 3 (1969) 127-232 .
76. H.WIDOM : 'Toeplitz matrices and Toeplitz operators' , pp.319-341 in Complex Analysis and its Applications , Vol.1 , IAEA , Vienna , 1976 .
77. J.WIMP Computation with Recurrence Relations , Pitman , Boston 1984.

MULTIDIMENSIONAL AND MULTIVARIATE PROBLEMS

Chairmen:

E. B. Saff

P. Graves-Morris

Invited communications:

P. Graves-Morris*
Vector-valued rational interpolation.

Short communications:

M. de Bruin
Simultaneous rational approximation to some q-hypergeometric functions.

D. Roberts*
Clifford algebras and vector-valued rational approximants.

M. Van Barel
Minimal Padé-sense matrix approximations around $s = 0$ and $s = \infty$.

J. Van Iseghem*
Convergence and divergence of columns of vector Padé approximants associated to meromorphic functions.

C. Chaffy
$(\text{Padé})_y$ of $(\text{Padé})_x$ approximants of $F(x,y)$.

C. Lutterodt
Rational approximants of hypergeometric series in $\mathbb{C}^n$.

W. Siemaszko*
Branched continued fractions and hypergeometric functions.

B. Verdonk
Different techniques for the construction of multivariate rational interpolants.

---

* Lecture notes are not included.

SIMULTANEOUS RATIONAL APPROXIMATION TO SOME q-HYPERGEOMETRIC FUNCTIONS

Marcel G. de Bruin
Department of Mathematics
University of Amsterdam
Roetersstraat 15
1018 WB Amsterdam, The Netherlands

ABSTRACT. The purpose of this paper is to show that it is possible to extend results that hold for the existence of the Padé-n-table for n-tuples of hypergeometric ${}_2F_1, {}_1F_1, {}_2F_0$ functions to so called basic q-hypergeometric functions of the same type, including explicit formulae for the last two types mentioned. Appropriate changes of the variable and/or taking the limit for $q \to 1$ show that the results include known results on both ordinary (cf. P. Wynn) and simultaneous Padé approximation.

1. INTRODUCTION

The Padé-n-table for an n-tuple of formal power series $f_1(z), f_2(z), \dots, f_n(z)$ over C, with $n \geq 1$

$$f_j(z) = \sum_{k=0}^{\infty} c_k^{(j)} z^k, \quad c_0^{(j)} \neq 0 \qquad (j = 1, 2, \dots, n), \tag{1}$$

is defined in the customary linearised form given below. For each (n+1)-tuple $(\rho_0, \rho_1, \dots, \rho_n)$ of nonnegative integers with $\sigma = \rho_0 + \rho_1 + \cdots + \rho_n$ we put the approximation problem of finding polynomials

$$P_j(z) = P_j(\rho_0, \rho_1, \dots, \rho_n; z) \qquad (j = 0, 1, \dots, n), \tag{2a}$$

satisfying

$$deg\ P_j(z) \leq \sigma - \rho_j \qquad (j = 0, 1, \dots, n), \tag{2b}$$

$$P_0(z) f_j(z) - P_j(z) = O(z^{\sigma+1}) \quad (j = 1, 2, \dots, n). \tag{2c}$$

It is outside the scope of this paper to go into details of existence and uniqueness of the solution of this problem: the reader is referred to [3] and the references therein. If the problem has a solution which is unique up to a multiplicative constant, the point $(\rho_0, \rho_1, \dots, \rho_n)$ will be called **regular**. Consider the "building blocks" $D^{(j)} = D^{(j)}(\rho_0, \rho_1, \dots, \rho_n)$ for $j = 1, 2, \dots, n$, defined by

$$D^{(j)} = \begin{pmatrix} c_{\sigma-\rho_j}^{(j)} & c_{\sigma-\rho_j-1}^{(j)} & \cdots & c_{\rho_0-\rho_j+1}^{(j)} \\ c_{\sigma-\rho_j+1}^{(j)} & c_{\sigma-\rho_j}^{(j)} & \cdots & c_{\rho_0-\rho_j+2}^{(j)} \\ \vdots & \vdots & \ddots & \vdots \\ c_{\sigma-1}^{(j)} & c_{\sigma-2}^{(j)} & \cdots & c_{\rho_0}^{(j)} \end{pmatrix}, \tag{3a}$$

A. Cuyt (ed.), Nonlinear Numerical Methods and Rational Approximation, 135–142.

(for $\rho_j = 0$ this block is empty) and the $(\sigma - \rho_0) \times (\sigma - \rho_0)$ determinant $D = D(\rho_0, \rho_1, \ldots, \rho_n)$ by

$$D = \det \begin{pmatrix} D^{(1)} \\ \vdots \\ D^{(n)} \end{pmatrix}. \tag{3b}$$

Throughout this paper it is tacitly assumed that $c_k^{(j)} = 0$ for $k < 0$ and that empty building blocks are omitted.
Then we have (cf. [3]):

Theorem 1. *If $D(\rho_0, \rho_1, \ldots, \rho_n) \neq 0$, then $(\rho_0, \rho_1, \ldots, \rho_n)$ is regular and the solution of (2), which is unique after the normalisation $P_0(\rho_0, \rho_1, \ldots, \rho_n; 0) = 1$, is given by*

$$P_0(z) = \frac{1}{D} \det \begin{pmatrix} 1 & z & z^2 & \ldots & z^{\sigma-\rho_0} \\ \cdots & \cdots & \cdots & \cdots & \cdots \\ & & \vdots & & \\ \cdots & \cdots & \cdots & \cdots & \cdots \\ c^{(j)}_{\sigma-\rho_j+1} & c^{(j)}_{\sigma-\rho_j} & c^{(j)}_{\sigma-\rho_j-1} & \ldots & c^{(j)}_{\rho_0-\rho_j+1} \\ c^{(j)}_{\sigma-\rho_j+2} & c^{(j)}_{\sigma-\rho_j+1} & c^{(j)}_{\sigma-\rho_j} & \ldots & c^{(j)}_{\rho_0-\rho_j+2} \\ \vdots & \vdots & \vdots & \ddots & \vdots \\ c^{(j)}_{\sigma} & c^{(j)}_{\sigma-1} & c^{(j)}_{\sigma-2} & \ldots & c^{(j)}_{\rho_0} \\ \cdots & \cdots & \cdots & \cdots & \cdots \\ & & \vdots & & \end{pmatrix} \tag{4}$$

and

$$P_0(z)f_i(z) - P_i(z) = \frac{1}{D} \sum_{k=1}^{\infty} \det \begin{pmatrix} c^{(i)}_{\sigma+k} & c^{(i)}_{\sigma+k-1} & c^{(i)}_{\sigma+k-2} & \ldots & c^{(i)}_{\rho_0+k} \\ \cdots & \cdots & \cdots & \cdots & \cdots \\ & & \vdots & & \\ \cdots & \cdots & \cdots & \cdots & \cdots \\ c^{(j)}_{\sigma-\rho_j+1} & c^{(j)}_{\sigma-\rho_j} & c^{(j)}_{\sigma-\rho_j-1} & \ldots & c^{(j)}_{\rho_0-\rho_j+1} \\ c^{(j)}_{\sigma-\rho_j+2} & c^{(j)}_{\sigma-\rho_j+1} & c^{(j)}_{\sigma-\rho_j} & \ldots & c^{(j)}_{\rho_0-\rho_j+2} \\ \vdots & \vdots & \vdots & \ddots & \vdots \\ c^{(j)}_{\sigma} & c^{(j)}_{\sigma-1} & c^{(j)}_{\sigma-2} & \ldots & c^{(j)}_{\rho_0} \\ \cdots & \cdots & \cdots & \cdots & \cdots \\ & & \vdots & & \end{pmatrix} z^{\sigma+k} \; (1 \leq i \leq n). \tag{5}$$

■

Now the restriction will be made that either all points $(\rho_0, \rho_1, \ldots, \rho_n)$ are regular or all points $(\rho_0, \rho_1, \ldots, \rho_n)$ with $\rho_0 \geq \rho_i - 1 \quad (i = 1, 2, \ldots, n)$. In these cases the n-tuples $f_1(z), f_2(z), \ldots, f_n(z)$ are referred to as *regular* respectively *semi-regular*. Just as in the ordinary Padé table we put at each point—here a point in $(n+1)$-space—its unique solution of problem (2); the resulting "table" is called the *Padé-n-table* for the n-tuple of functions given. As in the ordinary case we have the

Definition. *A point $(\rho_0, \rho_1, \ldots, \rho_n)$ in the Padé-n-table for a (semi-) regular n-tuple of power series will be called* **normal** *if the solution to problem (2) at that point does not appear at any other point in the Padé-n-table.*

Introducing building-blocks $\tilde{D}^{(j)}$ by adding a column and a row to the $D^{(j)}$ as follows

$$\tilde{D}^{(j)} = \begin{pmatrix} c^{(j)}_{\sigma-\rho_j+1} & c^{(j)}_{\sigma-\rho_j} & c^{(j)}_{\sigma-\rho_j-1} & \cdots & c^{(j)}_{\rho_0-\rho_j+1} \\ c^{(j)}_{\sigma-\rho_j+2} & c^{(j)}_{\sigma-\rho_j+1} & c^{(j)}_{\sigma-\rho_j} & \cdots & c^{(j)}_{\rho_0-\rho_j+2} \\ \vdots & \vdots & \vdots & \ddots & \vdots \\ c^{(j)}_{\sigma} & c^{(j)}_{\sigma-1} & c^{(j)}_{\sigma-2} & \cdots & c^{(j)}_{\rho_0} \\ c^{(j)}_{\sigma+1} & c^{(j)}_{\sigma} & c^{(j)}_{\sigma-1} & \cdots & c^{(j)}_{\rho_0+1} \end{pmatrix}, \tag{6a}$$

and the $(\sigma-\rho_0+n)\times(\sigma-\rho_0+1)$ matrix $\tilde{D}$ by

$$\tilde{D} = \begin{pmatrix} \tilde{D}^{(1)} \\ \vdots \\ \tilde{D}^{(n)} \end{pmatrix}, \tag{6b}$$

we can formulate the following theorem (for the proof the reader can consult the first reference in [3]; for $n=1$ the well known conditions for normality as given in [8] are recovered).

Theorem 2. *For a point $(\rho_0,\rho_1,\ldots,\rho_n)$ in the Padé-n-table for a (semi-) regular n-tuple of formal power series the following properties are equivalent:*
(a) $(\rho_0,\rho_1,\ldots,\rho_n)$ is normal.
(b) $deg\ P_j(z)=\sigma-\rho_j$ $(0\le j\le n)$, $P_0(z)f_j(z)-P_j(z)\ne O(z^{\sigma+2})$ for at least one $j\in\{1,2,\ldots,n\}$.
(c) The determinants $D(\rho_0,\rho_1,\ldots,\rho_n), D(\rho_0{+}1,\rho_1,\ldots,\rho_n), D(\rho_0,\rho_1,\ldots,\rho_{j-1},\rho_j{+}1,\rho_{j+1},\ldots,\rho_n)$ $(1\le j\le n)$ are different from zero and rank $\tilde{D}=\sigma-\rho_0+1$. ∎

In the last few years approximation in the vein given above—sometimes achieved via so-called *vector-valued interpolants*—has recieved some attention. For the ordinary Padé table many beautiful results have been obtained of which specifically [6], [7] deserve being mentioned. Whole classes of functions can be treated using these results and this paper is a first step into the direction of handling simultaneous approximation of q-hypergeometric functions.

The main results will be formulated in section 2 and the proofs will be given in section 3.

2. MAIN RESULTS

The n-tuples of power series to be considered will belong to certain classes of "basic" hypergeometric functions. Without wanting to enter into the dispute about the notation that *has been*, *is* and *will be* used for this type of function, this paper will use a "basic" generalisation of the ascending factorial (*Pochhammer's symbol*) for a complex number in the following form:

$$(A,\alpha;q)_0=1,(A,\alpha;q)_n=(A-q^{\alpha})(A-q^{\alpha+1})\cdots(A-q^{\alpha+n-1})\quad(n\ge1), \tag{7}$$

using the 'base' q, which—for the time being—will satisfy

$$q\in C\setminus\{0,1\},\ q^k\ne1\ (k\ge1).$$

Also the obvious generalisation of the binomial coefficient will appear:

$$\begin{bmatrix} k \\ s \end{bmatrix} = \frac{(1,1;q)_k}{(1,1;q)_s(1,1;q)_{k-s}}\quad(0\le s\le k) \tag{8}$$

Throughout this paper the principal value for the logarithm and exponentiation will be used:

$$\alpha^{\beta}=e^{\beta\ Log\ \alpha},\ Log\ \alpha=log|\alpha|+i\ Arg\ \alpha\ (-\pi<Arg\ \alpha\le\pi)\ for\ \alpha\ne0.$$

Now for the "basic" hypergeometric function:

$$ {}_r\Phi_s\left(\begin{matrix}(A_1,\alpha_1),\ldots,(A_r,\alpha_r)\\(C_1,\gamma_1),\ldots,(C_s,\gamma_s)\end{matrix};z\right)=\sum_{k=0}^{\infty}\frac{\prod_{i=1}^{r}(A_i,\alpha_i;q)_k}{\prod_{j=1}^{s}(C_j,\gamma_j;q)_k}\frac{z^k}{(1,1;q)_k} \tag{9} $$

The need for the condition prior to (8) becomes clear now as otherwise $(1,1;q)_k$ vanishes; furthermore the complex parameters A_i, α_i, C_j and γ_j should satisfy some trivial conditions to ensure that the function in (9) is well defined and does not reduce to a polynomial:

$$ A_i\neq q^{\alpha_i+k}\quad(1\le i\le r),\ C_j\neq q^{\gamma_j+k}\quad(1\le j\le s)\qquad \textit{for all } k\ge 0. \tag{10} $$

The main results are concerned with semi-regularity, normality (theorem 3) and explicit forms (theorem 4); the conditions (10) on the parameters will be included in the formulation of the theorem in order not to 'forget' any of the conditions.

Theorem 3. *Let n be an arbitrary natural number, $q\in C\setminus\{0,1\}$ and $q^k\neq 1$ for $k\ge 1$; $(\rho_0,\rho_1,\ldots,\rho_n)$ is an $(n+1)$-tuple of nonnegative integers. Then the point $(\rho_0,\rho_1,\ldots,\rho_n)$ is*
(a) regular for $\rho_0\ge\rho_i-1\quad(1\le i\le n)$, (b) normal for $\rho_0\ge\rho_i\quad(1\le i\le n)$,
for the following classes:
A. The n-tuple $\{{}_2\Phi_1(\begin{smallmatrix}(A,\gamma_j+\beta),(1,1)\\(C,\gamma_j)\end{smallmatrix};z)\}_{j=1}^{n}$ under the conditions

$$ A\neq q^{\gamma_j+\beta+k},\ C\neq q^{\gamma_j+k},\ Cq^{\beta}\neq Aq^k\quad \textit{for } k\ge 0,\qquad q^{\gamma_i-\gamma_j+k}\neq 1\quad(i\neq j)\qquad \textit{for } k\in Z. \tag{11} $$

B. The n-tuple $\{{}_1\Phi_1(\begin{smallmatrix}(1,1)\\(C,\gamma_j)\end{smallmatrix};z)\}_{j=1}^{n}$ under the conditions

$$ C\neq q^{\gamma_j+k}\quad \textit{for } k\ge 0,\qquad q^{\gamma_i-\gamma_j+k}\neq 1\quad(i\neq j)\qquad \textit{for } k\in Z. \tag{12} $$

C. The n-tuple $\{{}_2\Phi_0((A,\alpha_j),(1,1);z)\}_{j=1}^{n}$ under the conditions

$$ A\neq q^{\alpha_j+k}\quad \textit{for } k\ge 0,\qquad q^{\alpha_i-\alpha_j+k}\neq 1\quad(i\neq j)\qquad \textit{for } k\in Z. \tag{13} $$

Remark

(a) For $n=1$ we recover the results on normality from a paper by P. Wynn [9] (cf. also [2]), including the limitcases considered therein.
(b) Putting $A=C=1$ in part A of the theorem, $C=1$ and $z\mapsto(1-q)z$ in part B, $A=1$ and $z\mapsto(1-q)^{-1}z$ in part C, and letting q tend to unity, we recover the results in [1].
(c) Under the conditions given in (11), (12) and (13), we even find that $P_0(z)f_j(z)-P_j(z)$ is of the *exact order* $\sigma+1$ for *all* $j\in\{1,2,\ldots,n\}$

Furthermore we have

Theorem 4. *Let n be an arbitrary natural number and $(\rho_0,\rho_1,\ldots,\rho_n)$ an $(n+1)$-tuple of nonnegative integers satisfying $\rho_0\ge\rho_j-1\quad(1\le j\le n)$.*
A. For the n-tuple

$$ f_j(z)={}_1\Phi_1\left(\begin{matrix}(1,1)\\(C,\gamma_j)\end{matrix};z\right)=\sum_{k=0}^{\infty}\frac{z^k}{\prod_{p=0}^{k-1}(C-q^{\gamma_j+p})}\qquad(1\le j\le n), \tag{14} $$

satisfying the conditions (12) **and** $C \neq 0$, *the denominator polynomial is given by*

$$P_0(z) = P_0(\rho_0, \rho_1, \ldots, \rho_n; z) = \sum_{k=0}^{\sigma-\rho_0} d_k(-z)^k, \tag{15a}$$

with

$$d_k = \frac{1}{(1,1;q)_k C^k} \sum_{s=0}^{k} (-1)^s \begin{bmatrix} k \\ s \end{bmatrix} q^{s(\sigma-\rho_0)+\frac{1}{2}(k-s)(k-s-1)} \prod_{j=1}^{n} \frac{(C, \gamma_j + \sigma - \rho_j - s; q)_{\rho_j}}{(C, \gamma_j + \sigma - \rho_j; q)_{\rho_j}}. \tag{15b}$$

Furthermore

$$P_0(z) f_j(z) - P_j(z) = \sum_{k=1}^{\infty} d_k^{(j)} z^{\sigma+k} \quad (1 \leq j \leq n), \tag{16a}$$

with

$$d_k^{(j)} = \frac{(-1)^{\sigma-\rho_0}}{(C, \gamma_j; q)_{\sigma+k} q^{\sigma-\rho_0}} \prod_{i=1}^{n} \frac{(1, \gamma_j - \gamma_i + k; q)_{\rho_i}}{(C, \gamma_i + \sigma - \rho_i; q)_{\rho_i}} \prod_{r=1}^{\rho_i} q^{\gamma_i + \sigma - \rho_i + r}. \tag{16b}$$

B. For the n-tuple

$$f_j(z) = {}_2\Phi_0\big((A, \alpha_j), (1,1); z\big) = \sum_{k=0}^{\infty} \prod_{p=0}^{k-1} (A - q^{\alpha_j + p}) z^k \quad (1 \leq j \leq n), \tag{17}$$

satisfying the conditions (13) **and** $A \neq 0$, *the denominator polynomial is given by*

$$P_0(z) = P_0(\rho_0, \rho_1, \ldots, \rho_n; z) = \sum_{k=0}^{\sigma-\rho_0} d_k z^k, \tag{18a}$$

with

$$d_k = \frac{(-1)^{\sigma-\rho_0} q^{\frac{1}{2}(\sigma-\rho_0-k)(\sigma-\rho_0-k+1)}}{(1,1;q)_{\sigma-\rho_0-k} A^{\sigma-\rho_0-k}} \sum_{s=0}^{\sigma-\rho_0-k} (-1)^s \begin{bmatrix} \sigma-\rho_0-k \\ s \end{bmatrix} \times$$

$$\times q^{\frac{1}{2}(\sigma-\rho_0-s)(\sigma-\rho_0-s+1)} \prod_{j=1}^{n} (A, \alpha_j + \rho_0 - \rho_j + s + 1; q)_{\rho_j}. \tag{18b}$$

Furthermore

$$P_0(z) f_j(z) - P_j(z) = \sum_{k=1}^{\infty} d_k^{(j)} z^{\sigma+k} \quad (1 \leq j \leq n), \tag{19a}$$

with

$$d_k^{(j)} = (-1)^{\sigma-\rho_0} q^{\frac{1}{2}(\sigma-\rho_0)(\sigma+2\alpha_j+2k+\rho_0-1)} (A, \alpha_j; q)_{\rho_0+k} \times$$

$$\times \prod_{i=1}^{n} (1, \alpha_i - \alpha_j - \rho_i - k + 1; q)_{\rho_i}. \tag{19b}$$

Remark. Putting $C = 1$, $z \mapsto (1-q)z$ in A and $A = 1$, $z \mapsto z(1-q)^{-1}$ in B and letting q tend to unity, the results in [3] are recovered. The cases $A = 0$ and $C = 0$ can not be treated using the method of this paper; for $n = 1$ these reduce to the well-studied partial theta functions, cf. Lubinsky and Saff [7].

3. PROOFS

For the proofs we need two lemmas on the explicit evaluation of determinants.

Lemma 1. *Let the elements of the $N \times N$ determinant $|e_{r,s}|$ satisfy*

$$e_{r,s+1} = a_s(y_r - x_s)e_{r,s} \quad (1 \le r \le N,\ 1 \le s \le N-1), \tag{20}$$

where $x_1, x_2, \ldots, x_{N-1}, y_1, y_2, \ldots, y_N$ are complex numbers, x_N arbitrary. Let $\phi_s(y)$ be polynomials in y of degree $\deg \phi_s \le N - s$ $(1 \le s \le N)$ with $\phi_N(y) \equiv 1$, then

$$|e_{r,s}\phi_s(y_r)| = \prod_{r=1}^{N} e_{r,1} \cdot \prod_{1 \le s < r \le N} (y_r - y_s) \cdot \prod_{s=1}^{N} \phi_s(x_s). \tag{21}$$

Proof. See [4]. ∎

Lemma 2. *Let $|e_{r,s}; e_r|_k$ be the $N \times N$ determinant obtained by replacing the elements of the kth column with $e_1, e_2, \ldots, e_N$ respectively, where $|e_{r,s}|$ satisfies (20). Let a function $g(z)$ be given that is analytic with the exception of a finite number of singularities $z_1, z_2, \ldots, z_M$, but that is regular at $y_1, y_2, \ldots, y_N$ and moreover satisfies*

$$e_r = e_{r,1}g(y_r) \qquad (1 \le r \le N).$$

Take $q(z) = \prod_{r=1}^{N}(z - y_r)$ and let $p(z)$ be the coefficient of x^{-1} in the Laurent series for $q(x)/\{(x - z)\prod_{s=1}^{k}(x - x_s)\}$. Then

$$|e_{r,s}; e_r|_k = -|e_{r,s}|\left(\sum_{j=1}^{M} Res\left[\frac{g(z)p(z)}{q(z)}, z_j\right] + Res\left[\frac{g(z)p(z)}{q(z)}, \infty\right]\right). \tag{22}$$

Proof. See [5]. ∎

Proof of Theorem 3. This reduces to simply applying Lemma 1, to show that the determinants $D(\rho_0, \rho_1, \ldots, \rho_n)$ differ from zero for $\rho_0 \ge \rho_i - 1$ $(1 \le i \le n)$ under certain conditions on the parameters in the hypergeometric series. According to Theorem 1, this implies regularity, while Theorem 2 (c) then shows that the Padé-n-table under consideration is normal for $\rho_0 \ge (\rho_i + 1) - 1 = \rho_i$ (rank $\tilde{D} = \sigma - \rho_0 + 1$ follows by exhibiting a $(\sigma - \rho_0 + 1) \times (\sigma - \rho_0 + 1)$ minor of $\tilde{D}$, having determinant different from zero) without introducing subsequent conditions on the parameters.

A. For $D(\rho_0, \rho_1, \ldots, \rho_n)$ we introduce

$$\begin{cases} y_r = q^{\gamma_j + \sigma - \rho_j + r_j - 1} & r = \sum_{k=1}^{j-1} \rho_k + r_j \text{ with } 1 \le r_j \le \rho_j, 1 \le j \le n \\ x_s = Cq^s & 1 \le s \le \sigma - \rho_0 \\ a_s = -q^{-s} & 1 \le s \le \sigma - \rho_0 \\ \phi_s(y) = \prod_{u=1}^{\sigma - \rho_0 - s}(A - yq^{\beta - \sigma + \rho_0 + u}) & 1 \le s \le \sigma - \rho_0 \end{cases}$$

Now the denominators of the entries in the determinant are used to define the $e_{r,s}$ as follows

$$e_{r,s} = \frac{1}{\prod_{p=0}^{\sigma - \rho_j + r_j - s - 1} (C - q^{\gamma_j + p})}.$$

From the numerators in row number r we take out a factor $\prod_{p=0}^{\rho_0 - \rho_j + r_j - 1}(A - q^{\gamma_j + \beta + p})$, the entry in row number r, column number s becomes

$$\prod_{p=\rho_0 - \rho_j + r_j}^{\sigma - \rho_j + r_j - s - 1} (A - q^{\gamma_j + p}) \Big/ \prod_{p=0}^{\sigma - \rho_j + r_j - s - 1} (C - q^{\gamma_j + p}) = e_{r,s}\phi_s(y_r).$$

Lemma 1 leads to $D(\rho_0, \rho_1, \dots, \rho_n) \neq 0$ $(\rho_0 \geq \rho_j - 1)$ under the conditions (11). Using the same method for $k = 1$ on the determinant behind the summation sign in (5), we find that the coefficient of $z^{\sigma+1}$ in each of the remainders $P_0 f_i - P_i$ is different from zero and also that $\tilde{D}$ has rank $\sigma - \rho_0 + 1$ (for $k = 1$ the numerator determinant is a minor of $\tilde{D}$!).

B. In this case no factors are taken out of the determinant and we proceed as in A.

C. Now we first have to put the columns in reversed order and then, using

$$\begin{cases} y_r = q^{\alpha_j + \rho_0 - \rho_j + r_j - 1} & r = \sum_{k=1}^{j-1} \rho_k + r_j \text{ with } 1 \leq r_j \leq \rho_j, 1 \leq j \leq n \\ x_s = Aq^{-s} & 1 \leq s \leq \sigma - \rho_0 \\ a_s = -q^s & 1 \leq s \leq \sigma - \rho_0 \end{cases}$$

proceed as in A, without polynomials ϕ_s. ∎

Proof of Theorem 4. Again the choices to make, in order to be able to apply Lemma 2, will be given only; the simple details are left to the reader.

A. We use y_r, x_s, a_s as in the proof of Theorem 3 and $g(y) = (C - y)^{-1}$. From Theorem 1 we see that

$$P_0(z) = 1 + \sum_{k=1}^{\sigma-\rho_0} (-1)^k \frac{D_k}{D} z^k,$$

where D_k arises from D by removing the kth column and putting a new column in front. Thus

$$D_k = (-1)^k |e_{r,s}; g(y_r)|_k.$$

To finish the calculations with the aid of Lemma 2, we have to find the residu of $\frac{g(y)p_k(y)}{q(y)}$ in $y = C$ and $y = \infty$, where q(y) and $p_k(y)$ follow from the formulation of the lemma. The condition $C \neq 0$ implies $x_s \neq x_t$ for $s \neq t$ and we find $p_k(y)$ by a very simple partial fraction decomposition

$$\frac{A}{(x-y)} + \sum_{k=1}^{s} \frac{B_s}{(x - x_s)} + \text{ polynomial in x,}$$

leading to $p_k(y) = A + \sum_{k=1}^{s} B_s$ (the polynomial in x has to be omitted for $k = \sigma - \rho_0$).

B. This time we have to invert the order of the columns again and we find

$$P_0(z) = 1 + \sum_{k=1}^{\sigma-\rho_0} (-1)^k \frac{D_k}{D} z^k,$$

with, using $d_{r,s}$ for the elements of the order-inverted determinant and the same values for y_r, x_s, a_s as in the proof of Theorem 3 part C, the following abbreviations

$$\begin{cases} D & = (-1)^{\frac{1}{2}(\sigma-\rho_0)(\sigma-\rho_0-1)} |d_{r,s}| \\ D_k & = (-1)^{\frac{1}{2}(\sigma-\rho_0)(\sigma-\rho_0+1)+k-1} |d_{r,s}; g(y_r)|_{\sigma-\rho_0+1-k} \\ g(y) & = \prod_{t=1}^{\sigma-\rho_0} (A - yq^t) \end{cases}$$

Note that the number of the column that has been replaced is complementary to k with respect to the dimension of the determinant while we have reversed the order of the columns! Again the extra condition, here $A \neq 0$, makes way for a simple partial fraction decomposition, here of the form

$$\frac{A}{(x-y)} + \sum_{s=1}^{\sigma-\rho_0+1-k} \frac{B_s}{(x-x_s)},$$

and we only need the residue in ∞. ∎

4. CONCLUDING REMARKS

Due to space limitations, it is not possible to give an extensive treatment of convergence results including proofs. To give the reader a glimpse of what can happen, we consider the following situation in Theorem 4(A):

$$\rho_1 + \rho_2 + \ldots + \rho_n = M \ (M \ fixed), \ |q| < 1, \ \rho_0 \to \infty.$$

Then we find, amongst other things, that the denominators—having the fixed degree M—converge to the polynomial

$$\sum_{k=0}^{M} \frac{(-z)^k}{(1,1;q)_k C^k} \sum_{j=0}^{k} (-1)^j \begin{bmatrix} k \\ j \end{bmatrix} q^{jM + \frac{1}{2}(k-j)(k-j-1)},$$

which is nothing else than

$$\prod_{j=0}^{M-1} (1 - q^j \frac{z}{C}).$$

An analysis of this situation, and the study of other sequences of approximants, will be given in another paper.

5. REFERENCES

[1] M.G. de Bruin, 'Three new examples of generalised Padé tables which are partly normal,' *Report* **76-11** (1976), University of Amsterdam.

[2] M.G. de Bruin, 'Some classes of Padé tables whose upper halves are normal,' *Nw. Arch. v. Wisk.(3)* **25** (1977), 148–160.

[3] M.G. de Bruin, 'Some explicit formulae in simultaneous Padé approximation,' *Lin. Alg. Applics.* **63** (1984), 271–281.

[4] J.G. van der Corput, 'Over eenige determinanten,' *Kon. Akad. v. Wet. Verhandelingen Amsterdam* **14** (1930), 1–44 (Dutch).

[5] J.G. van der Corput and H.J.A Duparc, 'Determinants and quadratic forms I,' *Proc. Kon. Akad. v. Wet. Amsterdam* **49** (1946), 995–1002.

[6] D.S. Lubinsky, 'Uniform convergence of rows of the Padé table for functions with smooth MacLaurin series coefficients,' unpublished.

[7] D.S. Lubinsky and E.B. Saff, 'Convergence of Padé approximants of partial theta functions and the Rogers-Szegö polynomials,' *Report* **#ICM-86-001** (1986), University of South Florida, Tampa.

[8] O. Perron, 'Die Lehre von den Kettenbrüchen II,' Teubner Verlag (1975), Stuttgart.

[9] P. Wynn, 'A general system of orthogonal polynomials,' *Quart. J. Math. Oxford (2)* **18** (1967), 81–96.

MINIMAL PADE-SENSE MATRIX APPROXIMATIONS AROUND $s = 0$ AND $s = \infty$

Marc Van Barel and Adhemar Bultheel
Department of Computer Science
Katholieke Universiteit Leuven
Celestijnenlaan 200A
B-3030 Leuven (Heverlee)
Belgium

ABSTRACT. An algorithm is presented for the solution of the minimal Padé-sense matrix approximation problem around $s = 0$ and $s = \infty$ or what is called in linear system theory, the minimal partial realization problem with Markov parameters and time moments. It gives a solution in the form of a polynomial matrix fraction. The algorithm is a generalization of the Berlekamp-Massey algorithm for the scalar case [1,2,7]. Our approach is mainly based on the work of Dickinson, Morf and Kailath who treated the problem with Markov parameters only [6]. Y. Bistritz [3,4,5] gives a solution of the above problem in a state space setting and refers to [6] for a polynomial approach. However an extra condition should be imposed on the solution as will be shown in this paper. Our main tool is the construction of a nested basis for the null spaces of block Hankel matrices. The proof of the algorithm and other aspects of the minimal partial realization can be found in [8]. We only give the main result.

1. FORMULATION AND TRANSFORMATION OF THE PROBLEM

Suppose the following matrix power series are given:

$$M(s) = M_1 s^{-1} + M_2 s^{-2} + \cdots + M_\alpha s^{-\alpha} + + M_{\alpha+1} s^{-\alpha-1} + \cdots \quad \text{around } s = \infty \tag{1}$$

and

$$m(s) = m_0 + m_1 s^1 + \cdots + m_\beta s^\beta + m_{\beta+1} s^{\beta+1} + \cdots \quad \text{around } s = 0 \tag{2}$$

where

$$M_i, m_j \in \mathbb{R}^{p \times m}.$$

In this paper we want to construct a rational function matrix $T(s) = C(s) \cdot B(s)^{-1}$ with $C(s) \in \mathbb{R}^{p\times m}[s]$ and $B(s) \in \mathbb{R}^{m\times m}[s]$ and regular such that

A. Cuyt (ed.), Nonlinear Numerical Methods and Rational Approximation, 143–154.

a. the power series of $T(s)$ around $s = \infty$ corresponds with the given series (1) in the first α terms :

$$T(s) = M_1 s^{-1} + M_2 s^{-2} + M_\alpha s^{-\alpha} + M'_{\alpha+1} s^{-\alpha-1} + \cdots$$

(Padé-sense around $s = \infty$)

b. the power series of $T(s)$ around $s = 0$ corresponds with the given series (2) in the first β+1 terms:

$$T(s) = m_0 + m_1 s^{+1} + \cdots + m_\beta s^\beta + m'_{\beta+1} s^{\beta+1} + \cdots$$

(Padé-sense around $s = 0$)

c. the degree of the determinant of the polynomial matrix $B(s)$ is as small as possible (minimal).

This problem originates from the field of linear multi-input multi-output continuous time systems and is called there the minimal partial realization problem.

In a system theoretic terminology, M_k are Markov parameters and if the given system is stable the m_k are related with the time moments of that system because the time moments only exist for a stable system. The Markov parameters characterize the transient behaviour (at small time) and the time moments characterize the steady state behaviour (at large time for a stable system). By solving the previously formulated problem, we find the transfer function $T(s)$ of a system which has a minimal number of states and which models the given system behaviour for small as well as for large time (if $T(s)$ is stable). It is possible that for certain values of α and β none of the solutions of the above problem are stable. This means that the time moments of $T(s)$ do not exist. However the solutions $T(s)$ solve the mathematical problem described above but as said before the Padé-type approximation around $s = 0$ has no system theoretic interpretation anymore. Note that in certain cases the algorithm of Bistritz [3,4,5] does not give a solution of the mathematical problem of finding a rational matrix function whose denominator determinant has minimal degree and who approximates in a Padé-sense around $s = 0$ and $s = \infty$. This is the case when the matrix A of a state-space representation found by Bistritz is singular or equivalently when his approximation has a pole in $s = 0$. Of course this approximation is not stable. Note that by using the results of [9], our algorithm can easily be adapted to solve the above problem with an additional constraint :

d. The rational matrix function $T(s)$ has to be stable i.e. the poles of $T(z)$ have to lie in the left half plane or equivalent the zeros of the determinant of $B(z)$ have to be in the left half plane.

Definitions:

* K is an arbitrary field.
* A polynomial matrix

$$A(s) = \left[\sum_{k=1}^{d_1} a_{1k}s^k, \sum_{k=1}^{d_2} a_{2k}s^k, \ldots, \sum_{k=1}^{d_m} a_{m,k}s^k \right] \in K^{m \times m}[s]$$

with $a_{i,d_i} \neq 0, i = 1,2, \ldots, m$ is called *column reduced* iff the matrix of its highest degree coefficients denoted by A_{hc} is nonsingular. I.e. iff $\det[a_{1,d_1}, a_{2,d_2}, \ldots, a_{m,d_m}]$ is a nonzero element of K.

The number d_i is called the *degree* of the i-th column of $A(s)$. Note that for a column reduced polynomial matrix $A(s)$, $\det A(s)$ has degree $\sum_{i=1}^{m} d_i$.

* We set $M_{-k} = -m_{-k}, k \leq 0$.
* Based on the M_k the block Hankel matrices $H_{k,d}$ for given β are defined as:

$$H_{k,d} = \begin{bmatrix} M_{-\beta} & M_{-\beta+1} & \cdots & M_{-\beta+d} \\ M_{-\beta+1} & M_{-\beta+2} & \cdots & M_{-\beta+d+1} \\ \vdots & \vdots & & \vdots \\ M_{-\beta+k-d-1} & M_{-\beta+k-d} & \cdots & M_{-\beta+k-1} \end{bmatrix}.$$

* From this point on we use the notation $a(s) \in K^m[s]$ not only to denote a polynomial vector but also to denote the vector $[a_0^T a_1^T \cdots a_d^T]^T$ if $a(s) = a_0 + a_1 s + a_2 s^2 + \cdots + a_d s^d$.

With these definitions it is shown in [8] that the minimal Padé-sense matrix approximation problem around $s = 0$ (β+1 terms) and $s = \infty$ (α terms) can be reformulated as :

Find columns $b_i(s)$ of $B(s)$ of degree d_i that are the solutions of

$$H_{\alpha+\beta+1,d_i} b_i(s) = 0 \quad i = 1,2, \ldots, m$$

such that :

$$\begin{cases} \text{(a) } B_{hc} \text{ is regular.} \\ \text{(b) } B(0) \text{ is regular.} \\ \text{(c) } \sum_{i=1}^{m} d_i \text{ is minimal.} \end{cases}$$

A proof can be found in [8]. Note in particular condition (b) which is not required in [6] but which is necessary if $\beta \geq 0$. This means that a Markov parameter, time moment problem is equivalent with a Markov parameter problem with an additional condition (b). When the equivalent Markov parameter problem is solved, the solution of the original is found by shifting. Since Bistritz didn't include condition (b) this shift property doesn't work when his state transition matrix $\overline{A}$ of the equivalent Markov parameter problem is singular (see for example (2.8) in [5]).

2. AN ALGEBRAIC ALGORITHM TO SOLVE THE PROBLEM

To find these polynomial vectors $b_i(s)$ we give an algorithm that computes recursively a nested basis for the kernels of $H_{k,d}$ for $k = 0,1,2,\ldots,\alpha+\beta$. From these basis vectors we can get the final solution vectors $b_i(s)$ $i = 1,2,\ldots,m$. The proof of this algorithm and further aspects of the minimal partial realization problem can be found in [8].

We use the following notation in the algorithm:

* When we use the notion "level k" we refer to the fact that we are working with the block Hankel matrices $H_{k,d}$.
* The operator R denotes 'the residual of' and the operator L 'the lowest degree coefficient of'.
* The set $A_k = [a_1^k, a_2^k, \ldots, a_m^k]$ denotes the m basis vectors a_i^k of level k and degree α_i^k. Initially we take $A_0 = [e_1, \ldots, e_m]$ the m unit vectors of K^m.
* The elements of A_k of degree d are denoted by $\{k,d\}$ (possibly empty). The set A_k is equal to: $A_k = \{k,0\} \cup \{k,1\} \cup \cdots \cup \{k,k\}$.
* The residual of a basis vector a_i^k is defined as:
$$r_i^k = [M_{-\beta+k-d}\; M_{-\beta+k-d+1} \;\cdots\; M_{-\beta+k}] \cdot a_i^k \;.$$
The set of residuals of the basis vectors in A_k is denoted by RA_k.
* The set $B_k = [b_1^k, \ldots, b_m^k]$ denotes the set of solution vectors b_i^k of level k of degree β_i^k.
* The set $X_k = [x_1^k, \ldots, x_t^k]$ denotes the set of t auxiliary vectors x_i^k for level k. The set of residuals of the auxiliary vectors X_i^k is denoted by RX_k. The vectors in RX_k form a basis for the set of residuals of all previously computed basis vectors.
Every auxiliary vector x_i^k is also an element of one or more of the sets A_j, $j = 0,\ldots,k$. Suppose $x_i^k = a_j^r \in A_r$ with r as high as possible. Then the potential degree of x_i^k is defined as $\pi_i^k = d_j^r + (k-r)$. The set X_k is used to make the residuals r_i^k equal to zero.
* The set $G_k = [g_1^k, \ldots, g_m^k]$ denotes the generating vectors with respect to level k. They are used to generate the solution vectors b_i^{k+1} of the next level. These vectors g_i^k are basisvectors with linearly independent lowest degree coefficients or $g_i^k \in \bigcup_{j=1}^{k} B_j$. The potential degree of g_i^k is denoted by σ_i^k.

With these notations the algorithm looks as follows :

A.1 {Initialization}

A.1.1 Choose $A_0 = \{0,0\} = [e_1, e_2, \ldots, e_m]$ the m unit vectors of K^m. These basis vectors have degree $\alpha_1^0 = \alpha_2^0 = \cdots = \alpha_m^0 = 0$. The corresponding set of residuals $RA_0 = M_{-\beta}$ (the m columns of $M_{-\beta}$).

A.1.2 The set of auxiliary vectors X_0 and the corresponding set of residuals RX_0 is empty ($t = 0$).

A.1.3 The set G_0 can be initialized as $G_0 = A_0$ with potential degrees $\sigma_i^0 = 0$.

A.1.4 The solution vectors of B_0 can be chosen as: $B_0 = A_0$. These solution vectors have degree $\beta_1^0 = \beta_2^0 = \cdots = \beta_m^0 = 0$.

A.2 for $k = 0,1,2,\ldots,\alpha+\beta$ {loop to find the basis vectors and solution vectors of level $k+1$}
Suppose we know:

- $X_k = [x_1^k,\ldots,x_t^k]$ the set of auxiliary vectors with potential degrees $\pi_1^k, \pi_2^k, \cdots, \pi_t^k$.
- $RX_k = [Rx_1^k,\ldots,Rx_t^k]$ the set of corresponding residuals.
- X_{k+1} is at the beginning of this block equal to X_k.
- $A_k = [a_1^k,\ldots,a_m^k]$ the set of basisvectors for level k having degrees $\alpha_1^k \leqslant \alpha_2^k \leqslant \cdots \leqslant \alpha_m^k$.
- A_{k+1} is the set of basis vectors for level $k+1$ which is empty at the beginning of this block.
- $G_k = [g_1^k,\ldots,g_m^k]$ the set of generating vectors with potential degrees $\sigma_1^k \leqslant \sigma_2^k \leqslant \cdots \leqslant \sigma_m^k$.
- B_{k+1} is the set of solution vectors for level $k+1$ which is empty at the beginning of this block.

A.2.1 for $i = 1,2,\ldots,m$ {loop to find the basis vector of level $k+1$ based on a_i^k}

A.2.1.1 Compute the residual Ra_i^k of $a_i^k \in \{k,\alpha_i^k\}$

A.2.1.2 Try to write Ra_i^k as a linear combination of $Rx_1^{k+1},\ldots,Rx_t^{k+1}$:

$$Ra_i^k = -\sum_{j=1}^{t} \mu_j Rx_j^{k+1}$$

r is defined as : $\mu_r \neq 0$ and $\pi_r^k \geqslant \pi_j^k$ if $\mu_j \neq 0$.
Note that when all μ_j are zero, r is not defined.

A.2.1.3 {Find a_i^{k+1}, the basis vector of level $k+1$ based on a_i^k and update the set X_{k+1} of auxiliary vectors}

A.2.1.3.1 If $r \leqslant t$ does not exist then

A.2.1.3.1.1 $a_i^{k+1} = s^{k+1-\alpha_i^k} a_i^k$

A.2.1.3.1.2 $\alpha_i^{k+1} = k+1$

A.2.1.3.1.3 Add a_i^k of potential degree α_i^k to X_{k+1} $(t \to t+1)$.

A.2.1.3.1.4 Introduce the residual Ra_i^k at the corresponding place in RX_{k+1}.

A.2.1.3.2 If $r \leqslant t$ does exist and $\alpha_i^k < \pi_r^{k+1}$ then

A.2.1.3.2.1 $a_i^{k+1} = s^{\pi_r^{k+1}-\alpha_i^k} a_i^k + \sum_{j=1}^{t} \mu_j s^{\pi_r^{k+1}-\pi_j^{k+1}} x_j^{k+1}$

A.2.1.3.2.2 $\alpha_i^{k+1} = \pi_r^{k+1}$

A.2.1.3.2.3 Replace x_r^{k+1} in X_{k+1} by a_i^k of potential degree α_i^k

A.2.1.3.2.4 Replace Rx_r^{k+1} by the residual Ra_i^k at the corresponding position in RX_{k+1}.

A.2.1.3.3 If $r \leqslant t$ does exist and $\alpha_i^k \geqslant \pi_r^{k+1}$ then

A.2.1.3.3.1 $a_i^{k+1} = a_i^k + \sum_{j=1}^{t} \mu_j s^{\alpha_i^k-\pi_j^{k+1}} x_j^{k+1}$

A.2.1.3.3.2 $\alpha_i^{k+1} = \alpha_i^k$

A.2.2 {All basis vectors of level $k+1$ are found. Before we compute the solution vectors of level $k+1$, we do the following}

A.2.2.1 Reorder the vectors of A_{k+1} such that they have nondecreasing degree

A.2.2.2 Increase the potential degrees of the vectors in X^{k+1} with 1

A.2.3 for $i = 1,2,\ldots,m$ {loop to find the solution vector of level $k+1$ based on a_i^{k+1}}

A.2.3.1 If $La_i^{k+1} \in \operatorname{span}\{Lb_j^{k+1}, j = 1,2,\ldots,i-1\}$ then

A.2.3.1.1 Look for the first r such that $Lg_r^k \notin \operatorname{span}\{Lb_j^{k+1}, j = 1,2,\ldots,i-1\}$. It will always exist and r will be $\leqslant i$.

A.2.3.1.2 If $\sigma_r^k+1 \leqslant \alpha_i^{k+1}$ then

$$b_i^{k+1} = a_i^{k+1} + g_r^k \text{ of degree } \beta_i^{k+1} = \alpha_i^{k+1}$$

A.2.3.1.3 If $\sigma_r^k+1 > \alpha_i^{k+1}$ then

$$b_i^{k+1} = s^{\sigma_r^k+1-\alpha_i^{k+1}} a_i^{k+1} + g_r^k$$

$$\text{of degree } \beta_i^{k+1} = \sigma_r^k+1$$

A.2.3.2 If $La_i^{k+1} \notin \operatorname{span}\{Lb_j^{k+1}, j = 1,2,\ldots,i-1\}$ then

$$b_i^{k+1} = a_i^{k+1} \text{ of degree } \beta_i^{k+1} = \alpha_i^{k+1}$$

A.2.4 {All solution vectors of level $k+1$ are found. Before we go to the next level we do}

A.2.4.1 Increase the potential degrees of the vectors in G_k with 1

A.2.4.2 The set G_{k+1} is formed by m vectors of G_k and A_{k+1} whose lowest degree coefficients are linearly independent and who have lowest potential degree. These vectors are ordered according to increasing potential degree.

When we consider the solution vectors b_i^k as vector polynomial columns $b_i^k(s)$ of a polynomial matrix:

$$B_k = [b_1^k(s) \; b_2^k(s) \cdots b_m^k(s)] \quad k \geq \beta+1$$

then the matrix B_k can be seen as a denominator polynomial of a minimal Padé-sense matrix approximation with $\beta+1$ time moments and $k-\beta-1$ Markov parameters. The numerator polynomial matrices $C_k = [c_1^k(s) \; c_2^k(s) \cdots c_m^k(s)]$ can be found using the matrices T_d and M_d defined as:

$$T_d = \begin{bmatrix} m_{d-1} & & m_0 & 0 \\ & & & 0 \\ & & & \\ m_0 & 0 & & 0 \end{bmatrix}$$

$$M_d = \begin{bmatrix} 0 & & 0 & M_1 \\ & & & \\ 0 & & & \\ 0 & M_1 & & M_d \end{bmatrix} .$$

The numerator vector c_j^k has degree smaller than or equal to β_j^k-1.
If $k-\beta-1 \leq \beta$ then c_j^k is computed as:

$$T_d \cdot b_j^k = \begin{bmatrix} c_{j,\beta_j^k-1}^k \\ \vdots \\ c_{j,0}^k \end{bmatrix} \tag{3}$$

else

$$M_d \cdot a_j^k = c_j^k . \tag{4}$$

Note: The vectors a_i^k are called basis vectors because a basis for the solution space of $H_{r,d} \cdot a = 0$ is given by all the vectors

$$z^{d-p_i^k} a_i^k \text{ with } \begin{cases} k = 0,1,\ldots,r \text{ and} \\ p_i^k = \alpha_i^k + (r-k) \leqslant d \end{cases}$$

3. RELATED WORK OF Y. BISTRITZ [3,4,5]

The idea of putting time moments and Markov parameters together into one block Hankel matrix has originally been used by Bistritz [3,4,5] who gives a state space algorithm to fit both time moments and Markov parameters. His algorithm is designed for system theoretic applications. This implies that his algorithm is not applicable to solve our mathematical problem if the approximant has a pole in $s = 0$, which means that the shifting operation doesn't work. In our formulation this problem has been taken care of by imposing the condition that the lowest degree coefficient of the matrix polynomial denominator should be nonsingular (condition (b)). For example Bistritz's algorithm applied to the data

$$T(s) = \begin{bmatrix} 0 & 0 \\ 0 & 0 \end{bmatrix} s^0 - \begin{bmatrix} 0 & 0 \\ 1 & 1 \end{bmatrix} s^1 + \cdots$$

would generate the solution :

$$\bar{A} = [0]\,,\ \bar{B} = [1 \ \ 1]\,,\ \bar{C} = \begin{bmatrix} 0 \\ 1 \end{bmatrix}$$

which solves the equivalent Markov parameter problem :

$$T(s) = \begin{bmatrix} 0 & 0 \\ 1 & 1 \end{bmatrix} s^{-1} + \begin{bmatrix} 0 & 0 \\ 0 & 0 \end{bmatrix} s^{-2} + \cdots$$

but because $\bar{A}$ is singular he finds for the original problem the solution

$$C(sI-A)^{-1}B$$

with $C = \bar{C}\bar{A}^2 = [0 \ \ 0]^T$, $B = \bar{B} = [1 \ \ 1]$ and $A = \bar{A} = 0$ which is the only possible solution and this gives the approximation

$$\begin{bmatrix} 0 & 0 \\ 0 & 0 \end{bmatrix}$$

which doesn't match the data. Our algorithm would give the approximation $C(s)B(s)^{-1}$ with

$$B(s) = \begin{bmatrix} -1 & s^2+1 \\ 1 & 0 \end{bmatrix}$$

$$C(s) = \begin{bmatrix} 0 & 0 \\ 0 & -s \end{bmatrix}.$$

Some other solutions (see [9]) can be written down as

$$B_{a,b}(s) = \begin{bmatrix} -1 & s^2+bs+a \\ 1 & 0 \end{bmatrix}$$

$$C_{a,b}(s) = \begin{bmatrix} 0 & 0 \\ 0 & -as \end{bmatrix} \quad \text{with } 0 \neq a \in \mathbb{R}.$$

So when the zeros of s^2+bs+a are in the left half plane, the system $C_{a,b}(s)\,.\,B_{a,b}^{-1}(s)$ is not only a solution of the mathematical problem but it is even stable and therefore also solves the system modelling problem. Since Bistritz's solution is based on the equivalent Markov parameter problem without considering the additional condition (b), he can only find solutions with the number of states equal to the rank of the block Hankel matrix of the equivalent Markov parameter problem, which is 1. The stable solution above can only be found by Bistritz algorithm if the data are extended with a carefully chosen Markov parameter such that the rank of the block Hankel matrix is 2 and his matrix $\overline{A}$ is regular.

4. EXAMPLE

Our notation for the example is the following: By

$$\begin{array}{l} M_n = [\; \underset{(\alpha_1')}{x_1(s)} \;\cdots\; \underset{(\alpha_t')}{x_t(s)} \mid \underset{(\alpha_1)}{a_1^n(s)} \;\cdots\; \underset{(\alpha_m)}{a_m^n(s)} \;], \\ RM_n = [Rx_1 \cdots Rx_t \mid Ra_1^n \cdots Ra_m^n] \rightarrow A_{n+1} = [a_1^{n+1}(s) \cdots a_m^{n+1}(s)], \\ \qquad \rightarrow B_{n+1} = [\; b_1^{n+1}(s) \;\cdots\; b_m^{n+1}(s) \;], \end{array}$$

$$G_{n+1} = [\; \underset{(\sigma_1^{n+1})}{g_1^{n+1}(s)} \;\cdots\; \underset{(\sigma_m^{n+1})}{g_m^{n+1}(s)} \;]$$

we mean that $x_1(s), \ldots, x_t(s)$ are auxiliary vectors for level n and $a_1^n(s), \ldots, a_m^n(s)$ are the basis vectors (they form A_n). RM_n contains the corresponding residuals. A_{n+1} represents the basis vectors and B_{n+1} the solution vectors of level $n+1$. G_{n+1} represents the generating vectors. The numbers between brackets represent the potential degree of the vector immediately above them.

$$\begin{aligned} T(s) &= \begin{bmatrix} 0 & 0 \\ 1 & 1 \end{bmatrix} s^0 + \begin{bmatrix} 0 & 0 \\ 0 & 0 \end{bmatrix} s^1 + \begin{bmatrix} 1 & 1 \\ 1 & 1 \end{bmatrix} s^2 + \cdots \quad s \rightarrow 0 \\ &= \begin{bmatrix} 1 & 0 \\ 1 & 0 \end{bmatrix} s^{-1} + \begin{bmatrix} 0 & 0 \\ 1 & 1 \end{bmatrix} s^{-2} + \cdots \quad s \rightarrow \infty \end{aligned}$$

We should apply the algorithm to the data

$$\begin{bmatrix} -1 & -1 \\ -1 & -1 \end{bmatrix}, \begin{bmatrix} 0 & 0 \\ 0 & 0 \end{bmatrix}, \begin{bmatrix} 0 & 0 \\ -1 & -1 \end{bmatrix}, \begin{bmatrix} 1 & 0 \\ 1 & 0 \end{bmatrix}, \begin{bmatrix} 0 & 0 \\ 1 & 1 \end{bmatrix}, \ldots,$$

which gives :

$$A_0 = B_0 = \begin{bmatrix} 1 & 0 \\ 0 & 1 \end{bmatrix}, \quad G_0 = \left[\begin{array}{cc} 1 & 0 \\ 0 & 1 \\ (0) & (0) \end{array} \right]$$

$$M_0 = \begin{array}{c} \left[\begin{array}{c|cc} & 1 & 0 \\ & 0 & 1 \end{array}\right] \\ \begin{array}{cc} (0) & (0) \end{array} \end{array},$$

$$RM_0 = \left[\begin{array}{c|cc} & -1 & -1 \\ & -1 & -1 \end{array}\right] \rightarrow A_1 = \begin{bmatrix} -1 & s \\ 1 & 0 \end{bmatrix} \rightarrow B_1 = \begin{bmatrix} -1 & s+1 \\ 1 & 0 \end{bmatrix},$$

$$G_1 = \begin{array}{c} \begin{bmatrix} -1 & 1 \\ 1 & 0 \end{bmatrix} \\ \begin{array}{cc} (0) & (1) \end{array} \end{array}$$

$$M_1 = \begin{array}{c} \left[\begin{array}{c|cc} 1 & -1 & s \\ 0 & 1 & 0 \end{array}\right] \\ \begin{array}{c|cc} (1) & (0) & (1) \end{array} \end{array},$$

$$RM_1 = \left[\begin{array}{c|cc} -1 & 0 & 0 \\ -1 & 0 & 0 \end{array}\right] \rightarrow A_2 = \begin{bmatrix} -1 & s \\ 1 & 0 \end{bmatrix} \rightarrow B_2 = \begin{bmatrix} -1 & s^2+1 \\ 1 & 0 \end{bmatrix},$$

$$G_2 = \begin{array}{c} \begin{bmatrix} -1 & 1 \\ 1 & 0 \end{bmatrix} \\ \begin{array}{cc} (0) & (2) \end{array} \end{array}$$

$$M_2 = \begin{array}{c} \left[\begin{array}{c|cc} 1 & -1 & s \\ 0 & 1 & 0 \end{array}\right] \\ \begin{array}{c|cc} (2) & (0) & (1) \end{array} \end{array},$$

$$RM_2 = \left[\begin{array}{c|cc} -1 & 0 & 0 \\ -1 & 0 & -1 \end{array}\right] \rightarrow A_3 = \begin{bmatrix} -1 & s^3 \\ 1 & 0 \end{bmatrix} \rightarrow B_3 = \begin{bmatrix} -1 & s^3+1 \\ 1 & 0 \end{bmatrix},$$

$$G_3 = \begin{array}{c} \begin{bmatrix} -1 & 1 \\ 1 & 0 \end{bmatrix} \\ \begin{array}{cc} (0) & (3) \end{array} \end{array}$$

$$M_3 = \begin{array}{c} \left[\begin{array}{cc|cc} s & 1 & -1 & s^3 \\ 0 & 0 & 1 & 0 \end{array}\right] \\ \begin{array}{cc|cc} (2) & (3) & (0) & (3) \end{array} \end{array},$$

$$RM_3 = \left[\begin{array}{cc|cc} 0 & -1 & -1 & 1 \\ -1 & -1 & -1 & 1 \end{array}\right] \rightarrow A_4 = \begin{bmatrix} -s^3-1 & 0 \\ s^3 & s^3 \end{bmatrix} \rightarrow B_4 = \begin{bmatrix} -s^3-1 & -1 \\ s^3 & s^3+1 \end{bmatrix},$$

$$RM_3 = \begin{bmatrix} 0 & -1 & | & -1 & 1 \\ -1 & -1 & | & -1 & 1 \end{bmatrix} \to A_4 = \begin{bmatrix} -s^3-1 & 0 \\ s^3 & s^3 \end{bmatrix} \to B_4 = \begin{bmatrix} -s^3-1 & -1 \\ s^3 & s^3+1 \end{bmatrix},$$

$$G_4 = \begin{bmatrix} -1 & -s^3-1 \\ 1 & s^3 \\ (1) & (3) \end{bmatrix}$$

$$M_4 = \begin{bmatrix} -1 & s & | & -s^3-1 & 0 \\ 1 & 0 & | & s^3 & s^3 \\ (1) & (3) & & (3) & (3) \end{bmatrix}$$

$$RM_4 = \begin{bmatrix} -1 & 0 & | & 0 & 0 \\ 1 & -1 & | & 0 & 1 \end{bmatrix} \to A_5 = \begin{bmatrix} -s^3-1 & s \\ s^3 & s^3 \end{bmatrix} \to B_5 = \begin{bmatrix} -s^3-1 & s-1 \\ s^3 & s^3+1 \end{bmatrix},$$

$$G_5 = \begin{bmatrix} -1 & -s^3-1 \\ 1 & s^3 \\ (2) & (3) \end{bmatrix}.$$

The numerator coefficients can be found by using (3) which gives :

$$C_5(s) = \begin{bmatrix} -s^2 & 0 \\ -s^2-1 & s \end{bmatrix}.$$

$C_5(s)B_5(s)^{-1}$ will now fit 3 time moments and 2 Markov parameters. Indeed,

$$C_5(s)B_5(s)^{-1} = \begin{bmatrix} -s^2 & 0 \\ -s^2-1 & s \end{bmatrix} \begin{bmatrix} -s^3-1 & s-1 \\ s^3 & s^3+1 \end{bmatrix}^{-1}$$

$$= \begin{bmatrix} -s^2 & 0 \\ -s^2-1 & s \end{bmatrix} \begin{bmatrix} s^3+1 & -s+1 \\ -s^3 & -s^3-1 \end{bmatrix} / \left[-(s^3+1)^2 - s^4 + s^3\right]$$

$$= \begin{bmatrix} s^5+s^2 & -s^3+s^2 \\ s^5+s^4+s^3+s^2+1 & s^4-s^3+s^2+1 \end{bmatrix} / (s^6+s^4+s^3+1)$$

$$= \begin{bmatrix} 1 & 0 \\ 1 & 0 \end{bmatrix} s^{-1} + \begin{bmatrix} 0 & 0 \\ 1 & 1 \end{bmatrix} s^{-2} + \begin{bmatrix} 0 & -1 \\ 0 & -1 \end{bmatrix} s^{-3} + \cdots \qquad s \to \infty$$

$$= \begin{bmatrix} 0 & 0 \\ 1 & 1 \end{bmatrix} s^0 + \begin{bmatrix} 0 & 0 \\ 0 & 0 \end{bmatrix} s + \begin{bmatrix} 1 & 1 \\ 1 & 1 \end{bmatrix} s^2 + \begin{bmatrix} 0 & -1 \\ 0 & -2 \end{bmatrix} s^3 + \cdots \qquad s \to 0.$$

5. CONCLUSIONS

An algebraic algorithm was developed to solve the minimal Padé-sense matrix approximation problem around $s = 0$ and $s = \infty$ also called in linear system terminology the minimal partial realization problem with Markov parameters and time moments. It was shown that the algorithm given by Y. Bistritz does not solve the mathematical problem when $s = 0$ is a pole of the approximation. To clarify the theory an example was given.

6. REFERENCES

[1] E.R. BERLEKAMP : 'Nonbinary BCH decoding', *International symposium on Information Theory*, San Remo, Italy (1962).

[2] E.R. BERLEKAMP : *Algebraic coding theory*, McGraw-Hill, New York, chapts 7 and 10 (1968).

[3] Y. BISTRITZ : 'Nested bases of invariants for minimal realizations of finite matrix sequences', *SIAM J. Control Opt.*, 21, 804-821 (1983).

[4] Y. BISTRITZ, U. SHAKED : 'Discrete multivariable system approximations by minimal Padé type state models', *IEEE Trans. Circuits and Systems*, CAS-31, 382-390 (1984).

[5] Y. BISTRITZ, U. SHAKED : 'Minimal Padé model reduction for multivariable systems', *Journal of Dynamic Systems, Measurement, and Control*, 106, 293-299 (1984).

[6] B.W. DICKINSON, M. MORF, T. KAILATH : 'A minimal realization algorithm for matrix sequences', *IEEE Trans. Autom. Control*, AC-19, 31-38 (1974).

[7] J.L. MASSEY : 'Shift register synthesis and BCH decoding', *IEEE Trans. Information Theory*, IT-15, 122-127 (1969).

[8] M. VAN BAREL, A. BULTHEEL : *A minimal partial realization algorithm for MIMO systems*, Report TW79, Katholieke Universiteit Leuven, Department of Computer Science (July 1986, revised: April 1987).

[9] M. VAN BAREL, A. BULTHEEL : *A minimal partial realization algorithm for MIMO systems III. A parametrization of all solutions*, Report TW93, Katholieke Universiteit Leuven, Department of Computer Science (August 1987).

(PADE)y of (PADE)x APPROXIMANTS OF F(x,y)

C.CHAFFY
Laboratoire TIM3
Algorithmique parallèle et calcul formel
46 av. Félix Viallet
38041 Grenoble cedex. France

A BSTRACT. The computer algebra system "Reduce" allows us to define a new kind of multivariate Pade approximants. The method arises from a quite standard idea in multivariate theory. To construct these $(\text{Pade})_y$ of $(\text{Pade})_x$ approximants to a function of two variables f(x,y), the usual univariate algorithms must be applied twice (first with respect to x and then with respect to y), provided that the system accepts data depending on parameters. The coefficients of the double power series expansion of the rational functions constructed in this way match those of f(x,y) on a rectangular interpolation set and this process presents some invariance properties for functions such as g(x)h(y) or f(xy) . Contrary to other types of Pade approximants, the generalization to the n-variate case (n>2) is computationally easy. For g(x,y)/P(x,y) (g holomorphic, P polynomial...) numerical approximation results are quite good: they are supported by a de Montessus-type theorem.

0. INTRODUCTION

There have been many attempts to solve the multivariate Pade problem [6,7,8,9,10] but generally, for such approximants, no theorem of uniform convergence on compact subsets has been proved. Our definition tries to preserve as many univariate convergence results as possible. Consequently, the fractions that we construct are not uniquely defined by interpolation conditions.

Let us remember some classical properties in multivariate theory.

Integration: $\iint f(x,y)\,dx\,dy = \int \left(\int f(x,y)\,dx \right) dy$

The univariate process is first applied to the variable x , and then applied one more time to the variable y.

In this case, the order of these two operations doesn't matter.

Derivation:

$$\delta^{k+l}(f)/\delta y^l\,\delta x^k = \delta^l(\delta^k f/\delta x^k)/\delta y^l$$

Pade approximation: So, in the generic case, let us define **pade(f,x,l,m)** the operator that substitutes a rational function of x to f in the usual way:
the numerator degree is l, the denominator degree is m, and the power series in x of this fraction matches those of f, up to the order l+m .

By applying this operator twice to f(x,y), we compute

A. Cuyt (ed.), Nonlinear Numerical Methods and Rational Approximation, 155–166.

pade(pade(f,x,l,m), y,n,p)

This rational function of x,y has some "rectangular" coincidence properties with the double power series of f(x,y).

Let us explain the details of this construction.

1. FIRST STEP.

1.1. Definition

Let f be in $\mathbb{C}[[x,y]] = \mathbb{C}[[y]][[x]]$ and (FSx) be the power series in x:

$$\sum_{i\geq 0} c_i(y)\, x^i = f(x,y)$$

Problem (Padé,f,l,m,x): Assuming that $c_0(y) \neq 0$, we are looking for a function

$$\Big(\sum_{0\leq k\leq l} d_k(y)\, x^k\Big) / \Big(\sum_{0\leq k\leq m} e_k(y)\, x^k\Big) \quad \text{in} \quad \mathbb{C}[[y]](x), \quad \text{where } e_0(y) \neq 0$$

(ie: the $d_k(y)$ and $e_k(y)$ are power series in y), whose development coincides with (FSx) up to the order l+m.

If there exists a unique solution, it is denoted by **pade(f,x,l,m)** or $\mathbf{[l/m]_x}$ (f).

For example, $\mathbf{[l+m/0]_x}$ (f) always equals

$$\sum_{0\leq i\leq l+m} c_i(y)\, x^i .$$

This method is illustrated by the following examples.

1.2. Examples

* Consider **f(x,y) = cos(x+y²)**

The Mac-Laurin polynomial of order 2 of $x \to f(x,y)$ (y as parameter) is $\cos y^2 - (\sin y^2)\, x - (\cos y^2)\, x^2/2$ and denoted by pade(**f,x,2,0**) or $[2/0]_x$ (f)

The function

$$(-2 \cos y^2 \sin y^2 + (2(\sin y^2)^2 + (\cos y^2)^2)\, x) / (-2 (\sin y^2) + (\cos y^2) x)$$

is pade(**f,x,1,1**) or $\mathbf{[1/1]_x}$ (f) because it is the unique solution of the Problem (Pade,f,1,1,x) :

we have $\mathbf{[2/0]_x}$ (f) = $\mathbf{[2/0]_x}$ (pade(f,x,1,1))

* For $\mathbf{f(x,y) = 1 + e^{y^2} + Ln(1\text{-}x\text{-}y)}$

pade(**f,x,2,0**) = $1 + e^{y^2} + \mathrm{Ln}(1-y) + x/(y-1) - x^2/(2(y-1)^2)$ = $\mathbf{[2/0]_x}$ (f) and

pade(**f,x,1,1**) = $(2y-2 + (2y-2)\mathrm{Ln}(1-y) + (2y-1)e^{y^2} + (3+\mathrm{Ln}(1-y)+ e^{y^2})x)/(2y-2+x)$

We have $\mathbf{[2/0]_x}$ (f) = $\mathbf{[2/0]_x}$ (pade(f,x,1,1))

* Particularly, if f(x,y) = **g(x)** h(y)

pade(**f,x,l,m**) = h(y) pade(**g,x,l,m**)

$f(x,y) = e^{x+2y} = e^x\ e^{2y}$
$[2/0]x\ (f) = e^{2y}\ (1 + x + x^2/2) = e^{2y}\ [2/0]x\ (e^x)$
$[1/1]_x\ (f) = e^{2y}\ ((-x-2)/(x-2)) = e^{2y}\ [1/1]_x\ (e^x)$
Notice that these intermediate approximants are not yet rational functions of x,y.

1.3. Singularities

Let us present here some analytic properties of the functions constructed in this formal way. Suppose that $f(x,y) = g(x,y) / Q_M(x,y)$, with g holomorphic, Q_M polynomial and $Q_M(0,0) \neq 0$:
$Q_M(x,y) = a_0(y) + a(y)\ x + a(y)\ x^2 + ... + a_M(y)\ x^M$
The previous construction is applied to the power series of f(x,y).
For each value y_0 of y such that $a_M(y_0) \neq 0$ and $g(x,y_0) \neq 0$, the de Montessus de Ballore 's theorem applied to the univariate function $x \to f(x,y_0)$ gives uniform convergence on x of some column $[l/M_0]_x\ (x,y_0)$ when the numerator degree l tends to infinity, and the denominator degree $M_0 \leq M$ remains fixed.
Under some other conditions, we even get uniform convergence on compact subsets [5], as illustrated by the following examples.

* $f(x,y) = e^{xy} / (1-x-y)$
In this particular case, $[l/1]_x$ is a rational function of x,y. Its denominator defines x as a function $\varphi_l(y)$, that we compare to the straight line: x=1-y, as illustrated on Figure 1.
$\varphi_0(y) = (y-1)/(y^2-y-1)$
$\varphi_1(y) = 2\ (y^3-2y+1)/(y^3 -2y^2 -y +2y+2)$
$\varphi_2(y) = 3\ (y^5-3y^4 +y^3+3y^2 -2)/(y^6 -3y^5 +5y^3 +3y^2 -6y-6)$

* $f(x,y) = 1/(x^2 -2x+y^2-2y+1)$
Here, the poles of $[l/1]_y$ define y as a function $\varphi_l(x)$ that we compare to 1-sqrt(2x-x²), as illustrated on Figure 2. For example,
$\varphi_0(x) = (x^2 -2x +1)/2$
$\varphi_1(x) = 2\ (x^2 -2x +1)/(- x^2 +2x+3)$
$\varphi_2(x) = (x^4 -4x^3 +2x^2 +4x-3)/(4(x^2 -2x-1))$
$\varphi_3(x) = 4\ (-x^4 +4x^3 -4x^2 + 1)/(x^4 -4x^3 -6x^2 +20x +5)$

* $f(x,y) = 1/(e^x-y-2)$
Even for non-polynomial coefficients of the denominator, numerical results are quite good. Here, $[l/1]_y$ is exactly f. The poles of $[l/1]_x$ define x as a function $\varphi_l(y)$ that we compare to x=Ln(y+2), as illustrated on Figure 3.

Figure 1: $f(x,y) = e^{xy} / (1-x-y)$ Poles of $[1/1]_x$

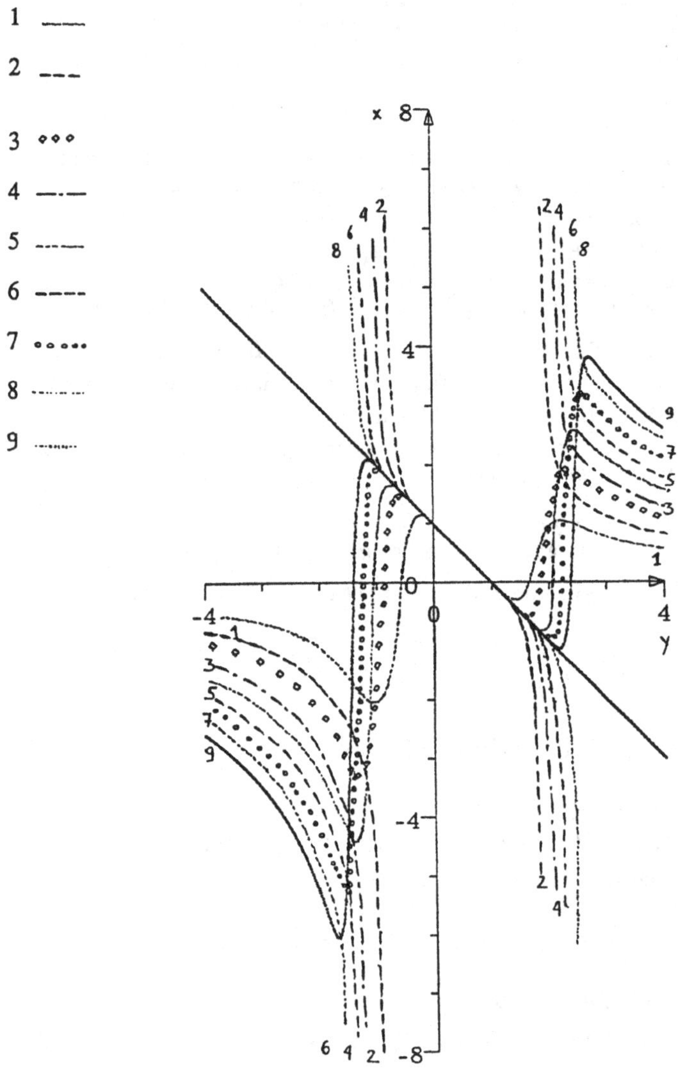

Figure 2:   f(x,y) = 1/(x^2 -2x+y^2-2y+1)   Poles of [l/1]$_y$

0 – – –

1 ....

2 ......

3 ○ ○ ○

4 ◇ ◇ ◇

5 •——•——•

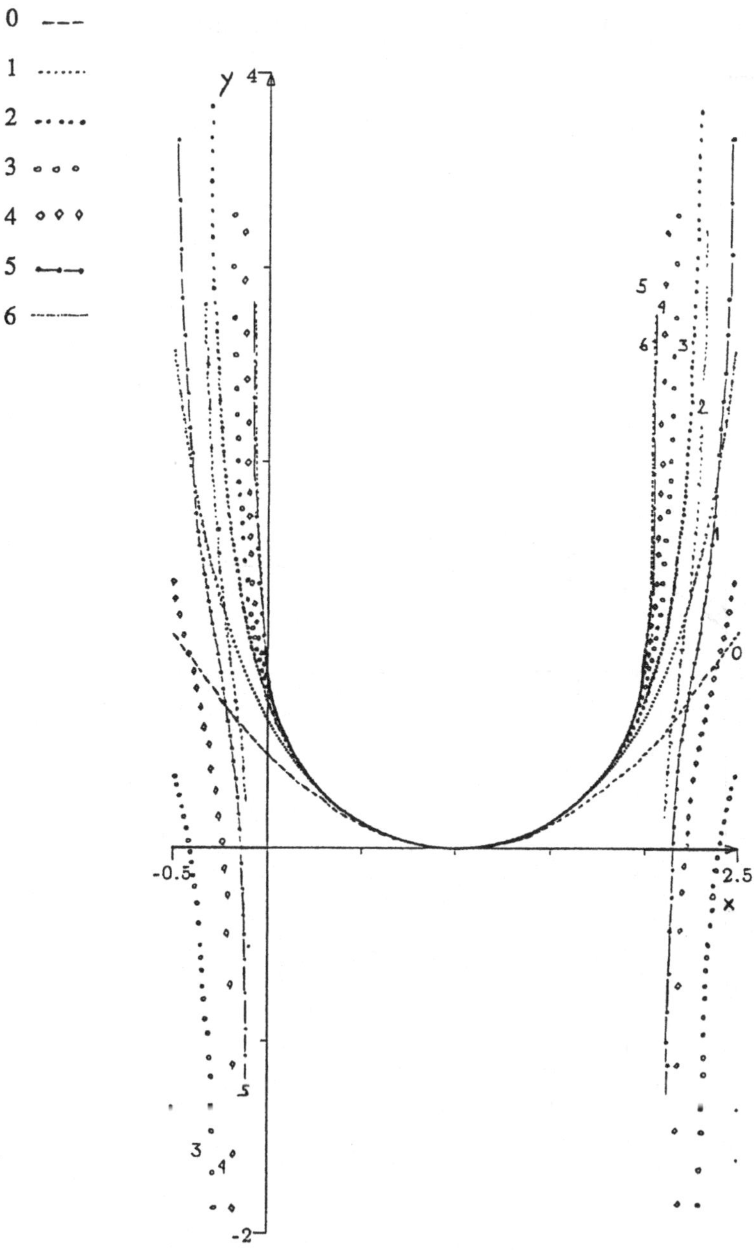

Figure 3: $f(x,y) = 1/(e^{x}-y-2)$ Poles of $[1/1]_x$

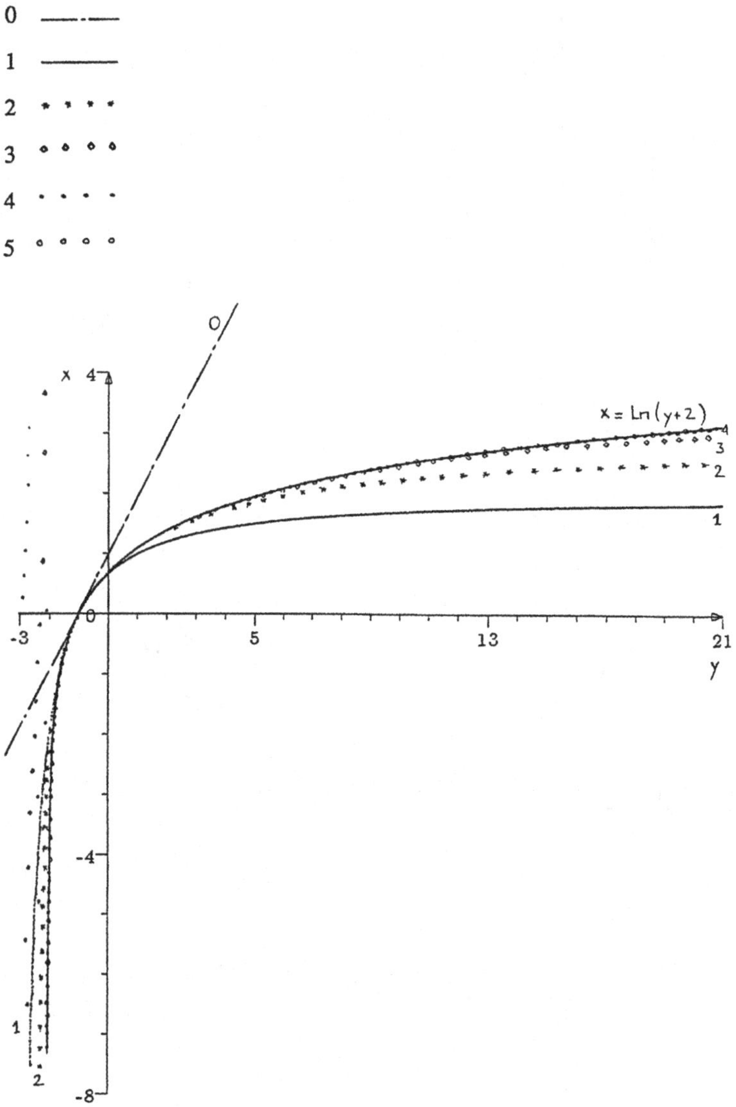

We now consider the second step of our construction.

2. SECOND STEP.

2.1. The method.

We start from the intermediate approximant g=pade(f,x,l,m) and consider it as a function of y , to calculate its Pade approximant , with respect to y, if it exists.

Therefore, we need to compute the power series expansion $\sum_{j\geq 0} b_j(x)\, y^j$ of g(x,y).

For example, $b_0(x) = g(x,0) = \left(\sum_{0\leq k\leq l} d_k(0)\, x^k\right) / \left(\sum_{0\leq k\leq m} e_k(0)\, x^k\right)$.

More generally, $j!\, b_j(x)$ formally equals $\left(\delta^j g(x,y)/\delta y^j\right)_{/y=0}$, and since g(x,y) depends rationally on x, all $b_j(x)$ are rational functions.

If there exists a unique solution to the problem (Padé,g,n,p,y), it is pade(g,y,n,p) or **pade**(pade(f,x,l,m),**y,n,p**),which is denoted by $[\mathbf{n/p}]_y$ o $[l/m]_x$ (f).

It is now a rational function of x,y; in the general case, the numerator (resp. the denominator) has degree n (resp. degree p) with respect to y, but the degrees with respect to x have been changed by the last transformation.

For example, in $[\mathbf{n/p}]_y$ o $[l/0]_x$ (f) ,

- the numerator has a degree/x $\leq l(p+1)$
- the denominator has a degree/x $\leq l\,p$

This increase of the degrees is the counterpart of the good numerical results that these fractions give, as is the case for the "homogeneous" Pade approximants. [1]

Prop: Let $f(x,y) = \sum_{i\geq 0}\sum_{j\geq 0} a_{i,j}\, x^i\, y^j$

The technique used here implies that the fractions satisfy a "rectangular-type" interpolation property: when there is no block at the first, nor at the second step, the approximant $h = [\mathbf{n/p}]_y$ o $[l/m]_x$ (f) verifies:

$[\mathbf{n+p/0}]_y$ o $[l+m/0]_x$ (h) $= [\mathbf{n+p/0}]_y$ o $[l+m/0]_x$ (f)

$$= \sum_{0\leq i\leq l+m}\ \sum_{0\leq j\leq n+p} a_{i,j}\, x^i\, y^j$$

Proof:

$[\mathbf{n+p/0}]_y$ o $[l+m/0]_x$ (h) $= [\mathbf{l+m/0}]_x$ o $[n+p/0]_y$ (h)

$= [\mathbf{l+m/0}]_x$ o $[n+p/0]_y$ (g) (definition of h)

$= [\mathbf{n+p/0}]_y$ o $[l+m/0]_x$ (g)

$= [\mathbf{n+p/0}]_y$ o $[l+m/0]_x$ (f) . (definition of g)

But such "rectangular" conditions are not sufficient to define (Pade)of(Pade) approximants.

$[0/1]_y \circ [1/0]_x$ (f) = $(a_{10}^2 x^2 + 2 a_{10} a_{00} x + a_{00}^2)/(-a_{11} xy + a_{10} x - a_{01} y + a_{00})$:

we have 7 (-1) coefficients and only 4 =2*2 equations: same coefficients a_{00}, a_{10}, a_{01}, a_{11} as f(x,y) in the double power series expansion of this last fraction.

But is there any connection between these (Padé) of (Pade) approximants and other kinds of multivariate Pade approximants? The approximants introduced here seem to depend more totally on the function $x \to f(x,y)$ than branched continued fractions because, as explained in [2,3], some of the branched continued fractions result from Pade approximation with respect to y of each reciprocal derivative c(f,x,k) (that depends on the variable y) in the x-continued fraction:

$$g = c(f,x,0) + \frac{x|}{|c(f,x,1)} + \frac{x|}{|c(f,x,2)} + \ldots + \frac{x|}{|c(f,x,m)}$$

The bivariate function g(x,y) belongs to a staircase of the Pade table of the intermediate approximants pade(f,x,l,m). Instead of interpolating each coefficient c(f,x,k), our process interpolates the whole function: $y \to g(x,y)$.

In the next paragraph, we present some examples of these $(\text{Pade})_y$ of $(\text{Pade})_x$ approximants.

2.2. Examples.

* $f(x,y) = \cos(x+y^2) + e^{x+y}$

At the first step, we compute

$$g = [1/1]_x (f) = (2 e^y (\cos y^2 - 2\cos y^2 \sin y^2 - \sin y^2) + 2 e^{2y} + ((\cos y^2)^2 + e^{2y} - 4 e^y \sin y^2 + 2 (\sin y^2)^2) x) / (2 e^y - 2 \sin y^2 + (\cos y^2 - e^y)x)$$

which verifies

$$[2/0]_x (f) = [2/0]_x (g) = (\cos y^2 + e^y) + (e^y - \sin y^2) x + (e^y - \cos y^2)x^2/2$$

Next, we find that $[1/1]_y$ (g) equals

$$(2x^3+ 8x^2+12 x+ 8+ (x^3+ 8x^2 +10x) y)/(2x^2+ 4 x+ 4 + (-x^3 - x^2+ 2x - 2)y)$$
$$= h = [1/1]_y \circ [1/1]_x (f)$$

because

$$[2/0]_y (h) = [2/0]_y (g) = (4x + 8 + (2x^2+ 4 x+ 4)y + (x^3+ x^2 -2x+ 2)y^2) /4$$

The double power series of f and h coincide on the rectangular set:

a_{02}	a_{12}	a_{22}
a_{01}	a_{11}	a_{21}
a_{00}	a_{10}	a_{20}

ie: $$[2/0]_y \circ [2/0]_x (h) = [2/0]_y \circ [2/0]_x (f) = (x + 2) + (x^2/2 + x + 1)y + (x^2 -2x + 2)y^2 /4$$

* $\mathbf{f(x,y) = 1 + Ln(1\text{-}x\text{-}y) + e^{y^2}}$

Here $[1/1]_x$ (f) is

$$(2\, e^{y^2} (y-1) + 2\,(y-1)\, Ln(1-y) + 2y - 2 + (e^{y^2} + Ln(1-y) + 3)\, x) / (2y - 2 + x) = g$$

Next, $h = \mathbf{[1/1]_y}\ (g) = \mathbf{[1/1]_y}\ o\ [1/1]_x\ (f)$

equals $2\,(4\,x^3 - 4\,x^2 + 16\,x - 16 + (x^3 - 12x^2 + 36x)\, y)$
$/ (2x^3 - 4x^2 + 8x - 16 + (x^3 - 6x^2 + 28x - 8)y)$

because

$\mathbf{[2/0]_y}\ (h) = \mathbf{[2/0]_y}\ (g) = (8x^3 - 40x^2 + 64x - 32 + (-2x^3 + 4x^2 - 8x + 4)y$
$+ (x^3 - 6x^2 + 28x - 8)y^2) / (2\,(x^3 - 6x^2 + 12x - 8)$

The rectangular interpolation property is

$\mathbf{[2/0]_y}\ o\ [2/0]_x\ (h) = \mathbf{[2/0]_y}\ o\ [2/0]_x\ (f)$
$= (-x^2/2 - x + 2) - (x^2 + x + 1)y + (-3x^2/2 - x + 1/2)y^2$

* $\mathbf{f(x,y) = e^{x+2y}}$

$[1/1]_x\ (f) = g = e^{2y}\ (x+2)/(-x+2)$

Then $h = \mathbf{[0/1]_y}\ (g) = \mathbf{[0/1]_y}\ o\ [1/1]_x\ (f) = (x+2)/(2-x+(2x-4)y)$

because

$\mathbf{[1/0]_y}\ (h) = \mathbf{[1/0]_y}\ (g) = (1+2y)\ (x+2)/(-x+2)$

The interpolation set is

a_{01}	a_{11}	a_{21}
a_{00}	a_{10}	a_{20}

ie: $\mathbf{[1/0]_y}\ o\ [2/0]_x\ (h) = \mathbf{[1/0]_y}\ o\ [2/0]_x\ (f) = (1+2y)\ (1+x+x^2/2)$

2.3. Generalisations.

The operator **pade(f,x,l,m)** is computed by using the classical univariate Pade algorithms, programmed in the computer algebra system Reduce. This single operator allows us to calculate multivariate Pade approximants also in the n-variate case $(n \geq 2)$, because they are computed as they are defined.

For example, **pade(pade(pade (f,x,l,m),y,n,p),z,q,r)** is the rational function:

$$\mathbf{[q/r]_z\ o\ [n/p]_y\ o\ [l/m]_x}\ (f)$$

of a 3-variate function f(x,y,z).

Moreover, the same process can be applied to the multipoint - multivariate case: with data at the points (x_i, y_j) of a grid: $0 \leq i \leq l+m \quad 0 \leq j \leq n+p$ (they can be distinct or not). The multipoint - univariate theory just needs to run under Reduce.

3. ALGEBRAIC AND ANALYTIC PROPERTIES.

3.1. Factorization property.

Consider $f(x,y) = g(x)\, h(y)$.
The factor $h(y)$ seems to be constant in the Pade/x approximation;
so **pade (f,x,l,m) = h(y) pade(g,x,l,m)**
At the next step, pade(g,x,l,m) seems to be constant in the Pade/y approximation, because this function only depends on x. Then,
pade (h(y)pade(g,x,l,m),y,n,p) = pade(g,x,l,m) pade(h,y,n,p).
Consequently, in such a case:

$$\text{pade}(\text{pade}(f,x,l,m),y,n,p) = \text{pade}(\text{pade}(f,y,n,p),x,l,m) = \underbrace{\text{pade}(f,x,l,m)}_{\substack{\text{univariate}\\ \text{Pade}\\ \text{approximant}}}\ \underbrace{\text{pade}(g,y,n,p)}_{\substack{\text{univariate}\\ \text{Pade}\\ \text{approximant}}}$$

3.2. Projection properties.

Our construction implies that for all values y_0 of y for which there exists a univariate Padé approximant for the function $f_{y0} : x \rightarrow f(x,y_0)$, the following holds:

$$[l/m]_x\,(f)\,(x, y_0) = [l/m](f_{y0})\,(x)\,.$$

If x=0 or y=0 , (Pade) of (Pade) approximants reduce to univariate Pade approximants.
Indeed, $[n/p]_y \circ [l/m]_x\,(f)\,(x,0) = [l/m]_x\,(f)\,(x,0) = [l/m]\,(f_0)\,(x)$.
Let us denote $g = [l/m]_x\,(f)$.
If $[n/p]_y\,(g)$ exists, it verifies $[n/p]_y\,(g)\,(x,0) = g(x,0)$
$[n/p]_y \circ [l/m]_x\,(f)\,(0,y)$ equals $[n/p]_y\,(f(0,y))$ too , which is the univariate Pade approximant [n/p] of $y \rightarrow f(0,y)$, because $[l/m]_x\,(f)\,(0,y) = f(0,y)$.

3.3. Invariance properties.

In the special case: $f(x,y) = g(xy)$, we have the following substitution result:

$$[n/p]_y \circ [n+p/0]_x\,(f)\,(x,y) = [n/p]_x \circ [n+p/0]_y\,(f)\,(x,y)$$
$$= [n/p]_y\,(f)\,(x,y) = [n/p]_x\,(f)\,(x,y)$$
$$= [n/p]_z\,(g(z))\Big/_{z=xy}$$

More generally, if $f(x,y) = g((a+x)y)$ then

$$[n/p]_y \circ [n+p/0]_x\,(f)\,(x,y) = [n/p]_y\,(f)\,(x,y) = [n/p]_z\,(g(z))\Big/_{z=(a+x)y}$$

Of course, if f(x,y) is a rational function, it equals a double infinity of its (Pade) of (Pade) approximants.

3.4. Convergence results.

Prop: Let us assume that $f(x,y) = g(x,y)/q(y)$, where q is a polynomial such that $q(0) \neq 0$ and where g is holomorphic .
Since $[l/0]_x\,(f)\,(x,y) = \{\, [l/0]_x\,(g)\,(x,y)\,\}/q(y)$, for all values y_0 of y such that $q(y_0) \neq 0$ and for all values x_0 of x such that $g(x_0,y)/q(y)$ has exactly p poles, the following holds:

$$f(x_0,y_0) = \lim_{l\to\infty} \left(\lim_{n\to\infty} [n/p]_y \circ [l/0]_x (f) (x_0,y_0) \right).$$

Proof: [4]

Theorem : Furthermore, if $f(x,y) = g(x,y)/Q(x,y)$, where Q is a bivariate polynomial such that $Q(0,0)\neq 0$ and g is still holomorphic (and non zero when $Q(x,y)=0$), we get results of uniform convergence to f of the $(\text{Padé})_y$ of $(\text{Pade})_x$ approximants $[n/p]_y \circ [l/M]_x (f)$ in some compact domains of $\mathbb{C}^2$, assuming that :

$Q_M(x,y) = q_0(y) + q_1(y)\, x + q_2(y)\, x^2 + ... + x^M$ and $y \to Q_M(x,y)$ has exactly p zeros in a disk depending on the least modulus root of $q_0(y)$ (x as parameter).

Proof : [5]

The result is based on the proof by Saff of the theorem of Montessus de Ballore,by studying in detail at each step the integral form of the errors, where y or x appear as parameters .The main idea is that no information about y is lost at the intermediate step,by using Pade approximation with respect to the variable x, because these approximants are exact for almost all values of y.

The next table presents some results for the function $f(x,y) = e^{xy}/(1-x-y)$, where the polynomials $[l+1/0]_y \circ [l+1/0]_x (f)$ are replaced by the corresponding (Pade) of (Pade) approximants $[l/1]_y \circ [l/1]_x (f)$: they happen to converge to the value of f, even when the power series expansion of f diverges.

Numerical results for $f(x,y) = e^{xy}/(1-x-y)$:

Point (x,y)	$[4/0]_y \circ [4/0]_x$	$[3/1]_y \circ [3/1]_x$	function
(1/4,1/4)	2.110	2.1289898	2.128989
(1/2,2/5)	6.4	12.2199	12.2140
(1/2,9/20)	7.6	25.09	25.05
(1/2,3/5)	13.0	-13.46	-13.50
(1/2,5/8)	14.2	-10.91	-10.93
(1/2,3)	2107.9	-1.64	-1.79
(2/3,1/4)	6.0	14.180	14.176
(2/3,2/3)	31.6	-4.66	-4.68
(2,0)	31.0	-1.000088	-1.000000
(2,1)	3640.3	-3.08	-3.69
(3,0)	121.0	-0.5000018	-0.5000000
(4,1/4)	1523.7	-0.817	-0.836
(i,i)	43.7 -i 62.3	0.06 +i 0.140	0.07 +i 0.147

To get a symmetric function, it would be better to use the arithmetic mean:
$\{ [l/1]_y \circ [l/1]_x (f)(x,y) + [l/1]_x \circ [l/1]_y (f)(x,y) \}/2$
Here, $[l/1]_x \circ [l/1]_y (f)(x,y) = [l/1]_y \circ [l/1]_x (f)(y,x)$.

In this particular case, let $g(z,t) = e^z/(1-t)$. The $(\text{Pade})_z$ of $(\text{Pade})_t$ approximants to $g(z,t)$ for $z = xy$ and $t = x+y$ would give more beautiful results, because g satisfies the factorization property and $1/(1-t)$ is a rational function.

Still, it seems more effective to choose the degree l of the numerator to be large at the first step, and the degree n to be small at the second step (which has been confirmed by graphic experiments) : the more wrong is the intermediate result, the more wrong is the final approximant.Even more strongly we remark that there is not much difference between various approximants when the intermediate approximant is kept fixed.

In fact, these theoretical results about (Padé) of (Pade) approximants have been confirmed by many experiments, and we feel very optimistic about their applications.

REFERENCES

[1] C.CHAFFY: "A homogeneous process for Pade approximants in two complex variables". Num.Math.vol 45 fasc 1 p. 149-144 1984

[2] C.CHAFFY: "Fractions continues à deux variables et calcul formel". RR 611M. TIM3. juin 1986.

[3] C.CHAFFY: " How to compute multivariate Pade approximants". ACM: Proceedings SYMSAC'86. Waterloo (Canada) . juillet 1986.

[4] C.CHAFFY: "Approximation de Padé à plusieurs variables:une nouvelle méthode".
I Théorie RR 631 M. TIM3.
II Programmation RR 632 M. TIM3.
octobre 1986.

[5] C.CHAFFY: "Convergence uniforme d'une nouvelle classe d'approximants de Padé à plusieurs variables". note à soumettre au CRAS en septembre 1987.

[6]J.S.R.CHISHOLM:"Rational approximants defined from double power series".Math. Comp. 27 p. 841-848 1973

[7]A.CUYT:"Abstract Padé approximants for operators: theory and applications". thèse Universiteit Antwerpen 1982

[8]P.R.GRAVES-MORRIS:"Generalisations of the theorem of de Montessus using Canterbury approximants" in "Padé and rational approximation" Saff-Varga eds. p. 73-82 1977

[9]J.KARLSSON and H.WALLIN:"Rational approximation by an interpolation procedure in several variables" in "Padé and rational approximation" Saff-Varga eds. p. 83-100 1977

[10]C.H.LUTTERODT:"Rational approximants to holomorphic functions in n dimensions". Lecture Notes in physics 47 . Springer-Verlag p. 33-54 1976

DIFFERENT TECHNIQUES FOR THE CONSTRUCTION OF MULTIVARIATE RATIONAL INTERPOLANTS

A. CUYT AND B. VERDONK

Department of Mathematics and Computer Science
University of Antwerp (UIA)
Universiteitsplein 1, B–2610 Wilrijk, Belgium

Abstract

One approach to multivariate rational interpolation is based on the accuracy through order principle. With this approach it is possible to write down explicit formulas for the multivariate rational interpolants, to construct a recursive scheme or to find the interpolant as the convergent of a multivariate continued fraction [9].

A second and equivalent approach does not rewrite the interpolation data in a Newton series. Immediately starting from the interpolation conditions determinant formulas, a recursive computation scheme and an interpolating continued fraction can be given. These algorithms are different from the ones in [9] but they compute the same multivariate rational interpolant, given the uniqueness of this interpolant.

In view of the large number of definitions an overview of these is given in the appendix.

1. Introduction

In this section we shall briefly go over these different approaches in the univariate case in order to show how these techniques generalize to the case of multivariate rational interpolation. Let the univariate function $f(x)$ be given in the non-coincident interpolation points $\{x_0, x_1, x_2, \ldots\}$. The rational interpolation problem of order (n, m) for f consists in finding polynomials

$$p(x) = \sum_{i=0}^{n} a_i x^i$$

$$q(x) = \sum_{i=0}^{m} b_i x^i$$

with $p(x)/q(x)$ irreducible and such that

$$f(x_i) = \frac{p}{q}(x_i) \qquad i = 0, \ldots, n+m \tag{1}$$

A. Cuyt (ed.), Nonlinear Numerical Methods and Rational Approximation, 167–190.

In order to solve (1) we rewrite it as

$$f(x_i)q(x_i) - p(x_i) = 0 \qquad i = 0, \ldots, n+m \tag{2}$$

Condition (2) is a homogeneous system of $n+m+1$ linear equations in the $n+m+2$ unknown coefficients a_i and b_i of p and q and hence it has at least one nontrivial solution. It is well-known that all the solutions of (2) have the same irreducible form and we shall therefore denote by

$$r_{n,m}(x) = \frac{p_0}{q_0}(x)$$

the irreducible form of p/q with p and q satisfying (2) where q_0 is normalized according to a chosen normalization. We say that $r_{n,m}$ "interpolates" the given function and by this we mean that p_0 and q_0 satisfy some of the interpolation conditions (1). This does not imply that $r_{n,m}$ actually interpolates the given function at all the data because, by constructing the irreducible form, a common factor and hence some interpolation conditions may be cancelled in the polynomials p and q that provided $r_{n,m}$. Since $r_{n,m}$ is the irreducible form, the rational functions p/q with p and q satisfying (2) are called "equivalent". If the rank of the system (2) is maximal then $r_{n,m} = p_0/q_0 = p/q$.

The next equivalence can be proved for representations of $r_{n,m}$ if the rank of (2) is maximal.

(A) from [13]:

$$r_{n,m}(x) = \frac{\begin{vmatrix} f_0 & \Delta_0 & \ldots & \Delta_0^n & f_0\Delta_0 & \ldots & f_0\Delta_0^m \\ \vdots & \vdots & & \vdots & \vdots & & \vdots \\ f_{n+m} & \Delta_{n+m} & \ldots & \Delta_{n+m}^n & f_{n+m}\Delta_{n+m} & \ldots & f_{n+m}\Delta_{n+m}^m \end{vmatrix}}{\begin{vmatrix} 1 & \Delta_0 & \ldots & \Delta_0^n & f_0\Delta_0 & \ldots & f_0\Delta_0^m \\ \vdots & \vdots & & \vdots & \vdots & & \vdots \\ 1 & \Delta_{n+m} & \ldots & \Delta_{n+m}^n & f_{n+m}\Delta_{n+m} & \ldots & f_{n+m}\Delta_{n+m}^m \end{vmatrix}}$$

where

$$f_i = f(x_i) \qquad \Delta_i = (x - x_i) \qquad \Delta_i^j = (x - x_i)^j$$

(B) from [6]:

$$r_{n,m}(x) = \frac{\begin{vmatrix} \sum_{\ell=0}^{n} f[x_0,\ldots,x_\ell]B_\ell(x) & \ldots & \sum_{\ell=0}^{n} f[x_m,\ldots,x_\ell]B_\ell(x) \\ f[x_0,\ldots,x_{n+1}] & \ldots & f[x_m,\ldots,x_{n+1}] \\ \vdots & & \vdots \\ f[x_0,\ldots,x_{n+m}] & \ldots & f[x_m,\ldots,x_{n+m}] \end{vmatrix}}{\begin{vmatrix} B_0(x) & \ldots & B_m(x) \\ f[x_0,\ldots,x_{n+1}] & \ldots & f[x_m,\ldots,x_{n+1}] \\ \vdots & & \vdots \\ f[x_0,\ldots,x_{n+m}] & \ldots & f[x_m,\ldots,x_{n+m}] \end{vmatrix}}$$

where

$$B_i(x) = \prod_{j=1}^{i} (x - x_{j-1}) \qquad B_0(x) = 1$$

(C) from [12]:

$$r_{n,m}(x) = r_{n,m}^{(0)}(x)$$

with for $k = 0, \ldots, n-m$

$$r_{k,0}^{(j)}(x) = f[x_j] + \sum_{i=1}^{k} f[x_j,\ldots,x_{j+i}](x - x_j)\ldots(x - x_{j+i-1})$$

$$r_{0,k}^{(j)}(x) = \left[(1/f)[x_j] + \sum_{i=1}^{k}(1/f)[x_j,\ldots,x_{j+i}](x - x_j)\ldots(x - x_{j+i-1})\right]^{-1}$$

$$r_{n,m}^{(j)}(x) = \frac{\begin{array}{c}(x - x_j)\left(r_{n-1,m-1}^{(j+1)} - r_{n-1,m}^{(j)}\right) r_{n-1,m}^{(j+1)}(x) \\ -(x - x_{j+n+m})\left(r_{n-1,m-1}^{(j+1)} - r_{n-1,m}^{(j+1)}\right) r_{n-1,m}^{(j)}(x)\end{array}}{(x - x_j)\left(r_{n-1,m-1}^{(j+1)} - r_{n-1,m}^{(j)}\right) - (x - x_{j+n+m})\left(r_{n-1,m-1}^{(j+1)} - r_{n-1,m}^{(j+1)}\right)}$$

$$r_{n,m}^{(j)}(x) = \frac{\begin{array}{c}(x - x_j)\left(r_{n-1,m-1}^{(j+1)} - r_{n,m-1}^{(j)}\right) r_{n,m-1}^{(j+1)}(x) \\ -(x - x_{j+n+m})\left(r_{n-1,m-1}^{(j+1)} - r_{n,m-1}^{(j+1)}\right) r_{n,m-1}^{(j)}(x)\end{array}}{(x - x_j)\left(r_{n-1,m-1}^{(j+1)} - r_{n,m-1}^{(j)}\right) - (x - x_{j+n+m})\left(r_{n-1,m-1}^{(j+1)} - r_{n,m-1}^{(j+1)}\right)}$$

(D) from [5]:

$$r_{n,m}(x) = \epsilon_{2m}^{(n-m)}$$

with

$$\begin{aligned}
\epsilon_{-1}^{(n)} &= 0 \qquad n = 0,1,\ldots \\
\epsilon_{2m}^{(-m-1)} &= 0 \qquad m = 0,1,\ldots \\
\epsilon_{0}^{(n)} &= \sum_{i=0}^{n} f[x_0,\ldots,x_i]B_i(x) \qquad n = 0,1,\ldots \\
\epsilon_{m+1}^{(n)} &= \epsilon_{m-1}^{(n+1)} + \frac{1}{(x - x_{n+m+1})(\epsilon_m^{(n+1)} - \epsilon_m^{(n)})}
\end{aligned}$$

$$n = -\lfloor \tfrac{m}{2} \rfloor - 1, -\lfloor \tfrac{m}{2} \rfloor, \ldots \qquad m = 0,1,\ldots$$

(E) from [2, p. 280]:

$$r_{n,m}(x) = t(x) + \frac{u(x)\,|}{|\,1} + \sum_{i=1}^{2m-1} \frac{x - x_{n-m+i}\,|}{|\,\varphi^{u/(f-t)}[x_{n-m+1},\ldots,x_{n-m+1+i}]}$$

for $n \geq m$ with for $\ell \geq 1$

$$\begin{aligned}
t(x) &= \sum_{i=0}^{n-m} f[x_0,\ldots,x_i]B_i(x) \\
u(x) &= f[x_0,\ldots,x_{n-m+1}]B_{n-m+1}(x) \\
\varphi_0^{u/(f-t)}[x_0] &= [u/(f-t)](x_0) \\
\varphi_\ell^{u/(f-t)}[x_0,\ldots,x_\ell] &= \frac{x_\ell - x_{\ell-1}}{\varphi_{\ell-1}^{u/(f-t)}[x_0,\ldots,x_{\ell-2},x_\ell] - \varphi_{\ell-1}^{u/(f-t)}[x_0,\ldots,x_{\ell-1}]}
\end{aligned}$$

and for $m > n$ the reciprocal covariance of rational interpolants is used.

(F) from [4]:

$$\begin{aligned}
r_{n,m}(x) = \sum_{i=0}^{n-m} f[x_0,\ldots,x_i]B_i(x) &+ \frac{f[x_0,\ldots,x_{n-m+1}]B_{n-m+1}(x)\,|}{|\,1} + \\
&\frac{-q_1^{(n-m+1)}(x - x_{n-m+1})\,|}{|\,1 + q_1^{(n-m+1)}(x_0 - x_{n-m+1})} + \\
\sum_{i=1}^{m-1} \Bigg(\frac{-e_i^{(n-m+1)}(x - x_{n-m+2i})\,|}{|\,1 + e_i^{(n-m+1)}(x_0 - x_{n-m+2i})} &+ \frac{-q_{i+1}^{(n-m+1)}(x - x_{n-m+2i+1})\,|}{|\,1 + q_{i+1}^{(n-m+1)}(x_0 - x_{n-m+2i+1})} \Bigg)
\end{aligned}$$

for $n \geq m$ with for $\ell \geq 1$ and $k \geq 1$

$$e_0^{(k)} = 0 \qquad q_1^{(k)} = \frac{f[x_0,\ldots,x_{k+1}]}{f[x_1,\ldots,x_{k+1}]}$$

$$e_\ell^{(k)} = \frac{q_\ell^{(k+1)} - q_\ell^{(k)} + e_{\ell-1}^{(k+1)}\left[1 + q_\ell^{(k+1)}(x_0 - x_{k+2\ell-1})\right]}{1 + q_\ell^{(k)}(x_0 - x_{k+2\ell-1})}$$

$$q_{\ell+1}^{(k)} = \frac{e_\ell^{(k+1)} q_\ell^{(k+1)}\left[1 + e_\ell^{(k)}(x_0 - x_{k+2\ell})\right]}{e_\ell^{(k)}\left[1 + q_\ell^{(k+1)}(x_0 - x_{k+2\ell-1})\right] + e_\ell^{(k+1)}(e_\ell^{(k)} - q_\ell^{(k+1)})(x_0 - x_{k+2\ell+1})}$$

and for $m > n$ the reciprocal covariance of the rational interpolants is used.

The expressions in (A) and (B) use determinant formulas for the numerator and denominator coefficients of the rational interpolant $r_{n,m}$ which are obtained by solving equivalent formulations of (2) using Cramer's rule: the size of the determinants in (B) is related to the degree of the denominator while the size of the determinants in (A) is related to the number of data and hence is larger. The expressions in (C) and (D) give $r_{n,m}$ as the result of two different recursive computation schemes. The expressions in (E) and (F) each write the rational function as the convergent of an interpolating continued fraction, either by computing inverse differences or by starting a qd-like algorithm. The next three paragraphs deal with the multivariate generalization of the formulas (A) and (B), (C) and (D), (E) and (F) respectively.

2. Determinant formulas

Let the bivariate function $f(x,y)$ be given in the points (x_i, y_j) with $(i,j) \in I \subset \mathbb{N}^2$. Let us assume that none of the x_i and none of the y_j coincide. Let N and D be two finite subsets of $\mathbb{N}^2$ with which we associate the bivariate polynomials

$$p(x,y) = \sum_{(i,j)\in N} a_{ij} x^i y^j \qquad N \text{ from "numerator"} \tag{3a}$$

$$\#N = n+1$$

$$q(x,y) = \sum_{(i,j)\in D} b_{ij} x^i y^j \qquad D \text{ from "denominator"} \tag{3b}$$

$$\#D = m+1$$

The multivariate rational interpolation problem consists in finding polynomials $p(x,y)$ and $q(x,y)$ with $p(x,y)/q(x,y)$ irreducible such that

$$f(x_k, y_\ell) = \frac{p}{q}(x_k, y_\ell) \qquad (k,\ell) \in I$$

$$\#I = n+m+1$$

In order to give a generalization of the formulas (A), we require that the sets N and D satisfy the inclusion property. We say that a set N satisfies the inclusion property if for all (i,j) belonging to N also (k,ℓ) belongs to N for all $k \le i$ and $\ell \le j$. The problem of interpolating the data in I ("interpolation set") by the rational function $[N/D]_I$ ("numerator set"/"denominator set") is reformulated as follows:

$$f(x_k, y_\ell)\left(\sum_{(i,j)\in D} b_{ij} x_k^i y_\ell^j\right) - \left(\sum_{(i,j)\in N} a_{ij} x_k^i y_\ell^j\right) = 0 \quad (k,\ell) \in I \qquad (3c)$$

We shall denote the elements in N by

$$(i_0, j_0), (i_1, j_1), \ldots, (i_n, j_n)$$

the elements in D by

$$(d_0, e_0), (d_1, e_1), \ldots, (d_m, e_m)$$

the elements in I by

$$(k_0, \ell_0), \ldots, (k_{n+m}, \ell_{n+m})$$

and the function value in the data point (x_{k_i}, y_{ℓ_i}) by

$$f_i = f(x_{k_i}, y_{\ell_i}) \qquad i = 0, \ldots, n+m$$

The numbering of the points within each set follows the numbering $r(i,j)$ chosen for the points (i,j) in $\mathbb{N}^2$ in the sense that the next point in the set is the next in line from its intersection with $\mathbb{N}$. We assume throughout the text that this numbering is such that $r(k,\ell) \le r(i,j)$ for all $k \le i$ and $\ell \le j$. Several numberings $r(i,j)$ for $\mathbb{N}^2$ satisfy this condition as for instance

$$(0,0),(1,0),(0,1),(2,0),(1,1),(0,2),\ldots$$

or

$$(0,0),(1,0),(0,1),(1,1),(2,0),(2,1),(0,2),(1,2),(2,2),\ldots$$

The assumption made about the numbering $r(i,j)$ implies the following two facts for a set N satisfying the inclusion property:

1) all the subsets $N_s^{(0)} = \{(i_0, j_0), \ldots, (i_s, j_s)\}$ of N also satisfy the inclusion property

2) the element with lowest rank number in N is the origin and so $(i_0, j_0) = (0,0)$

Using this notation and assuming that the rank of the system (3c) is maximal, a solution $p(x,y)/q(x,y)$ of the set of $n+m+1$ homogeneous equations in the

$n+m+2$ unknowns a_{ij} and b_{ij} is given by

$$
-\frac{\begin{vmatrix}
x^{i_0}y^{j_0} & \cdots & x^{i_n}y^{j_n} & 0 & \cdots & 0 \\
x_{k_0}^{i_0}y_{\ell_0}^{j_0} & \cdots & x_{k_0}^{i_n}y_{\ell_0}^{j_n} & f_0 x_{k_0}^{d_0}y_{\ell_0}^{e_0} & \cdots & f_0 x_{k_0}^{d_m}y_{\ell_0}^{e_m} \\
\vdots & & \vdots & \vdots & & \vdots \\
x_{k_{n+m}}^{i_0}y_{\ell_{n+m}}^{j_0} & \cdots & x_{k_{n+m}}^{i_n}y_{\ell_{n+m}}^{j_n} & f_{n+m} x_{k_{n+m}}^{d_0}y_{\ell_{n+m}}^{e_0} & \cdots & f_{n+m} x_{k_{n+m}}^{d_m}y_{\ell_{n+m}}^{e_m}
\end{vmatrix}}{\begin{vmatrix}
0 & \cdots & 0 & x^{d_0}y^{e_0} & \cdots & x^{d_m}y^{e_m} \\
x_{k_0}^{i_0}y_{\ell_0}^{j_0} & \cdots & x_{k_0}^{i_n}y_{\ell_0}^{j_n} & f_0 x_{k_0}^{d_0}y_{\ell_0}^{e_0} & \cdots & f_0 x_{k_0}^{d_m}y_{\ell_0}^{e_m} \\
\vdots & & \vdots & \vdots & & \vdots \\
x_{k_{n+m}}^{i_0}y_{\ell_{n+m}}^{j_0} & \cdots & x_{k_{n+m}}^{i_n}y_{\ell_{n+m}}^{j_n} & f_{n+m} x_{k_{n+m}}^{d_0}y_{\ell_{n+m}}^{e_0} & \cdots & f_{n+m} x_{k_{n+m}}^{d_m}y_{\ell_{n+m}}^{e_m}
\end{vmatrix}}
$$

For a proof of this formula in the special case $n = m$ we refer to [11, p. 45]. Since the sets $N_s^{(0)}$ satisfy the inclusion property it is easy to see that for $r = 0, \ldots, n+m$ the expression $(x-x_{k_r})^{i_s}(y-y_{\ell_r})^{j_s}$ is a linear combination of the $(s+1)$ first columns in the determinants above:

$$(x - x_{k_r})^{i_s}(y - y_{l_r})^{j_s} = \sum_{v=0}^{i_s}\sum_{w=0}^{j_s}(-1)^{v+w}\binom{i_s}{v}\binom{j_s}{w}x^{i_s-v}y^{j_s-w}x_{k_r}^{v}y_{l_r}^{w}$$

Since the sets $D_s^{(0)}$ also satisfy the inclusion property, we can take analogous linear combinations of the columns $n+2$ through $n+m+2$ and rewrite $p(x,y)/q(x,y)$ as

$$
\frac{\begin{vmatrix}
f_0 & \cdots & f_{n+m} \\
(x-x_{k_0})^{i_1}(y-y_{\ell_0})^{j_1} & \cdots & (x-x_{k_{n+m}})^{i_1}(y-y_{\ell_{n+m}})^{j_1} \\
\vdots & & \vdots \\
(x-x_{k_0})^{i_n}(y-y_{\ell_0})^{j_n} & \cdots & (x-x_{k_{n+m}})^{i_n}(y-y_{\ell_{n+m}})^{j_n} \\
f_0(x-x_{k_0})^{d_1}(y-y_{\ell_0})^{e_1} & \cdots & f_{n+m}(x-x_{k_{n+m}})^{d_1}(y-y_{\ell_{n+m}})^{e_1} \\
\vdots & & \vdots \\
f_0(x-x_{k_0})^{d_m}(y-y_{\ell_0})^{e_m} & \cdots & f_{n+m}(x-x_{k_{n+m}})^{d_m}(y-y_{\ell_{n+m}})^{e_m}
\end{vmatrix}}{\begin{vmatrix}
1 & \cdots & 1 \\
(x-x_{k_0})^{i_1}(y-y_{\ell_0})^{j_1} & \cdots & (x-x_{k_{n+m}})^{i_1}(y-y_{\ell_{n+m}})^{j_1} \\
\vdots & & \vdots \\
(x-x_{k_0})^{i_n}(y-y_{\ell_0})^{j_n} & \cdots & (x-x_{k_{n+m}})^{i_n}(y-y_{\ell_{n+m}})^{j_n} \\
f_0(x-x_{k_0})^{d_1}(y-y_{\ell_0})^{e_1} & \cdots & f_{n+m}(x-x_{k_{n+m}})^{d_1}(y-y_{\ell_{n+m}})^{e_1} \\
\vdots & & \vdots \\
f_0(x-x_{k_0})^{d_m}(y-y_{\ell_0})^{e_m} & \cdots & f_{n+m}(x-x_{k_{n+m}})^{d_m}(y-y_{\ell_{n+m}})^{e_m}
\end{vmatrix}} \tag{4}
$$

where $(i_0, j_0) = (0,0) = (d_0, e_0)$. This formula is a bivariate analogon of the representation (A).

In order to give a generalization of the formulas (B) we switch to the basis functions

$$B_{ij}(x,y) = \prod_{k=1}^{i}(x - x_{k-1})\prod_{\ell=1}^{j}(y - y_{\ell-1})$$

and to the situation where the interpolation set I satisfies the inclusion property. We can then construct the interpolating Newton series

$$f(x,y) = \sum_{(i,j)\in I} f[x_0,\ldots,x_i][y_0,\ldots,y_j]B_{ij}(x,y) + \sum_{(i,j)\in \mathbb{N}^2\setminus I} c_{ij}B_{ij}(x,y)$$

where the bivariate divided differences are computed as in [10] from the data $f(x_i, y_j)$. The problem of interpolating the data in I by the rational function $[N/D]_I$ is reformulated as follows. Find polynomials

$$\tilde{p}(x,y) = \sum_{(i,j)\in N} \tilde{a}_{ij}B_{ij}(x,y) \tag{5a}$$

$$\#N = n+1$$

$$\tilde{q}(x,y) = \sum_{(i,j)\in D} \tilde{b}_{ij}B_{ij}(x,y) \tag{5b}$$

$$\#D = m+1$$

such that

$$(f\tilde{q} - \tilde{p})(x,y) = \sum_{(i,j)\in \mathbb{N}^2\setminus I} d_{ij}B_{ij}(x,y) \tag{5c}$$

$$\#I = n+m+1$$

Condition (5c) immediately implies the interpolation conditions

$$(f\tilde{q} - \tilde{p})(x_k, y_\ell) = 0 \qquad (k,\ell) \in I$$

Assuming that I satisfies the inclusion property and that the numerator set N is a subset of I, we can give a multivariate generalization of the formulas (B). The condition $N \subset I$ is required in order to be able to decouple the set of equations (5c) in a homogeneous and a non-homogeneous set. In [10] is shown that if the subsystem of homogeneous equations from (5c) has maximal rank m, then the solution $p(x,y)/q(x,y)$ of (5) is unique and is given by

$$\frac{\begin{vmatrix} \sum_{(i,j)\in N} f[x_{d_0}\dots x_i][y_{e_0}\dots y_j]B_{ij} & \dots & \sum_{(i,j)\in N} f[x_{d_m}\dots x_i][y_{e_m}\dots y_j]B_{ij} \\ f[x_{d_0}\dots x_{k_{n+1}}][y_{e_0}\dots y_{\ell_{n+1}}] & \dots & f[x_{d_m}\dots x_{k_{n+1}}][y_{e_m}\dots y_{\ell_{n+1}}] \\ \vdots & & \vdots \\ f[x_{d_0}\dots x_{k_{n+m}}][y_{e_0}\dots y_{\ell_{n+m}}] & \dots & f[x_{d_m}\dots x_{k_{n+m}}][y_{e_m}\dots y_{\ell_{n+m}}] \end{vmatrix}}{\begin{vmatrix} B_{d_0e_0} & \dots & B_{d_me_m} \\ f[x_{d_0}\dots x_{k_{n+1}}][y_{e_0}\dots y_{\ell_{n+1}}] & \dots & f[x_{d_m}\dots x_{k_{n+1}}][y_{e_m}\dots y_{\ell_{n+1}}] \\ \vdots & & \vdots \\ f[x_{d_0}\dots x_{k_{n+m}}][y_{e_0}\dots y_{\ell_{n+m}}] & \dots & f[x_{d_m}\dots x_{k_{n+m}}][y_{e_m}\dots y_{\ell_{n+m}}] \end{vmatrix}} \quad (6)$$

We shall now show that for $N \subset I$, D and I satisfying the inclusion property, the conditions (3) and (5) are equivalent and hence that, if we assume unicity of the solution, formulas (4) and (6) are just two different ways to write the same multivariate rational interpolant. Since N and D satisfy the inclusion property we can write the polynomials $p(x,y)$ and $q(x,y)$ given by (3a) and (3b) that satisfy (3c) in the form

$$\sum_{(i,j)\in N} a_{ij}x^iy^j = \sum_{(i,j)\in N} \tilde{a}_{ij}B_{ij}(x,y) \quad (7a)$$

$$\sum_{(i,j)\in D} b_{ij}x^iy^j = \sum_{(i,j)\in D} \tilde{b}_{ij}B_{ij}(x,y) \quad (7b)$$

Vice versa, the polynomials $\tilde{p}(x,y)$ and $\tilde{q}(x,y)$ given by (5a) and (5b) that satisfy (5c) can be written in the form (3a) and (3b). Hence it is clear that the interpolation conditions (5c) imply the conditions (3c). It remains to be shown that polynomials $p(x,y)$ and $q(x,y)$ of "degree" N and D respectively which satisfy (3c) also satisfy (5c). Since I satisfies the inclusion property, it has a structure similar to the picture below

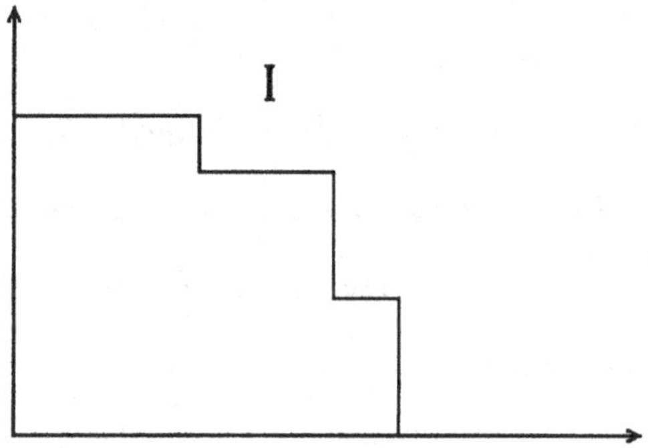

Let us now for simplicity, without loss of generality, assume that I has the structure

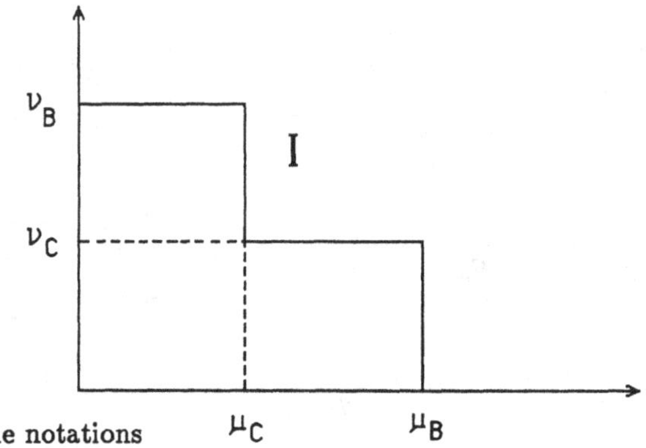

We can then introduce the notations

$$\nu_B = \max\{i \mid \exists j : (i,j) \in I\} \qquad \text{(B from "border")}$$
$$\mu_B = \max\{j \mid \exists i : (i,j) \in I\}$$
$$\nu_C = \max\{i \mid \forall\, 0 \leq j \leq \mu_B : (i,j) \in I\} \qquad \text{(C from "corner")}$$
$$\mu_C = \max\{j \mid \forall\, 0 \leq i \leq \nu_B : (i,j) \in I\}$$

The set I can be subdivided in the subsets

$$I_1 = \{(i,j) \in I \mid 0 \leq i \leq \nu_B \quad 0 \leq j \leq \mu_C\}$$

$$I_2 = \{(i,j) \in I \mid 0 \leq i \leq \nu_C \quad 0 \leq j \leq \mu_B\}$$

and we can consider the data set as being a union of data sets for the univariate functions $f(x, y_\ell)$ for $\ell = 0, \ldots, \mu_B$. For ℓ fixed and such that $0 \leq \ell \leq \mu_C$, the interpolation conditions

$$(fq - p)(x_k, y_\ell) = 0 \qquad 0 \leq k \leq \nu_B$$

are equivalent to [6]

$$(fq - p)[x_0, \ldots, x_k][y_\ell] = 0 \qquad 0 \leq k \leq \nu_B \tag{8a}$$

and for ℓ fixed and such that $\mu_C < \ell \leq \mu_B$, the interpolation conditions

$$(fq - p)(x_k, y_\ell) = 0 \qquad 0 \leq k \leq \nu_C$$

are equivalent to

$$(fq - p)[x_0, \ldots, x_k][y_\ell] = 0 \qquad 0 \leq k \leq \nu_C \tag{8b}$$

The conditions (8a) imply

$$(fq-p)[x_0,\dots,x_k][y_0,\dots,y_\ell]=0 \qquad (k,\ell)\in I_1$$

while the conditions (8a) and (8b) together imply

$$(fq-p)[x_0,\dots,x_k][y_0,\dots,y_\ell]=0 \qquad (k,\ell)\in I_2$$

In other words, the conditions (3c) imply

$$(fq-p)[x_0,\dots,x_k][y_0,\dots,y_\ell]=0 \qquad (k,\ell)\in I=I_1\cup I_2$$

which amounts to

$$(fq-p)(x,y)=\sum_{(i,j)\in \mathbb{N}^2\setminus I} d_{ij}B_{ij}(x,y)$$

Since p and q can be written in the form (5a) and (5b) respectively we also have

$$(f\tilde{q}-\tilde{p})(x,y)=\sum_{(i,j)\in \mathbb{N}^2\setminus I} d_{ij}B_{ij}(x,y)$$

3. Recursive formulas

The univariate recursive scheme (D) was developed especially for the calculation of the rational interpolants as formulated in (B). In [8] it is shown that the E-algorithm [3], of which (D) is a special case, is suitable for the computation of the multivariate generalization (6) of (B) which we gave in the previous section and which exists if $N\subset I$. We shall show here that the E-algorithm is also suitable for the computation of the multivariate generalization (4) of (A), which we gave at the beginning of the previous section and which can be written down also for $N\setminus I\neq\emptyset$. Remember that the E-algorithm computes a quotient of determinants of the following form

$$E_k^{(j)}=E_k(S_j)=\frac{\begin{vmatrix} S_j & \dots & S_{j+k}\\ g_1(j) & \dots & g_1(j+k)\\ \vdots & & \vdots\\ g_k(j) & \dots & g_k(j+k)\end{vmatrix}}{\begin{vmatrix} 1 & \dots & 1\\ g_1(j) & \dots & g_1(j+k)\\ \vdots & & \vdots\\ g_k(j) & \dots & g_k(j+k)\end{vmatrix}} \tag{9}$$

From the previous section we can see that $p/q(x,y)$ given by (4) is of this form if we define for $r = 0,\dots,n+m$ and $s = 0,\dots,\min(n,m)$

$$\begin{aligned} S_r &= f_r \\ g_{2s-1}(r) &= (x - x_{k_r})^{i_s}(y - y_{\ell_r})^{j_s} \\ g_{2s}(r) &= f_r(x - x_{k_r})^{d_s}(y - y_{\ell_r})^{e_s} \end{aligned} \tag{10a}$$

and for $s = \min(n,m)+1,\dots,\max(n,m)$

$$\begin{aligned} g_{m+s}(r) &= (x - x_{k_r})^{i_s}(y - y_{\ell_r})^{j_s} \qquad \text{if } n \geq m \\ g_{n+s}(r) &= f_r(x - x_{k_r})^{d_s}(y - y_{\ell_r})^{e_s} \qquad \text{if } n < m \end{aligned} \tag{10b}$$

where $(i_s, j_s) \in N$ and $(d_s, e_s) \in D$. Then

$$\frac{p}{q}(x,y) = [N/D]_{I_{n+m}^{(0)}} = E_{n+m}^{(0)}$$

We remark that the rows in the numerator and denominator determinants of (4) have been permuted before defining the functions $g_s(r)$. To compute the multivariate rational interpolants recursively, the E-algorithm can now be applied. For $j = 0,\dots,n+m$ we have

$$\begin{aligned} E_0^{(j)} &= f_j \\ g_{0,k}^{(j)} &= g_k(j) \qquad k = 1,\dots,n+m \\ E_k^{(j)} &= \frac{g_{k-1,k}^{(j+1)} E_{k-1}^{(j)} - g_{k-1,k}^{(j)} E_{k-1}^{(j+1)}}{g_{k-1,k}^{(j+1)} - g_{k-1,k}^{(j)}} \qquad k = 1,\dots,n+m \\ g_{k,i}^{(j)} &= \frac{g_{k-1,i}^{(j)} g_{k-1,k}^{(j+1)} - g_{k-1,i}^{(j+1)} g_{k-1,k}^{(j)}}{g_{k-1,k}^{(j+1)} - g_{k-1,k}^{(j)}} \qquad i = k+1, k+2, \dots \end{aligned}$$

The E-values are stored in a table as follows

$$\begin{array}{lll} f_0 = E_0^{(0)} = [N_0/D_0]_{I_0^{(0)}} & & \\ & E_1^{(0)} = [N_1/D_0]_{I_1^{(0)}} & \\ f_1 = E_0^{(1)} = [N_0/D_0]_{I_0^{(1)}} & & \ddots \\ & E_1^{(1)} = [N_1/D_0]_{I_1^{(1)}} & \\ f_2 = E_0^{(2)} = [N_0/D_0]_{I_0^{(2)}} & \vdots & E_{n+m}^{(0)} = [N_n/D_m]_{I_{n+m}^{(0)}} \\ \vdots & & \\ & & \cdot^{\cdot^{\cdot}} \\ & E_1^{(n+m-1)} = [N_1/D_0]_{I_1^{(n+m-1)}} & \\ f_{n+m} = E_0^{(n+m)} = [N_0/D_0]_{I_0^{(n+m)}} & & \end{array}$$

The auxiliary entries $g_{k,i}^{(j)}$ of the E-algorithm are also quotients of determinants

$$g_{k,i}^{(j)} = \frac{d_{k,i}^{(j)}}{d_{k,-1}^{(j)}} \qquad i = k+1, k+2, \ldots \tag{11}$$

with

$$d_{k,i}^{(j)} = \begin{vmatrix} g_i(j) & \ldots & g_i(j+k) \\ g_1(j) & \ldots & g_1(j+k) \\ \vdots & & \vdots \\ g_k(j) & \ldots & g_k(j+k) \end{vmatrix} \qquad i = -1, 0, k+1, k+2, \ldots \tag{12}$$

The starting values for the E-algorithm are the function values in the data points while intermediate E-values $E_s^{(j)}$ are rational functions interpolating on subsets

$$I_s^{(j)} = \{(k_j, l_j), \ldots, (k_{j+s}, l_{j+s})\}$$

Along a column the "degree" of numerator and denominator is constant and completely determined by the functions $g_s(r)$ appearing in the determinant expressions at that stage. Due to the row permutations performed above, advancing along a diagonal in the E-table alternatively increases the numerator and denominator index-sets until one of the sets is exhausted. Definition (10) of the functions $g_s(r)$ enables us to profit from the recursive scheme as much as possible when it has to be restarted for other sets N and D, since many intermediate values can usually be retained.

We shall now show the link between the E-algorithm for the recursive computation of the multivariate rational interpolants (4) and the univariate recursive scheme (C) for the computation of $r_{n,m}^{(j)}$. Let us assume for simplicity, but without loss of generality, that $n = m$. We remark that the univariate determinant formula (A) for $r_{n,n}^{(j)}$ is of the form (9) too:

$$r_{n,n} = r_{n,n}^{(0)} = E_{2n}^{(0)}$$

In this case the functions $g_s(r)$ appearing in formula (9) are univariate. We have for $r = 0, \ldots, 2n$ and $s = 0, \ldots, n$

$$\begin{aligned} S_r &= f_r = f(x_r) \\ g_{2s-1}(r) &= (x - x_r)^s \\ g_{2s}(r) &= f_r(x - x_r)^s \end{aligned} \tag{13}$$

which is analogous to the definition of the multivariate functions $g_s(r)$. Again we have first permuted the rows in the numerator and denominator determinants of (A). We can now apply the E-algorithm to compute $r_{n,n}^{(j)} = E_{2n}^{(j)}$ recursively:

$$r_{n,n}^{(j)} = \frac{g_{2n-1,2n}^{(j+1)} E_{2n-1}^{(j)} - g_{2n-1,2n}^{(j)} E_{2n-1}^{(j+1)}}{g_{2n-1,2n}^{(j+1)} - g_{2n-1,2n}^{(j)}}$$

Taking into account definition (13) of the univariate functions $g_s(r)$ we have that

$$E_{2n-1}^{(j)} = r_{n,n-1}^{(j)}$$

and so

$$r_{n,n}^{(j)} = \frac{g_{2n-1,2n}^{(j+1)} r_{n,n-1}^{(j)} - g_{2n-1,2n}^{(j)} r_{n,n-1}^{(j+1)}}{g_{2n-1,2n}^{(j+1)} - g_{2n-1,2n}^{(j)}}$$

This formula is similar in structure to the formulas (C) and the next theorem will show that applying the E-algorithm to $r_{n,n}^{(j)}$ is equivalent to using the Bulirsch-Stoer formula.

LEMMA 1:
With the functions $g_s(r)$ defined by (13) we have for $j \geq 0$ and $n \geq 1$

$$g_{2n-1,2n}^{(j)} = -\frac{d_{2n-2,-1}^{(j+1)}}{e_{2n-2,2n-3}^{(j+1)}} (x - x_j)(r_{n-1,n-1}^{(j+1)} - r_{n,n-1}^{(j)})$$

$$g_{2n-1,2n}^{(j+1)} = -\frac{d_{2n-2,-1}^{(j+1)}}{e_{2n-2,2n-3}^{(j+1)}} (x - x_{j+2n})(r_{n-1,n-1}^{(j+1)} - r_{n,n-1}^{(j+1)})$$

where

$$e_{k,i}^{(j)} = \begin{vmatrix} 1 & \dots & 1 \\ f_j & \dots & f_{j+k} \\ g_1(j) & \dots & g_1(j+k) \\ \vdots & & \vdots \\ g_{k-2}(j) & \dots & g_{k-2}(j+k) \\ g_i(j) & \dots & g_i(j+k) \end{vmatrix}$$

PROOF:
From the determinant expression for $d_{k,i}^{(j)}$ it is easy to see that, according to (A)

$$r_{\lfloor \frac{k+1}{2} \rfloor, \lfloor \frac{k}{2} \rfloor}^{(j)} = E_k^{(j)} = \frac{d_{k,0}^{(j)}}{d_{k,-1}^{(j)}}$$

where the functions $g_s(r)$ in $d_{k,i}^{(j)}$ are given by (13). From (13) we see that

$$g_s(r) = (x - x_r)g_{s-2}(r) \qquad s = 1, \ldots, n$$

and hence we can write

$$g_{2n-1,2n}^{(j)} = \frac{d_{2n-1,2n}^{(j)}}{d_{2n-1,-1}^{(j)}} = -\frac{(x-x_j)\ldots(x-x_{j+2n-1})}{d_{2n-1,-1}^{(j)}} e_{2n-1,2n-2}^{(j)}$$

Applying Sylvester's identity [1] to $e_{2n-1,2n-2}^{(j)}$ we find

$$g_{2n-1,2n}^{(j)} = \frac{(x-x_j)\ldots(x-x_{j+2n-1})}{d_{2n-1,-1}^{(j)} d_{2n-3,2n-2}^{(j+1)}} \left(d_{2n-2,0}^{(j+1)} d_{2n-2,-1}^{(j)} - d_{2n-2,0}^{(j)} d_{2n-2,-1}^{(j+1)}\right) \quad (14)$$

On the other hand we have

$$r_{n-1,n-1}^{(j+1)} - r_{n,n-1}^{(j)} = \frac{d_{2n-2,0}^{(j+1)} d_{2n-1,-1}^{(j)} - d_{2n-1,0}^{(j)} d_{2n-2,-1}^{(j+1)}}{d_{2n-2,-1}^{(j+1)} d_{2n-1,-1}^{(j)}}$$

Applying Sylvester's identity to $d_{2n-1,0}^{(j)}$ and $d_{2n-1,-1}^{(j)}$ we find

$$\begin{aligned} & r_{n-1,n-1}^{(j+1)} - r_{n,n-1}^{(j)} \\ &= -\frac{d_{2n-2,2n-1}^{(j+1)}}{d_{2n-2,-1}^{(j+1)} d_{2n-1,-1}^{(j)} d_{2n-3,2n-2}^{(j+1)}} \left(d_{2n-2,0}^{(j+1)} d_{2n-2,-1}^{(j)} - d_{2n-2,0}^{(j)} d_{2n-2,-1}^{(j+1)}\right) \\ &= -\frac{(x-x_{j+1})\ldots(x-x_{j+2n-1}) e_{2n-2,2n-3}^{(j+1)}}{d_{2n-2,-1}^{(j+1)} d_{2n-1,-1}^{(j)} d_{2n-3,2n-2}^{(j+1)}} \left(d_{2n-2,0}^{(j+1)} d_{2n-2,-1}^{(j)} - d_{2n-2,0}^{(j)} d_{2n-2,-1}^{(j+1)}\right) \end{aligned}$$

and so

$$(x - x_j)\left(r_{n-1,n-1}^{(j+1)} - r_{n,n-1}^{(j)}\right) = -\frac{e_{2n-2,2n-3}^{(j+1)}}{d_{2n-2,-1}^{(j+1)}} g_{2n-1,2n}^{(j)}$$

Applying an analogous reasoning to $r_{n-1,n-1}^{(j+1)} - r_{n,n-1}^{(j+1)}$ we find

$$\begin{aligned} & r_{n-1,n-1}^{(j+1)} - r_{n,n-1}^{(j+1)} \\ &= -\frac{d_{2n-2,2n-1}^{(j+1)}}{d_{2n-2,-1}^{(j+1)} d_{2n-1,-1}^{(j+1)} d_{2n-3,2n-2}^{(j+2)}} \left(d_{2n-2,0}^{(j+2)} d_{2n-2,-1}^{(j+1)} - d_{2n-2,0}^{(j+1)} d_{2n-2,-1}^{(j+2)}\right) \\ &= -\frac{(x-x_{j+1})\ldots(x-x_{j+2n-1}) e_{2n-2,2n-3}^{(j+1)}}{d_{2n-2,-1}^{(j+1)} d_{2n-1,-1}^{(j+1)} d_{2n-3,2n-2}^{(j+2)}} \left(d_{2n-2,0}^{(j+2)} d_{2n-2,-1}^{(j+1)} - d_{2n-2,0}^{(j+1)} d_{2n-2,-1}^{(j+2)}\right) \end{aligned}$$

Comparing this equality to (14) we find

$$(x - x_{j+2n})(r^{(j+1)}_{n-1,n-1} - r^{(j+1)}_{n,n-1}) = -\frac{e^{(j+1)}_{2n-2,2n-3}}{d^{(j+1)}_{2n-2,-1}} g^{(j+1)}_{2n-1,2n}$$

which concludes the proof. ■

In this section we have pointed out how the multivariate rational interpolants defined by the system of linear equations (3) can be computed recursively by means of the E-algorithm and how this relates to the univariate Bulirsch-stoer scheme. From the table of E-values it is clear that in order to compute $[N/D]_{I^{(0)}_{n+m}}$ given by (4) we need to compute $n + m + 1$ columns in the E-table. When the E-algorithm is applied to the quotient of determinants given by (6) only $m+1$ columns need to be computed [8] since the size of the determinants in (6) is related to the size of the denominator index set D. However, this does not imply that the E-algorithm applied to (6) takes less computation because its starting values are bivariate interpolating polynomials instead of function values.

In the next section we shall show that the multivariate rational interpolants (4) can also be obtained as the convergent of an ordinary interpolating continued fraction.

4. Interpolating continued fractions

The generalized qd-algorithm was again developed especially for the calculation of the rational interpolants as formulated in (B). In [7] is shown that a very similar algorithm is valid for the multivariate generalization of (B). We will now turn to a multivariate generalization of the formulas (E). The well-known explicit determinant formulas [2, p. 104] for the partial denominators in the univariate Thiele interpolating continued fraction

$$\varphi[x_j, \ldots, x_{j+k}] = \rho[x_j, \ldots, x_{j+k}] - \rho[x_j, \ldots, x_{j+k-2}]$$

can by a suitable combination of the rows be rewritten as

$\rho[x_j, \ldots, x_{j+2k-1}] =$

$$-\frac{\begin{vmatrix} 1 & f_j & (x-x_j) & f_j(x-x_j) & \ldots & (x-x_j)^{k-1} & (x-x_j)^k \\ \vdots & \vdots & & & & \vdots & \vdots \\ 1 & f_{j+2k-1} & \ldots & & & & \end{vmatrix}}{\begin{vmatrix} 1 & f_j & (x-x_j) & f_j(x-x_j) & \ldots & (x-x_j)^{k-1} & f_j(x-x_j)^{k-1} \\ \vdots & \vdots & & & & \vdots & \vdots \\ 1 & f_{j+2k-1} & \ldots & & & & \end{vmatrix}}$$

$\rho[x_j, \ldots, x_{j+2k}] =$

$$\frac{\begin{vmatrix} 1 & f_j & (x-x_j) & \ldots & (x-x_j)^{k-1} & f_j(x-x_j)^{k-1} & f_j(x-x_j)^k \\ \vdots & \vdots & & & & & \vdots \\ 1 & f_{j+2k} & (x-x_{j+2k}) & \ldots & & & f_{j+2k}(x-x_{j+2k})^k \end{vmatrix}}{\begin{vmatrix} 1 & f_j & (x-x_j) & \ldots & (x-x_j)^{k-1} & f_j(x-x_j)^{k-1} & (x-x_j)^k \\ \vdots & \vdots & & & & & \vdots \\ 1 & f_{j+2k} & (x-x_{j+2k}) & \ldots & & & (x-x_{j+2k})^k \end{vmatrix}}$$

If we consider the univariate functions $g_s(r)$ given by (13) then it is easy to see that the above formulas for the reciprocal differences can be joined into

$$\rho[x_j, \ldots, x_{j+r}] = (-1)^r \frac{\begin{vmatrix} g_{-1}(j) & \ldots & g_{-1}(j+r) \\ g_0(j) & \ldots & g_0(j+r) \\ g_1(j) & \ldots & g_1(j+r) \\ \vdots & & \vdots \\ g_{r-3}(j) & & \\ g_{r-2}(j) & & \\ g_r(j) & \ldots & g_r(j+r) \end{vmatrix}}{\begin{vmatrix} g_{-1}(j) & \ldots & g_{-1}(j+r) \\ g_0(j) & \ldots & g_0(j+r) \\ g_1(j) & \ldots & g_1(j+r) \\ \vdots & & \vdots \\ g_{r-3}(j) & & \\ g_{r-2}(j) & & \\ g_{r-1}(j) & \ldots & g_{r-1}(j+r) \end{vmatrix}} \tag{15}$$

Using the notation from section 3

$$e_{r,i}^{(j)} = \begin{vmatrix} g_{-1}(j) & \ldots & g_{-1}(j+r) \\ g_0(j) & \ldots & g_0(j+r) \\ g_1(j) & \ldots & g_1(j+r) \\ \vdots & & \vdots \\ g_{i-2}(j) & & \\ g_i(j) & \ldots & g_i(j+r) \end{vmatrix} \qquad i = r-1, r, \ldots \tag{16}$$

we have

$$\rho[x_j, \ldots, x_{j+r}] = (-1)^r \frac{e_{r,r}^{(j)}}{e_{r,r-1}^{(j)}}$$

A generalization of formula (15) to the multivariate case is now at hand if we take for the functions $g_s(r)$ appearing in (15) the bivariate analogon of the univariate functions (13). For $r = 0, 1, \ldots$ and $s = 0, 1, \ldots$ we let

$$\begin{aligned} g_{2s-1}(r) &= (x - x_{k_r})^{i_s}(y - y_{\ell_r})^{j_s} \qquad (i_s, j_s) \in N \\ g_{2s}(r) &= f_r(x - x_{k_r})^{d_s}(y - y_{\ell_r})^{e_s} \qquad (d_s, e_s) \in D \end{aligned} \tag{17}$$

With the functions $g_s(r)$ given as above we define a bivariate reciprocal difference as follows

$$\rho[(x_{k_0}, y_{\ell_0}), \ldots, (x_{k_r}, y_{\ell_r})] = (-1)^r \frac{e_{r,r}^{(0)}}{e_{r,r-1}^{(0)}} \tag{18}$$

One can easily see, by making a suitable combination of the rows, that the numerator and the denominator of the bivariate reciprocal difference are independent of x and y if the sets N and D satisfy the inclusion property. Definition (17) for the bivariate functions $g_s(r)$ is a special case of the general definition (10) in that we require

$$\#N = \#D \quad (r \text{ even}) \quad \text{or} \quad \#N = \#D + 1 \quad (r \text{ odd})$$

In order to give a multivariate generalization of the formulas (E) we shall therefore assume in the sequel of this section that

$$n = m \qquad \text{or} \qquad n = m + 1$$

which means that our multivariate rational interpolant $[N/D]_I$ is located on the main staircase

$$\begin{array}{lll} [N_0^{(0)}/D_0^{(0)}]_{I_0^{(0)}} & & \\ [N_1^{(0)}/D_0^{(0)}]_{I_1^{(0)}} & [N_1^{(0)}/D_1^{(0)}]_{I_2^{(0)}} & \\ & [N_2^{(0)}/D_1^{(0)}]_{I_3^{(0)}} & \cdots \\ & & \vdots \end{array}$$

in the "table" of multivariate rational interpolants. We would like to point out that this is not a restriction. As in the univariate case it is possible to compute rational interpolants on other staircases by constructing a rational interpolant on the main staircase for a function related to f. The reciprocal differences are, as in the univariate case, independent of the order of the points and they can be computed recursively. In the bivariate case the recurrence relations are

$$\rho[(x_{k_r}, y_{\ell_r})] = f_r \qquad r = 0,1,\ldots$$
$$\rho[(x_{k_0}, y_{\ell_0}),\ldots,(x_{k_r}, y_{\ell_r})] = (-1)^r h^{(0)}_{r,r} \qquad r = 1,2,\ldots$$

with

$$h^{(j)}_{r,i} = \frac{e^{(j)}_{r,i}}{e^{(j)}_{r,r-1}} = \frac{h^{(j)}_{r-1,i} - h^{(j+1)}_{r-1,i}}{h^{(j)}_{r-1,r-1} - h^{(j+1)}_{r-1,r-1}} \qquad i = r, r+1, \ldots \tag{19}$$

This last identity can easily be proved by applying Sylvester's identity to the numerator and denominator of $h^{(j)}_{r,i}$ where we discard the first and last column and the last two rows. The values $h^{(j)}_{r,i}$ take the same amount of storage as the values $g^{(j)}_{r,i}$ used in the E-algorithm. In the following lemma we give a relation between the values $g^{(j)}_{r,i}$, $h^{(j)}_{r,i}$ and $E^{(j)}_r$ which will turn out to be useful when generalizing the Thiele interpolating continued fraction (E) to the multivariate case.

LEMMA 2:
We have for $r \geq 2$ and $j \geq 0$

$$\frac{E^{(j)}_{r-1} - E^{(j)}_r}{E^{(j)}_r - E^{(j)}_{r-2}} = \frac{g^{(j)}_{r-1,r}}{g^{(j)}_{r-2,r-1} h^{(j)}_{r,r} - g^{(j)}_{r-2,r}}$$

PROOF:
Applying Schwein's identity [2, p. 43] to $e^{(j)}_{r,r}$ we find

$$e^{(j)}_{r,r} = (-1)^r \begin{vmatrix} g_r(j) & \cdots & g_r(j+r) \\ g_{-1}(j) & \cdots & g_{-1}(j+r) \\ g_0(j) & \cdots & g_0(j+r) \\ g_1(j) & \cdots & g_1(j+r) \\ \vdots & & \vdots \\ g_{r-2}(j) & \cdots & g_{r-2}(j+r) \end{vmatrix} \tag{20}$$
$$= \frac{1}{d^{(j)}_{r-1,0}} (d^{(j)}_{r,0}\, e^{(j)}_{r-1,r-2} - (-1)^r e^{(j)}_{r,r-1}\, f^{(j)}_{r-1,r})$$

where the values $d^{(j)}_{r,i}$ are given by (12) and

$$f^{(j)}_{r,i} = \begin{vmatrix} g_i(j) & \cdots & g_i(j+r) \\ g_0(j) & \cdots & g_0(j+r) \\ g_1(j) & \cdots & g_1(j+r) \\ \vdots & & \vdots \\ g_{r-1}(j) & \cdots & g_{r-1}(j+r) \end{vmatrix}$$

Again applying Schwein's identity, now to $f^{(j)}_{r-1,r}$, we have

$$e^{(j)}_{r,r} = \frac{1}{d^{(j)}_{r-1,0}} \left(d^{(j)}_{r,0}\, e^{(j)}_{r-1,r-2} - \frac{e^{(j)}_{r,r-1}}{d^{(j)}_{r-2,r-1}} \left(d^{(j)}_{r-1,r} d^{(j)}_{r-2,0} - d^{(j)}_{r-2,r} d^{(j)}_{r-1,0} \right) \right) \tag{21}$$

If we permute the second and third rows in the determinant expression (20) for $e^{(j)}_{r,r}$ we find in a completely analogous way

$$e^{(j)}_{r,r} = \frac{1}{d^{(j)}_{r-1,-1}} \left(d^{(j)}_{r,-1}\, e^{(j)}_{r-1,r-2} - \frac{e^{(j)}_{r,r-1}}{d^{(j)}_{r-2,r-1}} \left(d^{(j)}_{r-1,r} d^{(j)}_{r-2,-1} - d^{(j)}_{r-2,r} d^{(j)}_{r-1,-1} \right) \right) \tag{22}$$

Using the fact that

$$E^{(j)}_r = \frac{d^{(j)}_{r,0}}{d^{(j)}_{r,-1}}$$

we get from (21) and (22) that

$$E^{(j)}_r = \frac{d^{(j)}_{r-2,r-1}\, h^{(j)}_{r,r}\, d^{(j)}_{r-1,0} + d^{(j)}_{r-1,r} d^{(j)}_{r-2,0} - d^{(j)}_{r-2,r} d^{(j)}_{r-1,0}}{d^{(j)}_{r-2,r-1}\, h^{(j)}_{r,r}\, d^{(j)}_{r-1,-1} + d^{(j)}_{r-1,r} d^{(j)}_{r-2,-1} - d^{(j)}_{r-2,r} d^{(j)}_{r-1,-1}}$$

$$= \frac{g^{(j)}_{r-2,r-1}\, h^{(j)}_{r,r}\, E^{(j)}_{r-1} + g^{(j)}_{r-1,r} E^{(j)}_{r-2} - g^{(j)}_{r-2,r} E^{(j)}_{r-1}}{g^{(j)}_{r-2,r-1}\, h^{(j)}_{r,r} + g^{(j)}_{r-1,r} - g^{(j)}_{r-2,r}}$$

where $h^{(j)}_{r,i}$ is given by (19) and $g^{(j)}_{r,i}$ by (11). In this way

$$\begin{aligned} g^{(j)}_{r-2,r-1} h^{(j)}_{r,r} \left(E^{(j)}_r - E^{(j)}_{r-1}\right) &= g^{(j)}_{r-1,r} E^{(j)}_{r-2} - g^{(j)}_{r-2,r} E^{(j)}_{r-1} - \left(g^{(j)}_{r-1,r} - g^{(j)}_{r-2,r}\right) E^{(j)}_r \\ &= g^{(j)}_{r-2,r}\left(E^{(j)}_r - E^{(j)}_{r-1}\right) - g^{(j)}_{r-1,r}\left(E^{(j)}_r - E^{(j)}_{r-2}\right) \end{aligned}$$

which completes the proof of the lemma. ■

In order to generalize the formulas (E) we still have to define multivariate inverse differences:

$$\begin{aligned} &\varphi[(x_{k_r}, y_{\ell_r})] = f_r \\ &\varphi[(x_{k_0}, y_{\ell_0}), (x_{k_1}, y_{\ell_1})] = \rho[(x_{k_0}, y_{\ell_0}), (x_{k_1}, y_{\ell_1})] \\ &\varphi[(x_{k_0}, y_{\ell_0}), \ldots, (x_{k_r}, y_{\ell_r})] = \rho[(x_{k_0}, y_{\ell_0}), \ldots, (x_{k_r}, y_{\ell_r})] - \\ &\qquad\qquad \rho[(x_{k_0}, y_{\ell_0}), \ldots, (x_{k_{r-2}}, y_{\ell_{r-2}})] \qquad r \geq 2 \end{aligned} \tag{23}$$

Now consider the functions $g_s(r)$ given by (17) and the sequence of multivariate rational interpolants $[N_{\lfloor (r+1)/2 \rfloor}/D_{\lfloor r/2 \rfloor}]_{I^{(0)}_r} = E^{(0)}_r$. It is well-known that the continued fraction of which the rth convergent equals $E^{(0)}_r$ has its partial numerator and denominator coefficients given by

$$
\begin{aligned}
b_0 &= E_0^{(0)} \\
a_1 &= E_1^{(0)} - E_0^{(0)} \\
b_1 &= 1 \\
a_i &= \frac{E_{i-1}^{(0)} - E_i^{(0)}}{E_{i-1}^{(0)} - E_{i-2}^{(0)}} \qquad i \geq 2 \\
b_i &= 1 - a_i = \frac{E_i^{(0)} - E_{i-2}^{(0)}}{E_{i-1}^{(0)} - E_{i-2}^{(0)}} \qquad i \geq 2
\end{aligned}
$$

From the continued fraction

$$b_0 + \sum_{i=0}^{\infty} \left| \frac{a_i}{b_i} \right.$$

we construct an equivalent continued fraction of which the partial denominators are $B_i = \varphi_i = \varphi[(x_{k_0}, y_{\ell_0}), \ldots, (x_{k_i}, y_{\ell_i})]$. The partial numerators of the equivalent continued fraction then become

$$
\begin{aligned}
A_1 &= \frac{\varphi_1}{b_1} a_1 \\
A_i &= \frac{\varphi_i}{b_i} \frac{\varphi_{i-1}}{b_{i-1}} a_i \qquad i \geq 2
\end{aligned}
$$

More explicitly we have for $i = 1$

$$A_1 = h_{11}^{(0)} \left(E_0^{(0)} - E_1^{(0)}\right)$$

and for $i \geq 2$

$$
\begin{aligned}
A_i &= \varphi_i \varphi_{i-1} \frac{a_i}{b_i} \left(1 + \frac{a_{i-1}}{b_{i-1}}\right) \\
&= \varphi_i \varphi_{i-1} \left(\frac{E_{i-1}^{(0)} - E_i^{(0)}}{E_i^{(0)} - E_{i-2}^{(0)}} \right) \left(1 + \frac{E_{i-2}^{(0)} - E_{i-1}^{(0)}}{E_{i-1}^{(0)} - E_{i-3}^{(0)}} \right)
\end{aligned}
$$

Making use of lemma 2 and of the definition of the multivariate reciprocal and inverse differences we can summarize the multivariate generalization of the formulas (E) as follows. The successive convergents of the continued fraction

$$\varphi[(x_{k_0}, y_{\ell_0})] + \sum_{i=1}^{\infty} \left| \frac{A_i}{\varphi[(x_{k_0}, y_{\ell_0}), \ldots, (x_{k_i}, y_{\ell_i})]} \right.$$

with

$$A_1 = h_{11}^{(0)} (E_0^{(0)} - E_1^{(0)})$$

$$A_2 = -(h_{2,2}^{(0)} - h_{0,0}^{(0)}) h_{1,1}^{(0)} \frac{g_{1,2}^{(0)}}{g_{0,1}^{(0)} h_{2,2}^{(0)} - g_{0,2}^{(0)}}$$

$$A_i = -(h_{i,i}^{(0)} - h_{i-2,i-2}^{(0)}) (h_{i-1,i-1}^{(0)} - h_{i-3,i-3}^{(0)}) \times$$

$$\frac{g_{i-1,i}^{(0)}}{g_{i-2,i-1}^{(0)} h_{i,i}^{(0)} - g_{i-2,i}^{(0)}} \left(1 + \frac{g_{i-2,i-1}^{(0)}}{g_{i-3,i-2}^{(0)} h_{i-1,i-1}^{(0)} - g_{i-3,i-1}^{(0)}}\right) \qquad i \geq 2$$

where the functions $g_s(r)$ given by (17), are the multivariate rational interpolants on the "main staircase"

$$E_{2r}^{(0)} = \left[N_r^{(0)}/D_r^{(0)}\right]_{I_{2r}^{(0)}}$$

$$E_{2r+1}^{(0)} = \left[N_{r+1}^{(0)}/D_r^{(0)}\right]_{I_{2r+1}^{(0)}} \qquad r = 0, 1, \ldots$$

When we compare this technique with the one from the previous section we remark the following. The continued fraction of which the partial numerators and denominators are given above generates only rational interpolants $E_r^{(0)}$. The recursive scheme also explicitly computes intermediate values $E_r^{(j)}$ with $j \neq 0$. So different algorithms will serve different purposes.

Conclusion

In this paper a number of techniques have been presented for the construction and calculation of multivariate rational interpolants. These techniques generalize well-known univariate results. The next overview summarizes their similarities and differences.

Determinant formula	(A) N and D satisfy inclusion property	(B) $N \subset I$ I inclusion property
Recursive scheme	(C) input $f(x_i, y_j)$ $n + m + 1$ columns	(D) input $f[x_0, \ldots, x_i][y_0, \ldots, y_j]$ m columns
Continued fraction	(E) input $f(x_i, y_j)$ inverse differences	(F) input $f[x_0, \ldots, x_i][y_0, \ldots, y_j]$ q- and e-values

Appendix

This appendix lists the notations used for the determinants and quotients of determinants in the paper with the convention that $g_{-1}(j) = 1$ and $g_0(j) = f_j$.

$$d_{r,i}^{(j)} = \begin{vmatrix} g_i(j) & \dots & g_i(j+r) \\ g_1(j) & \dots & g_1(j+r) \\ \vdots & & \vdots \\ g_r(j) & \dots & g_r(j+r) \end{vmatrix} \qquad i = -1, 0, r+1, r+2, \dots$$

$$e_{r,i}^{(j)} = \begin{vmatrix} g_{-1}(j) & \dots & g_{-1}(j+r) \\ g_0(j) & \dots & g_0(j+r) \\ g_1(j) & \dots & g_1(j+r) \\ \vdots & & \vdots \\ g_{r-2}(j) & & \\ g_i(j) & \dots & g_i(j+r) \end{vmatrix} \qquad i = r-1, r, \dots$$

$$f_{r,i}^{(j)} = \begin{vmatrix} g_i(j) & \dots & g_i(j+r) \\ g_0(j) & \dots & g_0(j+r) \\ g_1(j) & \dots & g_1(j+r) \\ \vdots & & \vdots \\ g_{r-1}(j) & \dots & g_{r-1}(j+r) \end{vmatrix} \qquad i = r, r+1, \dots$$

$$E_r^{(j)} = E_r(f_j) = \frac{d_{r,0}^{(j)}}{d_{r,-1}^{(j)}}$$

$$g_{r,i}^{(j)} = \frac{d_{r,i}^{(j)}}{d_{r,-1}^{(j)}} \qquad i = r+1, r+2, \dots$$

$$h_{r,i}^{(j)} = \frac{e_{r,i}^{(j)}}{e_{r,r-1}^{(j)}} \qquad i = r, r+1, \dots$$

$$\rho[x_j, \dots, x_{j+r}] = (-1)^r h_{r,r}^{(j)} = (-1)^r \frac{e_{r,r}^{(j)}}{e_{r,r-1}^{(j)}} \text{ with } g_s(r) \text{ as in (13)}$$

$$\rho[(x_{k_j}, y_{\ell_j}), \dots, (x_{k_{j+r}}, y_{\ell_{j+r}})] = (-1)^r h_{r,r}^{(j)} = (-1)^r \frac{e_{r,r}^{(j)}}{e_{r,r-1}^{(j)}} \text{ with } g_s(r) \text{ as in (17)}$$

References

1. Aitken A., "Determinants and Matrices", Oliver and Boyd, Edingburgh and London, 1967.
2. Brezinski C., "Accélération de la convergence en Analyse Numérique", LNM 584, Springer Verlag, Berlin, 1977.
3. Brezinski C., *A general extrapolation algorithm*, Numer. Math. **35** (1980), 175-187.
4. Claessens G., *A generalization of the qd-algorithm*, J. Comput. Appl. Math. **7** (1981), 237-247.
5. Claessens G., *A useful identity for the rational Hermite interpolation table*, Numer. Math. **29** (1978), 227-231.
6. Claessens G., "Some aspects of the rational Hermite interpolation table and its applications", Ph. D., University of Antwerp, 1976.
7. Cuyt A., *A multivariate qd-like algorithm*, BIT (to appear).
8. Cuyt A., *A recursive computation scheme for multivariate rational interpolants*, SIAM J. Num. Anal. **24** (1987), 228-238.
9. Cuyt A., "General order multivariate rational Hermite interpolants", Habilitation, University of Antwerp, 1986.
10. Cuyt A. and Verdonk B., *General order Newton-Padé approximants for multivariate functions*, Numer. Math. **43** (1984), 293-307.
11. Cuyt A. and Verdonk B., *Multivariate rational interpolation*, Computing **34** (1985), 41-61.
12. Stoer J., *Über zwei Algorithmen zur Interpolation mit rationalen Funktionen*, Numer. Math. **3** (1961), 285-304.
13. Wynn P., *Über einen Interpolations-algorithmus und gewisse andere Formeln die in der Theorie der Interpolation durch rationale Funktionen bestehen*, Numer. Math. **2** (1960), 151-182.

RATIONAL APPROXIMANTS OF HYPERGEOMETRIC SERIES IN $\mathbb{C}^N$.

C.H. Lutterodt
Department of Mathematics
Howard University
Washington, D.C. 20059

ABSTRACT. Rational Approximants to two variable Appell hypergeometric series are discussed. A number of examples showing the first few elements of Padé-type table, which depends on a suitable choice of an index set $E^{\mu\nu}$, is given. An example is constructed to verify Montessus de Ballore theorem in $\mathbb{C}^2$.

§ INTRODUCTION.

The object of this paper is to open up an area rife in power series from which one can churn out quick examples of rational approximants. Hypergeometric series in two or more variables provide us with such an area. They were extensively studied by Bailey [1935] and later by his former student and co-worker Slater [1966]. Two of the four types of hypergeometric series in two variables known as Appell's series, that converge in the unit bidisc, are the ones discussed in this paper. The other Appell's series that converge in other circular domains in $\mathbb{C}^2$ as well as Lauricella series in $\mathbb{C}^n$, currently under study, will not be presented here. We have introduced a new two variable hypergeometric series that is entire in $\mathbb{C}^2$ and have used a candidate of that class in making up a locally meromorphic function used to verify Montessus result.

A. Cuyt (ed.), Nonlinear Numerical Methods and Rational Approximation, 191–210.

§1. NOTATIONS AND DEFINITIONS.

Let $\mathbb{N}$ be the set of natural numbers so that $\mathbb{N}^n := \mathbb{N}\times\ldots\times\mathbb{N}$ n-times. We introduce a partial ordering in $\mathbb{N}^n$ as follows: For $\gamma,\lambda \in \mathbb{N}^n$, $0 \preccurlyeq \gamma \preccurlyeq \lambda$ iff $0 \le \gamma_j \le \lambda_j$, $j = 1,\ldots,n$

Let $E_\lambda := \{\gamma \in \mathbb{N}^n : 0 \preccurlyeq \gamma \preccurlyeq \lambda\}$. We call $E^{\mu\nu} \subset \mathbb{N}^n$ an index set if

(i) $E^{\mu\nu} = E_\mu \cup A_\nu$ where $E_\mu \cap A_\nu = \phi$

(ii) $\lambda \in E^{\mu\nu} \Longrightarrow \gamma \in E^{\mu\nu}$ for all γ: $0 \preccurlyeq \gamma \preccurlyeq \lambda$.

(iii) $|E^{\mu\nu}| \le \prod_{j=1}^{n} (\mu_j + 1) + \prod_{j=1}^{n} (\nu_j + 1) - 1$

(iv) Each axis of $E^{\mu\nu}$ has a Padé index set.

Here $|E^{\mu\nu}|$ is the cardinality of $E^{\mu\nu}$. Now observe that even though E_μ and A_ν are disjoint there is no real gap between them since (ii) guarantees that for all $\lambda \in E^{\mu\nu}$, $E_\lambda \subset E^{\mu\nu}$.

A polynomial $P_\lambda(z) := \sum_{\gamma\in E_\lambda} a_\gamma z^\gamma$, where γ,λ are both multiple indices, is said to have a multiple degree at most λ.

For $\mu,\nu \in \mathbb{N}^n$, let $\mathcal{R}_{\mu\nu}$ be the class of rational functions of the form $R_{\mu\nu}(z) = P_\mu(z)/Q_\nu(z)$ in $\mathbb{C}^n$, with $Q_\nu(0) \ne 0$, where $P_\mu(z)$ and $Q_\nu(z)$ are polynomials in $\mathbb{C}^n$ with multiple degrees at most μ and ν respectively: furthermore, $P_\mu(z)$ and $Q_\nu(z)$ are coprime off sets of codimension at least 2. In other words, $P_\mu(z)$ and $Q_\nu(z)$ are relatively prime everywhere except at points where their non localizable zero sets intersect. The following illustration (figure 1) below shows the exceptional points corresponding to the points of intersection of zero sets of $P_\mu(z)$ and $Q_\nu(z)$; these points in $\mathbb{C}^{n-1}$ have been marked 1,2,3,4,5,6. For further details see Gunning and Rossi [1965] or Hörmander [1966].

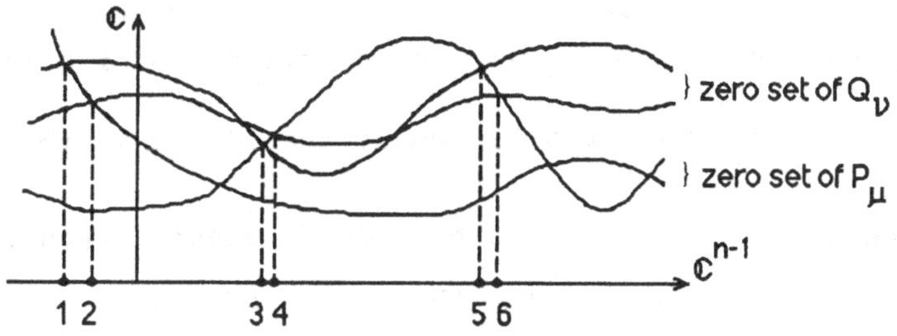

Figure 1

This illustrates the points in $\mathbb{C} \times \mathbb{C}^{n-1}$ that corresponds to zero sets of P_μ and Q_ν.

Definition 1.1: Let f be holomorphic in an 0-neighborhood A rational function $R_{\mu\nu}(z) \in \mathcal{R}_{\mu\nu}$ is called a rational approximant to f at 0 if

$$\text{(i)} \quad \partial^{|\lambda|}(Q_\nu f - P_\mu)\big|_{z=0} = 0 \qquad \text{for all } \lambda \in E_\mu \tag{1.1}$$

$$\text{(ii)} \quad \partial^{|\lambda|}_{A_\nu}(Q_\nu f)\big|_{z=0} = 0 \qquad \text{for all } \lambda \in A_\nu \tag{1.2}$$

where

$$\partial^{|\lambda|}_{A_\nu}(Q_\nu f) := \sum_{\substack{\gamma \le \lambda \\ \gamma, \lambda \in A_\nu}} \prod_{j=1}^{n} \binom{\lambda_j}{\gamma_j} \partial^{|\gamma|}(Q_\nu)\, \partial^{|\lambda-\gamma|}(f) \tag{1.3}$$

and

$$\partial^{|\lambda|} \equiv \partial^{|\lambda|}/\partial z_1^{\lambda_1} \ldots \partial z_n^{\lambda_n} .$$

As indicated earlier $E_\mu \subset E^{\mu\nu}$ and so trivially $|E^{\mu\nu}| \ge |E_\mu| = \prod_{j=1}^{n} (\mu_j + 1)$. Since in general, $E^{\mu\nu} = E_\mu \cup A_\nu$ and $E_\mu \cap A_\nu = \phi$, $|E^{\mu\nu}| = |E_\mu| + |A_\nu|$. Now $|A_\nu|$ provides the number of equations needed to solve the linear system involving the coefficients of $Q_\nu(z)$, with $Q_\nu(0) \neq 0$ normalized. The maximum number of coefficients of $Q_\nu(z)$ is $\prod_{j=1}^{n} (\nu_j + 1) - 1$. To solve uniquely the linear system involving $\prod_{j=1}^{n} (\nu_j + 1) - 1$ unknown coefficients, requires at least $|A_\nu|$ equations, meaning the matrix of these coefficients has rank $\prod_{j=1}^{n} (\nu_j + 1) - 1$.

Definition 1.2. An index set $E^{\mu\nu}$ is called maximal if $|E^{\mu\nu}| \geq \prod_{j=1}^{n} (\mu_j + 1) + \prod_{j=1}^{n} (\nu_j + 1) - 1$.

Proposition 1.1. If $E^{\mu\nu}$ is maximal, the associated (μ,ν)-rational approximant is uniquely determined.

We call rational approximants associated with maximal $E^{\mu\nu}$, that in addition have normalized denominator polynomials, unisolvent rational approximants (URA). We denote these URAs typically by

$$\pi_{\mu\nu}(z) = P_{\mu\nu}(z)/Q_{\mu\nu}(z). \tag{1.4}$$

We have already indicated that $|A_\nu|$ yields the number of equations necessary to solve the linear system of the coefficients of $Q_\nu(z)$, arising from equation (1.2). In a case in which $E^{\mu\nu}$ is maximal, we shall take $|A_\nu| = \prod_{j=1}^{n} (\nu_j + 1) - 1$. There are various expansions of $|A_\nu|$ that provide the building "blocks" for the linear system obtained from equation (1.2). One such form that has featured in our work Lutterodt [1976] [1976a] in the past is the following which consists of 2^{n-1} building "blocks":

$$\prod_{j=1}^{n}(\nu_j + 1) - 1 = \nu_1\nu_2\ldots\nu_{n-1}(\nu_n + 1) \qquad \text{1 block}$$

$$+ \nu_1\nu_2\ldots\nu_{n-2}(\nu_n + 1) + \nu_2\nu_3\ldots\nu_{n-1}(\nu_n + 1)$$

$$+\ldots \nu_{n-1}\nu_1\ldots\nu_{n-3}(\nu_n + 1) \qquad \text{(n-1) blocks}$$

$$+ \nu_1\nu_2\ldots\nu_{n-3}(\nu_n + 1) + \nu_2\nu_3\ldots\nu_{n-2}(\nu_n + 1)$$

$$+\ldots+ \nu_{n-2}\nu_1\ldots\nu_{n-3}(\nu_n + 1) \qquad \frac{(n-1)(n-2)}{2!} \quad \text{blocks}$$

$$\vdots \qquad\qquad \vdots$$

$$+ \nu_1(\nu_n + 1) + \nu_2(\nu_n + 1) \qquad (1.5)$$

$$+\ldots+ \nu_{n-1}(\nu_n + 1) \qquad \text{(n-1) blocks}$$

$$+ \nu_n \qquad \text{1 block}$$

To see how the blocks are used to determine the form of A_ν, we consider an example from two dimensions (see figure 2)

<u>n = 2</u>: $A_\nu = (\nu_1 + 1)(\nu_2 + 1) - 1 = \nu_1(\nu_2 + 1) + \nu_2$

Now take the "block" $\nu_1(\nu_2 + 1)$. This gives the following part of A_ν:

$$\mu_1 + 1 \le \lambda_1 \le \mu_1 + \nu_1; \quad 0 \le \lambda_2 \le \nu_2, \qquad (1.6a)$$

Take the "block" ν_2. This corresponds to the other part of A_ν

$$\lambda_1 = 0; \quad \mu_2 + 1 \le \lambda_2 \le \mu_2 + \nu_2 \qquad (1.6b)$$

The following diagram (figure 2) shows a maximal set of $E^{\mu\nu}$ in which we have used the A_ν specified by (1.6a) and (1.6b).

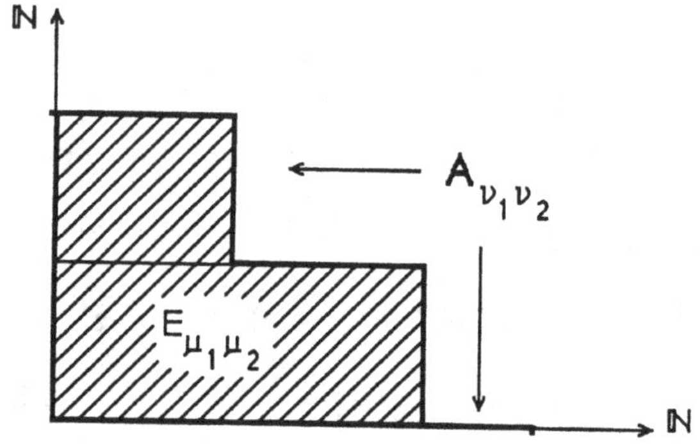

Figure 2

This illustrates the form of index set $E^{\mu\nu}$ using the pattern provided by the building blocks.

We shall use this in the next section to construct some examples of URAs to hypergeometric series in two dimensions.

There is a degenerate class of the index set $E^{\mu\nu}$ which is of some interest in this brief study of rational approximants to hypergeometric series. The degenerate $E^{\mu\nu}$'s are reduced to the skeleton of their Padé index sets in each axis of $\mathbb{N}^n$. For instance, in the j-th axis the projection of E_μ along that axis is E_{μ_j} and that of A_ν is A_{ν_j} where

$$E_{\mu_j} := \{\lambda_j \in N: 0 \le \lambda_j \le \mu_j\} \tag{1.7}$$

$$A_{\nu_j} := \{\lambda_j \in N: \mu_j + 1 \le \lambda_j \le \mu_j + \nu_j\} \tag{1.8}$$

Thus the j-th axis Padé index set is $E_{\mu_j} \cup A_{\nu_j}$ and a degenerate $E^{\mu\nu}$ is given by

$$E^{\mu\nu} = \bigcup_{j=1}^{n} (E_{\mu_j} \cup A_{\nu_j}). \tag{1.9}$$

This type of $E^{\mu\nu}$ is what is essential in handling the construction of rational approximants to holomorphic functions of several variables expressible as a product of n functions analytic in one and only one variable. Thus if $f_1(z_1)$, $f_2(z_2),\ldots,f_n(z_n)$ are analytic functions in the variables in which they are expressed then the holomorphic function $F(z)$ has the form

$$F(z) = f_1(z_1) \times \ldots \times f_n(z_n) \qquad (1.10)$$

Proposition 1.2: Let $F(z)$ be as in (1.10) and holomorphic in some neighborhood of $z = 0$ in $\mathbb{C}^n$. Then the (μ,ν) rational approximant to $F(z)$ at $z = 0$ is given by

$$r_{\mu\nu}(z) = \prod_{j=1}^{n} r_{\mu_j\nu_j}(z_j). \qquad (1.11)$$

where $r_{\mu_j\nu_j}(z_j)$ is the Padé approximant to $f_j(z_j)$ at $z_j = 0$.

Remark: We use the notation $r_{\mu\nu}(z)$ for the rational approximants of the degenerate class.

Proof of Proposition 1.2: The definition 1.1 provides the machinery. However since $E^{\mu\nu} = \bigcup_{j=1}^{n} (E_{\mu_j} \cup A_{\nu_j})$, each variable in the equations (1.1) and (1.2) gets separated, so that computations from (1.1) and (1.2) can be carried out successively over each variable. This requires the separation of the rational function $R_{\mu\nu}(z)$ into separated products of rational functions of one variable. The desired result is then immediate.

This form of degenerate $E^{\mu\nu}$ has a special case if each of its components $E_{\mu_j} \cup A_{\nu_j}$ is maximal. Since maximality of each $E_{\mu_j} \cup A_{\nu_j}$ guarantees uniqueness of $r_{\mu_j\nu_j}(z_j)$ Padé approximant, the special case $E^{\mu\nu}$ therefore implies the uniqueness of the $r_{\mu\nu}(z)$. If we normalize the denominator polynomial of $r_{\mu\nu}(z)$ in addition to its uniqueness then we call $r_{\mu\nu}(z)$ a unisolvent rational approximant (URA) just like $\pi_{\mu\nu}(z)$.

§2. URA to Appell series

There are four different types of Appell series in $\mathbb{C}^2$. Two of these series converge in the unit polydisc while the other two converge in some unit circular domains. Since our analysis is largely polydisc bound we shall focus mainly on those two series that converge in the unit bidisc.

On the unit bidisc $\Delta_1^2 := \{z \in \mathbb{C}^2 : |z_1| < 1, |z_2| < 1\}$ we have the following two series

$$F_1(\alpha;\beta,\beta';\gamma;z_1,z_2)$$

$$= \sum_{(\lambda_1,\lambda_2)\in\mathbb{N}^2} \frac{(\alpha)_{\lambda_1+\lambda_2}(\beta)_{\lambda_1}(\beta')_{\lambda_2}}{\lambda_1!\lambda_2!\ (\gamma)_{\lambda_1+\lambda_2}} z_1^{\lambda_1} z_2^{\lambda_2} \tag{2.1}$$

$$F_3(\alpha,\alpha';\beta,\beta';\gamma;z_1,z_2)$$

$$= \sum_{(\lambda_1,\lambda_2)\in\mathbb{N}^2} \frac{(\alpha)_{\lambda_1}(\alpha')_{\lambda_2}(\beta)_{\lambda_1}(\beta')_{\lambda_2}}{\lambda_1!\lambda_2!\ (\gamma)_{\lambda_1+\lambda_2}} z_1^{\lambda_1} z_2^{\lambda_2} \tag{2.2}$$

In both series $(k)_m = k(k+1)\ldots\ldots(k+m-1)$. We shall first consider the series $F_1(\alpha;\beta,\beta';\gamma;z_1,z_2)$ constructing its URA for certain choices of the parameters $\alpha,\beta,\beta',\gamma$. We shall write

$$c_{\lambda_1\lambda_2} = \frac{(\alpha)_{\lambda_1+\lambda_2}(\beta)_{\lambda_1}(\beta')_{\lambda_2}}{\lambda_1!\lambda_2!\ (\gamma)_{\lambda_1+\lambda_2}} \tag{2.3}$$

and let

$$P_{\mu_1\mu_2}(z_1,z_2) = \sum_{\gamma_1=0}^{\mu_1} \sum_{\gamma_2=0}^{\mu_2} a_{\gamma_1\gamma_2} z_1^{\gamma_1} z_2^{\gamma_2} \tag{2.4}$$

$$Q_{\nu_1\nu_2}(z_1,z_2) = \sum_{\gamma_1=0}^{\nu_1} \sum_{\gamma_2=0}^{\nu_2} b_{\gamma_1\gamma_2} z_1^{\gamma_1} z_2^{\gamma_2} \tag{2.5}$$

where $Q_\nu(0,0) \neq 0$ and in fact, we normalize it to unity. Using the definition of a rational approximant given in Section 1, with respect to $E^{\mu\nu} = E_{\mu_1\mu_2} \cup A_{\nu_1\nu_2}$ maximal where we take $|E_{\mu_1\mu_2}|=(\mu_1+1)(\mu_2+1)$ and $|A_{\nu_1\nu_2}| = (\nu_1 + 1)\nu_2 + \nu_2$ the equations (1.1) and (1.2) in the definition yield

$$\sum_{\gamma_1=0}^{\min(\nu_1,\lambda_1)} \sum_{\gamma_2=0}^{\min(\nu_2,\lambda_2)} b_{\gamma_1\gamma_2}\, c_{\lambda_1-\gamma_1,\lambda_2-\gamma_2} = a_{\lambda_1\lambda_2};\quad (\lambda_1,\lambda_2) \in E_{\mu_1,\mu_2} \tag{2.6}$$

$$\sum_{\gamma_1=0}^{\min(\nu_1,\lambda_1)} \sum_{\gamma_2=0}^{\min(\nu_2,\lambda_2)} b_{\gamma_1\gamma_2}\, c_{\lambda_1-\gamma_1,\lambda_2-\gamma_2}=0;\quad (\lambda_1,\lambda_2) \in A_{\nu_1\nu_2} \tag{2.7}$$

Refer to figure 2 for the sets $E_{\mu_1\mu_2}$ and $A_{\nu_1\nu_2}$.

In a simple case when $\mu_1=1$, $\mu_2=1$, $\nu_1=0$, $\nu_2=1$, we get from (2.6), (2.7) together with block inequalities for $A_{\nu_1\nu_2}$ from (1.6a) and 1.6b) that the URA $\pi_{1101}(z_1,z_2)$ to $F_1(\alpha;\beta,\beta';\gamma;z_1,z_2)$ at (0,0) is

$$\pi_{1101}(z_1,z_2) = \frac{1 + \frac{\alpha\beta}{\gamma} z_1 + \left(\frac{\alpha\beta}{\gamma} - \frac{(\alpha+1)(\beta'+1)}{2(\gamma+1)}\right)z_2 + \left(\frac{\alpha(\alpha+1)\beta\beta'}{2\gamma(\gamma+1)} - \frac{\alpha(\alpha+1)\beta}{2\gamma(\gamma+1)}\right)z_1z_2}{\frac{\alpha+1)(\beta'+1)}{\gamma+1}\, z_2} \tag{2.8}$$

For arbitrary $\alpha,\beta,\beta',\gamma$ the formulas for the URAs get unbearably heavy so as we said earlier, we shall compute URA's for special values of α,β,β' and γ. The first case we consider is when $\alpha = \gamma$ so that

$$F_1(\alpha;\beta,\beta';\gamma;z_1,z_2) = F_1(1;\beta,\beta';1;z_1,z_2)$$

$$= (1 - z_1)^{-\beta} (1 - z_2)^{-\beta'} \qquad (2.9)$$

This function is rational if $\beta,\beta' \in Z^+$ and after a finite number of steps (not necessarily the same for each variable) the Padé approximants in the separate variables coincide with the rational functions in those variable. For example, $F_1(1;\beta,\beta,1;z_1,z_2)$ has the following samples of rational approximants according to (1.11).

$$r_{1101}(z_1,z_2) = (1 + \beta z_1)\left[\frac{1 + (\frac{\beta' - 1}{2})z_2}{1 - (\frac{\beta' + 1}{2})z_2}\right] \qquad (2.10)$$

$$r_{1111}(z_1,z_2) = \frac{(1 + (\frac{\beta - 1}{2})z_1)(1 + (\frac{\beta' - 1}{2})z_2)}{(1 - (\frac{\beta + 1}{2})z_1)(1 - (\frac{\beta' + 1}{2})z_2)} \qquad (2.11)$$

$$r_{2111}(z_1,z_2) = \left[\frac{1+\frac{2}{3}(\beta-1)z_1+\frac{\beta(\beta-1)}{3!}\ z_1^2}{1 - (\frac{\beta + 1}{2})z_1}\right]\left[\frac{1 + (\frac{\beta'-1}{2})z_2}{1 - (\frac{\beta'+1}{2})z_2}\right] \qquad (2.12)$$

When $\beta=\beta'=1$, $r_{1111}(z_1,z_2) = (1 - z_1)^{-1}(1 - z_2)^{-1}$. Thus $r_{1111}(z_1,z_2) = F_1(1;1,1;1,z_1,z_2)$ and for μ_1, $\nu_1 \geq 1$ and $\mu_2,\nu_2 \geq 1$, the individual Padé approximants in the separate variables z_1,z_2, have non-normal Padé tables. It turns out that for $\beta,\beta' \in Z^+$, $\beta,\beta' > 1$ the rational approximant $r_{\beta\beta',\beta\beta'}(z_1,z_2) = F_1(1;\beta,\beta';1,z_1,z_2)$. The cases when β,β' are non integral rationals or irrationals do not lead to this property of the rational approximants reproducing $F_1(1;\beta,\beta';1;z_1,z_2)$ after any number of steps. This suggests a definition of normality for the tables (see Lutterodt 1974) of rational approximants of functions of several variables that separate into products of functions of single variable. The following definition does not apply to the tables of rational approximants $\pi_{\mu\nu}$ of the non separable functions, in general.

Definition 2.1. Suppose $F(z) = f_1(z_1) \times \ldots \times f_n(z_n)$ is holomorphic in a neighborhood of $z = 0$. Suppose $r_{\mu\nu}(z) = r_{\mu_1\nu_1}(z_1) \times \ldots \times r_{\mu_n\nu_n}(z_n)$ is a unisolvent rational approximant to $f(z)$ at $z = 0$, then the table of the sequences $\{r_{\mu\nu}(z)\}_{\mu\nu}$ is normal if each $\{r_{\mu_j\nu_j}(z_j)\}_{\mu_j\nu_j}$ has a normal Padé table.

Remark: Clearly each $r_{\mu\nu} = \prod_{j=1}^{n} r_{\mu_j\nu_j}$ occurs once and only once in the multidimensional configuration of this generalized Padé table.

Next we return to $F_1(\alpha;1,1;\gamma;z_1,z_2)$. Unlike $F_1(1;\beta,\beta';1;z_1,z_2)$, the hypergeometric function $F_1(\alpha;1,1;\gamma;z_1,z_2)$ does not separate into product of functions of its individual variables. Its expansion is given by

$$F_1(\alpha;1,1;\gamma;z_1,z_2) = \sum_{\lambda_1=0}^{\infty} \sum_{\lambda_2=0}^{\infty} \frac{(\alpha)_{\lambda_1+\lambda_2}}{(\gamma)_{\lambda_1+\lambda_2}} z_1^{\lambda_1} z_2^{\lambda_2} \qquad (2.13)$$

We compute a URA with $\mu_1=1$, $\mu_2=1$, $\nu_1=1$, $\nu_2=0$ and using the equations (2.6) and (2.7) we obtain

$$\pi_{1110}(z_1,z_2) = \frac{1 + \frac{\alpha - \gamma}{\gamma(\gamma + 1)} z_1 + \frac{\alpha}{\gamma} z_2}{1 - \frac{(\alpha + 1)}{(\gamma + 1)} z_1} \qquad (2.14)$$

For $\mu_1=\mu_2=\nu_1=\nu_2=1$, using equation (2.6) and (2.7) and A_ν as in (1.6a and (1.6b) we obtain

$$\pi_{1111}(z_1,z_2) =$$

$$\frac{1 + \frac{\alpha-\gamma}{\gamma(\gamma+1)}z_1 + \frac{\alpha-\gamma}{\gamma(\gamma+1)}z_2 + \frac{\alpha+1}{\gamma+1}\left[\frac{\alpha(\gamma+1) + 2(\alpha+1)}{(\gamma+1)(\gamma+2)} - \frac{\alpha}{\gamma}\right]z_1 z_2}{1 + \frac{\alpha+1}{\gamma+1}z_1 + \frac{\alpha+1}{\gamma+1}z_2 - \frac{\alpha+1}{\gamma+1}\left[\frac{\alpha(\gamma+1) + 2(\alpha+1)}{(\gamma+1)(\gamma+2)}\right]z_1 z_2} \qquad (2.15)$$

Again these get rapidly complicated using arbitrary real α and γ, so we shall consider a special case in which $\alpha=1$ and $\gamma=2$. A few of the URA's to $F_1(1;1,1;2;z_1,z_2)$ at $(0,0)$ are:

$$\pi_{1110}(z_1,z_2) = \frac{1 - \frac{1}{6} z_1 + \frac{1}{2} z_2}{1 - \frac{2}{3} z} \tag{2.16}$$

$$\pi_{1111}(z_1,z_2) = \frac{1 - \frac{1}{6} z_1 - \frac{1}{6} z_2 - \frac{1}{18} z_1 z_2}{1 + \frac{2}{3} z_1 + \frac{2}{3} z_2 - \frac{7}{18} z_1 z_2} \tag{2.17}$$

$$\pi_{2110}(z_1,z_2) = \frac{1 - \frac{1}{4}z_1+\frac{1}{2}z_2 - \frac{1}{24}z_1 z_2 - \frac{1}{24}z_1^2 - \frac{1}{3}z_2^2}{1 - \frac{3}{4} z_1} \tag{2.18}$$

We now move on to the other hypergeometric function $F_3(\alpha,\alpha';\beta,\beta';\gamma;z_1,z_2)$ in the bidisc Δ_1^2 where the ensuing series converges.

$$F_3(\alpha,\alpha';\beta,\beta';\gamma;z_1,z_2) = \sum_{(\lambda_1,\lambda_2)\in\mathbb{N}^2} \frac{(\alpha)_{\lambda_1}(\alpha')_{\lambda_2}(\beta)_{\lambda_1}(\beta')_{\lambda_2}}{\lambda_1!\lambda_2!(\gamma)_{\lambda_1+\lambda_2}} z_1^{\lambda_1} z_2^{\lambda_2} \tag{2.19}$$

Using equation (2.6), (2.7) and (1.6a) and (1.6b) we compute $\pi_{1110}(z_1,z_2)$ URA to $F_3(\alpha,\alpha';\beta,\beta';\gamma;z_1,z_2)$ at $(0,0)$ to get

$$\pi_{1110}(z_1,z_2) = \frac{1+\left(\frac{\alpha\beta}{\gamma} - \frac{(\alpha+1)(\beta+1)}{2!(\gamma+1)}\right)z_1+\frac{\alpha'\beta'}{\gamma}z_2+\left[\frac{\alpha\alpha'\beta\beta'}{\gamma(\gamma+1)} - \frac{(\alpha+1)\alpha'(\beta+1)\beta'}{2!\gamma(\gamma+1)}\right]z_1 z}{1 - \frac{(\alpha + 1)(\beta + 1)}{2!(\gamma + 1)} z_1} \tag{2.20}$$

Calculation involving all five parameters $\alpha,\alpha',\beta,\beta',\gamma$ gets very involved right from the start so we shall now examine special cases of F_3 and their rational approximants. Now suppose $\alpha=\alpha'=1$, then $F_3(1,1;\beta,\beta';\gamma;z_1,z_2)$ has the following first few URAs at $(0,0)$

$$\pi_{1110}(z_1,z_2)$$

$$= \frac{1 - \frac{(\beta - \gamma)}{\gamma(\gamma + 1)} z_1 + \frac{\beta'}{\gamma} z_2 - \frac{\beta'}{\gamma(\gamma + 1)} z_1 z_2}{1 - \frac{(\beta + 1)}{(\gamma + 1)} z_1} \quad (2.22)$$

$$\pi_{1101}(z_1,z_2)$$

$$= \frac{1 + \frac{\beta}{\gamma} z_1 - \frac{(\beta' - \gamma)}{\gamma(\gamma + 1)} z_2 - \frac{\beta}{\gamma(\gamma + 1)} z_1 z_2}{1 - \frac{(\beta' + 1)}{(\gamma + 1)} z_2} \quad (2.23)$$

$$\pi_{1111}(z_1,z_2)$$

$$= \frac{1+\frac{\beta-\gamma}{\gamma(\gamma+1)}z_1+\frac{\beta'-\gamma}{\gamma(\gamma+1)}z_2+[\frac{(\gamma-1)(\beta+1)\beta'+\gamma-\beta}{\gamma(\gamma+1)^2} - \frac{(\beta+1)\beta'}{\gamma(\gamma+1)(\gamma+2)}]z_1z_2}{1 - \frac{(\beta+1)}{\gamma+1}z_1 - \frac{(\beta'+1)}{\gamma+1}z_2+\frac{(\beta+1)}{\gamma(\gamma+1)}[\frac{2\beta'+1}{\gamma+1} - \frac{\beta'}{\gamma+2}]z_1z_2} \quad (2.24)$$

If we take $\alpha=\alpha'=\beta=\beta'=\gamma=1$ in F_3 then its first few URAs at (0,0) are

$$\pi_{1110}(z_1,z_2) = \frac{1 + z_2 - \frac{1}{2} z_1z_2}{1 - z_1} \quad (2.25)$$

$$\pi_{1111}(z_1,z_2) = \frac{1 - \frac{1}{3} z_1z_2}{1 - z_1 - z_2 + \frac{7}{6} z_1z_2}$$

$$\pi_{1201}(z_1,z_2) = \frac{1 + z_1 - \frac{1}{2} z_1z_2 + z_1^2 - \frac{1}{6} z_1z_2^2}{1 - z_2} \quad (2.26)$$

Now for $F_3(1,1;\frac{1}{2},\frac{1}{3};1;z_1,z_2)$ we compute the following few URAs with respect to the same maximal $E^{\mu\nu}$'s as above.

$$\pi_{1110}(z_1,z_2) = \frac{1 - \frac{1}{4} z_1 + \frac{1}{3} z_2 - \frac{1}{6} z_1 z_2}{1 - \frac{3}{4} z_1}$$

$$\pi_{1101}(z_1,z_2) = \frac{1 + \frac{1}{2} z_1 - \frac{1}{3} z_2 - \frac{1}{4} z_1 z_2}{1 - \frac{2}{3} z_2}$$

$$\pi_{1111}(z_1,z_2) = \frac{1 - \frac{1}{4} z_1 - \frac{1}{4} z_2 + \frac{1}{24} z_1 z_2}{1 - \frac{3}{4} z_1 - \frac{2}{3} z_2 + \frac{13}{24} z_1 z_2} \tag{2.27}$$

In this section we have made no attempt to use the various URAs that we have computed in any numerical approximation. Our main goal is to introduce Appells series especially the ones that converge on a bidisc disc Δ_1^2 and simply compute their URA's for the series $F_1(\alpha;\beta,\beta';\gamma;z_1,z_2)$ and $F_3(\alpha,\alpha';\beta,\beta';\gamma;z_1,z_2)$. In the next section we consider a specific case of the Montessus result in relation to certain meromorphic hypergeometric functions.

§3. Montessus Convergence

We begin this section by considering a certain class of hypergeometric series that are entire, unlike the ones discussed in the preceding section. This class of hypergeometric series is new in the sense that we are introducing them for the first time. (We have used a member in our earlier work (see Lutterodt [1974], [1984a]). We shall call these hypergeometric functions of two variables

$$F_{L1}(\alpha;\beta,\beta';\gamma;\delta;z_1,z_2) = \sum_{(\lambda_1,\lambda_2)\in N^2} c_{\lambda_1\lambda_2} z_1^{\lambda_1} z_2^{\lambda_2} \tag{3.1}$$

where

$$c_{\lambda_1\lambda_2} = \frac{(\alpha)_{\lambda_1+\lambda_2} (\beta)_{\lambda_1} (\beta')_{\lambda_2}}{\lambda_1!\lambda_2!\ (\gamma)_{\lambda_1+\lambda_2} (\delta)_{\lambda_1+\lambda_2}} \tag{3.2}$$

<u>Proposition 3.1</u>. $F_{L1}(\alpha;\beta,\beta';\gamma;\delta;z_1,z_2)$ is entire in $\mathbb{C}^2$.

<u>Proof</u>: We follow Bailey's [1935] technique by considering the asymptotic behaviour of the general term in (3.1). We shall write in place λ_1 and λ_2, m and n and assume that m and n are large positive numbers. Then

$$c_{mn}z_1^m z_2^n = \left[\frac{\Gamma(\gamma)\Gamma(\delta)}{\Gamma(\alpha)\Gamma(\beta)(\beta')}\right]\frac{\Gamma(\alpha+m+n)\Gamma(\beta+m)\Gamma(\beta'+n)z_1^m z_2^n}{\Gamma(\gamma+m+n)\Gamma(\delta+m+n)\Gamma(m+1)\Gamma(n+1)} \tag{3.3}$$

Using Stirling's formula $\Gamma(n) \sim (2\pi)^{1/2}n^{n+1/2}e^{-n}$ (3.4)

$$c_{mn}z_1^m z_2^n \sim \frac{\Gamma(\gamma)\Gamma(\delta)}{\Gamma(\alpha)\Gamma(\beta)\Gamma(\beta')}\frac{m^{\beta-1}\ n^{\beta'-1}}{(m+n)^{\gamma+\delta-\alpha}}\left(\frac{z_1}{m+n}\right)^m\left(\frac{z_2}{m+n}\right)^n \tag{3.5}$$

Now for fixed parameters $\alpha,\beta,\beta',\gamma,\delta \in \mathbb{C}$, we can find a large enough positive number N that dominates $|\Gamma(\gamma)\Gamma(\delta)/\Gamma(\alpha)\Gamma(\beta)\Gamma(\beta')|$ so that asymptotically speaking we have

$$(1/N)\ \frac{m^{\mathrm{Re}(\beta)-1}\ n^{\mathrm{Re}(\beta')-1}}{(m+n)^{\mathrm{Re}(\gamma+\delta-\alpha)}}\left[\frac{|z_1|}{m+n}\right]^m\left[\frac{|z_2|}{m+n}\right]^n$$

$$\le |c_m z_1^m z_2^n| \le \frac{Nm^{\mathrm{Re}(\beta)-1}\ n^{\mathrm{Re}(\beta')-1}}{(m+n)^{\mathrm{Re}(\gamma+\delta-\alpha)}}\left[\frac{|z_1|}{m+n}\right]^m\left[\frac{|z_2|}{m+n}\right]^n \tag{3.6}$$

From (3.6) it is evident that $F_{L1}(\alpha_1;\beta,\beta';\gamma,\delta;z_1,z_2)$ is entire in $\mathbb{C}^2$.

The F_{L1} series is somewhat related to the Appell's F_1 series whose rational approximants we discussed in the preceding section. The difference between them is that the F_{L1} has additional δ-factors in the denominator of its coefficients. As with the F_1 series, when $\alpha = \gamma$ we get

$$F_{L1}(\alpha;\beta,\beta';\alpha,\delta;z_1,z_2) = F_{L1}(1;\beta,\beta';1,\delta;z_1,z_2) \tag{3.7}$$

Other properties of F_{L1} will be reported elsewhere.

We now consider the following function:

$$F_{L1}(1;1,1;1,1;z_1,z_2) \cdot F_1(1;1,1;1;\frac{z_1}{2},\frac{z_2}{3})$$

$$= \left[\frac{z_1 e^{z_1} - z_2 e^{z_2}}{z_1 - z_2}\right] / \left[1 - \frac{z_1}{2} - \frac{z_2}{3} + \frac{1}{6} z_1; z_2)\right] \quad (3.8)$$

This function is locally meromorphic in $\mathbb{C}^2$, however, it is holomorphic in the bicylinder $\{z \in \mathbb{C}^2: z_1 < 2, z_2 < 3\}$.

We now state the Montessus theorem in $\mathbb{C}^2$ and then use the above function to verify the result. For details about the Montessus theorem see Lutterodt [1984], [1987].

<u>Theorem 3.2</u> (Montessus). Let $\nu=(\nu_1,\nu_2) \in \mathbb{N}^2$ and $\rho>0$ be fixed. Suppose $f(z_1,z_2)$ is meromorphic in Δ_ρ^2 but holomorphic in $\mathfrak{U}$ ($\subset \Delta_\rho^2$ but $\neq \Delta_\rho^2$) neighborhood of $(0,0)$. Suppose the polar set of $f(z_1,z_2)$ in Δ_ρ^2 is specified by a polynomial $q_{\nu_1\nu_2}(z_1 z_2) = 0$ with the multiple degree of $q_{\nu_1\nu_2}$ being minimal, also $q_{\nu_1\nu_2} f \in C(\bar{\Delta}_\rho^2)$. Suppose $\pi_{\mu_1\mu_2\nu_1\nu_2}(z_1,z_2)$ is a URA to $f(z_1,z_2)$ at $(0,0)$ then as $\min(\mu_1,\mu_2) \to \infty$ we get

(i) $Q^{-1}_{\mu_1\mu_2\nu_1\nu_2}(0) \cap \Delta_\rho^2 \to q^{-1}_{\nu_1\nu_2}(0) \cap \Delta_\rho^2$

(ii) $\pi_{\mu_1\mu_2\nu_1\nu_2}(z_1,z_2) \to f(z_1,z_2)$ uniformly on $\Delta_\rho^2 \backslash q^{-1}_{\nu_1\nu_2}(0)$.

$$\text{Let } f(z_1,z_2) = \sum_{(\lambda_1,\lambda_2)\in\mathbb{N}^2} c_{\lambda_1\lambda_2} z_1^{\lambda_1} z_2^{\lambda_2} \quad (3.9)$$

be the power series representation of the locally meromorphic $F_{L_1}(1;1,1;1,1;z_1,z_2) \times F(1;1,1;1;z_1,z_2)$

restricted to the bicylinder $\Delta_2^1 \times \Delta_3^1 = \{(z_1,z_2) \in \mathbb{C}^2$: $z_1 <2$, $z_2 <3\}$
where it is holomorphic with coefficients given by

$$c_{\lambda_1\lambda_2} = \frac{1}{2^{\lambda_1}} \cdot \frac{1}{3^{\lambda_2}} \sum_{r=0}^{\lambda_1} \sum_{s=0}^{\lambda_2} \frac{2^r 3^s}{(r+s)!} \tag{3.10}$$

We now construct a unisolvent rational approximant $\pi_{\mu_1\mu_2 11}(z_1,z_2)$ from the equations (2.6) and (2.7) together with (1.6a) and (1.6b) to have

$$\sum_{r=0}^{\min(1,\lambda_1)} \sum_{s=0}^{\min(1,\lambda_2)} b_{rs}c_{\lambda_1-r,\lambda_2-s} = a_{\lambda_1\lambda_2};$$

$$(\lambda_1,\lambda_2) \in E_{\mu_1\mu_2} \tag{3.11}$$

and

$$b_{00}c_{\mu_1+1,0} + b_{10}^{(\mu)}c_{\mu_1,0} = 0$$

$$b_{00}c_{\mu_1+1,1} + b_{10}^{(\mu)}c_{\mu_1,1} + b_{01}^{(\mu)}c_{\mu_1+1,0} + b_{11}^{(\mu)}c_{\mu_1,0} = 0 \tag{3.12}$$

$$b_{00}c_{0,\mu_2+1} + b_{01}^{(\mu)}c_{0,\mu_2} = 0.$$

Using the normalization $b_{00} = 1$, we readily solve the system of equations for the $b^{(\mu)}$'s obtaining

$$b_{01}^{(\mu)} = -\frac{1}{2}\frac{\sum_{r=0}^{\mu_1+1}\frac{2^r}{r!}}{\sum_{r=0}^{\mu_1}\frac{2^r}{r!}} \; ; \; b_{01}^{(\mu)} = -\frac{1}{3}\frac{\sum_{s=0}^{\mu_2+1}\frac{3^s}{s!}}{\sum_{s=0}^{\mu_2}\frac{3^s}{s!}} \tag{3.13}$$

$$b_{11}^{(\mu)} = -\frac{1}{6}\,\frac{\sum_{r=0}^{\mu_1+1}\sum_{s=0}^{1}\frac{2^r\,3^s}{(r+s)!}}{\sum_{r=0}^{\mu_1}\frac{2^r}{r!}} - $$

$$- b_{10}^{(\mu)}\,\frac{1}{3}\,\frac{\sum_{r=0}^{\mu_1}\sum_{s=0}^{1}\frac{2^r\,3^s}{(r+s)!}}{\sum_{r=0}^{\mu_1}\frac{2^r}{r!}} - b_{01}^{(\mu)}\,\frac{1}{2}\,\frac{\sum_{r=0}^{\mu+1}\frac{2^r}{r!}}{\sum_{r=0}^{\mu_1}\frac{2^r}{r!}}\ .$$

The bracket (μ) indicates the dependence of the b's on $\mu = (\mu_1, \mu_2)$. We obtain the following estimates on the coefficients $b^{(\mu)}$'s:

$$|b_{10}^{(\mu)} - (-\tfrac{1}{2})| \le \frac{2^{\mu_1+1}}{(\mu_1+1)!} \quad \text{i.e. } b_{10}^{(\mu)} \to -\frac{1}{2} \quad \text{as} \quad \mu \to (\infty,\infty)$$

$$|b_{01}^{(\mu)} - (-\tfrac{1}{3})| \le \frac{3^{\mu_2+1}}{(\mu_2+1)!} \quad \text{i.e. } b_{01}^{(\mu)} \to -\frac{1}{3} \quad \text{as} \quad \mu \to (\infty,\infty)$$

$$|b_{11}^{(\mu)} - \tfrac{1}{6}| \le \tfrac{1}{3}|\ b_{10}^{(\mu)} + \tfrac{1}{2}| \times$$

$$\times \left[\frac{\sum_{r=0}^{\mu_1+1}\sum_{s=0}^{1}\frac{2^r\,3^s}{(r+s)!}}{\sum_{r=0}^{\mu_1}\frac{2^r}{r!}}\right] + \tfrac{1}{2}|b_{01}^{(\mu)} + \tfrac{1}{3}| \left[\frac{\sum_{r=0}^{\mu_1+1}\frac{2^r}{r!}}{\sum_{r=0}^{\mu_1}\frac{2^r}{r!}}\right] \qquad (3.14)$$

from which we get

$$|b_{11}^{(\mu)} - \tfrac{1}{6}| \le \tfrac{5}{3}\ |b_{10}^{(\mu)} + \tfrac{1}{2}| + \tfrac{3}{2}|b_{01}^{(\mu)} + \tfrac{1}{3}| \qquad (3.15)$$

Thus as $\mu \to (\infty,\infty)$ $b_{11}^{(\mu)} \to \frac{1}{6}$. This basically establishes that $Q_{\mu 11}(z_1,z_2)$ the denominator polynomial of the URA candidate considered, tends to $q_{11}(z_1,z_2)$ where

$$q_{11}(z_1,z_2) = \frac{1}{F_1(1;1,1;1;\frac{1}{2}z_1\ \frac{1}{3}z_2)}$$

$$= 1 - \frac{1}{2} z_1 - \frac{1}{3} z_2 + \frac{1}{6}z_1z_2 \qquad (3.16)$$

In fact the convergence of $Q_{\mu 11}(z_1,z_2)$ to $q_{11}(z_1,z_2)$ is uniform on every compact subset of $\mathbb{C}^2$. Thus to show that the URA $\pi_{\mu 11}(z_1,z_2)$ constructed earlier tends uniformly to $f(z_1,z_2)$ on compact subsets of $\mathbb{C}^2\backslash q_{11}^{-1}(0)$ (not necessarily limited to $\Delta_4^2\backslash q_{11}^{-1}(0)$). We check that coefficients of the numerator of the rational approximant $\pi_{\mu 11}(z_1,z_2)$ tend to that of $F_{L1}(1;1;1;1,1;z_1,z_2)$ the numerator entire function of $f(z_1,z_2)$.

Let $P_{\mu 11}(z_1,z_2) = \sum_{\lambda_1=0}^{\mu_1} \sum_{\lambda_2=0}^{\mu_2} a^{(\mu)}_{\lambda_1\lambda_2} z^{\lambda_1} z_2^{\lambda_2}$ be the numerator of the URA candidate then

$$a^{(\mu)}_{\lambda_1\lambda_2} = \sum_{r=0}^{\min(\lambda_1,1)} \sum_{s=0}^{\min(\lambda_2,1)} b^{(\mu)}_{rs} c_{\lambda_1-r,\lambda_2-s} \qquad (3.17)$$

$$\to c_{\lambda_1\lambda_2} - \frac{1}{2} c_{\lambda_1-1,\lambda_2} - \frac{1}{3} c_{\lambda_1,\lambda_2-1} + \frac{1}{6} c_{\lambda_1-1,\lambda_2-1}$$

$$= \frac{1}{(\lambda_1+\lambda_2)!} \qquad \text{as } \mu \to (\infty,\infty)$$

on simplifying (cf. Lutterodt [1984a)).This concludes our discussion of an example to verify the Montessus' result.

Acknowledgement: This research was supported by HU grant FY87FRP and the author is grateful to the referee for his comments.

References

1. W.N. Bailey, Generalized Hypergeometric Series, Cambridge University Press, Cambridge, UK, 1935.

2. R.C. Gunning & H. Rossi, Analytic Functions of Several Complex Variables, Prentice-Hall, Englewood Cliffs, N.J., 1965.

3. L. Hörmander, An Introduction to Complex Analysis in Several Variables, D. Van Nostrand, Princeton, N.J., 1966.

4. C.H. Lutterodt, 'A Two Dimensional Analogue of Padé Approximant Theory', J. Phys. A Math. Gen. 7, (1974), 1027-1037.

5. C.H. Lutterodt, 'Rational Approximants to Holomorphic Functions is n-Dimensions', J. Math. Anal. & Applic. 53, (1976), 89-98.

6. C.H. Lutterodt, 'Rational Approximation by Approximants in $\mathbb{C}^n$', Complex Analysis & Applic. III, (1976a), 25-35, IAEA.

7. C.H. Lutterodt, 'On a Partial Converse of Montessus de Ballores Theorem in $\mathbb{C}^n$', J. Approx. Theory 40, (1984), 216-225.

8. C.H. Lutterodt, 'Meromorphic Functions, Maps and Their Rational Approximants in $\mathbb{C}^n$' Approx. Theory & Spline Functions, (1984a), 379-396, NATO.

9. C.H. Lutterodt, 'A Generalized Oka-Weil Approximation in a Polynomial Polyhedron in $\mathbb{C}^n$,' to appear in Complex Variables & Applic., Int. J., 1987.

10. L.J. Slater, 'Generalized Hypergeometric Functions', Cambridge University Press, Cambridge, UK., 1966.

ORTHOGONAL POLYNOMIALS AND THE MOMENT PROBLEM

Chairmen:

C. Brezinski

W. B. Gragg

Invited communications:

W. B. Gragg*
Computational aspects of polynomials orthogonal on the unit circle.

Short communications:

E. Hendriksen
Some orthogonal systems of ${}_{p+1}F_p$-type Laurent polynomials.

F. Marcellán
The moment problem on equipotential curves.

D. R. Masson
Difference equations, continued fractions, Jacobi matrices and orthogonal polynomials.

O. Njåstad
Multipoint Padé approximation and orthogonal rational functions.

W. J. Thron
L-Polynomials orthogonal on the unit circle.

W. Van Assche*
Orthogonal polynomials on several intervals.

---

* Lecture notes are not included.

SOME ORTHOGONAL SYSTEMS OF ${}_{p+1}F_p$-TYPE LAURENT POLYNOMIALS

E. Hendriksen

University of Amsterdam

Department of Mathematics

Roetersstraat 15, 1018 WB Amsterdam

The Netherlands

ABSTRACT. Starting from the Jacobi Laurent polynomials (see [1]), related to the ordinary polynomials ${}_2F_1(-n,-a;-c-n+1;z)$ and the orthogonality of these Laurent polynomials on the unit circle in **C** with respect to the weight function $w(z) = \text{const.}(-z)^{-c}(1-z)^{c-a-1}$, $a,c \in \mathbf{R}\backslash\mathbf{Z}$, $c>a$, we get orthogonal systems of ${}_{p+3}F_{p+2}$-type Laurent polynomials by adding linear combinations of the delta function δ_1 at the point 1 and its first p derivatives to the weight function w.

If only a scalar multiple of δ_1 is added to w we give for the resulting ${}_3F_2$-type system

(i) a second order linear differential equation with polynomial coefficients of bounded degrees which is satisfied by the corresponding ordinary polynomials.

(ii) the three term recurrence relation.

1.INTRODUCTION

In the more recent past it has been rather popular to add delta functions to the weight functions of orthogonal polynomial systems to get new orthogonal systems. For instance in Koornwinders paper [3] an orthogonal system of ${}_4F_3$-type polynomials is obtained by adding scalar multiples of the delta functions at -1 and 1 to the weight of the Jacobi polynomials on [-1,1]. In the present paper we add a linear combination of the delta function δ_1 at the point 1 and its first p derivatives $\delta_1^{(1)}, \delta_1^{(2)}, \dots, \delta_1^{(p)}$ to the weight functions of certain systems of Laurent polynomials which are orthogonal with respect to a weight function w on the unit circle in C. We consider only orthogonal systems of Laurent polynomials $(Q_n)_{n=0}^{\infty}$ of the following special form

A. Cuyt (ed.), Nonlinear Numerical Methods and Rational Approximation, 213–227.

$$Q_{2n}(z) = \alpha_{-n}^{(2n)} z^{-n} + \alpha_{-n+1}^{(2n)} z^{-n+1} + \ldots + \alpha_{n}^{(2n)} z^{n} \ ,$$

$$Q_{2n+1}(z) = \alpha_{-n-1}^{(2n+1)} z^{-n-1} + \alpha_{-n}^{(2n+1)} z^{-n} + \ldots + \alpha_{n}^{(2n+1)} z^{n}$$

with

$$\alpha_{-n}^{(2n)} = \alpha_{-n-1}^{(2n+1)} = 1 \quad \text{and} \quad \alpha_{n}^{(2n)} \neq 0 \ , \quad \alpha_{n}^{(2n+1)} \neq 0 \ ,$$

$$\alpha_{-n}^{(2n)}, \ldots, \alpha_{n}^{(2n)}, \alpha_{-n-1}^{(2n+1)}, \ldots, \alpha_{n}^{(2n+1)} \in \mathbf{C} \ , \quad n = 0, 1, 2, \ldots \ .$$

The system $(Q_n)_{n=0}^{\infty}$ is orthogonal with respect to a moment functional Φ (i.e. Φ is linear and $\Phi(Q_nQ_k) = 0$ if $n \neq k$ and $\Phi(Q_n^2) \neq 0$, $n,k = 0, 1, 2, \ldots$) if and only if $(Q_n)_{n=0}^{\infty}$ satisfies the recurrence relations

$$\begin{aligned} Q_{2n+1} &= (z^{-1} + g_{2n+1})Q_{2n} + f_{2n+1}Q_{2n-1} \ , \\ Q_{2n+2} &= (1 + g_{2n+2}\, z)Q_{2n+1} + f_{2n+2}Q_{2n} \end{aligned} \tag{1.1}$$

with $f_{2n+1}, f_{2n+2}, g_{2n+1}, g_{2n+2} \neq 0$, $n = 0, 1, 2, \ldots$, and $Q_{-1} = 0$ and $Q_0 = 1$. (See [1], Th.1.1).
Using the corresponding ordinary polynomials $V_{2n}(z) = z^n Q_{2n}(z)$ and $V_{2n+1} = z^{n+1}Q_{2n+1}(z)$ $n = 0, 1, 2, \ldots$ we see that $(Q_n)_{n=0}^{\infty}$ is orthogonal if and only if $(V_n)_{n=0}^{\infty}$ satisfies

$$V_n = (1 + g_n z)V_{n-1} + f_n z V_{n-2} \tag{1.2}$$

with $f_n, g_n \neq 0$, $n = 0, 1, 2, \ldots$ and $V_{-1} = 0$, $V_0 = 1$.

Let $(V_n)_{n=0}^{\infty}$ satisfy (1.2). Then the sequence $(V_n)_{n=0}^{\infty}$ is easily extended to a biorthogonal system $\{V_n, U_n\}_{n=0}^{\infty}$ with respect to Φ if we put

$$U_n = z^{-n-1}(V_n + f_{n+1} z V_{n-1}) \ , \quad n = 0, 1, 2, \ldots \ .$$

For this reason we say that the polynomial system $(V_n)_{n=0}^{\infty}$ is "biorthogonal with respect to Φ ".

In terms of the moment sequence $(c_n)_{n\in \mathbf{Z}}$, $c_n = \Phi(z^{-n})$, $n \in \mathbf{Z}$, and with the usual notation for Hankel determinants

$H_{n+1}^{(k)} = \det (c_{k+s+t})_{s,t=0}^{n}$, $n = 0, 1, 2, \ldots$, $H_0^{(k)} = 1$, $k \in \mathbf{Z}$, we have

$$V_n = \frac{1}{H_n^{(-n+1)}} \begin{vmatrix} c_{-n+1} \cdots c_1 \\ \cdots\cdots \\ c_0 \cdots\cdots c_n \\ z^n \cdots\cdots 1 \end{vmatrix} \quad \text{and} \quad U_n = \frac{z^{-n-1}}{H_n^{(-n+2)}} \begin{vmatrix} c_{-n+2} \cdots c_2 \\ \cdots\cdots \\ c_1 \cdots\cdots c_{n+1} \\ z^n \cdots\cdots 1 \end{vmatrix} .$$

Clearly, given a moment sequence $(c_n)_{n\in \mathbf{Z}}$, there exists an orthogonal system of Laurent polynomials Q_n of the prescribed form with moment functional $\Phi(z^{-n}) = c_n$, $n\in \mathbf{Z}$, if and only if $H_n^{(-n+1)} \neq 0$ and $H_n^{(-n+2)} \neq 0$, $n = 0, 1, 2, \ldots$. In this case, i.e. if $H_n^{(-n+1)} \neq 0$ and $H_n^{(-n+2)} \neq 0$, $n = 0, 1, 2, \ldots$, the moment sequence $(c_n)_{n\in \mathbf{Z}}$ will be called **normal**.

Applying shift and difference operator to the moment sequence $(c_n)_{n\in \mathbf{Z}}$ we get new sequences which, in case of normality, lead to new orthogonal systems of Laurent polynomials.

Let $\Delta x_n = x_{n+1} - x_n$ for any sequence $(x_n)_{n\in \mathbf{Z}}$.

For $j = 0, 1, 2, \ldots$ and $k\in \mathbf{Z}$ we define $\Phi_k^{(j)}$ by $\Phi_k^{(j)}(z^{-n}) = \Delta^j c_{n+k}$, $n\in \mathbf{Z}$.

For the moment sequence $(\Delta^j c_{n+k})_{n\in \mathbf{Z}}$ we use the notation

$H_{n+1,j}^{(k)} = \det(\Delta^j c_{k+s+t})_{s,t=0}^{n}$, $n = 0, 1, 2, \ldots$, and $H_{0j}^{(k)} = 1$, $j = 0, 1, 2, \ldots$; $k\in \mathbf{Z}$,

to indicate the corresponding Hankel determinants and in case of normality of $(\Delta^j c_{n+k})_{n\in \mathbf{Z}}$ the corresponding biorthogonal polynomials are written as $V_{nk}^{(j)}$. For negative integers n we put $V_{nk}^{(j)} = 0$.

Suppose now that w is a weight function on the unit circle Γ in $\mathbf{C}$ such that

$\int_\Gamma z^{-n}\, w\, dz = \Phi(z^{-n})$, $n\in \mathbf{Z}$, and suppose that $w^* = w + \Sigma_{k=0}^{p} \rho_k\, \delta_1^{(k)}$ where $\delta_1^{(k)}$ is the k-th derivative of the delta function at the point 1 and $\rho_k \in \mathbf{C}$, $\rho_p \neq 0$.

Let new moments c_n^* and the functional Ω be defined by

$$\Omega(z^{-n}) = c_n^* = \int_\Gamma z^{-n}\, w^*\, dz , \quad n\in \mathbf{Z} .$$

Then it is easily verified that

$$c_n^* = c_n + \Sigma_{k=0}^{p} \rho_k\, (n)_k , \quad n\in \mathbf{Z} .$$

Here the Pochhammer symbol $(x)_n = \dfrac{\Gamma(x+n)}{\Gamma(x)}$ for $x\in \mathbf{C}\backslash\mathbf{N}$ and $n\in \mathbf{Z}$.

If $(c_n^*)_{n\in Z}$ is again a normal sequence, then there is a system $(W_n)_{n=0}^{\infty}$ of polynomials which is biorthogonal with respect to Ω.

Using shift and difference operator on $(c_n^*)_{n\in Z}$ we obtain $W_{n,k}^{(j)}$, $\Omega_k^{(j)}$ from W_n, Ω in the same way as $V_{n,k}^{(j)}$, $\Phi_k^{(j)}$ are obtained from V_n and Φ. The corresponding Hankel determinants are

$$K_{n+1,j}^{(k)} = \det(\Delta^j c^*_{k+s+t})_{s,t=0}^{n}, \quad n = 0, 1, 2, \dots \text{ and } K_{0,j}^{(k)} = 1, \quad k\in Z, \quad j = 0, 1, 2, \dots .$$

In section 2 we give, under some additional normality conditions, an expression for W_n as a linear combination of $z^j V_{n-j,-j}^{(j)}$, $j = 0, 1, 2, \dots, p+1$, and an expression for W_n as a $_{p+3}F_{p+2}$-type hypergeometric polynomial if we start from the Jacobi Laurent polynomials. Section 3 is devoted to the case $p = 0$. Only a scalar multiple of δ_1 is added to the weight of the Jacobi Laurent polynomials. The new parameters in the resulting $_3F_2$-type polynomials are calculated and for these polynomials we give a second order differential equation and we give the parameters of the three term recurrence relation.

2. GENERAL FORM OF W_n

If $(V_n)_{n=0}^{\infty}$ is biorthogonal with respect to Φ and the moment sequence $(c_n)_{n\in Z}$ with $c_n = \Phi(z^{-n})$, $n\in Z$, is normal and f_n and g_n are as in (1.2), then (see [1]),

$$\Phi(z^{-n-1}V_n) = \frac{H_{n+1}^{(-n+1)}}{H_n^{(-n+1)}} = (-1)^n f_1 f_2 \dots f_{n+1} \neq 0 \tag{2.1}$$

and

$$\Phi(V_n) = (-1)^n \frac{H_{n+1}^{(-n)}}{H_n^{(-n+1)}} = (-1)^{n+1} \frac{f_1 f_2 \dots f_{n+1}}{g_1 g_2 \dots g_{n+1}} \neq 0 \quad . \tag{2.2}$$

PROPOSITION 2.1. If $(c_n)_{n\in Z}$ and $(\Delta c_{n-1})_{n\in Z}$ are normal, then there are nonzero scalars A_n such that

$$V_n = V_{n,1}^{(1)} + A_n z V_{n\,1,1}^{(1)}, \quad n = 0, 1, 2, \dots . \tag{2.3}$$

Proof. By the convention $V^{(j)}_{m,k} = 0$ if $m < 0$, this is obvious if $n = 0$. Let $n \geq 1$. From $\Phi^{(1)}_{-1}(z^{-j}) = \Phi((1-z)z^{-j})$, $j \in \mathbf{Z}$, we get $\Phi^{(1)}_{-1}(z^{-1}V_n) = -\Phi(V_n) \neq 0$ and $\Phi^{(1)}_{-1}(z^{-1}zV^{(1)}_{n-1,-1}) \neq 0$ by normality, hence there is $A_n \neq 0$ such that

$$\Phi^{(1)}_{-1}(z^{-1}(V_n - A_n z V^{(1)}_{n-1,-1})) = 0.$$

If $n \geq 2$ we have moreover

$$\Phi^{(1)}_{-1}(z^{-j} V_n) = \Phi((1-z) z^{-j} V_n) = 0, \quad j = 2, 3, \dots, n$$

and

$$\Phi^{(1)}_{-1}(z^{-j} z V^{(1)}_{n-1,-1}) = 0, \quad j = 2, 3, \dots, n .$$

Hence

$$\Phi^{(1)}_{-1}(z^{-j}(V_n - A_n z V^{(1)}_{n-1,-1})) = 0, \quad j = 1, 2, \dots, n.$$

Since $(V^{(1)}_{n,-1})^{\infty}_{n=0}$ is biorthogonal (2.3) follows.

COROLLARY 2.1. If $j, k \in \mathbf{Z}$, $j \geq 0$ are such that $(\Delta^j c_{n+k})_{n\in \mathbf{Z}}$ and $(\Delta^{j+1} c_{n+k-1})_{n\in \mathbf{Z}}$ are normal, then there exist nonzero scalars $A^{(j)}_{n,k}$ such that

$$V^{(j)}_{n,k} = V^{(j+1)}_{n,k-1} + A^{(j)}_{n,k} z V^{(j+1)}_{n-1,k-1}, \quad n = 0, 1, 2, \dots .$$

PROPOSITION 2.2. Let $(c_n)_{n\in \mathbf{Z}}$ and $(\Delta c_{n-1})_{n\in \mathbf{Z}}$ be normal. Suppose $c^*_n = c_n + \rho$, $n \in \mathbf{Z}$, where $\rho \in \mathbf{C}\backslash\{0\}$ is such that also $(c^*_n)_{n\in \mathbf{Z}}$ is normal. Let $(W_n)^{\infty}_{n=0}$ be biorthogonal with respect to Ω. Then there are nonzero $\lambda_n \in \mathbf{C}$ such that

$$W_n = V_n + \lambda_n z V^{(1)}_{n-1,-1}, \quad n = 0, 1, 2, \dots .$$

Proof. Since $\Delta c^*_{n-1} = \Delta c_{n-1}$ we have, by proposition (2.1),

$$V_n = V^{(1)}_{n,-1} + A_n z V^{(1)}_{n-1,-1}, \quad n = 0, 1, 2, \dots$$

and

$$W_n = W^{(1)}_{n,-1} + B_n z W^{(1)}_{n-1,-1} = V^{(1)}_{n,-1} + B_n z V^{(1)}_{n-1,-1}, \quad n = 0, 1, 2, \dots$$

for certain $A_n, B_n \in \mathbf{C}$. Elimination of $V^{(1)}_{n,-1}$ gives

$$W_n = V_n + \lambda_n z V^{(1)}_{n-1,-1}, \quad n = 0, 1, 2, \dots .$$

In order to show that $\lambda_n \neq 0$ we put $\Omega(z^{-n}) = c^*_n$. If $\lambda_n = 0$ for some n, then $V_n = W_n$ and

$$0 = \Omega(z^{-n} W_n) = \Phi(z^{-n} W_n) + (\Omega - \Phi)(z^{-n} W_n) = \Phi(z^{-n} V_n) + (\Omega - \Phi)(z^{-n} V_n) = \rho V_n(1).$$

As $H_n^{(-n+1)} V_n(1) = (-1)^n H_{n,1}^{(-n+1)}$ and $\rho \neq 0$, it follows that $H_{n,1}^{(-n+1)} = 0$. This is a contradiction with the normality of $(\Delta c_{n-1})_{n\in \mathbf{Z}}$. So $\lambda_n \neq 0$, $n = 0, 1, 2, \ldots$.

PROPOSITION 2.3. Let $(\Delta^j c_{n-j})_{n\in \mathbf{Z}}$ be normal moment sequences for $j = 0, 1, \ldots, p+1$. Let

$$c_n^* = c_n + \Sigma_{k=0}^{p} \rho_k (n)_k \,, \quad n \in \mathbf{Z}, \quad \text{with} \quad \rho_p \neq 0,$$

and suppose that $(\Delta^j c_{n-j}^*)_{n\in \mathbf{Z}}$ are normal for $j = 0, 1, \ldots, p$. If $(W_n)_{n=0}^{\infty}$ is biorthogonal with respect to Ω, then there are $\lambda_{n,0}, \lambda_{n,1}, \ldots, \lambda_{n,p+1}$ with $\lambda_{n,0} = 1$ and $\lambda_{n,p+1} \neq 0$ such that

$$W_n = \Sigma_{j=0}^{p+1} \lambda_{n,j} z^j V_{n-j,-j}^{(j)} \,, \quad n = 0, 1, 2, \ldots . \tag{2.4}$$

$(V_{n,k}^{(j)} = 0$ if $n<0)$

Proof. Mathematical induction on p. If $p = 0$ we have just Proposition 2.2. Now let $p \geq 1$ and assume that the proposition is true for $p-1$. Then, since

$$\Delta c_{n-1}^* = \Delta c_{n-1} + \Sigma_{k=0}^{p-1} \rho_{k+1} (k+1)\,(n)_k \,, \quad n \in \mathbf{Z},$$

for the polynomial sequence $(W_{n,-1}^{(1)})_{n=0}^{\infty}$, biorthogonal with respect to $\Omega_{-1}^{(1)}$, $\Omega_{-1}^{(1)}(z^{-k}) = \Delta c_{k-1}^*$, $k \in \mathbf{Z}$, we have

$$W_{n,-1}^{(1)} = \Sigma_{j=0}^{p} \alpha_{n,j} z^j V_{n-j,-j-1}^{(j+1)} \tag{2.5}$$

with $\alpha_{n,0} = 1$, $n = 0, 1, 2, \ldots$.

By Proposition 2.1 there are $A_n \in C$ such that

$$W_n = W_{n,-1}^{(1)} + A_n z W_{n-1,-1}^{(1)} \,, \quad n = 0, 1, 2, \ldots . \tag{2.6}$$

By Corollary 2.1 there are also $B_{n,j} \in C$ such that

$$V_{n-j,-j-1}^{(j+1)} = V_{n-j,-j}^{(j)} - B_{n,j} z V_{n-j-1,-j-1}^{(j+1)} \,, \; n = 0, 1, 2, \ldots ; \; j = 0, 1, \ldots, p. \tag{2.7}$$

Replacing n by n-1 in (2.5) yields

$$W_{n-1,-1}^{(1)} = \Sigma_{j=0}^{p} \alpha_{n-1,j} z^j V_{n-j-1,-j-1}^{(j+1)} \,, \quad n = 1, 2, \ldots, \tag{2.8}$$

and with (2.7) we get from (2.5)

$$W^{(1)}_{n,-1} = \alpha_{n,0} V^{(0)}_{n,0} + \Sigma^{p}_{j=1} (\alpha_{n,j} - \alpha_{n,j-1} B_{n,j-1}) z^j V^{(j)}_{n-j,-j} - \alpha_{n,p} B_{n,p} z^{p+1} V^{(p+1)}_{n-p-1,-p-1} \tag{2.9}$$

$n = 0, 1, 2, \dots$.

Combination of (2.6), (2.8) and (2.9) gives (2.4) with

$$\lambda_{n,j} = \begin{cases} \alpha_{n,0} = 1 , & j = 0 \\ \alpha_{n,j} - \alpha_{n,j-1} B_{n,j-1} + \alpha_{n-1,j-1} A_n , & j = 1, \dots , p \\ -\alpha_{n,p} B_{n,p} + \alpha_{n-1,p} A_n , & j = p + 1. \end{cases}$$

Since $V^{(j)}_{n,k} = 0$ if $n < 0$ we only have to show that $\lambda_{n,p+1} \neq 0$ if $n>p$. With

$$\Omega(z^{-n}) = c^*_n \text{ and } \Phi^{(j)}_{-j}(z^{-n}) = \Delta^j c_{n-j} , \quad j = 0, 1, \dots , p ; \; n\in \mathbf{Z} ,$$

we clearly have

$$\Phi^{(j)}_{-j}(z^{-n}) = \Phi((1-z)^j z^{-n}) , \quad n\in \mathbf{Z} , \; j = 0, \dots , p$$

and

$$\Omega(z^{-n}) = \Phi(z^{-n}) + \Sigma^{p}_{k=0} (-1)^k \rho_k (D^k z^{-n})_{z=1} , \quad n\in \mathbf{Z} ,$$

where $D = \dfrac{d}{dz}$.

Suppose $\lambda_{n,p+1} = 0$ and $n > p$. Then

$$W_n = \Sigma^{p}_{j=0} \lambda_{n,j} z^j V^{(j)}_{n-j,-j} \text{ and } \Omega(z^{-n} (1-z)^p W_n) = 0 .$$

On the other hand

$$\Phi(z^{-n}(1-z)^p z^j V^{(j)}_{n-j,-j}) = \Phi^{(j)}_{-j}(z^{-n+j}(1-z)^{p-j} V^{(j)}_{n-j,-j}) = 0 , \quad j = 0, 1, \dots , p ,$$

and

$$(\Omega - \Phi)(z^{-n}(1-z)^p z^j V^{(j)}_{n-j,-j}) = \Sigma^{p}_{k=0} (-1)^k \rho_k (D^k((1-z)^p z^{-n+j} V^{(j)}_{n-j,-j}))_{z=1}$$
$$= (-1)^p \rho_p (-1)^p p! \, V^{(j)}_{n-j,-j}(1) = p! \, \rho_p V^{(j)}_{n-j,-j}(1) , \quad j = 0, 1, \dots , p ,$$

hence

$$\Omega(z^{-n}(1-z)^p W_n) = p! \, \rho_p W_n(1) ,$$

and it follows that

$$\rho_p W_n(1) = 0 . \tag{2.10}$$

As

$$K_n^{(-n+1)} W_n(1) = (-1)^n K_{n,1}^{(-n+1)} \quad \text{and} \quad \rho_p \neq 0$$

it follows from (2.10) that $K_{n,1}^{(-n+1)} = 0$, contradicting the normality of $(\Delta c^*_{n-1})_{n\in \mathbf{Z}}$.

Hence $\lambda_{n,p+1} \neq 0$.

In the remaining part of this paper we assume that

$$V_n = {}_2F_1(-n, -a; -c-n+1; z), \quad n = 0, 1, 2, \dots,$$

where $a,c \in \mathbf{C}$ and $a, -c, a-c \neq 0, 1, 2, \dots$.

As in [1], where only real values of a and c are considered, it can be shown that $(V_n)_{n=0}^{\infty}$ is biorthogonal with respect to the moment functional

$$\Phi(z^{-n}) = c_n = \frac{(a)_n}{(c)_n}, \quad n \in \mathbf{Z}.$$

In [1] the corresponding orthogonal Laurent polynomials Q_n were called "Jacobi Laurent polynomials", and it has been shown that these Laurent polynomials are orthogonal with respect to the weight function

$$w(z) = \frac{-1}{2\pi i} \frac{\Gamma(c)\Gamma(1-a)}{\Gamma(c-a)} (-z)^{-c}(1-z)^{c-a-1}, \quad 0 \leq \arg z < 2\pi,$$

on the unit circle in $\mathbf{C}$ if a and c satisfy the additional conditions $a,c \in \mathbf{R}\backslash\mathbf{Z}$ and $c > a$.

In the present paper the weight function is not essential. Therefore we assume only that $a,c \in \mathbf{C}$ and $a, -c, a-c \neq 0, 1, 2, \dots$. Under this assumption we have

$$\Phi^{(j)}_{-j}(z^{-n}) = \Delta^j c_{n-j} = \frac{(c-a)_j}{(-a+1)_j} \cdot \frac{(a-j)_n}{(c)_n}, \quad n \in \mathbf{Z}, \ j = 0, 1, 2, \dots,$$

while $(\Delta^j c_{n-j})_{n\in \mathbf{Z}}$ is normal, $j = 0, 1, 2, \dots$, and

$$V^{(j)}_{n-j,-j} = {}_2F_1(-n+j,-a+j;-c-n+j+1; z), \quad j = 0, 1, 2, \dots.$$

Moreover we have

$$D^j V_n = D^j {}_2F_1(-n,-a;-c-n+1; z) = \frac{(-n)_j(-a)_j}{(-c-n+1)_j} \cdot {}_2F_1(-n+j,-a+j;-c-n+j+1; z)$$

$$= \frac{(-n)_j(-a)_j}{(-c-n+1)_j} V^{(j)}_{n-j,-j}, \quad n,j = 0, 1, 2, \dots. \tag{2.11}$$

For use in the proof of the next theorem we mention the following simple relation between hypergeometric functions

$$ {}_pF_q(a_1,\dots,a_p;b_1,\dots,b_q;\ z) + \frac{1}{r}\cdot\frac{a_1\dots a_p}{b_1\dots b_q}\, z\, {}_pF_q(a_1+1,\dots,a_p+1;b_1+1,\dots,b_q+1;\ z) $$

$$ = {}_{p+1}F_{q+1}(a_1,\dots,a_p,r+1;b_1,\dots,b_q,r;\ z) $$

or, equivalently,

$$ (I + \frac{1}{r} zD)\, {}_pF_q(a_1,\dots,a_p;b_1,\dots,b_q;\ z) = {}_{p+1}F_{q+1}(a_1,\dots,a_p,r+1;b_1,\dots,b_q,r;\ z) \tag{2.12} $$

THEOREM 2.1. Let $a,c \in \mathbf{C}$; $a, -c, a-c \neq 0, 1, 2, \dots$. Let $c_n = \frac{(a)_n}{(c)_n} = \Phi(z^{-n})$, $n \in \mathbf{Z}$, and $c_n^* = c_n + \Sigma_{k=0}^{p} \rho_k\, (n)_k = \Omega(z^{-n})$, $n \in \mathbf{Z}$, with $\rho_p \neq 0$. Suppose that $\rho_0, \rho_1, \dots, \rho_p$ are chosen such that the sequences $(\Delta^j c^*_{n-j})_{n\in Z}$ are normal as $j = 0, 1, \dots , p$, and assume that $(W_n)_{n=0}^{\infty}$ is biorthogonal with respect to Ω.

Then there are $s_{1,n}, s_{2,n}, \dots, s_{p+1,n} \in \mathbf{C}\setminus\{0\}$ such that

$$ W_n = {}_{p+3}F_{p+2}(-n,-a,s_{1,n}+1, \dots ,s_{p+1,n}+1;\ -c-n+1,s_{1,n}, \dots ,s_{p+1,n};\ z)\ , \quad n = 0, 1, 2, \dots . $$

Proof. By proposition 2.3 and (2.11) there are $\mu_{n,0}, \mu_{n,1}, \dots, \mu_{n,p+1} \in \mathbf{C}$ with $\mu_{n,0}=1$ and $\mu_{n,p+1} \neq 0$ such that

$$ W_n = \Sigma_{j=0}^{p+1} \mu_{n,j}\, z^j D^j V_n\, , \quad n = 0, 1, 2, \dots . $$

Since for each $m \in \mathbf{N}$ there are $\beta_{m,1}, \beta_{m,2}, \dots ,\beta_{m,m} \in \mathbf{C}$ with $\beta_{m,1}=\beta_{m,m}=1$ so that

$$ (zD)^m = \beta_{m,1}\, zD + \beta_{m,2}\, z^2 D^2 + \dots + \beta_{m,m} z^m D^m, $$

there is a polynomial P of the form

$$ P(t) = \mu_{n,p+1}(s_{1,n}+ t)\, (s_{2,n}+ t) \dots (s_{p+1,n}+ t) $$

with

$$ \Sigma_{j=0}^{p+1} \mu_{n,j}\, z^j D^j = P(zD)\, . $$

Clearly we have $\mu_{n,p+1}\, s_{1,n}\, s_{2,n} \dots s_{p+1,n} = 1$, so $s_{1,n}, s_{2,n}, \dots , s_{p+1,n} \neq 0$,

and

$$P(t) = (1+\frac{t}{s_{1,n}})(1+\frac{t}{s_{2,n}}) \dots (1+\frac{t}{s_{p+1,n}}) .$$

Hence

$$W_n = (I+\frac{1}{s_{p+1,n}} zD)(I+\frac{1}{s_{p,n}} zD) \dots (I+\frac{1}{s_{1,n}} zD)V_n, \quad n = 0, 1, 2, \dots . \tag{2.13}$$

It follows with mathematical induction from (2.12) and (2.13) that

$$W_n = {}_{p+3}F_{p+2}(-n,-a,s_{1,n}+1,s_{2,n}+1, \dots ,s_{p+1,n}+1; -c-n+1,s_{1,n},s_{2,n}, \dots ,s_{p+1,n}; z) .$$

REMARK 2.1. The fact that the weight function it self is not essential as long as we know the moments, is also illustrated by the fact that every moment sequence of the form $(\gamma_n)_{n\in\mathbf{Z}}$ is a difference of two "strong Stieltjes sequences",(see [2] for definition of strong Stieltjes sequence). This implies that there exists a real (signed) measure of bounded variation on the interval $[0,\infty)$ such that

$$\int_0^\infty z^{-n}\, d\mu(z) = \gamma_n, \quad n\in \mathbf{Z} .$$

REMARK 2.2. Starting from the (ordinary) Jacobi polynomials ${}_2F_1(-n,c+n;a+1; z)$, $n = 0, 1, \dots,$ on the interval $(0,1)$, $c>a>-1$, we obtain also a ${}_{p+3}F_{p+2}$-type orthogonal polynomial system if we add a linear combination of the delta function δ_1 at 1 and its derivatives $\delta_1^{(1)}, \dots , \delta_1^{(p)}$ to the Jacobi weight function on (0,1). We get as well a ${}_{p+3}F_{p+2}$-type orthogonal system if we add such a linear combination of the delta function δ_0 at 0 and its derivatives to the Jacobi weight function.

3.THE CASE $p = 0$.

In this section we consider the special case $p = 0$. We have

$$\Omega(z^{-n}) = c_n^* = c_n + \rho, \; n\in \mathbf{Z}, \text{ and } W_n = V_n + \lambda_n z V^{(1)}_{n-1,-1}, \quad n = 0, 1, 2, \dots .$$

Assuming $\Phi(z^{-n}) = c_n = \dfrac{(a)_n}{(c)_n}$, $n\in \mathbf{Z}$, (a, -c, a-c $\neq$ 0, 1, 2, ...), so that

$$V_n = {}_2F_1(-n,-a;-c-n+1; z) \quad\text{and}\quad V^{(1)}_{n,-1} = {}_2F_1(-n,-a+1;-c-n+1; z) ,$$

we calculate λ_n from $0 = \Omega(z^{-1} W_n)$. As V_n satisfies the recurrence relation (1.2) with (see [1])

$$f_1 = \frac{a}{c} \quad \text{and} \quad f_n = -\frac{(c-a+n-2)(n-1)}{(c+n-2)(c+n-1)}, \quad n = 2, 3, \ldots \tag{3.1}$$

and

$$g_n = \frac{-a+n-1}{c+n-1}, \quad n = 1, 2, \ldots, \tag{3.2}$$

(2.2) implies

$$\frac{H_{n+1}^{(-n)}}{H_n^{(-n+1)}} = (-1)^n \Phi(V_n) = (-1)^n \frac{n!\,(c-a)_n}{(c)_n(-a+1)_n}, \quad n = 0, 1, 2, \ldots . \tag{3.3}$$

Furthermore we have

$$\frac{H_{n,1}^{(-n+1)}}{H_n^{(-n+1)}} = (-1)^n V_n(1) = (-1)^n \frac{(c-a)_n}{(c)_n}, \quad n = 0, 1, 2, \ldots, \tag{3.4}$$

by elementary manipulations with determinants, and

$$\frac{H_{n,1}^{(-n+1)}}{H_{n,1}^{(-n)}} = (-1)^n \frac{(-a+1)_n}{(c)_n}, \quad n = 0, 1, 2, \ldots, \tag{3.5}$$

(leading coefficient of $V_{n,-1}^{(1)}$).

Since $\Omega(V) = \Phi(V) + \rho V(1)$ for every Laurent polynomial V we get

$$0 = \Omega(z^{-1}W_n) = \Phi(z^{-1}V_n) + \rho V_n(1) + \lambda_n[\Phi(V_{n-1,-1}^{(1)}) + \rho V_{n-1,-1}^{(1)}(1)] \tag{3.6}$$

where $\Phi(z^{-1}V_n) = 0$ and

$$\Phi(V_{n-1,-1}^{(1)}) = (-1)^{n-1} \frac{H_n^{(-n+1)}}{H_{n-1,1}^{(-n+1)}} = (-1)^{n-1} \frac{H_n^{(-n+1)} H_{n-1}^{(-n+2)} H_{n-1,1}^{(-n+2)}}{H_{n-1}^{(-n+2)} H_{n-1,1}^{(-n+2)} H_{n-1,1}^{(-n+1)}}$$

$$= (-1)^{n-1} (-1)^{n-1} \frac{(n-1)!\,(c-a)_{n-1}}{(c)_{n-1}(-a+1)_{n-1}} (-1)^{n-1} \frac{(c)_{n-1}}{(c-a)_{n-1}} (-1)^{n-1} \frac{(-a+1)_{n-1}}{(c)_{n-1}} = \frac{(n-1)!}{(c)_{n-1}} \tag{3.7}$$

by (3.3), (3.4) and (3.5).

With $V_n(1) = \frac{(c-a)_n}{(c)_n}$ and $V_{n-1,-1}^{(1)}(1) = \frac{(c-a+1)_{n-1}}{(c)_{n-1}}$ it follows from (3.6) and (3.7) that

$$\lambda_n = -\frac{\dfrac{c-a}{c+n-1}}{1 + \dfrac{1}{\rho} \cdot \dfrac{(n-1)!}{(c-a+1)_{n-1}}}, \quad n = 1, 2, \ldots .$$

With $r_n = \frac{na}{c-a}(1+\frac{1}{\rho}\cdot\frac{(n-1)!}{(c-a+1)_{n-1}})$, $n = 1, 2, \ldots$, we get

$$W_n = V_n + \frac{1}{r_n} z\,DV_n, \quad n = 1, 2, \ldots, \tag{3.8}$$

so with (2.12) we obtain

$$W_n = {}_3F_2(-n,-a,r_n+1;-c-n+1,r_n;\, z), \quad n = 1, 2, \ldots .$$

SECOND ORDER DIFFERENTIAL EQUATION:

In order to derive a second order differential equation for W_n we write in this subsection

$$r = r_n, \ \phi = r_n W_n \ \text{and} \ y = V_n .$$

Then (3.8) becomes

$$\phi = ry + zy' \tag{3.9}$$

Using the hypergeometric differential equation

$$z(1-z)y''+[(a+n-1)z-(c+n-1)]y'-nay = 0$$

we obtain from (3.9) by differentiation

$$(1-z)\phi' = nay - [(a+n+r)z - (c+n+r)]y'. \tag{3.10}$$

Elimination of y' from (3.9) and (3.10) gives

$$z(1-z)\phi' + [(a+n+r)z - (c+n+r)]\phi = - Ay \tag{3.11}$$

where $A = r(c+n+r) - (r+n)(r+a)z$.

Differentiation of (3.11) and elemination of y' with the help of (3.9) yields

$$z^2(1-z)\phi''+[(a+n+r-2)z - (c+n+r-1)]z\phi'+[A+(a+n+r)z]\phi = (rA - A'z)y . \tag{3.12}$$

Elimination of y from (3.11) and (3.12) gives

$$z^2(1-z)A\phi''+ \{[(a+n+r-2)z - (c+n+r-1)]zA+z(1-z)(rA - A'z)\}\phi'+$$
$$+\{[(a+n+r)z+A]A+[(a+n+r)z - (c+n+r)](rA - A'z)\}\phi = 0. \tag{3.13}$$

Let B_j be the the coefficient of the jth derivative of ϕ in (3.13). Then, using $-A'z = r(c+n+r) - A$, we get

$$B_0 = \{(a+n+r)z+A+r[(a+n+r)z - (c+n+r)]\}A+[(a+n+r)z - (c+n+r)][r(c+n+r) - A]$$
$$= \{A+r[(a+n+r)z - (c+n+r)]\}A+(c+n+r)\{A+r[(a+n+r)z - (c+n+r)]\}$$
$$= z[r(a+n+r) - (r+n)(r+a)][A+c+n+r]$$

$$= z[r(a+n+r) - (r+n)(r+a)][A+c+n+r]$$

$$= - naz[c+n+r+A],$$

and with $\beta = (a+n - 1)z - (c+n - 1)$,

$$z^{-1} B_1 = \beta A + [(r - 1)z - r]A + (1-z)[rA+r(c+n+r) - A]$$

$$= \beta A + [(r - 1)z - r + (1-z)(r - 1)]A + (1-z)r(c+n+r)$$

$$= \beta A - A + r(c+n+r) - r(c+n+r)z$$

$$= \beta A - A(1)z.$$

Thus we have

$$z^{-1} B_0 = - na(c+n+r+A),$$

$$z^{-1} B_1 = \beta A - A(1)z,$$

$$z^{-1} B_2 = z(1-z)A,$$

and (3.13) reduces to

$$z(1-z)A\phi''+\{[(a+n-1)z - (c+n-1)]A - A(1)z\}\phi' - na(c+n+r+A)\phi = 0 \tag{3.14}$$

Clearly (3.14) is a second order linear differential equation with polynomial coefficients depending on n but of degrees at most 3, 2 and 1 .

RECURRENCE RELATION:

The normality of $(c_n^*)_{n\in Z}$ means that the W_n satisfy a recurrence relation

$$W_n = (1+g_n^* z)W_{n-1} + f_n^* zW_{n-2}, \quad n = 1, 2, ... \tag{3.15}$$

with $f_n^*, g_n^* \neq 0$, $n = 1, 2, \dots$, and $W_{-1} = 0$ and $W_0 = 1$.

Let $r_0 = \dfrac{a}{\rho}$. Then $W_n = {}_3F_2(-n,-a,r_n+1;-c-n+1,r_n;z)$ $n = 0, 1, 2, \dots$, and for the leading coefficient of W_n, denoted as $\text{l.c.}W_n$, we have

$$\text{l.c.}W_n = \frac{(-a)_n}{(c)_n} \cdot \frac{r_n+n}{r_n}, \quad n = 0, 1, 2, ...$$

and also, by (3.15),

$$\text{l.c.}W_n = g_1^* g_2^* \dots g_n^*, \quad n = 1, 2, \dots .$$

Hence with (3.2) we get

$$g_n^* = \frac{-a+n-1}{c+n-1} \cdot \frac{r_{n-1}(r_n+n)}{r_n(r_{n-1}+n-1)} = g_n \frac{r_{n-1}(r_n+n)}{r_n(r_{n-1}+n-1)} \tag{3.16}$$

For the calculation of f_n^* we use the analogue of (2.1).

$$\Omega(z^{-n-1}W_n) = (-1)^n f_1^* f_2^* \dots f_{n+1}^*, \quad n = 0, 1, 2, \dots . \tag{3.17}$$

Since by (2.1) and (3.1)

$$\Phi(z^{-n-1}V_n) = (-1)^n f_1 f_2 \dots f_{n+1} = \frac{n!\, a\, (c-a)_n}{(c)_{n+1}\,(c)_n}, \quad n = 0, 1, 2, \dots$$

and similarly

$$\Phi_{-1}^{(1)}(z^{-n}V_{n-1,-1}^{(1)}) = -\frac{(n-1)!\,(c-a)_n}{(c)_{n-1}\,(c)_n}, \quad n = 1, 2, \dots .$$

and

$$\lambda_n = \frac{-na}{r_n(c+n-1)}, \quad n = 1, 2, \dots ,$$

we have

$$\Omega(z^{-n-1}W_n) = \Omega((1-z)z^{-n-1}W_n) = \Phi((1-z)\, z^{-n-1}(V_n+\lambda_n z V_{n-1,-1}^{(1)})) = \Phi(z^{-n-1}V_n)+\lambda_n \Phi_{-1}^{(1)}(z^{-n}V_{n-1,-1}^{(1)})$$

$$= \frac{n!\, a\, (c-a)_n}{(c)_{n+1}\,(c)_n} + \frac{n\, a}{r_n(c+n-1)} \cdot \frac{(n-1)!\,(c-a)_n}{(c)_{n-1}\,(c)_n} = \frac{n!\, a\, (c-a)_n}{(c)_{n+1}\,(c)_n} \cdot \frac{r_n+n+c}{r_n}, \quad n = 1, 2, \dots .$$

With (3.17) this implies

$$(-1)^n f_1^* f_2^* \dots f_{n+1}^* = \frac{n!\, a\, (c-a)_n}{(c)_{n+1}\,(c)_n} \cdot \frac{r_n+n+c}{r_n}, \quad n = 1, 2, \dots .$$

Since this also holds for $n = 0$ as $f_1^* = \Omega(z^{-1}) = \frac{a}{c} + \rho$ and $r_0 = \frac{a}{\rho}$, it follows that

$$f_{n+1}^* = \frac{-n\,(c-a+n-1)}{(c+n)\,(c+n-1)} \cdot \frac{r_{n-1}(r_n+n+c)}{r_n(r_{n-1}+n-1+c)} = f_{n+1} \frac{r_{n-1}(r_n+n+c)}{r_n(r_{n-1}+n-1+c)}, \quad n = 1, 2, \dots . \tag{3.18}$$

Furthermore, using

$$(c-a+n)\, r_{n+1} = (n+1)\,(r_n+a), \quad n = 0, 1, 2, \dots , \tag{3.19}$$

which follows easily from the definition of the r_n, (3.16) and (3.18) yield

$$\frac{f^*_{n+1}}{g^*_{n+1}} = \frac{f_{n+1}}{g_{n+1}} \cdot \frac{r_{n-1}(r_n+a)}{r_n(r_{n-1}+a)}, \quad n = 1, 2, \dots . \tag{3.20}$$

It is obvious that with (3.19) other expressions than (3.16), (3.18) and (3.20) for g^*_n, f^*_n and f^*_n / g^*_n can be given. Moreover it follows directly that for the polynomial A in (3.14) we have $\deg A = 1$.

REFERENCES

[1] Hendriksen,E. and H.van Rossum, *Orthogonal Laurent polynomials*, Proc. Kon. Ned. Akad. v.Wet., Amsterdam, ser. A, **89** (1) (1986),17 - 36 ≡ Indag. Math. **48** (1) (1986).

[2] Jones,W.B., W.J.Thron and H.Waadeland, *A strong Stieltjes moment problem*, Trans. Amer. Math. Soc.(1980),503 - 528.

[3] Koornwinder,T.H., *Orthogonal polynomials with weight function* $(1-x)^\alpha (1+x)^\beta + M\delta(x+1) + N\delta(x-1)$, Canad. Math. Bull. **27** (2), (1984), 205 - 214.

THE MOMENT PROBLEM ON EQUIPOTENTIAL CURVES

F. Marcellán
Dep. Matemática Aplicada
ETS Ingenieros Industriales
28006 Madrid, Spain

I. Pérez-Grasa
Dep. Análisis Matemático
Fac. Ciencias Económicas y Empresariales
50005 Zaragoza, Spain

ABSTRACT. Some types of Hermitian and positive definite matrices whose elements satisfy a linear recurrence relation have been studied in connection with the theory of orthogonal polynomials on algebraic curves; more precisely, for the lemniscates ([1]) and harmonic algebraic curves.

Through a linear recurrence relation for the moments of a Gram matrix, a necessary and sufficient condition for the extension is obtained. We obtain the moments in terms of a family of parameters, generalising the one's introduced by Geronimus for the unit circle [4] .

1. INTRODUCTION

Let Γ be an equipotential curve $|A(z)| = 1$, where $A(z) = \sum_{k=0}^{m} a_k z^k$ is an algebraic polynomial with complex coefficients.

If μ is a finite and positive Borel measure defined on Γ , it is well known ([10]) that the elements c_{ij} $(i, j \in \mathbb{N})$ of the Gram matrix for the inner product associated with μ, $\langle f(z), g(z)\rangle_\mu = \int_\Gamma f(z)\,\overline{g(z)}\,d\mu$, satisfy a linear relation

$$\sum_{k,h=0}^{m} a_k \bar{a}_h c_{i+k,j+h} = c_{ij} \quad (i, j \in \mathbb{N}) \qquad (1)$$

The linear dependencies satisfied by the moments associated with a real linear functional are of great interest in the study of semiclassical orthogonal polynomials (see [7] and [9]).

On the other hand, the above mentioned linear relation has been profusely used in the study of orthogonal polynomials on equipotential curves, as it is well known, that condition is equivalent to the isometric character of the multiplication operator $A(z)$ in the pre-Hilbert space $\mathbb{P}$ of the algebraic polynomials. More specifically, the recurrence relation that satisfy the orthogonal polynomials associated with μ , and the problems of density of $\mathbb{P}$ in L^2_μ are intimately linked to the isometric character of $A(z)$ (see [1]).

In the present paper, we consider two types problems:

1. Given a Hermitian positive definite matrix $(c_{ij})_{i,j \in \mathbb{N}}$, that

A. Cuyt (ed.), Nonlinear Numerical Methods and Rational Approximation, 229–238.

satisfies (1), we study the existence of a finite and positive Borel measure defined on Γ such that

$$c_{ij} = \int_\Gamma z^i \bar{z}^j \, d\mu .$$

that is the generalization of the moment problem for the real line $\mathbb{R}$ or the unit circle $\mathbb{T}$.

2. Given a Hermitian positive definite matrix $(c_{ij})_{i,j=0}^{n}$ whose elements satisfy the relation (1), we characterise its extensions in such a way that $(c_{ij})_{i,j=0}^{n+1}$ is Hermitian positive definite and its elements satisfy (1). A recursive formula for the moments is provided.

So as to facilitate the presentation of the expressions, we consider the case of Bernoulli's lemniscate, $BL = \{ z \in \mathbb{C} : |z^2-1| = 1\}$.

2. PROBLEM 1

Definition 1.- A Hermitian matrix $(c_{ij})_{i,j \in \mathbb{N}}$ is said to be relative to BL, if it is positive definite and if its elements satisfy a relation of the type

$$c_{i+2,j+2} = c_{i+2,j} + c_{i,j+2} \qquad (i,j \in \mathbb{N}) ,$$

or, equivalently, in the matrix there exists a relation

$$\begin{matrix} \cdot & \cdot & + \\ \cdot & \cdot & \cdot \\ + & \cdot & = \end{matrix}$$

(see [2]).

It is evident that, for every finite positive Borel measure μ defined on BL, its Gram matrix is relative to BL.

On the other hand, if we consider the positive definite sesquilinear form $\Psi(z^i, z^j) = c_{ij}$ associated with a matrix relative to BL, the formal monic orthogonal polynomials are given by

$$\tilde{P}_n(z) = \frac{1}{m_{n-1}} \begin{vmatrix} c_{00} & c_{10} & \cdots & c_{n0} \\ c_{01} & c_{11} & \cdots & c_{n1} \\ \cdots & \cdots & \cdots & \cdots \\ c_{0,n-1} & c_{1,n-1} & \cdots & c_{n,n-1} \\ 1 & z & \cdots & z^n \end{vmatrix}$$

where

$$m_{n-1} = \det M_{n-1} \quad \text{and} \quad M_{n-1} = (c_{ij})_{i,j=0}^{n-1}$$

If instead of applying the Gram-Schmidt orthogonalization process to the sequence $(z^k)_{k \in \mathbb{N}}$, we apply it to

$$\{1, z, z^2-1, z(z^2-1), \ldots (z^2-1)^k, z(z^2-1)^k \ldots\}$$

a block Toeplitz matrix for the generalized matrix of moments is obtained:

$$N_{2n} = \begin{pmatrix} D_0 & D_1 & \cdots & D_{n-1} \\ D_1^* & D_0 & \cdots & D_{n-2} \\ \cdots & \cdots & \cdots & \cdots \\ D_{n-1}^* & D_{n-2}^* & & D_0 \end{pmatrix} \in \mathbb{C}^{2n\times 2n}, \qquad n \in \mathbb{N}$$

where D_k^* means the transposed conjugate of D_k

$$D_n = \begin{pmatrix} d_{2n,0} & d_{2n+1,0} \\ d_{2n,1} & d_{2n+1,1} \end{pmatrix} \quad \text{and}$$

$$d_{2n,i} = \sum_{k=0}^{n} \binom{n}{k} c_{2k,i} (-1)^{n-k} \qquad (i=0,1);$$

$$d_{2n+1,i} = \sum_{k=0}^{n} \binom{n}{k} c_{2k+1,i} (-1)^{n-k} \qquad (i=0,1);$$

so

$$D_n = \sum_{k=0}^{n} \binom{n}{k} (-1)^{n-k} C_k,$$

where

$$C_k = \begin{pmatrix} c_{2k,0} & c_{2k+1,0} \\ c_{2k,1} & c_{2k+1,1} \end{pmatrix}$$

If

$$N = \begin{pmatrix} D_0 & D_1 & \cdots & D_n & \cdots \\ D_1^* & D_0 & \cdots & D_{n-1} & \cdots \\ \cdots & \cdots & \cdots & \cdots & \cdots \end{pmatrix}$$

is a 2x2 block Toeplitz infinite matrix, it is well known (see [3]) that a matrix valued measure Ω , hermitian positive definite, over T exists, such that

$$D_n = \frac{1}{2\pi} \int_{-\pi}^{\pi} e^{in\theta}\, d\,\Omega(\theta)$$

where

$$d\,\Omega(\theta) = \begin{vmatrix} d\mu_{11}(\theta) & d\mu_{12}(\theta) \\ d\bar{\mu}_{12}(\theta) & d\mu_{22}(\theta) \end{vmatrix}$$

Thus, since

$$C_n = \sum_{k=0}^{n} \binom{n}{k} D_k = \int_{-\pi}^{\pi} \sum_{k=0}^{n} \binom{n}{k} e^{ik\theta} \, d\Omega(\theta)$$

$$= \int_T (1+w)^n \, d\Omega_1(w) = \int_{BL} z^{2n} \, d\tilde{\Omega}(z)$$

$$= \int_{BL} \begin{bmatrix} z^{2n} \, d\tilde{\mu}_{11}(z) & \bar{z}^{2n} \, d\tilde{\mu}_{12}(z) \\ z^{2n} \, d\overline{\tilde{\mu}}_{12}(z) & z^{2n} \, d\tilde{\mu}_{22}(z) \end{bmatrix}$$

we have

$$c_{2n,0} = \int_{BL} z^{2n} \, d\tilde{\mu}_{11}(z) \quad ; \quad c_{2n,1} = \int_{BL} z^{2n} \, d\overline{\tilde{\mu}_{12}}(z) \; ,$$

$$c_{2n+1,0} = \int_{BL} z^{2n} \, d\tilde{\mu}_{12}(z) \; ; \quad c_{2n+1,1} = \int_{BL} z^{2n} \, d\tilde{\mu}_{22}(z),$$

and thus

Proposition 1

The moment problem for a matrix relative to BL can be solved if and only if

$$d\tilde{\mu}_{22}(z) = |z|^2 \, d\tilde{\mu}_{11}(z) \qquad z \, d\overline{\tilde{\mu}}_{12}(z) = \bar{z} \, d\tilde{\mu}_{12}(z) \qquad (z \in BL).$$

An example of a matrix relative to BL for which the moment problem can not be solved, would be:

$$D_0 = \begin{bmatrix} 1 & 0 \\ 0 & 1 \end{bmatrix} \quad ; \quad D_n = \begin{bmatrix} 0 & 0 \\ 0 & 0 \end{bmatrix} \qquad (n \geqslant 1).$$

Then, $C_n = D_0 \quad (\forall\, n \in \mathbb{N})$.

3. PROBLEM 2

In a recent paper ([5]), we have proved the following result: given two sequences $(a_n^{(i)})_{n \in \mathbb{N}}$ $(i = 1, 2)$, with

$$a_0^{(i} = 1 \quad (i = 1, 2) \qquad \text{and} \qquad a_1^{(1} - a_1^{(2} = 2,$$

and satisfying the condition

$$e_{n-2} - \sum_{i,j=1}^{2} a_n^{(i} \, \bar{a}_n^{(j} \, M_{ji}^{(n-1} = e_n > 0 \; , \qquad (2)$$

where $\{M_{ji}^{(n-1}\}_{i,j=1}^{2}$ and e_{n-2} are obtained from $(a_k^{(i})_{0\leq k\leq n-1}$ (i = 1, 2), there exists a sequence of monic orthogonal polynomials relative to BL, such that

$$\tilde{P}_n(1) = a_n^{(1}, \quad \tilde{P}_n(-1) = a_n^{(2} \qquad (n \in \mathbb{N}).$$

Condition (2) is similar to the one used by Geronimus in the unit circle ([4]).

Proposition 2.

Given the sequences $(a_n^{(i})_{n\in\mathbb{N}}$ (i=1,2) that satisfy the above mentioned conditions, there is one and only one family $(c_{ni})_{n\in\mathbb{N}}$, i=0,1, of moments that define a matrix relative to BL.

More precisely:

$$c_{n0} = -\sum_{k=0}^{n-3} b_k^{(n-2} c_{k+2,0} - \sum_{j=1}^{2} a_n^{(j} (M_{1j}^{(n-1} + M_{2j}^{(n-1})$$

$$c_{n1} = -\sum_{k=0}^{n-3} b_k^{(n-2} c_{k+2,1} - \sum_{j=1}^{2} a_n^{(j} (-M_{1j}^{(n-1} + M_{2j}^{(n-1})$$

where

$$\tilde{P}_{n-2}(z) = \sum_{k=0}^{n-2} b_k^{(n-2} z^k. \quad \text{and} \quad b_{n-2}^{(n-2} = 1.$$

Proof. Considering the basis $\{z^2, z^3, z^2-1, z(z^2-1), \ldots, z^k(z^2-1), \ldots\}$ of $\mathbb{P}$, and applying the Gram-Schmidt orthogonalization process, then:

$$\tilde{P}_n(z) = \frac{1}{m_{n-1}} \begin{vmatrix} c_{22} & c_{32} & c_{20} & c_{30} & \cdots & c_{n-1,0} & c_{n0} \\ c_{23} & c_{33} & c_{21} & c_{31} & \cdots & c_{n-1,1} & c_{n1} \\ c_{02} & c_{12} & c_{00} & c_{10} & \cdots & c_{n-3,0} & c_{n-2,0} \\ c_{03} & c_{13} & c_{01} & c_{11} & \cdots & c_{n-3,1} & c_{n-2,1} \\ \cdots & \cdots & \cdots & \cdots & \cdots & \cdots & \cdots \\ c_{0,n-1} & c_{1,n-1} & c_{0,n-3} & c_{1,n-3} & \cdots & c_{n-3,n-3} & c_{n-2,n-3} \\ z^2 & z^3 & z^2-1 & z(z^2-1) & \cdots & z^{n-3}(z^2-1) & z^{n-2}(z^2-1) \end{vmatrix}$$

and

$$c_{ni} = \Psi(z^n, z^i) = -\sum_{k=0}^{n-3} b_k^{(n-2} c_{k+2,0} - \sum_{j=1}^{2} a^{(j} \Psi(\phi_{n-1}^{(j}(z), z^i)$$

(i= 0,1). Since

$$(z^2 - 1)\,\tilde{P}_{n-2}(z) = \tilde{P}_n(z) - \sum_{j=1}^{2} a_n^{(j} \phi_{n-1}^{(j}(z)$$

and

$$\phi^{(1}_{n-1}(z) = \frac{\begin{vmatrix} K_{n-1}(z,1) & K_{n-1}(z,-1) \\ K_{n-1}(-1,1) & K_{n-1}(-1,-1) \end{vmatrix}}{\begin{vmatrix} K_{n-1}(1,1) & K_{n-1}(1,-1) \\ K_{n-1}(-1,1) & K_{n-1}(-1,-1) \end{vmatrix}}$$

where

$$K_{n-1}(z,y) = \sum_{j=0}^{n-1} \hat{P}_j(z)\, \overline{\hat{P}_j(y)}$$

$$\phi^{(2}_{n-1}(z) = \frac{\begin{vmatrix} K_{n-1}(z,-1) & K_{n-1}(z,1) \\ K_{n-1}(1,-1) & K_{n-1}(1,1) \end{vmatrix}}{\begin{vmatrix} K_{n-1}(1,1) & K_{n-1}(1,-1) \\ K_{n-1}(-1,1) & K_{n-1}(-1,-1) \end{vmatrix}}$$

It can be easily shown

$$\Psi(\phi^{(i}_{n-1}(z), 1) = M^{(n-1}_{1i} + M^{(n-1}_{2i}, \quad \Psi(\phi^{(i}_{n-1}(z), z) = M^{(n-1}_{1i} - M^{(n-1}_{2i}$$

$(i = 1,2)$.

So

$$\begin{pmatrix} c_{n0} \\ \\ c_{n1} \end{pmatrix} = \begin{pmatrix} -M^{(n-1}_{11} - M^{(n-1}_{21} & -M^{(n-1}_{12} - M^{(n-1}_{22} \\ \\ -M^{(n-1}_{11} + M^{(n-1}_{21} & -M^{(n-1}_{12} + M^{(n-1}_{22} \end{pmatrix} \begin{pmatrix} a^{(1}_{n} \\ \\ a^{(2}_{n} \end{pmatrix} - \begin{pmatrix} \sum_{k=0}^{n-3} b^{(n-2}_{k} c_{k+2,0} \\ \sum_{k=0}^{n-3} b^{(n-2}_{k} c_{k+2,1} \end{pmatrix} \tag{3}$$

The determinant of the matrix (3) is

$$\begin{vmatrix} -2\, M^{(n-1}_{11} & -2\, M^{(n-1}_{12} \\ M^{(n-1}_{21} & M^{(n-1}_{22} \end{vmatrix} = - \frac{1}{\det\left[K_{n-1}(\alpha_i, \alpha_j)\right]^{2}_{i,j=1}} \neq 0. \quad \begin{matrix} \alpha_1 = 1 \\ \alpha_2 = -1 \end{matrix}$$

and the result follows.

<u>Remarks</u>: 1. The elements c_{ni} $(i = 0, 1)$ are linearly dependent on the above moments (c_{ki}) with $k < n$.

2. $[M_{ij}^{(n-1)}]_{i,j=1}^{2} = \{[K_{n-1}(\alpha_i, \alpha_j)]_{i,j=1}^{2}\}^{-1}$

Given the matrix

$$R_{n-1} = \left(\begin{array}{cc|ccc} c_{22} & c_{32} & c_{20} & \cdots & c_{n-1,0} \\ c_{23} & c_{33} & c_{21} & \cdots & c_{n-1,1} \\ \hline c_{02} & c_{12} & & & \\ \vdots & \vdots & & M_{n-3} & \\ c_{0,n-1} & c_{1,n-1} & & & \end{array}\right) \in \mathbb{C}^{(n,n}$$

we try to obtain the elements c_{n0} and c_{n1} for which the matrix

$$R_n = \left(\begin{array}{cc|ccc} c_{22} & c_{23} & c_{20} & \cdots & c_{n0} \\ c_{23} & c_{33} & c_{21} & \cdots & c_{n1} \\ \hline c_{02} & c_{12} & & & \\ \vdots & \vdots & & M_{n-2} & \\ c_{0n} & c_{1n} & & & \end{array}\right) \in \mathbb{C}^{n+1,n+1}$$

is Hermitian and positive definite.

It can be easily shown that the possible values will be found within a quadratic variety of an elliptic type in $\mathbb{C}^2$.

Moreover, in relation to [2] and [5] , we have:

$$e_{n-2} - (\bar{a}_n^{(1}, \bar{a}_n^{(2}) \begin{pmatrix} M_{11}^{(n-1} & M_{12}^{(n-1} \\ M_{21}^{(n-1} & M_{22}^{(n-1} \end{pmatrix} \begin{pmatrix} a_n^{(1} \\ a_n^{(2} \end{pmatrix} > 0 \qquad (4)$$

and using (3),

$$\begin{pmatrix} a_n^{(1} \\ a_n^{(2} \end{pmatrix} = \begin{pmatrix} -M_{11}^{(n-1} - M_{21}^{(n-1} & -M_{12}^{(n-1} - M_{22}^{(n-1} \\ -M_{11}^{(n-1} + M_{21}^{(n-1} & -M_{12}^{(n-1} + M_{22}^{(n-1} \end{pmatrix}^{-1} \begin{pmatrix} c_{n0} + \sum_{k=0}^{n-3} b_k^{(n-2} c_{k+2,0} \\ c_{n1} + \sum_{k=0}^{n-3} b_k^{(n-2} c_{k+2,1} \end{pmatrix}$$

By substitution in (4) , we obtain:

$$e_{n-2} - (\bar{v}_n, \bar{w}_n) \begin{pmatrix} A_{n-1} & B_{n-1} \\ C_{n-1} & D_{n-1} \end{pmatrix} \begin{pmatrix} v_n \\ w_n \end{pmatrix} > 0 \qquad (5)$$

where

$$v_n = c_{n0} + \sum_{k=0}^{n-3} b_k^{(n-2} c_{k+2,0}$$

$$w_n = c_{n1} + \sum_{k=0}^{n-3} b_k^{(n-2} c_{k+2,1}$$

and the matrix

$$\begin{pmatrix} A_{n-1} & B_{n-1} \\ C_{n-1} & D_{n-1} \end{pmatrix}$$

is

$$\begin{pmatrix} -M_{11}^{(n-1}-M_{21}^{(n-1} & -M_{11}^{(n-1}+M_{21}^{(n-1} \\ -M_{12}^{(n-1}-M_{22}^{(n-1} & -M_{12}^{(n-1}+M_{22}^{(n-1} \end{pmatrix}^{-1} \begin{pmatrix} M_{11}^{(n-1} & M_{12}^{(n-1} \\ M_{21}^{(n-1} & M_{22}^{(n-1} \end{pmatrix} \begin{pmatrix} -M_{11}^{(n-1}-M_{21}^{(n-1} & -M_{12}^{(n-1}-M_{22}^{(n-1} \\ -M_{11}^{(n-1}+M_{21}^{(n-1} & -M_{12}^{(n-1}+M_{22}^{(n-1} \end{pmatrix}^{-1},$$

On the other hand, we can write:

$$\begin{pmatrix} c_{n0} \\ c_{n1} \end{pmatrix} = \begin{pmatrix} -1 & -1 \\ -1 & 1 \end{pmatrix} \begin{pmatrix} M_{11}^{(n-1} & M_{12}^{(n-1} \\ M_{21}^{(n-1} & M_{22}^{(n-1} \end{pmatrix} \begin{pmatrix} a_n^{(1} \\ a_n^{(2} \end{pmatrix} -$$

$$- \begin{pmatrix} \sum_{k=0}^{n-3} b_k^{(n-2} c_{k+2,0} \\ \sum_{k=0}^{n-3} b_k^{(n-2} c_{k+2,1} \end{pmatrix}$$ and finally

$$\begin{pmatrix} a_n^{(1} \\ a_n^{(2} \end{pmatrix} = \frac{1}{2} \begin{pmatrix} K_{n-1}(1,1) & K_{n-1}(1,-1) \\ K_{n-1}(-1,1) & K_{n-1}(-1,-1) \end{pmatrix} \begin{pmatrix} -1 & -1 \\ -1 & -1 \end{pmatrix} \begin{pmatrix} c_{n0} + \sum_{k=0}^{n-3} b_k^{(n-2} c_{k+2,0} \\ c_{n1} + \sum_{k=0}^{n-3} b_k^{(n-2} c_{k+2,1} \end{pmatrix}$$

Thus, the elements $a_n^{(1}$ and $a_n^{(2}$ are explicitly expressed as a function of the parameters c_{n0}, c_{n1} and

$$\{\hat{P}_j(1)\}_{j=0}^{n-1}, \qquad \{\hat{P}_j(-1)\}_{j=0}^{n-1},$$

where $\{\hat{P}_n(z)\}_{n \in \mathbb{N}}$ donotes the orthonormal polynomial sequence.

4. EXAMPLES

As occurs for the unit circle, if $\{\hat{P}_n(1)\}_{n \in \mathbb{N}}$ and $\{\hat{P}_n(-1)\}_{n \in \mathbb{N}}$ are known, the moments can be determined.

As an application of this, we shall consider the following examples:

1) Let

$$\{\hat{P}_n(1)\} = \{1,a,0,0,\ldots\} \qquad \text{and} \qquad \{\hat{P}_n(-1)\} = \{1,b,0,0,\ldots\}$$

with $a \neq b$. In these conditions, by a direct application of (3) we obtain

$$c_{ni} = -\sum_{k=0}^{n-3} b_k^{(n-2} c_{k+2,i} \qquad (i=0,1)$$

2) If we consider the following sequences

$$\{\hat{P}_n(1)\} = \{1,1,1,1,\ldots\} \qquad \text{and} \qquad \{\hat{P}_n(-1)\} = \{1,0,1,0,1,\ldots\} ,$$

then, if n is even:

$$M_{11}^{(n} = \frac{2}{n} \ ; \qquad M_{12}^{(n} = -\frac{2}{n} = M_{21}^{(n} \ ; \qquad M_{22}^{(n} = \frac{4(n+1)}{n(n+2)} ,$$

so

$$c_{n0} = -\frac{2}{n+1} - \sum_{k=0}^{n-3} b_k^{(n-2} c_{k+2,0}$$

$$c_{n1} = \frac{2}{n+1} - \sum_{k=0}^{n-3} b_k^{(n-2} c_{k+2,1}$$

If n is odd,

$$M_{11}^{(n} = \frac{2}{n+1} \ ; \qquad M_{12}^{(n} = -\frac{2}{n+1} = M_{21}^{(n} \ ; \qquad M_{22}^{(n} = \frac{n}{n+1}$$

so

$$c_{n0} = -\sum_{k=0}^{n-3} b_k^{(n-2} c_{k+2,0} \ ; \qquad c_{n1} = -\frac{4}{n} - \sum_{k=0}^{n-3} b_k^{(n-2} c_{k+2,1}$$

and the extension is determined.

For other examples, see [8] in connection with problems in approximation theory.

REFERENCES

[1] ALFARO, M. and MARCELLAN, F.: "Recurrence relations for orthogonal polynomials on algebraic curves". Portugaliae Mathematica 42, pp. 41-52. 1984.

[2] ATENCIA, E.: "Polinomios ortogonales relativos a la lemniscata de Bernouilli". Ph. D. Thesis. Zaragoza, 1974.

[3] DELSARTE & Al.: "Orthogonal Polynomials matrices on the unit circle". IEEE Trans. on Circuits and Systems. Cas. 25,3, pp. 149-160. 1978.

[4] GERONIMUS, Ya.: "Orthogonal Polynomials". Consultants Bureau. New York, 1961.

[5] MARCELLAN, F. and MORAL, L.: "Minimal recurrence formulas for orthogonal polynomials on Bernoulli's Lemniscate". Lecture Notes in Mathematics, 1171, pp. 211-220. 1985.

[6] MARCELLAN, F. and MORAL, L.: "Polinomios Ortogonales sobre la lemniscata de Bernoulli : Una interpretación matricial del problema de momentos". XI Jornadas Hispano-Lusas de Matemáticas, pp. 292-302. Murcia, 1984.

[7] MARONI, P.: "Prolegomènes à l'étude des polynômes orthogonaux semiclassique". Public. Laboratoire d'Analyse Numérique Univ. P. et M. Curie, 85013. París, 1985.

[8] PEREZ-GRASA, I.: "Propiedades formales de polinomios ortogonales sobre arcos de Jordan". Ph. D. Thesis. Zaragoza, 1986.

[9] RONVEAUX, A.: "Semiclassical weight $(-\infty, +\infty)$: semi-Hermite orthogonal polynomials". Proc. II Simpos. de Polinomios Ortogonales. Segovia, 1986. (To appear).

[10] VIGIL, L.: "Polinomios ortogonales sobre curvas algebraicas". Act. XI R.A.M.E. Murcia, 1970. pp. 58-70. Univ. Complutense. Madrid, 1973.

DIFFERENCE EQUATIONS, CONTINUED FRACTIONS, JACOBI MATRICES AND ORTHOGONAL POLYNOMIALS*

David R. Masson
Department of Mathematics
University of Toronto
Toronto M5S 1A1
Canada

ABSTRACT: The linear coefficient, linear, second order difference equation is solved in terms of the $_2F_1$ hypergeometric function and its limits $_1F_1$, Ψ , $_0F_1$ and D_λ. This provides a generic link between two families of Jacobi matrices and their associated moment problems, continued fractions and orthogonal polynomials. The parent case contains the associated Meixner-Pollaczek polynomials whose weight function is calculated with minimal restrictions on the parameters. A neglected but related case contains what we call Bessel order polynomials. We emphasize throughout the role played by the subdominant boundary value solution to a difference equation and show how it determines the resolvents of the associated Jacobi matrix and its abbreviates.

1. INTRODUCTION

We solve the second order, linear coefficient, linear difference equation in terms of the hypergeometric function $_2F_1$ and its limits $_1F_1$, Ψ , $_0F_1$ and D_λ. The number of essential complex parameters is five. As a standard equation to work with we choose

$$X_{n+1} - (z - dn)X_n + (an^2 + bn + c)X_{n-1} = 0, \quad n \geq 0. \tag{1.1}$$

With suitable ranges of values for the parameters this provides a generic link between two families of Jacobi matrices, their associated moment problems and their corresponding classical orthogonal polynomials [4]. This method of cataloging also yields a new set of polynomials orthogonal with respect to a discrete measure determined by Bessel functions. We call them "Bessel order" polynomials.

*Partially supported by the Natural Sciences and Engineering Research Council (Canada)

A. Cuyt (ed.), Nonlinear Numerical Methods and Rational Approximation, 239–257.

If $an^2 + bn + c \neq 0$, $n = 1,2,\ldots$, this difference equation is related to the continued fraction ([16], § 81, § 82)

$$CF = z + \mathop{K}_{n=1}^{\infty} \left[\frac{-(an^2 + bn + c)}{z - dn} \right] \tag{1.2}$$

through the formulas [(1)]

$$1/CF = \lim_{n\to\infty} Q_n(z) / P_n(z) = X_0^{(s)}(z) / c\, X_{-1}^{(s)}(z) . \tag{1.3}$$

Here P_n, Q_n are initial value solutions to (1.1). We make the standard choice $Q_0 = 0$, $Q_1 = 1$, $P_{-1} = 0$, $P_0 = 1$.

$X_n^{(s)}$ is more elusive [10], [14]. It is a subdominant boundary value solution with the property

$$\lim_{n\to\infty} X_n^{(s)} / X_n^{(d)} = 0 \tag{1.4}$$

whenever $X_n^{(s)}$, $X_n^{(d)}$ are linearly independent solutions to (1.1).

The second equality in (1.3) is a simple but important, yet often neglected theorem due to Pincherle [10], [17]. Thus for a given z, the existance of $X_n^{(s)}(z)$ is a necessary and sufficient condition for the convergence of CF and if $X_{-1}^{(s)}(z_0) = 0$ then CF "converges" to infinity at $z = z_0$.

Pincherle's theorem points to the necessity of examining the large n behaviour of the solutions to such a difference equation since $X_n^{(s)}$ is the key to not only the convergence but also the value of the CF and, as we show later, the resolvent and spectral measure of the associated Jacobi matrix.

The solutions to (1.1) tie together a rich variety of classical orthogonal polynomials $P_n(x)$ and their associated positive measures $d\sigma(x)$. In this connection there are actually two distinct families which can be obtained by taking the parameters a,b,c,d real and having either

I. $an^2 + bn + c > 0$, $n = 1,2,\ldots$ or

II. $(an^2 + bn + c) / (e - dn)(e + d - dn) > 0$, $n = 1,2,\ldots$

where for II, $z - dn$ in (1.1) has been replaced by $z(e - dn) + f$ with e, f real and d, $e - dn$, $e + d - dn \neq 0$.

The positivity condition in I and II is known as Favard's Theorem ([3] p.866). It reflects the fact that the associated Jacobi matrix should be symmetric [2].

The measures for family II all have bounded support. Included are the $\alpha = \beta$ Jacobi polynomials, their Pollaczek-Szegö generalizations [18], [20], [22], the more recently examined, discretely orthogonal polynomials of Chihara and Ismail [7], and the Bessel related polynomials of Ismail [11], [24].

(1) If $c = 0$ then $cX_{-1}^{(s)}(z)$ should be replaced by $zX_0^{(s)}(z) - X_1^{(s)}(z)$.

Family I is more familiar and dictates our notation. All our results can however be translated over to family II. Family I contains the Meixner [15], Meixner-Pollaczek [15], [19], Laguerre, Hermite and Charlier polynomials [9]. Also included are the recently examined associated Laguerre [6] and associated Hermite polynomials [6], [13] as well as the neglected case of Bessel order polynomials.

Note that our viewpoint connects not only the polynomials $P_n(x)$ but also their associated measures $d\sigma(x)$ since one has the Cauchy transform of $d\sigma$ from the relation [(1)]

$$1/CF = \int_{-\infty}^{\infty} \frac{d\sigma(x)}{z-x} = X_0^{(s)}(z) \,/\, c\, X_{-1}^{(s)}(z), \ \mathrm{Im}\, z \neq 0. \tag{1.5}$$

Thus from a boundary value solution $X_n^{(s)}$ to the difference equation (1.1) we can obtain $d\sigma$ using (1.5) while the associated orthogonal polynomials P_n are determined as initial value solutions to the same difference equation.

In Section 2. we find general, subdominant and polynomial solutions to (1.1) in terms of ${}_2F_1$ and its limits.

In Section 3. we apply the parent ${}_2F_1$ formula to obtain the associated Pollaczek polynomial measure for the most general possible range of parameters.

In Section 4. we show that the resolvent of a Jacobi matrix of type D is determined in terms of its associated $X_n^{(s)}$. The subdominant solution thus contains all the spectral information and determines the associated projection valued measure.

2. SOLUTIONS

We derive general, subdominant and polynomial solutions to (1.1) for five different cases interconnected by limits. Each case is labeled with the name of the family I classical polynomial associated with it. All parameters are assumed to be complex except n which takes on values 0, ±1, ±2,....

The transformations A and R defined by

$$Af(n) = (-1)^n f(n) \tag{2.1}$$

$$Rf(n) = f(-n) \tag{2.2}$$

(which we call the alternating and reflection transformation respectively) are used throughout to derive new solutions from established ones.

2.1 Meixner-Pollaczek ($a \neq 0$, $d^2 - 4a \neq 0$):

(1) In the case of (1.2), $d\sigma(x)$ is unique and can be obtained from (1.5). However, one can have indeterminant cases where the associated continued fraction may be divergent or convergent and the measure nonunique [2], [23].

Theorem 1. *If* $a \neq 0,\ d^2 - 4a \neq 0$ *then (1.1) has :*

a) *linearly independent solutions*

$$X_{n-1}^{(1),\pm}\begin{bmatrix} a,b,c \\ d,z \end{bmatrix} = \left[\pm \frac{a}{\mu}\right]^n \frac{\Gamma(n+\alpha)\Gamma(n+\beta)}{\Gamma(n+\gamma^\pm)}\ {}_2F_1\,(n+\alpha,\ n+\beta;\ n+\gamma^\pm;\ \delta^\pm) \tag{2.3}$$

where

$$\mu = \sqrt{d^2 - 4a},\ -\pi/2 < \arg \mu \leq \pi/2$$

$$\delta^\pm = \tfrac{1}{2}\,(1 \pm d/\mu)$$

$$\gamma^\pm = \left[\frac{a+b}{a}\right]\delta^\pm \pm z/\mu \tag{2.4}$$

$$an^2 + bn + c = a(n+\alpha)(n+\beta)$$

b) *a subdominant solution if and only if*

$$|\ \mathrm{Re}(d/\mu)\ | + |\ \mathrm{Re}\ ((\frac{a+b}{2a})\ \frac{d}{\mu} + \frac{z}{\mu}\ )\ | \neq 0$$

given by

$$X_n^{(s)} = \begin{cases} X_n^{(1),+} & \text{if } \mathrm{Re}(\frac{d}{\mu}) < 0 \text{ or if } \mathrm{Re}(\frac{d}{\mu}) = 0 \text{ and } \mathrm{Re}(\gamma^+ - \gamma^-) > 0 \\ X_n^{(1),-} & \text{if } \mathrm{Re}(\frac{d}{\mu}) > 0 \text{ or if } \mathrm{Re}(\frac{d}{\mu}) = 0 \text{ and } \mathrm{Re}(\gamma^+ - \gamma^-) < 0 \end{cases} \tag{2.5}$$

with

$$|X_n^{(s)}/X_n^{(d)}| = \mathrm{const.} \left[\frac{(1-|\mathrm{Re}(\frac{d}{\mu})|)^2 + (\mathrm{Im}(\frac{d}{\mu}))^2}{(1+|\mathrm{Re}(\frac{d}{\mu})|)^2 + (\mathrm{Im}(\frac{d}{\mu}))^2}\right]^{n/2} n^{-2|\mathrm{Re}(\gamma^+ - \gamma^-)|}(1+O(\frac{1}{n})) \tag{2.6}$$

c) *additional solutions*

$$X_n^{(2),\pm}\begin{bmatrix} a,b,c \\ d,z \end{bmatrix} = (\pm\mu)^n \frac{{}_2F_1(-n-\alpha,\ -n-\beta;\ 1-n-\gamma^\pm;\ \delta^\pm)}{\Gamma(1-n-\gamma^\pm)} \tag{2.7}$$

which are linearly dependent and proportional to $P_n(z)$ *if* $c = 0$ *and* $n \geq 0$.

Proof: (See [14] for the case d = 0)

a) Define

$$u_n(w) = \frac{\Gamma(\alpha_n)\Gamma(\beta_n)}{\Gamma(\gamma_n)}\ {}_2F_1(\alpha_n,\ \beta_n;\ \gamma_n; w)$$

with $\alpha_n = n + \alpha$, $\beta_n = n + \beta$, $\gamma_n = n + \gamma^+$, $w = \delta^+$. One has

$$X_{n+1}^{(1),+} - (z - dn)X_n^{(1),+} + (an^2 + bn + c)X_{n-1}^{(1),+} =$$

$$(\frac{a}{\mu})^{n+2}(-\frac{\mu^2}{a})\,[w(1 - w)u_n''(w) + (\gamma_n - (1 + \alpha_n + \beta_n)w)u_n'(w) - \alpha_n\beta_n u_n(w)] = 0.$$

Note that in going from the difference equation to the differential equation we have used the identities $w(1 - w) = -a/\mu^2$, $\gamma_n - (1 + \alpha_n + \beta_n)w = (z - dn)/\mu$, $a\alpha_n\beta_n = an^2 + bn + c$ and $u_n'(w) = u_{n+1}(w)$. One may redo the calculation with $X_n^{(1),-}$ or use the alternating transformation applied to (1.1). Thus $AX_n^{(1),+}\begin{bmatrix}a,b,c\\-d,-z\end{bmatrix} = X_n^{(1),-}\begin{bmatrix}a,b,c\\d,z\end{bmatrix}$ is also a solution to (1.1). The linear independence follows from the large n behaviour which we calculate next.

b) Using the transformation

$${}_2F_1(a, b; c; z) = (1 - z)^{c-a-b}\,{}_2F_1(c - a, c - b; c; z),$$

the estimate ([8], 2.3.2 (10))

$${}_2F_1(a, b, c, z) = 1 + O(\frac{1}{\mathrm{Re}\ c})\ ,\ z \notin [1,\infty),\ \mathrm{Re}\ c \to \infty$$

and Stirling's formula one has

$$X_{n-1}^{(1),\pm} = (\pm\frac{a}{\mu})^n(\delta^\pm)^{\gamma^\pm-(\alpha+\beta+n)}\, n^{\alpha+\beta-½-\gamma^\pm}\, e^{-n}\, \sqrt{2\pi}\, (1+O(\frac{1}{n})). \tag{2.8a}$$

Thus

$$X_{n-1}^{(1),+} / X_{n-1}^{(1),-} = (-1)^n(\delta^-)^{\gamma^+}(\delta^+)^{\gamma^-}(\frac{\delta^+}{\delta^-})^n\, n^{\gamma^- - \gamma^+}(1 + O(\frac{1}{n})). \tag{2.8b}$$

This ratio is dominated by $(\delta^+ / \delta^-)^n$ if $\mathrm{Re}(d/\mu) \neq 0$ with $(\delta^+ / \delta^-)^{\pm n} \to 0$ if $\pm\mathrm{Re}(d / \mu) < 0$. If $\mathrm{Re}(\frac{d}{\mu}) = 0$ then the term $n^{\gamma^- - \gamma^+}$ takes charge with $n^{\pm(\gamma^- - \gamma^+)} \to 0$ if $\pm\mathrm{Re}(\gamma^+ - \gamma^-) = \pm 2\mathrm{Re}\left[\frac{(a+b)}{2a}\,\frac{d}{\mu} + \frac{z}{\mu}\right] > 0$. Finally if $\mathrm{Re}(d/\mu) = \mathrm{Re}\left[\frac{(a+b)d}{2a\mu} + \frac{z}{\mu}\right] = 0$ then $\lim_{n \to \infty} X_n^{(1),+} / X_n^{(1),-}$ diverges by oscillation.

c) We use the reflection transformation $Y_n = RX_n = X_{-n}$. Thus if X_n is a solution to (1.1) then Y_n is a solution to $(an^2 - bn + c)Y_{n+1} - (z + dn)Y_n + Y_{n-1} = 0$. The multiplicative transformation $Z_{n+1} = a^n\,\Gamma(n - \alpha + 1)\Gamma(n - \beta + 1)Y_{n+1}$ then yields

$$Z_{n+1} - (z + dn)Z_n + a(n - \alpha - 1)(n - \beta - 1)Z_{n-1} = 0.$$

It follows that if $X_n\begin{bmatrix}a,b,c\\d,z\end{bmatrix}$ is a solution to (1.1) then so is

$$a^n\Gamma(n + \alpha + 1)\Gamma(n + \beta + 1)X_{-n-1}\begin{bmatrix}a,-b,c\\-d,z+d\end{bmatrix}.$$

Applying this to the solutions in a), assuming that $c \neq 0$ and discarding factors that are n independent yields the solutions (2.7). Note that this calculation uses the identity ([8], 1.2(6)).

$$\Gamma(1 + n + x)\Gamma(- n - x) = (-1)^{n+1}\,\pi / \sin \pi x \tag{2.9}$$

for x equal to α and β. Hence the requirement $c = a\alpha\beta \neq 0$. After discarding a factor of $(\sin \pi \alpha \ \sin \pi \beta)^{-1}$ one may then take $c \to 0$. For $c = 0$ finds that $\Gamma(1 - \gamma^{\pm})X_n^{(2)\pm} = P_n(z)$, $n = 0,1,2,...$ by comparing the $n = 0,1$ initial values. □

2.2 Charlier ($a = 0$, b, $d \neq 0$):

Theorem 2. *If* $a = 0$, b, $d \neq 0$ *then (1.1) has:*

a) *linearly independent solutions*

$$X_{n-1}^{(1)}\begin{bmatrix} b,c \\ d,z \end{bmatrix} = (-\frac{b}{d})^n \frac{\Gamma(n + \frac{c}{b})}{\Gamma(n - \frac{b}{d^2} - \frac{z}{d})} {}_1F_1(n + \frac{c}{b}; n - \frac{b}{d^2} - \frac{z}{d}; -\frac{b}{d^2}) \tag{2.10}$$

$$X_{n-1}^{(2)}\begin{bmatrix} b,c \\ d,z \end{bmatrix} = (\frac{b}{d})^n \Gamma(n + \frac{c}{b}) \Psi(n + \frac{c}{b}, n - \frac{b}{d^2} - \frac{z}{d}; -\frac{b}{d^2}) \tag{2.11}$$

b) *a subdominant solution* $X_n^{(s)} = X_n^{(1)}$ *with*

$$X_n^{(s)} / X_n^{(d)} = \text{const.}(\frac{be}{d^2})^n \, n^{-n+\frac{1}{2}+\frac{c}{b}+\frac{2b}{d^2}+\frac{2z}{d}} (1 + O(\frac{1}{n})) \tag{2.12}$$

c) *additional solutions*

$$X_n^{(3)}\begin{bmatrix} b,c \\ d,z \end{bmatrix} = d^n \frac{{}_1F_1(-n-\frac{c}{b}; 1-n+\frac{b}{d^2} + \frac{z}{d}; \frac{b}{d^2})}{\Gamma(1-n+\frac{b}{d^2}+\frac{z}{d})} \tag{2.13}$$

$$X_n^{(4)}\begin{bmatrix} b,c \\ d,z \end{bmatrix} = (-d)^n \Psi(-n - \frac{c}{b}, 1 - n + \frac{b}{d^2} + \frac{z}{d}; \frac{b}{d^2}) \tag{2.14}$$

which are linearly dependent and proportional to $P_n(z)$ *if* $c = 0$ *and* $n \geq 0$.

Proof:

a) We take the $a \to 0$ limit of (2.3) to obtain (2.10). If $-\pi/2 < \arg d \leq \pi/2$ then $\mu = d + O(a)$, $\delta^- = -a/d^2 + O(a^2)$ and $\gamma^- = -\frac{b}{d^2} - \frac{z}{d} + O(a)$ and for definiteness we may take $\alpha = \frac{b}{a} + O(1)$, $\beta = \frac{c}{b} + O(a)$. Hence

$$\Gamma(n + \alpha) = (\frac{b}{a})^{\frac{b}{a}+n-\frac{1}{2}} e^{-b/a} \sqrt{2\pi}(1 + O(a))$$

while

$$\lim_{a\to 0} {}_2F_1(n + \alpha, n + \beta; n + \gamma^-; \delta^-) = {}_1F_1 (n + \frac{c}{b}; n - \frac{b}{d^2} - \frac{z}{d}, -\frac{b}{d^2}).$$

Therefore $\lim_{a\to 0} \left(\frac{b}{a}\right)^{-\frac{b}{a}+\frac{1}{2}} e^{b/a} X_{n-1}^{(1),-}\begin{bmatrix}a,b,c\\ d,z\end{bmatrix} /\sqrt{2\pi}$ yields (2.10). For $\pi/2 < \arg d \le 3\pi/2$ one replaces $X_{n-1}^{(1),-}$ by $X_{n-1}^{(1),+}$ to obtain the same limit. To obtain the solution (2.11) we compare the recurrence formulas for ${}_1F_1$ and Ψ. Thus from [1] (13.4.8, 13.4.14, 13.4.21, 13.4.27 on p. 507) and (2.10) one obtains (2.11).

b) If $\frac{b}{d^2} + \frac{z}{d} \neq 0, \pm 1, \cdots$ one may take the Kummer transformation ([8], 6.3(7)) of (2.12) and the Γ identity (2.9) to obtain

$$X_n^{(3)}\begin{bmatrix}b,c\\ d,z\end{bmatrix} = (-d)^n \frac{\Gamma(n - \frac{b}{d^2} - \frac{z}{d})}{\pi \sin(\frac{b}{d^2} + \frac{z}{d})\pi} e^{b/d^2} \times \tag{2.15}$$

$${}_1F_1(1 + \frac{b}{d^2} + \frac{z}{d} + \frac{c}{b}; 1 - n + \frac{b}{d^2} + \frac{z}{d}; \frac{b}{d^2}).$$

From [1], p. 65 one then has

$$X_n^{(1)} / X_n^{(3)} = \text{const.} \left(\frac{b}{d^2}\right)^n \frac{\Gamma(1 + n + \frac{c}{b})}{\Gamma(1 + n - \frac{b}{d^2} - \frac{z}{d})\Gamma(n - \frac{b}{d^2} - \frac{z}{d})} (1 + O(\frac{1}{n}))$$

from which (2.12) follows using Stirling's formula. For the exceptional values of $\frac{b}{d^2} + \frac{z}{d}$ one can compare (2.10) and (2.11) using [21], p.8, (1.5.24).

c) We use the reflection transformation. Thus if $X_n\begin{bmatrix}b,c\\ d,z\end{bmatrix}$ is a solution to (1.1) then so is

$$b^n\, \Gamma(1 + n + \frac{c}{b})\, X_{-n-1}\begin{bmatrix}-b,c\\ -d,z+d\end{bmatrix}.$$

This applied to (2.10) and (2.11) yields (2.13) and (2.14) with factors independent of n discarded. If $c = 0$, the proportionality to $P_n(z)$ follows from the $n = 0,1$ initial values. □

2.3 Laguerre $(d \neq 0,\ d^2 - 4a = 0)$ [6]:

Theorem 3. *If* $d \neq 0$, $d^2 - 4a = 0$, $2zd + 4b \neq -d^2$ *then (1.1) has:*

a) *linearly independent solutions*

$$X_{n-1}^{(1)}\begin{bmatrix}\frac{d^2}{4},b,c\\ d,z\end{bmatrix} = (-\frac{d}{2})^n\, \Gamma(n+\alpha)\, \Gamma(n+\beta)\Psi(n+\alpha; \alpha-\beta+1; -\frac{2z}{d} - \frac{4b}{d^2} - 1) \tag{2.16}$$

$$X_{n-1}^{(2)}\begin{bmatrix}\frac{d^2}{4},b,c\\ d,z\end{bmatrix} = (-\frac{d}{2})^n\, \Gamma(n + \alpha)\, {}_1F_1(n + \alpha; \alpha - \beta + 1; -\frac{2z}{d} - \frac{4b}{d^2} - 1), \tag{2.17}$$

b) *a subdominant solution if and only if*

$$0 < \arg(1 + \frac{4b}{d^2} + \frac{2z}{d}) < 2\pi$$

given by $X_n^{(s)} = X_n^{(1)}$ *with*

$$| X_n^{(s)} / X_n^{(d)} | = \text{const.} \exp(-2\sqrt{n}\,\text{Im}(1 + \frac{4b}{d^2} + \frac{2z}{d})^{1/2})(1 + O(\frac{1}{\sqrt{n}})). \qquad (2.18)$$

c) *additional solutions*

$$X_n^{(3)} \begin{bmatrix} \frac{d^2}{4},b,c \\ d,z \end{bmatrix} = (\frac{d}{2})^n \Psi(-n-\alpha, \beta-\alpha+1; \frac{2z}{d} + \frac{4b}{d^2} + 1) \qquad (2.19)$$

$$X_n^{(4)} \begin{bmatrix} \frac{d^2}{4},b,c \\ d,z \end{bmatrix} = (-\frac{d}{2})^n \Gamma(1+n+\beta)\ {}_1F_1(-n-\alpha; \beta-\alpha+1; \frac{2z}{d} + \frac{4b}{d^2} + 1) \qquad (2.20)$$

which are linearly dependent and proportional to $P_n(z)$ *if* $c = 0$, $\alpha = (b \pm \sqrt{b^2 - 4ac}) / 2a$ *is chosen to be zero in (2.20) and* $n \geq 0$.

Proof.

a) We take the $\mu \to 0$ limit of (2.3) to obtain (2.16). Similar calculations have been done by Askey and Wimp [6]. The essential formula we use is a generalization of [8], p.262, 6.8(1). Namely

$$\lim_{|c| \to \infty} {}_2F_1(a,b; c+d; 1-c/x) = x^a \Psi(a, a-b+1, x) \qquad (2.21)$$

$$\text{if} \quad |\arg(\frac{c}{x})| < \pi, \quad |\arg x| < \pi.$$

This equation follows from [8], 2.10(3), 6.5(7) and the normal confluence. The application of (2.21) to (2.3) yields (2.16). By comparing the recursion relations for ${}_1F_1$ and Ψ ([21], (2.2.1), (2.2.8)) we obtain (2.17) from (2.16).

b) As noted by Askey and Wimp [6], $X_n^{(1)}$ is subdominant and (2.18) follows from comparing the large n behaviour of $X_n^{(1)}$ and $X_n^{(2)}$ ([21], (4.6.42), (4.6.43)).

c) The difference equation is symmetric under an interchange of α and β. One can use this in conjunction with the reflection transformation to obtain additional solutions. Thus (2.19) follows from applying the reflection transformation to (2.16) (see the proof of Theorem 1 c)) and then interchanging α and β while (2.20) follows from the same procedure applied to (2.17). Alternatively one could take the $\mu \to 0$ limit of (2.7) using (2.21) to obtain (2.19). The proportionality to P_n follows from the $n = 0,1$ initial values. □

Note that for the special case $2zd + 4b = -d^2$ one has, for $\alpha \neq \beta$, linearly independent solutions $X_{n-1}^{(2)} = (-\frac{d}{2})^n \Gamma(n+\alpha)$, $X_{n-1}^{(4)} = (-\frac{d}{2})^n \Gamma(n+\beta)$ and for $\alpha = \beta$ one may replace one of these Γ functions by Γ'. One thus has $X_{n-1}^{(s)} = (-\frac{d}{2})^n \Gamma(n+\alpha)$ for $\text{Re}\alpha < \text{Re}\beta$ or

$\alpha = \beta$ with

$$X_n^{(s)}/X_n^{(d)} = \begin{cases} O(n^{\mathrm{Re}(\alpha-\beta)}), & \mathrm{Re}\alpha < \mathrm{Re}\beta \\ O(\ln n)^{-1}, & \alpha = \beta \end{cases} .$$

2.4 Hermite ($a = d = 0$, $b \neq 0$) [6], [13]:

Theorem 4: *If* $a = d = 0$, $b \neq 0$ *then (1.1) has:*

a) *linearly independent solutions*

$$X_{n-1}^{(1)\pm}\begin{bmatrix} b,c \\ z \end{bmatrix} = (\pm i\sqrt{b})^n\, \Gamma(n + \frac{c}{b})\, D_{-n-\frac{c}{b}}(\pm iz/\sqrt{b}) \tag{2.22}$$

where $-\pi/2 < \arg\sqrt{b} \leq \pi/2$,

b) *a subdominant solution if and only if* $\mathrm{Im}(z/\sqrt{b}) \neq 0$ *given by*

$$X_n^{(s)} = \begin{cases} X_n^{(1),-} & \text{if } \mathrm{Im}(z/\sqrt{b}) > 0 \\ X_n^{(1),+} & \text{if } \mathrm{Im}(z/\sqrt{b}) < 0 \end{cases} \tag{2.23}$$

with

$$|\, X_n^{(s)} / X_n^{(d)} \,| = \text{const.} \exp(-\, 2\sqrt{n}\, |\, \mathrm{Im}(z/\sqrt{b})\, |)(1 + O(\frac{1}{\sqrt{n}})). \tag{2.24}$$

c) *additional solutions*

$$X_n^{(2)\pm}\begin{bmatrix} b,c \\ z \end{bmatrix} = (\pm\sqrt{b})^n\, D_{n+\frac{c}{b}}(\pm z / \sqrt{b}) \tag{2.25}$$

which are linearly dependent and proportional to $P_n(z)$ *if* $c = 0$ *and* $n \geq 0$.

Proof: (See also [13])

a) That these are solutions to the difference equation (1.1) follows directly from the recurrence relation for parabolic cylinder functions $D_\lambda(z)$ ([9], 8.3(14)). The linear dependence follows from examining their large n behaviour (see below). These solutions can also be derived:

i) from (2.3) by putting $d = 0$ and taking the $a \to 0$ limit (see [6], (4.6)),

ii) from (2.10), (2.11) or (2.16), (2.17) by taking the $d \to 0$ limit and using [6], (5.7), (5.8).

b) This follows from [9], 8.4(5). See also [13].

c) The reflection transformation applied to (2.22) yields (2.25) (see the proof of theorem 2c). For $c = 0$ one has ([9], 8.2(9))

$$D_n(\pm z/\sqrt{b}) = e^{-z^2/4b}\, 2^{-n/2}\, H_n(\pm z/\sqrt{2b}),\ n \geq 0$$

whereas $P_n(z) = (\pm\sqrt{b})^n 2^{-n/2} H_n(\pm z/\sqrt{2b})$. □

2.5 Case a = b = 0, c, d ≠ 0 (Bessel order):

Theorem 5: *If* $a = b = 0, c, d \neq 0$ *then (1.1) has:*

a) *linearly independent solutions for* $z/d \neq 0, \pm 1, \ldots$ *given by*

$$X_n^{(1)} \begin{bmatrix} c \\ d,z \end{bmatrix} = (-\frac{c}{d})^n \frac{{}_0F_1(\,;\, n+1-z/d;\, -c/d^2)}{\Gamma(n+1-z/d)} \tag{2.26}$$

$$X_n^{(2)} \begin{bmatrix} c \\ d,z \end{bmatrix} = d^n \frac{{}_0F_1(\,;\, 1-n+z/d;\, -c/d^2)}{\Gamma(1-n+z/d)}. \tag{2.27}$$

b) *a subdominant solution* $X_n^{(s)} = X_n^{(1)}$ *with*

$$X_n^{(s)} / X_n^{(d)} = \text{const.}\ c^n (\frac{e}{dn})^{2n} n^{2z/d} (1 + O(\frac{1}{n})). \tag{2.28}$$

Proof:

a) The $b \to 0$ limit of (2.10) and (2.13) yields (2.26) and (2.27).

b) Using ${}_0F_1(\,;\, c; z) = 1 + O(\frac{1}{c})$, $c \neq 0, -1, \ldots$ and the Γ identity (2.9) one has

$$X_n^{(1)} / X_n^{(2)} = -(\frac{c}{d^2})^n \frac{[\Gamma(n+1-z/d)\Gamma(n-z/d)]^{-1}}{\pi \sin(\pi z/d)} (1 + O(\frac{1}{n}))$$

which yields (2.28) using Stirling's formula. For the exceptional values $z/d = 0, \pm 1, \ldots$ one must compare $X_n^{(1)}$ with $X_n^{(3)}$ given by (2.29) below. □

Note that the solutions (2.26), (2.27) may be written in terms of Bessel functions with z *appearing in the order* since ([9], 7.2.1(3))

$$J_\nu(x) = (\frac{x}{2})^\nu \frac{{}_0F_1(\,;\, 1+\nu;\, -\frac{1}{4}x^2)}{\Gamma(1+\nu)}.$$

The lack of linear independence of $X_n^{(1)}, X_n^{(2)}$ for $z/d = 0, \pm 1, \ldots$ is a nuisance which may be overcome by replacing $X_n^{(2)}$ by

$$X_n^{(3)} \begin{bmatrix} c,d \\ z \end{bmatrix} = \frac{(\frac{c}{d^2})^{z/2d} X_n^{(2)} \begin{bmatrix} c,d \\ z \end{bmatrix} - (\frac{c}{d^2})^{-z/2d} X_n^{(1)} \begin{bmatrix} c,d \\ z \end{bmatrix} \cos(\frac{z\pi}{d})}{\sin(\frac{z\pi}{d})} \tag{2.29}$$

This is equivalent to the use of the pair J_ν, Y_ν instead of $J_\nu, J_{-\nu}$.

2.6 Remarks.

There is a final case (Chebyshev) when $a = b = d = 0$, $c \neq 0$ where the difference equation (1.1) has constant coefficients. Although the solutions are then well known one can still obtain some interesting relations by considering the Chebyshev case as the $b \to 0$ limit of the § 2.4 Hermite case (see [6], (5.9)) or the $d \to 0$ limit of the § 2.5 Bessel order case.

The § 2.1 Meixner-Pollaczek case is actually two separate cases. One has the Meixner case $\mathrm{Re}(\frac{d}{\mu}) \neq 0$ when (2.5) yields a discrete spectrum of singularities for (1.3) and the Pollaczek case $\mathrm{Re}(\frac{d}{\mu}) = 0$ when (2.5) yields a continuous spectrum of singularities for (1.3) along the line $\mathrm{Re}((\frac{a+b}{2a})\frac{d}{\mu} + \frac{z}{\mu}) = 0$. The latter case is also referred to as the Meixner-Pollaczek case because of its history [6].

The analytic properties of the solutions in the z variable are interesting. Except for the § 2.3 Laguerre case one has the subdominant solution as either an entire function of z (Meixner, Charlier and Bessel order) or the half-plane restriction of an entire function of z (Pollaczek and Hermite). Our formulas for $X_n^{(s)}$ allow one to easily obtain both the discontinuity and the analytic continuation across a continuous line of singularities of (1.3). In the Pollaczek and Hermite cases the analytic continuation is meromorphic in z.

The large n estimates for $X_n^{(s)} / X_n^{(d)}$ which were obtained allow one to make estimates on the rate of convergence of (1.3), its modified approximants and its perturbations (see for example [13] and [14]).

3. POSITIVE MEASURES

The subdominant formulas of § 2 allow one to exactly calculate the positive continuous real line measures associated with the orthogonal polynomials of family I and II with no restrictions on the parameters except the basic positivity condition. For convenience we will assume that $c \neq 0$. The $c = 0$ case (which is actually simpler to calculate) will then be recovered as a limit.

From the imaginary part of (1.5) with $z = x + io$ and x an interior point of the continuous spectrum one obtains, for the family I cases of § 2.1, 2.3 and 2.4,

$$\frac{d\sigma}{dx}(x) = \frac{1}{2\pi i} \frac{W(X_{-1}^{(s)}(x+io), X_{-1}^{(s)}(x-io))}{c\,|X_{-1}^{(s)}(x+io)|^2} \tag{3.1}$$

where $X_{-1}^{(s)}(x-io) = \overline{X}_{-1}^{(s)}(x+io)$ and the Wronskian W is defined by

$$W(X_n, Y_n) = X_n Y_{n+1} - X_{n+1} Y_n. \tag{3.2}$$

Equation (3.1) follows from (1.5) by using the fact that $X_{-1}^{(s)}(x+io) \neq 0$ (since $X_n^{(s)}(x+io)$, $X_n^{(s)}(x-io)$ are linearly independent solutions to (1.1)) and the fact that $X_n^{(s)}(x+io)$ is continuous in x (see Theorems 1, 3 and 4). Thus $|X_{-1}^{(s)}(x+io)|^2$ is bounded away from zero and the right side of (3.1) is bounded and continuous on any closed interior subinterval of the continuous spectrum. This proves that there are no imbedded eigenvalues and that the continuous spectrum is absolutely continuous.

The Wronskian in (3.1) may be simplified by calculating it in terms of the large n constant Wronskian solutions of a related difference equation. As an example we calculate the parent Pollaczek case of § 2.1 where a,b,c,d are real, $d^2 - 4a < 0$ and $an^2 + bn + c > 0$, $n = 1,2,\ldots$.

Let X_n be a solution to (1.1) with $a > 0$ and define

$$U_n = (4a)^{-(n+1)/2}\, X_n/\Gamma(\tfrac{(n+1+\alpha)}{2})\,\Gamma(\tfrac{(n+\beta+1)}{2}). \tag{3.3}$$

One then has

$$U_{n+1} - \frac{(z-dn)}{2\sqrt{a}}\,\frac{\Gamma(\frac{n+\alpha+1}{2})\Gamma(\frac{n+\beta+1}{2})}{\Gamma(\frac{n+\alpha+2}{2})\Gamma(\frac{n+\beta+2}{2})}\,U_n + U_{n-1} = 0. \tag{3.4}$$

and it follows that the Wronskian of two solutions to (3.4) is constant.

From (3.3) it then follows that if $X_n^{(1)}, X_n^{(2)}$ are solutions to (1.1) then

$$W(X_n^{(1)}, X_n^{(2)}) = \frac{(4a)^{n+3/2}\pi}{2^{2n+\alpha+\beta}}\,\Gamma(n+\alpha+1)\Gamma(n+\beta+1)C \tag{3.5}$$

where we have used the Γ identity ([8], 1.2(15)).

With the choice $X_n^{(1)} = X_n^{(1),+}$, $X_n^{(2)} = X_n^{(1),-}$ one has using (2.8) and (3.5) the value

$$C = \frac{-\sqrt{a}}{16\pi\mu}\left[\frac{-4\mu^2}{a}\right]^{\frac{(1+\alpha+\beta)}{2}}\left[\frac{1-d/\mu}{1+d/\mu}\right]^{[\frac{(1+\alpha+\beta)d}{2}+z]/\mu} \tag{3.6}$$

For the Pollaczek case a,b,c,d real and $d^2 - 4a < 0$ one has $X_n^{(s)}(x+io) = X_n^{(1),+}(x)$ and from (3.6), (3.5) and (3.1)

$$\frac{d\sigma(x)}{dx} = -\frac{1}{2\pi i\mu}\,\frac{(-\frac{\mu^2}{a})^{\frac{1+\alpha+\beta}{2}}\,|\Gamma(\gamma^+)|^2\,(\frac{1-d/\mu}{1+d/\mu})^{[\frac{(1+\alpha+\beta)d}{2}+x]/\mu}}{\Gamma(\alpha+1)\Gamma(\beta+1)|\,{}_2F_1(\alpha,\beta,\gamma^+,\delta^+)|^2} \tag{3.7}$$

where $\mu = i\sqrt{4a-d^2}$, $\delta^+ = \frac{1}{2}(1 + d/\mu)$, $\gamma^+ = (1+\alpha+\beta)\delta^+ + x/\mu$, $\alpha+\beta = b/a$, $\alpha\beta = c/a$.

We have thus proved:

Theorem 6. *If* a,b,c,d *are real,* $4a - d^2 > 0$ *and* $an^2 + bn + c > 0$, $n = 1,2,\ldots,$ *then* $P_n(z)$ *the polynomial solutions to (1.1) with initial values* $P_{-1} = 0$, $P_0 = 1$ *satisfy the orthogonality relation*

$$\int_{-\infty}^{\infty} P_n(x)\,P_m(x)\,d\sigma(x) = \frac{\Gamma(1+n+\alpha)\Gamma(1+n+\beta)}{\Gamma(1+\alpha)\Gamma(1+\beta)}\,\delta_{nm},\quad n,m = 0,1,\ldots \tag{3.8}$$

with $d\sigma(x)$ *the positive measure given by (3.7).*

The standard difference equation for the associated Pollaczek polynomial $p_n(x)$ is

$$(n+c+1)p_{n+1}(x) - 2(x\sin\phi + (n+c+\lambda)\cos\phi)p_n(x) + (n+c+2\lambda-1)p_{n-1}(x) = 0 \tag{3.9}$$

(note that [6], (1.8) has a misprint).

Comparing this with (1.1) we have the connecting formulas

$$P_n(x) = (2\sin\phi)^n \frac{\Gamma(1+n+c)}{\Gamma(1+c)}\, p_n(x + (c+\lambda)\cot\phi) \tag{3.10}$$

$$a = (4\sin^2\phi)^{-1},\ d = -\cot\phi,\ \mu = i,\ \alpha = c,\ \beta = c + 2\lambda - 1$$

where λ, c, ϕ are the Pollaczek parameters.

With (3.10) one may rewrite (3.7) as

$$\frac{d\sigma(x)}{dx} = \frac{(2\sin\phi)^{2(\lambda+c)}\, |\,\Gamma(\lambda + c - ix')\,|^2\, e^{(\pi-2\phi)x'}}{\Gamma(c+1)\Gamma(c+2\lambda)\ |\ {}_2F_1(c, c+2\lambda-1; c+\lambda-ix'; \dfrac{e^{i(\frac{\pi}{2}-\phi)}}{2\sin\phi})\ |^2} \tag{3.11}$$

$$x' = x - (c+\lambda)\cot\phi,\quad 0 < \phi < \pi,$$

which agrees with [6], (1.10) after an Euler transformation ([9], 2.14(22)) of ${}_2F_1$.

Note that our derivation makes no restrictions on the Pollaczek parameters c, λ except for the basic positivity condition $(n+c)(n+c+2\lambda-1) > 0$, $n = 1,2,\ldots$ Although it suffices to have $c > -1$ and $c + 2\lambda > 0$ one could also have c and $c + 2\lambda - 1$ complex conjugates or $-k-1 < c,\ c + 2\lambda - 1 < -k$, $k = 1,2,\ldots$

We may use (3.7) to connect the large n behaviour of the moments of $d\sigma(x)$ to the parameters of the difference equation. Thus from Stirling's formula applied to $\Gamma(\gamma^+)$ and [8], 2.3.2(10) one has

$$\frac{d\sigma(x)}{dx} = \text{const.}\ |x|^{b/a}\, e^{-|x|/r_\pm}\left(1 + \frac{(a+b)d}{2a|x|}\right)\left(1 + O\left(\frac{1}{x^2}\right)\right) \tag{3.12}$$

$$r_\pm = \sqrt{4a - d^2}\ /\ [\pi \pm 2\arctan\left(\frac{d}{\sqrt{4a-d^2}}\right)]$$

for $x \to \pm\infty$ respectively provided that either $c = 0$ or $a > d^2/3$. Thus from (3.12) one has

$$\int_{-\infty}^{\infty} x^n d\sigma(x) = \text{const.}\Gamma(1+n+b/a)\left[r_-^{n+1}\left(1+\frac{(a+b)d}{2ar_-n}\right) + (-1)^n r_+^{n+1}\left(1+\frac{(a+b)d}{2ar_+n}\right)\right]\left(1+O\left(\frac{1}{n^2}\right)\right) \tag{3.13}$$

which for $d = 0$ reduces to

$$\int_{-\infty}^{\infty} x^{2n} d\sigma(x) = \text{const.}\Gamma(1 + 2n + b/a)\left(2\frac{\sqrt{a}}{\pi}\right)^{2n}\left(1 + O\left(\frac{1}{n^2}\right)\right), \tag{3.14}$$

$$\int_{-\infty}^{\infty} x^{2n+1}\, d\sigma(x) = 0.$$

The discrete measures may be calculated exactly only for the standard $c = 0$ Meixner and Charlier cases where the zeros of $X_{-1}^{(s)}(x)$ are given by the poles of the Γ function. For

$c \neq 0$ one must resort to perturbation type information.

4. RESOLVENTS AND PROJECTIVE MEASURES [2]

Associated with the symmetric Jacobi matrix

$$A = \begin{bmatrix} a_0 & b_1 & 0 & . \\ b_1 & a_1 & b_2 & . \\ 0 & b_2 & \ddots & . \\ & \cdots & & \end{bmatrix}, \quad b_n > 0, \ a_n \text{ real}, \tag{4.1}$$

one has the difference equation (we choose $b_0 = 1$)

$$X_{n+1} - (z - a_n)X_n + b_n^2 X_{n-1} = 0, \ n \geq 0 \tag{4.2}$$

and the continued fraction

$$CF = z - a_0 + \mathop{K}_{n=1}^{\infty} \left[\frac{-b_n^2}{z - a_n} \right]. \tag{4.3}$$

Let A be considered as the closed symmetric operator in the Hilbert space $H = l^2(0, \infty) =$ closed span $\{e_n\}_{n=0}^{\infty}$ where $e_n = (0,0,\ldots,1,0,\ldots)^t$, ($e_n$ has a one in its nth component) with minimal domain $D(A) \supset \text{span}\{e_n\}_{n=0}^{<\infty}$. One then has the following fundamental connection between (4.1), (4.2) and (4.3).

If A is self-adjoint then

$$(e_0,(zI - A)^{-1} e_0) = 1/CF = X_0^{(s)}(z) \ / \ b_0^2 X_{-1}^{(s)}(z) \tag{4.4}$$

for Im $z \neq 0$ or more specifically for $z \notin sp(A)$ the spectrum of A.

The first equality in (4.4) is implicit in [2]. The second equality is the Pincherle theorem mentioned in § 1.

From (4.4) and the spectral theorem for self-adjoint operators one also has

$$\sigma(x) = (e_0, E_A(x)e_0) \tag{4.5}$$

where

$$1/CF = \int_{-\infty}^{\infty} \frac{d\sigma(x)}{z - x}$$

and $E_A(x)$ is the family of orthogonal projections (spectral family) associated with A.

Thus the measure $\sigma(x)$ which we calculated in § 3 from $X_n^{(s)}$ was the matrix element of the spectral family for the self-adjoint Jacobi matrix (4.1) having $a_n = dn$ and $b_n^2 = an^2 + bn + c$.

We wish to generalize the basic connection (4.4) to *all matrix elements* of the resolvent $(zI - A)^{-1}$ so that from $X_n^{(s)}$ one can obtain *all matrix elements* of $E_A(x)$.

Theorem 7: *If* A *is self-adjoint then for* $n \geq m \geq 0$ *and* $z \notin sp(A)$ *one has*

$$(e_m, (zI - A)^{-1} e_n) = \tilde{P}_m(z)\, X_n^{(s)}(z) \,/\, (b_0 b_1 \cdots b_n) X_{-1}^{(s)}(z) \tag{4.6}$$

where $\tilde{P}_m(z) = P_m(z) \,/\, (b_0 b_1 \cdots b_n)$ *is the normalized orthogonal polynomial satisfying* $\int \tilde{P}_n(x) \tilde{P}_m(x) d\sigma(x) = \delta_{nm}$.

Before proving this theorem we establish some notation and lemmas.

We denote by $A^{(k)}$ the kth abbreviated Jacobi matrix obtained from A by crossing out the first k rows and columns. Thus

$$A^{(k)} = \begin{pmatrix} a_k & b_{k+1} & 0 & . \\ b_{k+1} & a_{k+1} & b_{k+2} & \cdot \\ 0 & b_{k+2} & \ddots & \cdot \\ & \cdots & & \end{pmatrix}, \quad k \geq 0 \tag{4.7}$$

and (4.4) can be generalized to

$$(e_0, (zI - A^{(k)})^{-1} e_0) = 1/CF^{(k)} = X_k^{(s)}(z) \,/\, b_k^2\, X_{k-1}^{(s)}(z) \tag{4.8}$$

where

$$CF^{(k)} = z - a_k + \mathop{K}_{n=k+1}^{\infty} \left[\frac{-b_n^2}{z - a_n} \right]. \tag{4.9}$$

We denote by $A_N^{(k)}$ the finite $(N + 1) \times (N + 1)$ matrix obtained from $A^{(k)}$ by ignoring all elements beyond the N + 1 st row and column. Thus

$$A_N^{(k)} = \begin{pmatrix} a_k & b_{k+1} & 0 & . \\ b_{k+1} & a_{k+1} & b_{k+2} & . \\ 0 & b_{k+2} & \ddots & . \\ & \cdots & & a_{k+N} \end{pmatrix}, \quad N \geq 0 \tag{4.10}$$

and one has the standard connection with (4.9) given by ([23], (16.2) and [2], p. 28)

$$1/CF_N^{(k)} = Q_N^{(k)}(z) \,/\, P_N^{(k)}(z) \tag{4.11}$$

where

$$CF_N^{(k)} = z - a_k + \mathop{K}_{n=k+1}^{k+N} \left[\frac{-b_n^2}{z - a_n} \right], \tag{4.12}$$

$$Q_N^{(k)}(z) = P_{N-1}^{(k+1)}(z)\,, \quad N > 0\,, \tag{4.13}$$

$$P_N^{(k)}(z) = \det (zI_N - A_N^{(k)}),\ N > 0\,. \tag{4.14}$$

That is to say, the Nth approximant (4.11) of (4.8) is given by a ratio of determinants (4.14) of finite matrices associated with $A^{(k)}$.

Finally we introduce the infinite Jacobi matrix

$$A^{(k),N} = \begin{bmatrix} A_N^{(k)} & 0 \\ 0 & 0 \end{bmatrix} \tag{4.15}$$

which can also be written as

$$A^{(k),N} = P_N A^{(k)} P_N \tag{4.16}$$

where P_N is the orthogonal projection onto the subspace $H_N = \text{span } \{e_n\}_{n=0}^{N}$.

Lemma 1: If $j,k \le N$ *then*

$$[(zI_N - A_N^{(i)})^{-1}]_{j,k} = (e_j\,,\,(zI - A^{(i),N})^{-1} e_k). \tag{4.17}$$

Proof: One has the decomposition

$$zI - A^{(i),N} = (zI_N - A_N^{(i)}) \oplus zI$$

corresponding to $H = H_N \oplus (I - P_N)H$ where $e_j, e_k \in H_N$ and for any B acting in H_N, $(e_j, Be_k) = [B]_{j,k}$ the j,k matrix element of the matrix representation of B. □

Lemma 2: *If* A *is self-adjoint then* $A^{(i)}$ *is self-adjoint and*

$$\lim_{N\to\infty} (e_j,(zI - A^{(i),N})^{-1} e_k) = (e_j,(zI - A^{(i)})^{-1} e_k),\ \operatorname{Im} z \ne 0. \tag{4.18}$$

Proof: From (4.13) one deduces that $A^{(k+1)}$ is self-adjoint if $A^{(k)}$ is self-adjoint ([2], p.28). To prove (4.18) one uses the fact that $A^{(i),N} \to A^{(i)}$ in the generalized strong sense since span $\{e_n\}_{n=0}^{<\infty}$ is a core for $A^{(i)}$ and $A^{(i),N} e_k = A^{(i)} e_k$ for $k < N$. This last equality holds because $A^{(i),N} = P_N A^{(i)} P_N$, $P_N e_k = e_k$ for $k \le N$ and $A^{(i)} e_k \in H_N$ for $k < N$. Thus (4.18) follows from [12], p. 429, Corollary 1.6. □

Lemma 3: *One has*

$$\lim_{N\to\infty} \frac{\det (zI_{N-k-1} - A_{N-k-1}^{(k+1)})}{\det(zI_N - A_N)} = \frac{X_k^{(s)}(z)}{b_0^2 b_1^2 \cdots b_k^2\, X_{-1}^{(s)}(z)} \tag{4.19}$$

if and only if $X_n^{(s)}(z)$ *exists.*

Proof: From (4.11) - (4.14) and Pincherle's Theorem one obtains

$$\lim_{N\to\infty} \frac{\det(zI_{N-j-1} - A_{N-j-1}^{(j+1)})}{\det(zI_{N-j} - A_{N-j}^{(j)})} = \frac{X_j^{(s)}(z)}{b_j^2 X_{j-1}^{(s)}(z)} \tag{4.20}$$

and from k products of (4.20) with $j = 0,1,...k$, one obtains (4.19). □

Lemma 4: *Let* $0 \le j \le k \le N$. *Then*

$$[(zI_N - A_N)^{-1}]_{j,k} = \frac{P_j(z) b_j b_{j+1} \cdots b_k \det(zI_{N-k-1} - A_{N-k-1}^{(k+1)})}{b_j \det(zI_N - A_N)}. \tag{4.21}$$

Proof: One calculates the elements of the inverse of the finite $(N+1) \times (N+1)$ matrix $zI_N - A_N$ in terms of ratios of determinants. The special tridiagonal form (4.10) allows one to simplify the result to (4.21). □

Proof of Theorem 7: From (4.17), (4.18) one obtains

$$(e_m, (zI - A)^{-1} e_n) = \lim_{N\to\infty} [(zI_N - A_N)^{-1}]_{m,n}, \ \mathrm{Im}\, z \neq 0$$

and from (4.20), (4.21)

$$\lim_{N\to\infty} [(zI_N - A_N)^{-1}]_{m,n} = \frac{\tilde{P}_m(z)\, X_n^{(s)}(z)}{b_0 b_1 \cdots b_n\, X_{-1}^{(s)}(z)}.$$

This establishes (4.6) for $\mathrm{Im}\, z \neq 0$ and by analytic continuation for $z \notin \mathrm{sp}(A)$. □

Corollary 1: *If* A *is self-adjoint then for* $n \ge m \ge 0$ *and* $z \notin \mathrm{sp}(A^{(i)})$ *one has*

$$(e_m, (zI - A^{(i)})^{-1} e_n) = \tilde{P}_m^{(i)} X_{n+i}^{(s)}(z) \,/\, (b_i b_{i+1} \cdots b_{i+n})\, X_{i-1}^{(s)}(z)$$

where $\tilde{P}_m^{(i)} = P_m^{(i)}(z) / (b_i b_{i+1} \cdots b_{i+m})$ *is the normalized polynomial satisfying*

$$\int \tilde{P}_n^{(i)}(x)\, \tilde{P}_m^{(i)}(x) d\sigma^{(i)}(x) = \delta_{nm}, \quad \sigma^{(i)}(x) = (e_0, E_{A^{(i)}}(x) e_0).$$

Proof: In Lemma 3 and 4 and Theorem 7 we replace A by $A^{(i)}$. Note that $(A^{(i)})^{(j)} = A^{(i+j)}$ and the subscripts on the b's and $X^{(s)}$'s must be advanced by i. □

Corollary 2: *If* A *is self-adjoint then* A *and all* $A^{(i)}$ *have the same continuous spectrum.*

Proof: The continuous spectrum singularities of $(zI - A)^{-1}$ and $(zI - A^{(i)})^{-1}$ are both determined by the nonexistence of $X_n^{(s)}(z)$. □

In summary, one has associated with a self-adjoint Jacobi matrix (4.1) a family of positive measures

$$d\sigma_n^{(k)}(x) = d(e_n, E_{A^{(k)}}(x) e_n)\ , \ n,k = 0,1,\ldots \tag{4.22}$$

all related through ratios of the subdominant solution to (4.2), which is equivalent to products of the continued fraction (4.3) and its tails (4.9).

For the examples of § 2 one can concentrate on the $k = 0$ case since $k > 0$ corresponds to a change in the parameters for $k = 0$.

Although the $n > 0$ measures (4.22) are simply related to the basic $n = 0$ measure through the formula

$$d\sigma_n^{(k)} = (\tilde{P}_n^{(k)}(x))^2\, d\sigma^{(k)}(x) \tag{4.23}$$

it is an interesting problem to determine Jacobi matrices B(k,n), n > 0 such that

$$d\sigma_n^{(k)}(x) = d(e_0, E_B(x)e_0). \tag{4.24}$$

ACKNOWLEDGEMENT

The author is thankful to L. Lorch for pointing out that the name "Bessel" polynomial was already in use[1] and to M.E. Muldoon for pointing out reference [11].

Note added: After completing this work the author noticed that P.C. Ojha[2] in a recent issue of J. Math. Phys. has also solved the c = 0 Pollaczek case of (1.1) and that the Bessel order case has occurred previously in an article by D. Maki [3].

The referee also brought to my attention the book by P.M. Batchelder[4] where there is a detailed development of the solutions to (1.1) using the Laplace transform method.

REFERENCES

1. M. Abramowitz and I. A. Stegun, Eds., *Handbook of Mathematical Functions* (National Bureau of Standards, Washington, DC, 1965).
2. N.I. Akhiezer, *The Classical Moment Problem* (Oliver and Boyd, Edinburgh and London, 1965).
3. R. Askey, Ed., *Gabor Szegö: Collected Papers*, Vol.3, 1945-1972 (Birkhauser, Boston, 1982).
4. R. Askey and M. Ismail, 'Recurrence relations, continued fractions and orthogonal polynomials', Memoirs Amer. Math. Soc. #300 (1984).
5. R. Askey and J. Wilson, 'Some basic hypergeometric orthogonal polynomials that generalize Jacobi polynomials', Memoirs Amer. Math. Soc. #319 (1985).
6. R. Askey and J. Wimp, 'Associated Laguerre and Hermite polynomials', Proc. Roy. Soc. Edinburgh, 96 A (1984), pp. 15-37.
7. T.S. Chihara and M.E.H. Ismail, 'Orthogonal polynomials suggested by a Queueing model', Advances in Applied Mathematics 3 (1982), pp. 441-462.

---

(1) E. Grosswald, *Bessel Polynomials*, Lecture Notes in Mathematics 698 (Springer-Verlag, Berlin, 1978).

(2) P.C.Ojha, The Jacobi-matrix method in parabolic coordinates: Expansion of Coulomb functions in parabolic Sturmians, J. Math. Phys., **28** (1987), pp. 392-396.

(3) D. Maki, On constructing distribution functions with applications to Lommel polynomials and Bessel functions, Trans. Amer. Math. Soc. **130** (1968), pp. 281-297.

(4) P.M. Batchelder, *An Introduction to Linear Difference Equations* (Harvard Univ. Press, Cambridge, 1927).

8. A. Erdélyi, Ed., *Higher Transcendental Functions*, Vol. 1 (McGraw-Hill, New York, 1953).

9. A. Erdélyi, Ed., *Higher Transcendental Functions*, Vol. 2 (Robert E. Krieger, Malabar, Florida, 1981).

10. W. Gautschi, 'Computational aspects of three-term recurrence relations', SIAM Review, 9 (1967), pp. 24-82.

11. M.E.H. Ismail, 'The zeros of basic Bessel functions, the functions $J_{v+ax}(x)$, and associated orthogonal polynomials', J. Math. Anal. Appl. 86 (1982), pp. 1-19.

12. T. Kato, *Perturbation Theory for Linear Operators* (Springer- Verlag, New York, 1966).

13. D. Masson, 'The rotating harmonic oscillator eigenvalue problem. I. Continued fractions and analytic continuation', J. Math. Phys. 24 (1983), pp. 2074-2088.

14. D. Masson, 'Convergence and analytic continuation for a class of regular C-fractions', Canad. Math. Bull. 28 (1985), pp. 411-421.

15. J. Meixner, 'Orthogonale Polynomsysteme mit einem besonderen Gestalt der erzeugenden Funcktion', J. London Math. Soc. 9 (1934), pp 6-13.

16. O. Perron, *Die Lehre von den Kettenbrüchen* (B.G. Teubner, Leipzig und Berlin, 1929).

17. S. Pincherle, 'Delle funzioni ipergeometriche e di varie questioni ad esse attinenti', Giorn. Mat. Battaglini, 32 (1894), pp. 209-291.

18. F. Pollaczek, 'Sur un généralisation des polynomes de Legendre', C.R. Acad. Sci. Paris 228 (1949), pp. 1363-1365.

19. F. Pollaczek, 'Sur une famille de polynômes orthogonaux à quatre parametres', C.R. Acad. Sci. Paris 230 (1950), pp. 2254-2256.

20. F. Pollaczek, 'Sur une généralisation des polynomes de Jacobi', Mémoir. Sci. Math. 131 (Gauthier-Villars, Paris, 1956).

21. L.J. Slater, *Confluent Hypergeometric Functions* (Cambridge Univ. Press, Cambridge, 1960).

22. G. Szegö, 'On certain special sets of orthogonal polynomials', Proc. Amer. Math. Soc. 1 (1950), pp. 731-737.

23. H.S. Wall, *Analytic Theory of Continued Fractions* (Van Nostrand, Princeton, 1948).

24. J. Wimp, 'Some explicit Padé approximants for the function Φ' / Φ and a related quadrature formula involving Bessel functions', SIAM J. Math. Anal. 16 (1985), pp. 887-895.

MULTIPOINT PADÉ APPROXIMATION AND ORTHOGONAL RATIONAL FUNCTIONS

Olav Njåstad
Department of Mathematics
University of Trondheim-NTH
N-7034 Trondheim
Norway

ABSTRACT. Let $\{a_1,\dots,a_p\}$ be fixed points in the complex plane **C**, and denote by $\mathcal{R}$ the space of all functions of the form $R(z) = \alpha_0 + \sum_{i=1}^{p} \sum_{j=1}^{N_i} \frac{\alpha_{ij}}{(z-a_i)^j}$, $\alpha_0, \alpha_{ij} \in \mathbf{C}$. Let the series $\frac{c_0}{z}$ and , $\sum_{j=1}^{\infty} c_j^{(i)}(z-a_i)^{j-1}$, $i = 1,2,\dots,p$, be given, and define the linear functional Φ on $\mathcal{R}$ by $\Phi((z-a_i)^j) = c_j^{(i)}$, $j=1,2,\dots$, $i=1,\dots,p$,

$\Phi(1) = c_0$. . Define the bilinear form $\langle\ ,\ \rangle$ on $\mathcal{R}\times\mathcal{R}$ by $\langle A,B\rangle = \Phi(A\cdot B)$, and assume that $\Phi(R^2) \neq 0$ when $R(t) \not\equiv 0$. The Gram-Schmidt process applied to the sequence $\left\{1, \frac{1}{(z-a_1)}, \dots, \frac{1}{(z-a_p)}, \frac{1}{(z-a_1)^2}, \dots\right\}$ gives an orthogonal system $\{Q_n(z)\}$ of functions in $\mathcal{R}$. These orthogonal functions can be used to construct multipoint Padé approximants for the given series.

1. INTRODUCTION

In accordance with e.g. [3], [8], [13], [14] we may describe a general multi-point Padé approximation problem as follows. Let m,n be non-negative integers, and let $\beta_k^{(m,n)}$, $k = 0,1,\dots,m+n$, be $m+n+1$ not necessarily distinct points in the extended complex plane. Let $\rho(k)$ denote the multiplicity of the point $\beta_k^{(m,n)}$. For each

A. Cuyt (ed.), Nonlinear Numerical Methods and Rational Approximation, 259–270.

of these points let a formal power series $L_{m,n,k}(z)$ be given as follows:

$$(1.1)\qquad L_{m,n,k}(z) = \sum_{j=0}^{\infty} \alpha_j^{(m,n,k)} \cdot (z-\beta_k^{(m,n)})^j,$$

$$k = 0,1,\dots,m+n, \quad \alpha_j^{(m,n,k)} \in \mathbf{C}.$$

(Obvious modifications are to be made when ∞ is among the points $\beta_k^{(m,n)}$.) It is assumed that $L_{m,n,k}(z) \equiv L_{m,n,t}(z)$ if $\beta_s^{(m,n)} = \beta_t^{(m,n)}$.

A special case of this situation occurs when the series $L_{m,n,k}(z)$ are Taylor series expansions of a given (analytic) function at the points $\beta_k^{(m,n)}$. This is the situation discussed in [8], [13], [14].

We write $F_{m,n,k,h}(z)$ for the h-th partial sum of $L_{m,n,k}(z)$, i.e.:

$$(1.2)\qquad F_{m,n,k,h}(z) = \sum_{j=0}^{h} \alpha_j^{(m,n,k)} \cdot (z-\beta_k^{(m,n)})^j.$$

The rational function $\dfrac{A_m(z)}{B_n(z)}$ is called the multi-point Padé approximant of type (m,n) for the series (1.1) if $A_m(z)$ and $B_n(z)$ are polynomials such that $\deg A_m \leqslant m$, $\deg B_n \leqslant n$, and

$$(1.3)\qquad A_m(z)-B_n(z)\cdot L_{m,n,k}(z) = \prod_{s=0}^{m+n} (z-\beta_k^{(m,n)}) \cdot \sum_{j=0}^{\infty} d_j^{(m,n,k)} \cdot (z-\beta_k^{(m,n)})^j$$

for all the points $\beta_k^{(m,n)}$.

An equivalent condition is

$$(1.4)\qquad A_m(z)-B_n(z)\cdot F_{m,n,k,\rho(k)}(z) = \sum_{j=\rho(k)}^{\infty} e_j^{(m,n,k)} \cdot (z-\beta_k^{(m,n)})^j.$$

A special case of the situation above occurs when the points $\{\beta_0^{(m,n)},\dots,\beta_{m+n}^{(m,n)}\}$ are the first $m+n+1$ members of a given fixed sequence $\{\beta_k : k = 0,1,2,\dots\}$ of not necessarily distinct points in the extended complex plane. The multi-point Padé approximants in this case are called Newton-Padé approximants. See [1], [4].

When all the points β_k are equal to ∞ and $m < n$, the Newton-Padé approximants are the classical Padé approximants to the series $\sum_{j=1}^{\infty} \frac{\alpha_j}{z^j}$.

The Padé approximants of type $(n-1,n)$ for a series of the form $\sum_{j=0}^{\infty} \frac{c_j}{z^{j+1}}$ are closely related to the sequence of orthogonal polynomials determined by the linear functional defined by the sequence $\{c_j\}$, when this functional is definite in the sense of [2]. If $Q_n(z)$ is the n-th orthogonal polynomial and $P_n(z)$ is its associated polynomial, then $\frac{P_n(z)}{Q_n(z)}$ is the Padé approximant of type $(n-1,n)$. See [2] for details. Thus

(1.5) $$P_n(z)-Q_n(z)\cdot \sum_{j=0}^{\infty} \frac{c_j}{z^{j+1}} = \sum_{j=2n}^{\infty} \frac{d_j}{z^j}.$$

When the points β_k are alternately equal to ∞ and 0, i.e. $\beta_{2s+1} = 0$, $\beta_{2s} = \infty$, $s = 0,1,2,\ldots$, then the Newton-Padé approximants are two-point approximants to the series $\sum_{j=0}^{\infty} \frac{\alpha_j^{(1)}}{z^{j+1}}$, $\sum_{j=1}^{\infty} \alpha_j^{(2)} z^{j-1}$. The two-point Padé approximants to series of the form $\sum_{j=0}^{\infty} \frac{c_j}{z^{j+1}}$, $\sum_{j=1}^{\infty} -c_{-j} z^{j-1}$ are closely related to the sequence of orthogonal Laurent polynomials determined by the linear functional defined by the sequence $\{c_j: j = 0,1,2,\ldots\}$, $\{c_j: j = -1,-2,\ldots\}$, when this functional is definite. If $Q_n(z)$ is the n-th orthogonal Laurent polynomial and $P_n(z)$ is the associated Laurent polynomial, then $\frac{P_n(z)}{Q_n(z)}$ is the two-point Padé approximant of type $(n-1,n)$. Thus

(1.6a) $$U_n(z)-V_n(z)\cdot \sum_{j=0}^{\infty} \frac{c_j}{z^{j+1}} = \sum_{j=n+1}^{\infty} d_j z^{n-j}$$

$$(1.6b)\qquad U_n(z)-V_n(z)\cdot\sum_{j=1}^{\infty}-c_{-j}z^{j-1} = \sum_{j=n}^{\infty}c_jz^j.$$

where $U_n(z)$ and $V_n(z)$ are the numerators of $P_n(z)$ and $Q_n(z)$. This situation has been treated in detail in [12] for the case that all c_j are real and the corresponding functional is positive definite. For the theory of orthogonal Laurent polynomials in this case, see also [5]. See also the discussion in [6], [7] for the case that $c_{-j} = \overline{c_j}$ for all j and the corresponding functional is positive definite. In this case the orthogonal Laurent polynomials are closely related to orthogonal polynomials on the unit circle.

Note that the Newton-Padé approximants to the series $\sum_{j=1}^{\infty}\frac{c_j}{z^{j+1}}$, $\sum_{j=1}^{\infty}-c_{-j}z^{j-1}$ of type $(n-1,n)$ are the same if the points β_k are defined as follows: $\beta_0 = \infty$, $\beta_1 = \beta_2 = 0$, $\beta_3 = \beta_4 = \infty$, and in general $\beta_{4m-3} = \beta_{4m-2} = 0$, $\beta_{4m-1} = \beta_{4m} = \infty$, for $m = 1,2,\ldots$.

Let $\{a_1,\ldots,a_p\}$ be given distinct points in the complex plane. Set $\beta_0 = \infty$, $\beta_1 = \beta_2 = a_1,\ldots,\beta_{2p-1} = \beta_{2p} = a_p$, and in general $\beta_{2qp+2r-1} = \beta_{2qp+2r} = a_r$, $r = 1,\ldots,p$, $q = 0,1,2,\ldots$. Every natural number n can be written in a unique way as $n = pq_n+r_n$, $1 \leqslant r_n \leqslant p$. The points $\beta_i^{(n-1,n)}$, $i = 0,1,\ldots,2n$, constituting an initial section of the sequence $\{\beta_k\}$ is then determined by: $\beta_0^{(n-1,n)} = \infty$, $\beta_{2si+1}^{(n-1,n)} = \beta_{2si+2}^{(n-1,n)} = a_i$ for $s = 0,\ldots,q_n$. $i < r_n$, $\beta_{2si+1}^{(n-1,n)} = \beta_{2si+2}^{(n-1,n)} = a_i$ for $s = 0,\ldots,q_{n-1}$, $i > r_n$, $\beta_{2si+1}^{(n-1,n)} = \beta_{2si+2}^{(n-1,n)} = \beta_{2q_nr_n+1}^{(n-1,n)} = a_{r_n}$ for $s = 0,\ldots,q_{n-1}$, $i = r_n$. (Note that the total number of points here is $1 + (r_{n-1})2(q_n+1) + (p-r_n)2q_n + 2q_n+1 = 2(pq_n+r_n) = 2n = (n-1)+n+1$.)

Let $\{c_j^{(i)}: j = 1,2,\ldots\}$, $i = 1,\ldots,p$, be given sequences of complex numbers, and let c_0 be a given complex number. Let the power series $L_{n-1,n,k}(z)$ which is associated with the points $\beta_k^{(n-1,n)}$ that are equal to a_i be the series

$$(1.7a)\qquad \Lambda_i(z) = \sum_{j=0}^{\infty}c_{j+1}^{(i)}(z-a_i)^j$$

and let the series associated with $\beta_0^{(n-1,n)} = \beta_0 = \infty$ be

$$(1.7b) \qquad \Lambda_0(z) = -\frac{c_0}{z}.$$

We are going to show that the multi-point Padé approximants of type $(n-1,n)$ for the series (1.7) are related to a system of orthogonal functions in the same way as the Padé approximants and the two-point Padé approximants are related to such systems. The functions involved are called R-functions. (They are rational functions having no poles in extended complex plane outside the set $\{a_1,\ldots,a_p\}$.) A study of this relationship was initiated in [11] for the case that all the points $\{a_1,\ldots,a_p\}$ are on the real line, the sequences $\{c_j^{(i)}\}$ are real and the linear functional defined by these sequences is positive definite. (The interpolation result as stated in that paper is incomplete, in so far as the total degree of interpolation can be shown to be one better.) In this special case, orthogonal R-functions have been used to investigate the extended Hamburger moment problem, see [9], [10].

A similar interpolation result can be obtained for the problem which arises when the interpolation condition at ∞ is dropped. This may seem to be a more natural problem. However, we have chosen to concentrate on the problem with the condition at ∞, since the orthogonal functions connected with this problem can be used to treat a moment problem. The situation is not completely analogous when the condition at ∞ is dropped.

2. ORTHOGONAL R-FUNCTIONS

As in Section 1, let $\{a_1,\ldots,a_p\}$ be given distinct points in the complex plane. Let $\mathcal{R}$ denote the linear space consisting of all functions of the form

$$(2.1) \qquad R(z) = \alpha_0 + \sum_{i=1}^{p} \sum_{j=1}^{N_i} \frac{\alpha_{ij}}{(z-a_i)^j}, \quad \alpha_0,\alpha_{ij} \in \mathbf{C}.$$

The elements of $\mathcal{R}$ are called R-functions. We note that a function $R(z)$ belongs to $\mathcal{R}$ iff it can be written in the form $R(z) = \frac{A(z)}{B(z)}$, where $B(z)$ is a polynomial with all its zeros among the points $\{a_1,\ldots,a_p\}$ and $A(z)$ is a polynomial such that $\deg A \leqslant \deg B$. It follows that the space $\mathcal{R}$ is closed under multiplication.

Let n be a natural number. In the decomposition $n = q_n \cdot p + r_n$,

$1 \leq r_n \leq p$ (cf. Section 1) we write q for q_n, r for r_n. We denote by R_n the space of all R-functions of the form

$$(2.2) \qquad R(z) = \alpha_0 + \sum_{i=1}^{r} \sum_{j=1}^{q+1} \frac{\alpha_{ij}}{(z-a_i)^j} + \sum_{i=r+1}^{p} \sum_{j=1}^{q} \frac{\alpha_{ij}}{(z-a_i)^j} .$$

As in Section 1, let $\{c_j^{(i)}: j = 1,2,\ldots\}$, $i = 1,\ldots,p$, be given sequences of complex numbers, and let c_0 be a given complex number. Let Φ be the linear functional defined on R by

$$(2.3) \qquad \Phi(1) = c_0 , \quad \Phi((z-a_i)^j) = c_j^{(i)},$$

$j = 1,2,\ldots,$ $i = 1,\ldots,p.$ This functional gives rise to a bilinear form $\langle\ ,\ \rangle$ on $R \times R$, defined by $\langle A,B\rangle = \Phi(A\cdot B)$. We shall assume that the sequences are such that this form is definite, in the sense that $\langle R,R\rangle \neq 0$ when $R(z) \not\equiv 0$.

The Gram-Schmidt process applied to the sequence
$$\left\{1 , \frac{1}{(z-a_1)}, \ldots , \frac{1}{(z-a_p)} , \frac{1}{(z-a_1)^2}, \ldots , \frac{1}{(z-a_p)^2} , \frac{1}{(z-a_1)^3}, \ldots\right\}$$
gives an orthonormal sequence $\{Q_n(z)\}$ of R-functions. Obviously $Q_n \in R_n - R_{n-1}$, and $\{Q_0,\ldots,Q_{n-1}\}$ is a base for R_{n-1}, so that $\langle Q_n,A\rangle = 0$ for every $A \in R_{n-1}$. We also note that $Q_n(z)$ may be written in the form

$$(2.4) \qquad Q_n(z) = \frac{V_n(z)}{N_n(z)} ,$$

where

$$(2.5) \qquad N_n(z) = (z-a_1)^{q+1} \ldots (z-a_r)^{q+1}(z-a_{r+1})^q \ldots (z-a_p)^q,$$

and $V_n(z)$ is a polynomial of degree at most n. (The meaning of statements about a_{r+1} when $r = p$ and a_{r-1} when $r = 1$ will be clear from the context.)

The function $t \to \frac{Q_n(t)-Q_n(z)}{t-z}$ is an R-function. We define the R-function $P_n(z)$ associated with $Q_n(z)$ by

$$P_n(z) = \Phi\left(\frac{Q_n(t)-Q_n(z)}{t-z}\right) \tag{2.6}$$

where Φ operates on $\frac{Q_n(t)-Q_n(z)}{t-z}$ as a function of t. We can then write

$$P_n(z) = \frac{U_n(z)}{N_n(z)} , \tag{2.7}$$

where $U_n(z)$ is a polynomial such that $\deg U_n \leqslant \deg V_n - 1$.

A detailed study of R-functions is given in [9], [10] for the case that the points $\{a_1,\ldots,a_p\}$ are on the real axis, the sequences $\{c_j^{(i)}\}$ are real, and the bilinear form $\langle\ ,\ \rangle$ determined by the sequences in positive definite.

3. MAIN RESULT

Our aim is to show that $-\frac{U_n(z)}{V_n(z)}$ (or equivalently: $-\frac{P_n(z)}{Q_n(z)}$) is the multi-point Padé approximant of type $(n-1,n)$ for the series (1.7). The argument is essentially a modification of that used in [11], and is related to the proof of a result in [7] concerning interpolation at ∞ and 0.

Let $\frac{A_{n-1}(z)}{B_n(z)}$ be the multi-point Padé approximant. According to (1.4), this means that the following conditions must be satisfied:

$$A_{n-1}(z)-B_n(z)\cdot\sum_{j=0}^{2q+1} c_{j+1}^{(i)}(z-a_i)^j = \sum_{j=2q+2}^{\infty} \gamma_j^{(i)}(z-a_i)^j \tag{3.1a}$$

for $i = 1,\ldots,r-1$,

$$A_{n-1}(z)-B_n(z)\cdot\sum_{j=0}^{2q} c_{j+1}^{(r)}(z-a_r)^j = \sum_{j=2q+1}^{\infty} \gamma_j^{(r)}(z-a_r)^j \tag{3.1b}$$

(3.1c) $$A_{n-1}(z)-B_n(z)\cdot\sum_{j=0}^{2q-1} c_{j+1}^{(i)}(z-a_i)^j = \sum_{j=2q}^{\infty}\gamma_j^{(i)}(z-a_i)^j$$

for $i = r+1,\ldots,p$

(3.1d) $$A_{n-1}(z)-B_n(z)\cdot\left(-\frac{c_0}{z}\right) = \sum_{j=2}^{\infty}\gamma_{n-j}^{(0)}\, z^{n-j}.$$

THEOREM. The rational function $R_n(z) = -\frac{U_n(z)}{V_n(z)}$ is the multi-point Padé approximant of type $(n-1,n)$ for the series (1.7) in the following sense:

(3.2a) $$U_n(z)+V_n(z)\cdot\sum_{j=0}^{2q+1} c_{j+1}^{(i)}(z-a_i)^j = \sum_{j=2q+2}^{\infty}\gamma_j^{(i)}(z-a_i)^j$$

for $i = 1,\ldots,r-1$,

(3.2b) $$U_n(z)+V_n(z)\cdot\sum_{j=0}^{2q} c_{j+1}^{(r)}(z-a_r)^j = \sum_{j=2q+1}^{\infty}\gamma_j^{(r)}(z-a_r)^j$$

(3.2c) $$U_n(z)+V_n(z)\cdot\sum_{j=0}^{2q-1} c_j^{(i)}(z-a_i)^j = \sum_{j=2q}^{\infty}\gamma_j^{(i)}(z-a_i)^j$$

(3.2d) $$U_n(z) - \frac{c_0}{z}V_n(z) = \sum_{j=2}^{\infty}\gamma_{n-j}^{(0)}\, z^{n-j}.$$

Proof: We set $\delta_i = q+1$ when $i < r$, $\delta_i = q$ when $i \geq r$, $\varepsilon_i = q+1$ when $i \leq r$, $\varepsilon_i = q$ when $i > r$. We shall make use of the following expressions:

(3.3a) $$\Delta_i(z) = P_n(z)+Q_n(z)\cdot\sum_{j=0}^{\delta_i+\varepsilon_i-1} c_{j+1}^{(i)}(z-a_i)^j \quad \text{for } i = 1,\ldots,p$$

(3.3b) $$\Delta_0(z) = P_n(z) - \frac{c_0}{z}Q_n(z),$$

(3.4) $$\psi_i(t) = \frac{1}{(t-z)}\left[\frac{(z-a_i)^{\delta_i}-(t-a_i)^{\delta_i}}{(t-a_i)^{\delta_i}}\right] \quad \text{for } i = 1,\ldots,p$$

$$(3.5a)\qquad \phi_i(t) = \frac{1}{(t-z)}\left[\frac{(z-a_i)^{\delta_i}(t-a_i)^{\varepsilon_i}Q_n(t) - (z-a_i)^{\delta_i+\varepsilon_i}Q_n(z)}{(t-a_i)^{\delta_i+\varepsilon_i}}\right]$$

for $i = 1,\dots,p$

$$(3.5b)\qquad \phi_0(t) = \frac{t[Q_n(t)-Q_n(z)]}{z(t-z)} .$$

The function $\psi_i(t)$ belongs to R_{n-1}. Therefore we may write

$$\Phi\left(\frac{1}{(t-z)}\left[\frac{(z-a_i)^{\delta_i}}{(t-a_i)^{\delta_i}}Q_n(t)-Q_n(z)\right]\right) = \Phi\left(\frac{1}{(t-z)}[Q_n(t)-Q_n(z)]\right)+\Phi(\psi_i(t)Q_n(t)) =$$

$P_n(z) + \langle\psi_i,Q_n\rangle = P_n(z)$, and thus

$$(3.6a)\qquad P_n(z) = \Phi\left(\frac{1}{(t-z)}\left[\frac{(z-a_i)^{\delta_i}}{(t-a_i)^{\delta_i}}Q_n(t)-Q_n(z)\right]\right) \text{ for } i = 1,\dots,p.$$

Similarly we get $\Phi\left(\frac{tQ_n(t)-zQ_n(z)}{z(t-z)}\right) = \Phi\left(\frac{Q_n(t)-Q_n(z)}{t-z}\right) + \Phi\left(\frac{1}{z}Q_n(t)\right) = P_n(z)$, and thus

$$(3.6b)\qquad P_n(z) = \Phi\left(\frac{tQ_n(t)-zQ_n(z)}{z(t-z)}\right) .$$

By using the expression (3.6a) for $P_n(z)$, the definition of Φ and of $\Delta_i(z)$ (formula (3.3a)) and summation of finite geometric series we get for $i = 1,\dots,p$:

$$\Delta_i(z) = \Phi\left(\frac{1}{(t-z)}\left[\frac{(z-a_i)^{\delta i}}{(t-a_i)^{\delta_i}}Q_n(t)-Q_n(z)\right]\right) + Q_n(z)\cdot\sum_{j=0}^{\delta_i+\varepsilon_i-1}(z-a_i)^j\,\Phi\left(\frac{1}{(t-a_i)^{j+1}}\right) =$$

$$\Phi\left(\frac{1}{(t-z)}\left[\frac{(z-a_i)^{\delta_i}}{(t-a_i)^{\delta_i}}Q_n(t)-Q_n(z)\right]\right) + \Phi\left(\frac{1}{(t-z)}Q_n(z)\left[\frac{(t-a_i)^{\delta_i+\varepsilon_i}-(z-a_i)^{\delta_i+\varepsilon_i}}{(t-a_i)^{\delta_i+\varepsilon_i}}\right]\right)$$

$$= \Phi\left(\frac{1}{(t-z)}\left[\frac{(z-a_i)^{\delta_i}(t-a_i)^{\varepsilon_i} Q_n(t) - (z-a_i)^{\delta_i+\varepsilon_i} Q_n(z)}{(t-a_i)^{\delta_i+\varepsilon_i}}\right]\right). \text{ Similarly}$$

$$\Delta_0(z) = \Phi\left(\frac{tQ_n(t)-zQ_n(z)}{z(t-z)}\right) - \frac{Q_n(z)}{z}\Phi(1) = \Phi\left(\frac{t[Q_n(t)-Q_n(z)]}{z(t-z)}\right).$$

It follows that we may write

(3.7) $\Delta_i(z) = \Phi(\phi_i(t)),$ for $i = 0,1,\dots,p.$

The function $\phi_i(t)$ is an R-function. It follows from (2.4), (2.5) and the definition of δ_i, ε_i that we may write

(3.8a) $\phi_i(t) = (z-a_i)^{q+1} S_i(t,z)$ for $i = 1,\dots,r-1,$

(3.8b) $\phi_r(t) = (z-a_r)^q S_r(t,z),$

(3.8c) $\phi_i(t) = (z-a_i)^q S_i(t,z)$ for $i = r+1,\dots,p,$

(3.8d) $\phi_0(t) = \frac{1}{z^2} S_0(t,z).$

Here $S_i(t,z)$ is an R-function as a function of t and an R-function as a function of z. For $i = 1,\dots,p$, $S_i(t,z)$ has no factors $(z-a_i)$ in the denominator (recall that $N_n(z)$ has $q+1$ factors $(z-a_i)$ when $i = 1,\dots,r$, and q factors when $i = r+1,\dots,p$). As an R-function in z, $S_0(t,z)$ has degree of the numerator at most equal to the degree of the denominator.

It follows from (3.7) and (3.8) that $\Delta_i(z)$ has Taylor series expansions of the following form:

(3.9a) $\Delta_i(z) = \sum_{j=q+1}^{\infty} g_j^{(i)}(z-a_i)^j,$ for $i = 1,\dots,r-1$

(3.9b) $\Delta_r(z) = \sum_{j=q}^{\infty} g_j^{(r)}(z-a_r)^j$

$$(3.9c) \qquad \Delta_i(z) = \sum_{j=q}^{\infty} g_j^{(i)} (z-a_i)^j, \qquad \text{for } i = r+1,\ldots,p$$

$$(3.9d) \qquad \Delta_0(z) = \sum_{j=2}^{\infty} g_{j-2}^{(0)} \frac{1}{z^j} .$$

By multiplying the expressions (3.9) for $\Delta_i(z)$ on both sides by $N_n(z)$, again taking into account that the factor $(z-a_i)$ occurs $q+1$ times for $i = 1,\ldots,r$, and q times for $i = r+1,\ldots,p$, we get formulas of of the form (3.2).

REFERENCES

1. Baker, George A. Jr. and Peter Graves-Morris, Pade Approximants I, II, Encyclopedia of Mathematics and its Applications 13, 14, Addison-Wesley (1980).

2. Brezinski, Claude, Padé type approximation and general orthogonal polynomials, Birkhäuser Verlag (1980).

3. Chui, K.C., 'Recent results on Padé approximants and related problems', Approximation Theory II, Editors: C.C. Lorentz, C.K. Chui and L.L. Schumaker, Academic Press (1976) 79-115.

4. Gallucci, M.A. and William B. Jones, 'Rational approximations corresponding to Newton series (Newton-Padé approximants)', J. Approximation Theory 17 (1976) 366-392.

5. Jones, William B., Olav Njåstad and W.J. Thron, 'Orthogonal Laurent polynomials and the strong Hamburger moment problem', J. Math. Anal. Appl. 98 (1984) 528-554.

6. Jones, William B., Olav Njåstad and W.J. Thron, 'Continued fractions associated with the trigonometric and other strong moment problems', Constructive Approximation 2 (1986) 197-211.

7. Jones, William B., Olav Njåstad and W.J. Thron, 'Hermitian PC-fractions and their relation to Szegö polynomials and Gaussian quadrature on the unit circle', submitted.

8. Karlsson, J., 'Rational interpolation and best rational approximation', J. Math. Anal. Appl. 52 (1976) 38-52.

9. Njåstad, Olav, 'An extended Hamburger moment problem', Proc. Edinb. Math. Soc. (Series II) 28 (1985) 167-183.

10. Njåstad, Olav, 'Unique solvability of an extended Hamburger moment problem', J. Math. Anal. Appl., to appear.

11. Njåstad, Olav, 'A multi-point Padé approximation problem', Analytic Theory of Continued Fractions II, Editor: W.J. Thron, Springer Lecture Notes 1199 (1986) 263-268.

12. Njåstad, Olav and W.J. Thron, 'The theory of sequences of orthogonal L-polynomials', Padé approximants and continued fractions, Editors: Haakon Waadeland and Hans Wallin, Det Kongelige Norske Videnskabers Selskab, Skrifter, No. 1 (1983) 54-91.

13. Saff, E.B., 'An extension of Montessus de Ballore's theorem on the convergence of interpolating rational functions', J. Approximation Theory 6 (1972) 63-67.

14. Wallin, H., 'Rational interpolation to meromorphic functions', Padé Approximation and its Applications, Editors: M.G. de Bruin and H. van Rossum, Springer Lecture Notes 888 (1981) 371-382.

L-POLYNOMIALS ORTHOGONAL ON THE UNIT CIRCLE

W. J. Thron
Campus Box 426
University of Colorado
Boulder, Colo 80309-0426, U.S.A.

ABSTRACT. Two sequences $\{\sigma_n\}$ and $\{\tau_n\}$ of L-polynomials are derived which are orthogonal with respect to a linear functional M on Λ and span Λ. The sequences occur as denominators of approximants of a modified PC-fraction. Three term recursion relations for the sequence $\{\sigma_n\}$ as well as for $\{\tau_n\}$ are derived.

1. INTRODUCTION

Let a double sequence of moments $\{\mu_n\}_{-\infty}^{\infty}$ satisfying

$$\mu_{-n} = \bar{\mu}_n, \quad n \geq 0,$$

be given. In terms of this sequence one can define a linear functional on Λ by the rule

$$M(z^n) := \mu_{-n}, \quad n = 0, \pm 1, \pm 2, \ldots .$$

Here Λ is the set of all L-polynomials $L(z) = \sum_{\nu=p}^{q} a_\nu z^\nu$, $p \leq q$. By $\Lambda_{-m,n}$ we shall mean the subset of Λ consisting of those $L(z)$ for which $-m \leq p,q \leq n$.

We also define a functional $\langle\ ,\ \rangle$ on $\Lambda \times \Lambda$ by the requirement

$$\langle P,Q\rangle := M(P(z)\bar{Q}(1/z)).$$

We note that

$$(1.1) \qquad \overline{M(L(z))} = M(\bar{L}(1/z))$$

This research was supported in part by the U.S. National Science Foundation under grant no DMS-8401717

A. Cuyt (ed.), Nonlinear Numerical Methods and Rational Approximation, 271–278.

and hence

(1.2) $\overline{\langle P,Q\rangle} = \langle Q,P\rangle$.

If one requires that the Toeplitz determinants

$$\Delta_n := \begin{vmatrix} \mu_0 & \mu_1 & \cdots\cdots & \mu_n \\ \mu_{-1} & \mu_0 & \cdots\cdots & \mu_{n-1} \\ \vdots & & \ddots & \vdots \\ \mu_{-n} & \cdots\cdots & & \mu_0 \end{vmatrix}, \quad n \geq 0,$$

do not vanish then the Szegö polynomials

$$\rho_n(z) := \frac{1}{\Delta_{n-1}} \begin{vmatrix} \mu_0 & \cdots\cdots & \mu_{n-1} & \mu_n \\ \vdots & \ddots & \vdots & \vdots \\ \mu_{-(n-1)} & \cdots & \mu_0 & \mu_1 \\ 1 & \cdots\cdots & z^{n-1} & z^n \end{vmatrix}, \quad n \geq 1, \ \rho_0(z) = 1,$$

are orthogonal with respect to $\langle\ ,\ \rangle$ in the sense that they satisfy

(1.3) $\langle \rho_n, z^m\rangle = 0,\ 0 \leq m \leq n-1,\ \langle \rho_n, z^n\rangle = \Delta_n/\Delta_{n-1}$.

The reciprocal polynomials

$$\rho_n^x(z) := z^n \overline{\rho}_n(1/z)$$

satisfy

(1.4) $\langle \rho_n^x, z^m\rangle = 0,\ 1 \leq m \leq n,\ \langle \rho_n^x, 1\rangle = \Delta_n/\Delta_{n-1}$.

In addition the ρ_n are monic polynomials of degree n and hence the reciprocal polynomials satisfy $\rho_n^x(0) = 1$. For proofs and further details see [1,2].

Since both the ρ_n and the ρ_n^x are ordinary polynomials the sequences $\{\rho_n\}$ and $\{\rho_n^x\}$ cannot span Λ. It thus becomes of interest to find sequences of L-polynomials which are both orthogonal and span Λ. Such sequences do indeed exist. Define

$$(1.5)\qquad \begin{aligned}\sigma_{2n}(z) &:= z^{-n}\rho^{x}_{2n}(z), \quad n \geq 0,\\ \sigma_{2n+1}(z) &:= z^{-n}\rho_{2n+1}(z), \quad n \geq 0.\end{aligned}$$

Then $\{\sigma_n\}$ spans Λ in the order

$$\Lambda_{0,0}, \Lambda_{0,1}, \Lambda_{-1,1}, \Lambda_{-1,2}, \Lambda_{-2,2}, \ldots .$$

This is so because the leading coefficients of the L-polynomials σ_n, which are the coefficients of z^{n+1} in σ_{2n+1} and z^{-n} in σ_{2n}, do not vanish. As a matter of fact they equal 1, so that $\{\sigma_n\}$ can be thought of as a sequence of monic L-polynomials.

Next we observe that

$$\langle\sigma_{2n}, z^{m}\rangle = M(z^{-m-n}\rho^{x}_{2n}) = \langle\rho^{x}_{2n}, z^{n+m}\rangle$$

so that

$$(1.6a)\qquad \langle\sigma_{2n}, z^{m}\rangle = 0, \ -(n-1) \leq m \leq n, \ \langle\sigma_{2n}, z^{-n}\rangle = \Delta_{2n}/\Delta_{2n-1}.$$

Also

$$\langle\sigma_{2n+1}, z^{m}\rangle = M(z^{-m-n}\rho_{2n+1}) = \langle\rho_{2n+1}, z^{m+n}\rangle$$

from which

$$(1.6b)\qquad \langle\sigma_{2n+1}, z^{m}\rangle = 0, \ -n \leq m \leq n, \ \langle\sigma_{2n+1}, z^{n+1}\rangle = \Delta_{2n+1}/\Delta_{2n}$$

follows. Thus $\{\sigma_n\}$ is orthogonal with respect to M.

If one wants to span Λ in the order

$$\Lambda_{0,0}, \Lambda_{-1,0}, \Lambda_{-1,1}, \Lambda_{-2,1}, \Lambda_{-2,2}, \ldots$$

one can choose

$$(1.7)\qquad \tau_n(z) := \bar{\sigma}_n(1/z).$$

The sequence $\{\tau_n\}$ is a sequence of monic L-polynomials with leading coefficients determined by the order indicated above. From (1.2) it now follows that

$$\langle \tau_{2n}, z^m \rangle = 0, \quad -n \leq m \leq n-1, \ \langle \tau_{2n}, z^n \rangle = \Delta_{2n}/\Delta_{2n-1},$$
$$\langle \tau_{2n+1}, z^m \rangle = 0, \ -n \leq m \leq n, \quad \langle \tau_{2n+1}, z^{-(n+1)} \rangle = \Delta_{2n+1}/\Delta_{2n}.$$

Here the fact that $\mu_{-n} = \bar{\mu}_n$, $n \geq 0$ implies $\Delta_n \in \mathbf{R}$, $n \geq 0$ has been used.

THEOREM 1. Given a sequence $\{\mu_n\}_{-\infty}^{\infty}$ satisfying $\mu_{-n} = \bar{\mu}_n$ and $\Delta_n \neq 0$, $n \geq 0$, there exist two sequences $\{\sigma_n\}$ and $\{\tau_n\}$ of monic L-polynomials which are orthogonal with respect to M and span Λ. We have for all $n \geq 0$

$$\sigma_{2n}(z) = z^{-n}\rho_{2n}^{x}(z), \qquad \tau_{2n}(z) = z^{-n}\rho_{2n}(z),$$
$$\sigma_{2n+1}(z) = z^{-n}\rho_{2n+1}(z), \ \tau_{2n+1}(z) = z^{-(n+1)}\rho_{2n+1}^{x}(z),$$

where ρ_n, ρ_n^x are the Szegö polynomials with respect to M and their reciprocals.

2. THE SEQUENCES $\{\sigma_n\}$ AND $\{\tau_n\}$ AS PARTIAL DENOMINATORS OF A HERMITIAN PC-FRACTION

We recall that the hermitian PC-fraction $K(c_n/d_n)$ defined by

$$\begin{aligned}
&c_1 := -2\beta_0 && d_0 := \beta_0, \ d_1 := 1 \\
(2.1) \qquad &c_{2n} := 1 && d_{2n} := \bar{\delta}_n z \\
&c_{2n+1} := (1-|\delta_n|^2)z && d_{2n+1} := \delta_n, \qquad n \geq 1
\end{aligned}$$

has as denominators D_n of its approximants

$$(2.2) \qquad D_{2n} = \rho_n^x, \quad D_{2n+1} = \rho_n, \ n \geq 0,$$

provided one sets

$$\delta_n = \rho_n(0).$$

Since

$$\rho_n(0) = \frac{1}{\Delta_{n-1}} \begin{vmatrix} \mu_{-1} & & \mu_{-n} \\ & \cdot \ \cdot \ \cdot & \\ \mu_{n-2} & & \mu_{-1} \end{vmatrix}$$

and by the Jacobi identity

$$\Delta_{n=1}^2 = \Delta_n\Delta_{n-2} + \left\| \begin{matrix} \mu_{-1} & \mu_{-n} \\ \mu_{n-2} & \mu_{-1} \end{matrix} \right\|^2$$

one has

$$|\delta_n|^2 = |\rho_n(0)|^2 = 1 - \frac{\Delta_n\Delta_{n-2}}{\Delta_{n-1}^2} \neq 1$$

so that (2.1) is well defined.

A continued fraction $K(a_n/b_n)$ equivalent to the PC-fraction $K(c_n/d_n)$ is determined by a sequence $\{r_n\}$, $r_n \neq 0$. One must have

$$a_n = r_n r_{n-1} c_n,\ n \geqslant 1,\ b_n = r_n d_n,\ n \geqslant 0,\ r_0 = 1.$$

The denominators Q_n of the nth approximants of $K(a_n/b_n)$ satisfy

$$Q_n = r_0 \ldots r_n D_n, \quad n \geqslant 0.$$

We want to choose the r_n (if possible) so that

$$\begin{aligned} \sigma_{2n} &= z^{-n} D_{4n} = Q_{4n}, \\ \tau_{2n} &= z^{-n} D_{4n+1} = Q_{4n+1}, \\ \tau_{2n+1} &= z^{-(n+1)} D_{4n+2} = Q_{4n+2}, \\ \sigma_{2n+1} &= z^{-n} D_{3n+3} = Q_{4n+3}. \end{aligned} \tag{2.3}$$

It is easily verified that (2.3) holds if

$$r_{4n+1} := 1,\ r_{4n+2} := 1/z,\ r_{4n+3} := z,\ r_{4n+4} := 1/z, \quad n \geqslant 0 \tag{2.4}$$

This leads to

$$\begin{aligned} a_1 &= -2\beta_0, & b_0 &= \beta_0,\ b_1 = 1, \\ a_{4n-2} &= 1/z, & b_{4n-2} &= \overline{\delta}_{2n-1}, \\ a_{4n-1} &= (1-|\delta_{2n-1}|^2)z, & b_{4n-1} &= \delta_{2n-1}z, \\ a_{4n} &= 1, & b_{4n} &= \overline{\delta}_{2n}, \\ a_{4n+1} &= 1-|\delta_{2n}|^2, & b_{4n+1} &= \delta_{2n}, \qquad n \geqslant 1. \end{aligned} \tag{2.5}$$

We thus have proved:

THEOREM 2. The continued fraction

(2.6) $$\beta_0 + \frac{-2\beta_0}{1} + \frac{1}{\bar{\delta}_1 z} + \frac{(1-|\delta_1|^2)z}{\delta_1} + \frac{1/z}{\bar{\delta}_2} + \frac{1-|\delta_2|^2}{\delta_2} + \frac{1}{\bar{\delta}_3 z} + \cdots$$

which is equivalent to the PC-fraction (2.1) has as denominators Q_n of its nth approximants the values

$$Q_{4n} = \sigma_{2n},\ Q_{4n+1} = \tau_{2n},\ Q_{4n+2} = \tau_{2n+1},\ Q_{4n+3} = \sigma_{2n+1}.$$

3. THREE TERM RECURRENCE RELATIONS FOR THE SEQUENCES $\{\sigma_n\}$ AND $\{\tau_n\}$, RESPECTIVELY.

From the wellknown recursion relations for the denominators Q_n of the nth approximants of $K(a_n/b_n)$ one obtains

$$b_{m-2}(b_{m-1}Q_{m-2} + a_{m-1}Q_{m-3}) = b_{m-2}Q_{m-1}$$

$$a_{m-1}Q_{m-2} = a_{m-1}b_{m-2}Q_{m-3} + a_{m-1}a_{m-2}Q_{m-4}$$

and hence

$$(b_{m-1}b_{m-2} + a_{m-1})Q_{m-2} = b_{m-2}Q_{m-1} + a_{m-1}a_{m-2}Q_{m-4}.$$

Substituting this into $Q_m = b_mQ_{m-1} + a_mQ_{m-2}$ (provided $b_{m-1}b_{m-2} + a_{m-1} \neq 0$ which it will be in the cases we are interested in) yields

(3.1) $$Q_m = \left(b_m + \frac{a_mb_{m-2}}{b_{m-1}b_{m-2}+a_{m-1}}\right)Q_{m-1} + \frac{a_ma_{m-1}a_{m-2}}{b_{m-1}b_{m-2}+a_{m-1}}\, Q_{m-4}.$$

Next

$$Q_m = b_m(b_{m-1}(b_{m-2}Q_{m-3}+a_{m-2}Q_{m-4}) + a_{m-1}Q_{m-3})$$
$$+ a_m(b_{m-2}Q_{m-3}+a_{m-2}Q_{m-4}).$$

Hence

(3.2) $$Q_m = (b_mb_{m-1}b_{m-2}+b_ma_{m-1}+a_mb_{m-2})Q_{m-3} + (b_mb_{m-1}+a_m)a_{m-2}Q_{m-4}.$$

Setting $m = 4n$ in (3.1) we have the recursion formula for σ_{2n} in terms of $\sigma_{2n-1}, \sigma_{2n-2}$

$$\sigma_{2n} = \left(b_{4n} + \frac{a_{4n}b_{4n-2}}{b_{4n-1}b_{4n-2}+a_{4n-1}}\right)\sigma_{2n-1} + \frac{a_{4n}a_{4n-1}a_{4n-2}}{b_{4n-1}b_{4n-2}+a_{4n-1}}\sigma_{2n-2}.$$

Substituting from (2.5) and simplifying leads to

$$\text{(3.3)} \qquad \sigma_{2n} = \left(\overline{\delta}_{2n} + \frac{\overline{\delta}_{2n-1}}{z}\right)\sigma_{2n-1} + \frac{1-|\delta_{2n-1}|^2}{z}\sigma_{2n-2}, \quad n \geq 1.$$

Using $m = 4n+3$ in (3.2) one arrives at the formula

$$\sigma_{2n+1} = (b_{4n+3}b_{4n+2}b_{4n+1}+b_{4n+3}a_{4n+2}+a_{4n+3}b_{4n+1})\sigma_{2n}$$
$$+ (b_{4n+3}b_{4n+2}+a_{4n+3})a_{4n+1}\sigma_{2n-1}, \quad n \geq 1.$$

From this one obtains

$$\text{(3.4)} \qquad \sigma_{2n+1} = (\delta_{2n+1}+\delta_{2n}z)\sigma_{2n} + (1-|\delta_{2n}|^2)z\sigma_{2n-1}, \quad n \geq 1.$$

An application of (1.7) to (3.3) and (3.4) yields

$$\text{(3.5)} \qquad \tau_{2n} = (\delta_{2n}+\delta_{2n-1}z)\tau_{2n-1} + (1-|\delta_{2n-1}|^2)z\tau_{2n-2}, \quad n \geq 1$$

and

$$\text{(3.6)} \qquad \tau_{2n+1} = \left(\overline{\delta}_{2n+1} + \frac{\overline{\delta}_{2n}}{z}\right)\tau_{2n} + \frac{1-|\delta_{2n}|^2}{z}\tau_{2n-1}, \quad n \geq 1.$$

For the initial conditions one computes

$$\sigma_0 = 1, \quad \sigma_1 = \delta_1 + z,$$

and

$$\tau_0 = 1, \quad \tau_1 = \overline{\delta}_1 + \frac{1}{z}.$$

THEOREM 3.
(A) The sequence $\{\sigma_n\}$ satisfies the recursion relations

$$\sigma_0 = 1, \quad \sigma_1 = \delta_1 + z,$$

$$\sigma_{2n} = \left(\bar{\delta}_{2n} + \frac{\bar{\delta}_{2n-1}}{z}\right)\sigma_{2n-1} + \frac{1-|\delta_{2n-1}|^2}{z}\sigma_{2n-2}, \quad n \geq 1,$$

$$\sigma_{2n+1} = (\delta_{2n+1}+\delta_{2n}z)\sigma_{2n} + (1-|\delta_{2n}|^2)z\sigma_{2n-1}, \quad n \geq 1.$$

(B) <u>The sequence</u> $\{\tau_n\}$ <u>satisfies the recursion relations</u>

$$\tau_0 = 1, \quad \tau_1 = \bar{\delta}_1 + \frac{1}{z},$$

$$\tau_{2n} = (\delta_{2n}+\delta_{2n-1}z)\tau_{2n-1} + (1-|\delta_{2n-1}|^2)z\tau_{2n-2}, \quad n \geq 1,$$

$$\tau_{2n+1} = \left(\delta_{2n+1} + \frac{\bar{\delta}_{2n}}{z}\right)\tau_{2n} + \frac{1-|\delta_{2n}|^2}{z}\tau_{2n-1}, \quad n \geq 1.$$

REFERENCES

1) William B. Jones, Olav Njåstad and W.J. Thron, 'Continued fractions associated with trigonometric and other strong moment problems', Constr. Approx. (1986) 2, 197-211.

2) Gabor Szegö, 'Orthogonal Polynomials', Amer. Math. Soc. Providence, R.I (1939).

CONTINUED FRACTIONS

Chairmen:

G. A. Baker Jr.

Invited communications:

W. B. Jones
Schur's algorithm extended and Schur continued fractions.

H. Waadeland
Some recent results in the analytic theory of continued fractions.

Short communications:

L. Jacobsen*
Uniform convergence of limit periodic generalized continued fractions.

A. Lembarki*
Connection and convergence acceleration of continued fractions $K(a_n/1) : a_n \to \infty$ and $a_n \to -1/4$.

P. Levrie
Convergence acceleration for Miller's algorithm.

W. Reid
Best a posteriori truncation error estimates for continued fractions $K(a_n/1)$ with twin element regions.

Lecture notes are not included.

SCHUR'S ALGORITHM EXTENDED AND SCHUR CONTINUED FRACTIONS

William B. Jones
Department of Mathematics
University of Colorado
Boulder, CO 80309-0426 U.S.A.

Abstract. A generalization of Schur's algorithm is given which provides rational interpolants at a sequence of (not necessarily distinct) points in the complex plane. The algorithm can easily be extended to functions of several variables. It is also shown that Schur continued fractions with $\gamma_0 \neq 0$ are equivalent to Perron-Carathéodory (PC-) fractions. This connection is used to obtain new formulas for the Schur parameters γ_n and a new characterization of positive Schur continued fractions. Continued fraction methods are used to prove convergence and obtain truncation error bounds for Schur approximants.

1. Introduction.

Schur's algorithm and approximants have received much recent attention as a result of their applications in the field of digital signal processing and operator theory (see, for example, [1] and [5] and references contained therein). The use of Schur continued fractions for testing the stability and locating the poles of digital filters has been described in [9], [10], and [11]. Other recent work on continued fractions related to the Schur method is given in [6], [13], and [15]. This paper provides some further new results on Schur continued fractions and the Schur algorithm as well as new proofs of some known results.

We let $\mathcal{J}$ denote the family of all **Schur functions**

$$\mathcal{J} := \{f:\ f(z) \text{ is analytic and } |f(z)| \leq 1 \text{ for } |z| < 1\},$$

let $\mathscr{J}$ denote the subfamily of **normalized Schur functions**

$$\mathscr{J} := \{f \in \mathcal{J}:\ -1 < f(0) < 1\}$$

and let

$$\mathscr{J}_0 := \{f \in \mathscr{J}:\ f(0) \neq 0\}.$$

Research supported in part by the U.S. National Science Foundation under grants No. DMS-8401717 and DMS-8700498.

A. Cuyt (ed.), Nonlinear Numerical Methods and Rational Approximation, 281–298.

In [14] Schur described the following:

Schur Algorithm. For $f \in \mathcal{J}$ define $\{f_n\}$ and $\{\gamma_n\}$ as follows: **Initialization**. Set

$$f_0(z) := f(z), \ \gamma_0 := f_0(0) \text{ and } t_0(z,w) := \frac{zw+\gamma_0}{\bar{\gamma}_0 zw+1}.$$

Iteration nth step. For each $n = 0,1,2, \cdots$,

(A) If $|\gamma_n| < 1$, then set

$$f_{n+1}(z) := t_n^{-1}(z, f_n(z)) := \frac{1}{z} \frac{f_n(z) - \gamma_n}{1 - \bar{\gamma}_n f_n(z)}, \quad \gamma_{n+1} := f_{n+1}(0). \tag{1.1}$$

(B) If $|\gamma_n| = 1$, then stop, since $f_n(z) \equiv f_n(0)$.

The sequences $\{f_n\}$ and $\{\gamma_n\}$ can be finite or infinite. γ_n is called the nth **Schur coefficient** for f. Associated with each $f \in \mathcal{J}$ are sequences of linear fractional transformations (*l*.f.t.'s) $\{t_n\}$ and $\{T_n\}$ defined by

$$t_n(z,w) := \frac{zw+\gamma_n}{\bar{\gamma}_n zw+1} \tag{1.2a}$$

and

$$T_0(z,w) := t_0(z,w), \ T_n(z,w) := T_{n-1}(z, t_n(z,w)), \ n \geq 1. \tag{1.2b}$$

Thus

$$T_n(z,w) = t_0 \circ t_1 \circ \cdots \circ t_n(z,w),$$

where $\circ$ denotes functional composition with respect to the second variable w. It follows that

$$f(z) = T_n(z, f_{n+1}(z)), \quad n = 0,1,2, \cdots. \tag{1.3}$$

If $f_n \in \mathcal{J}$ and $|\gamma_n| < 1$, then by Schwarz's lemma $f_{n+1} \in \mathcal{J}$. If for some n, $|\gamma_k| < 1$ for $0 \leq k \leq n$ and $|\gamma_{n+1}| = 1$, then $f_{n+1}(z) \equiv f_{n+1}(0) = \gamma_{n+1}$, the process stops and

$$f(z) = T_n(z, \gamma_{n+1}). \tag{1.4}$$

If $|\gamma_n| < 1$ for $n = 0,1,2, \cdots$, then it was proved by Schur that

$$f(z) = \lim_{n\to\infty} T_n(z, w_n) \text{ for } |z| < 1 \text{ and } |w_n| \leq 1. \tag{1.5}$$

It can also be shown that

$$g(z) := \frac{1}{\overline{f(1/\bar{z})}} = \lim_{n\to\infty} T_n(z, e^{i\alpha}) \text{ for } |z| < 1, \ \alpha \in \mathbb{R}. \tag{1.6}$$

Since the *l*.f.t. $t_n(z,w)$ can be written in the form

$$t_n(z,w) = \gamma_n + \frac{(1 - |\gamma_n|^2)z}{\bar{\gamma}_n z + w^{-1}}, \tag{1.7}$$

it is natural to consider the continued fraction

(1.8a) $$S\{\gamma_n\} := \gamma_0 + \frac{(1-|\gamma_0|^2)z}{\overline{\gamma}_0 z} + \frac{1}{\gamma_1} + \frac{(1-|\gamma_1|^2)z}{\overline{\gamma}_1 z} + \frac{1}{\gamma_2} + \cdots$$

where

(1.8b) $$\gamma_0 \in \mathbb{R} \text{ and } |\gamma_n| < 1, \quad n = 1,2,3,\cdots .$$

These continued fractions, called **positive Schur fractions**, were first studied by Wall [16]. The class of all positive Schur fractions is denoted by Σ^+. If (1.8b) is replaced by the weaker condition

$$|\gamma_n| \neq 1, \quad \gamma_n \in \mathbb{C}, \quad n = 0,1,2,\cdots ,$$

then (1.8a) is called a **Schur fraction**. The class of all these continued fractions is denoted by Σ.

In two papers [3], [4] Frank investigated continued fractions of the form

$$k_0\gamma_0 + \frac{k_0(1-|\gamma_0|^2)z}{\overline{\gamma}_0 z} - \frac{1}{k_1\gamma_1} + \frac{k_1(1-|\gamma_1|^2)z}{\overline{\gamma}_1 z} - \frac{1}{k_2\gamma_2} + \cdots ,$$

where the k_j and γ_j are complex constants with $|\gamma_j| \neq 1$. She obtained results on correspondence and convergence for this extension of the Schur fractions.

In Section 2 we describe an extension of the Schur algorithm that gives rise to a family of rational approximants that subsumes Schur approximants, Thiele continued fraction approximants and certain Padé and Newton-Padé approximants. It is evident that the algorithm can be adapted to give rational interpolants in more than one complex variable. One new subfamily of these approximants (in one variable) gives finite Blaschke products that solve an interpolation problem in which the interpolation points and their images lie on the unit circle. These approximants are applied in a separate paper [2] to solve the frequency transformation problem for the design of digital filters.

In Section 3 it is shown, by equivalence transformations, that Schur fractions with $\gamma_0 \neq 0$ can be expressed as PC-fractions. We use this fact to give a new proof of correspondence (Theorem 3), and to obtain new formulas for the Schur parameters γ_n in Theorem 4. In particular, Schur fraction approximants are shown to correspond to a pair of formal power series (fps) (L_0, L_∞). A relation between these series is given in (3.33). Positive Schur fractions are studied in Section 4. Among the new results are a characterization of positive Schur fractions in terms of Toeplitz determinants (Theorem 5) and best a posteriori truncation error bounds (Theorem 10). For completeness we include (Theorem 6) a result from [6, Theorem 2.2(C)] on convergence of positive Schur fraction approximants.

2. Extended Schur Algorithm.

Let $\{z_k\}$ be a given sequence of complex numbers (not necessarily distinct). Let $f(z)$ be a complex-valued function defined at each z_k and let

$$w_k := f(z_k) \text{ for all } k \geq 0 .$$

For each k let $a_k(z)$, $b_k(z)$, $c_k(z)$, $d_k(z)$ be complex-valued functions defined at the z_k such that

(2.1a) $$\Delta_k(z) := a_k(z)d_k(z) - b_k(z)c_k(z) \not\equiv 0 ,$$

and let

(2.1b) $$t_k(z,w) := \frac{a_k(z)w + b_k(z)}{c_k(z)w + d_k(z)} .$$

For $k \geq 0$ let $f_k(z)$ and γ_k be defined recursively by

(2.2a) $$f_0(z) := f(z), \;\; f_{k+1}(z) := t_k^{-1}(z, f_k(z)) ,$$

(2.2b) $$\gamma_k := f_k(z_k) \;\; \textbf{(extended Schur coefficients)}.$$

Let

(2.3) $$T_0(z,w) := t_0(z,w); \;\; T_k(z,w) := T_{k-1}(z, t_k(z,w)), \;\; k \geq 1 .$$

For suitably chosen $\rho_n \in \mathbb{C}$ we call $T_n(z, \rho_n)$ the **extended nth Schur approximant.**

Theorem 1. *If for some* $n \geq 0$,

(2.4a) $$t_k(z_k, \lambda_k^{(n)}) = \gamma_k , \;\; k = 0,1, \cdots ,n ,$$

where

(2.4b) $$\lambda_k^{(n)} := t_{k+1} \circ t_{k+2} \circ \cdots \circ t_n(z_k, \rho_n), \;\; 0 \leq k \leq n-1, \;\; \lambda_n^{(n)} := \rho_n ,$$

then

(2.5) $$T_n(z_k, \rho_n) = w_k, \;\; k = 0,1, \cdots ,n .$$

Proof. Let $n \geq 0$ be given. By (2.2) $f_k(z) = t_k(z, f_{k+1}(z))$ for $k = 0,1, \cdots ,n$ and hence

(2.6) $$f(z) = T_{k-1}(z, f_k(z)), \;\; k = 1,2, \cdots ,n .$$

Therefore

(2.7a) $$w_0 = f(z_0) = f_0(z_0) = \gamma_0$$

and

(2.7b) $$w_k = f(z_k) = T_{k-1}(z_k, f_k(z_k)) = T_{k-1}(z_k, \gamma_k), \;\; k = 1,2, \cdots ,n .$$

Hence by (2.3), (2.4) and (2.7)

$$T_n(z_0, \rho_n) = t_0(z_0, \lambda_0^{(n)}) = \gamma_0 = w_0$$

and, for $k = 1,2, \cdots ,n$,

$$T_n(z_k, \rho_n) = T_{k-1}(z_k, t_k(z_k, \lambda_k^{(n)})) = T_{k-1}(z_k, \gamma_k) = w_k . \;\; \square$$

If $d_k(z) \not\equiv 0$ for all $k \geq 0$, then without loss of generality we can set $d_k(z) = 1$ and write

$$(2.8)\qquad t_k(z,w) = \frac{a_k(z)w + b_k(z)}{c_k(z)w + 1} = b_k(z) + \frac{a_k(z) - b_k(z)c_k(z)}{c_k(z) + w^{-1}} .$$

If we set $a_k := a_k(z)$, $b_k := b_k(z)$, $c_k := c_k(z)$, then it follows from (2.8) and (2.3) that

$$(2.9)\quad T_n(z,w) = b_0 + \frac{a_0-b_0c_0}{c_0} \;\underset{+}{}\; \frac{1}{b_1} \;\underset{+}{}\; \frac{(a_1-b_1c_1)}{c_1} \;\underset{+}{}\; \frac{1}{b_2} \;\underset{+}{}\; \cdots \;\underset{+}{}\; \frac{1}{b_n} \;\underset{+}{}\; \frac{a_n-b_nc_n}{c_n+w^{-1}} ,$$

which leads us to consider the continued fraction

$$(2.10)\qquad b_0 + \frac{a_0-b_0c_0}{c_0} \;\underset{+}{}\; \frac{1}{b_1} \;\underset{+}{}\; \frac{a_1-b_1c_1}{c_1} \;\underset{+}{}\; \frac{1}{b_2} \;\underset{+}{}\; \frac{a_2-b_2c_2}{c_2} \;\underset{+}{}\; \cdots .$$

We now define sequences of *l.f.t.*'s $\{s_n\}$ and $\{S_n\}$ by

$$(2.11a)\qquad s_0(z,w) := b_0 + w,\;\; s_1(z,w) := \frac{a_0-b_0c_0}{c_1+w} ,$$

$$(2.11b)\qquad s_{2m}(z,w) := \frac{1}{b_m+w} ,\quad s_{2m+1}(z,w) := \frac{a_m-b_mc_m}{c_m+w} ,\quad m \geq 1$$

$$(2.11c)\qquad S_0(z,w) := s_0(z,w),\quad S_n(z,w) := S_{n-1}(z, s_n(z,w)),\;\; n \geq 1 .$$

It follows that $S_n(z,0)$ is the nth approximant of (2.10) and

$$(2.12)\qquad S_{2m}(z,0) = T_m(z,0),\;\; S_{2m+1}(z,0) = T_m(z,\infty),\;\; m = 0,1,2,\cdots .$$

In the special case in which $a_k = z$, $b_k = \gamma_k$, $c_k = \overline{\gamma}_k z$ and $d_k = 1$, the continued fraction (2.10) reduces to the Schur fraction (1.8a).

Another continued fraction can be associated with extended Schur approximants. If $c_k(z) \not\equiv 0$ for all $k \geq 0$, then, without loss of generality, we can set $c_k := c_k(z) = 1$ and write

$$(2.13)\qquad t_k(z,w) = a_k(z) + \frac{b_k(z) - a_k(z)d_k(z)}{d_k(z) + w} .$$

Thus by (2.13) and (2.3) (setting $a_k := a_k(z)$, $b_k := b_k(z)$, $d_k := d_k(z)$)

$$(2.14)\quad T_n(z,w) = a_0 + \frac{b_0-a_0d_0}{d_0+a_1} \;\underset{+}{}\; \frac{b_1-a_1d_1}{d_1+a_2} \;\underset{+}{}\; \cdots \;\underset{+}{}\; \frac{b_{n-1}-a_{n-1}d_{n-1}}{d_{n-1}+a_n} \;\underset{+}{}\; \frac{b_n-a_nd_n}{d_n+w} .$$

This leads us to consider the continued fraction

$$(2.15)\qquad a_0 + \frac{b_0-a_0d_0}{d_0+a_1} \;\underset{+}{}\; \frac{b_1-a_1d_1}{d_1+a_2} \;\underset{+}{}\; \frac{b_2-a_2d_2}{d_2+a_3} \;\underset{+}{}\; \cdots .$$

With this continued fraction are associated the sequences of *l.f.t.*'s $\{s_n\}$ and $\{S_n\}$ defined by

$$(2.16a)\quad s_0(z,w) := a_0 + w,\;\; s_n(z,w) := \frac{b_{n-1}-a_{n-1}d_{n-1}}{d_{n-1}+a_n+w} ,\;\; n = 1,2,3,\cdots ,$$

(2.16b) $S_0(z,w) := s_0(z,w), \quad S_n(z,w) := S_{n-1}(z, s_n(z,w)), \quad n = 1,2,3, \cdots .$

From (2.16) it is clear that $S_n(z,0)$ is the nth approximant of (2.15) and

$$S_n(z,w) = T_n(z, a_n+w), \quad n = 1,2,3, \cdots .$$

In the special case in which $a_k(z) = \gamma_k$, $b_k(z) = z - z_k$, $c_k(z) = 1$ and $d_k(z) = 0$, (2.15) reduces to the Thiele continued fraction

$$\gamma_0 + \frac{z-z_0}{\gamma_1} + \frac{z-z_1}{\gamma_2} + \frac{z-z_2}{\gamma_3} + \cdots, \tag{2.17}$$

and $T_n(z, \rho_n)$ is the nth modified approximant

$$T_n(z, \rho_n) = \gamma_0 + \frac{z-z_0}{\gamma_1} + \frac{z-z_1}{\gamma_2} + \cdots + \frac{z-z_{n-1}}{\gamma_n} + \frac{z-z_n}{\rho_n}. \tag{2.18}$$

We conclude this section with a table of the transformations $t_k(z,w)$ for four well known special cases subsumed by the extended Schur algorithm.

Table 1. Special cases of the extended Schur algorithm

$T_n(z, \rho_n)$	$t_k(z,w)$	$\gamma_k = t_k(z_k, \lambda_k^{(n)})$	$\Delta_k(z)$
Schur $\rho_n = 0$ $z_k = 0$	$\dfrac{zw+\gamma_k}{\bar{\gamma}_k zw+1}$	γ_k	$(1-\lvert\gamma_k\rvert^2)z$
Thiele $\rho_n \neq 0$ $z_k \in \mathbb{C}$	$\dfrac{\gamma_k w+(z-z_k)}{w}$	$\dfrac{\gamma_k \lambda_k^{(n)}}{\lambda_k^{(n)}} = \gamma_k$ if $\lambda_k^{(n)} \neq 0$	$-(z-z_k)$
Newton Series $\rho_n = 0$ $z_k \in \mathbb{C}$	$(z-z_k)w + \gamma_k$	$[f(z_0), \cdots, f(z_k)]$ Divided difference	$z - z_k$
Taylor Series $\rho_n = 0$ $z_k = 0$	$zw + \gamma_k$	$\dfrac{f^{(k)}(0)}{k!}$	z

3. Schur Fractions and PC-fractions.

We let Σ_0 denote the subclass of all Schur fractions

$$S\{\gamma_n\} := \gamma_0 + \frac{(1-|\gamma_0|^2)z}{\bar{\gamma}_0 z} + \frac{1}{\gamma_1} + \frac{(1-|\gamma_1|^2)z}{\bar{\gamma}_1 z} + \frac{1}{\gamma_2} + \cdots, \quad \gamma_n \in \mathbb{C} \tag{3.1a}$$

satisfying

(3.1b) $\gamma_0 \neq 0$ and $|\gamma_n| \neq 1$, $n = 0,1,2,\cdots$

It is readily seen that (3.1a) is equivalent to

$$\gamma_0 + \frac{(1-|\gamma_0|^2)}{\overline{\gamma}_0} + \frac{z^{-1}}{\gamma_1} + \frac{(1-|\gamma_1|^2)}{\overline{\gamma}_1} + \frac{z^{-1}}{\gamma_2} + \cdots$$

which is equivalent to

$$\gamma_0 + \frac{(1-|\gamma_0|^2)}{\overline{\gamma}_0} + \frac{1}{\gamma_1 z} + \frac{(1-|\gamma_1|^2)z}{\overline{\gamma}_1} + \frac{1}{\gamma_2 z} + \frac{(1-|\gamma_2|^2)z}{\overline{\gamma}_2} + \cdots .$$

By further transformations this can be shown to be equivalent to

$$(3.2) \quad \gamma_0 + \frac{(1-|\gamma_0|^2)/\overline{\gamma}_0}{1} + \frac{1}{\overline{\gamma}_0\gamma_1 z} + \frac{(1-|\gamma_1|^2)z}{\overline{\gamma}_1/\overline{\gamma}_0} + \frac{1}{\overline{\gamma}_0\gamma_2 z} + \frac{(1-|\gamma_2|^2)z}{\overline{\gamma}_2/\overline{\gamma}_0} + \cdots .$$

We have shown that a Schur fraction (3.1) with $\gamma_0 \neq 0$ is equivalent to (3.2) which is a PC-fraction. We now use this fact to derive some new correspondence properties of Schur fractions. To make this discussion self-contained we first recall pertinent properties of PC-fractions.

A continued fraction of the form

$$(3.3a) \quad \beta_0 + \frac{\alpha_1}{1} + \frac{1}{\beta_2 z} + \frac{\alpha_3 z}{\beta_3} + \frac{1}{\beta_4 z} + \frac{\alpha_5 z}{\beta_5} + \frac{1}{\beta_6 z} + \cdots$$

where the α_n and β_n are complex constants satisfying

(3.3b) $\alpha_1 \neq 0$, $\alpha_{2n+1} = 1 - \beta_{2n}\beta_{2n+1} \neq 0$, $n = 1,2,3,\cdots$,

is called a **Perron-Carathéodory continued fraction** (or **PC-fraction**). PC-fractions were introduced in [8, 1986]; their connection to the Wiener linear prediction method was described in [7]. It is now clear that (3.2) is a PC-fraction. If P_n and Q_n denote the nth numerator and denominator, respectively, of (3.3a), then the difference equations [12, (2.1.6)] are given by

$$(3.4a) \quad P_0 = \beta_0,\ P_1 = \beta_0 + \alpha_1,\ Q_0 = Q_1 = 1 ,$$

$$(3.4b) \quad \begin{bmatrix} P_{2n} \\ Q_{2n} \end{bmatrix} = \beta_{2n} z \begin{bmatrix} P_{2n-1} \\ Q_{2n-1} \end{bmatrix} + \begin{bmatrix} P_{2n-2} \\ Q_{2n-2} \end{bmatrix}, \quad n = 1,2,3,\cdots ,$$

$$(3.4c) \quad \begin{bmatrix} P_{2n+1} \\ Q_{2n+1} \end{bmatrix} = \beta_{2n+1} \begin{bmatrix} P_{2n} \\ Q_{2n} \end{bmatrix} + \alpha_{2n+1} z \begin{bmatrix} P_{2n-1} \\ Q_{2n-1} \end{bmatrix}, \quad n = 1,2,3,\cdots .$$

From (3.4) it can be seen that P_n and Q_n are polynomials in z of the form

$$(3.5a) \quad P_{2n}(z) = \sum_{j=0}^{n} p_{2n,j} z^j = \beta_0 + \cdots + \beta_{2n}(\beta_0+\alpha_1)z^n, \quad n = 1,2,3,\cdots ,$$

$$(3.5b) \quad Q_{2n}(z) = \sum_{j=0}^{n} q_{2n,j} z^j = 1 + \cdots + \beta_{2n} z^n, \quad n = 1,2,3,\cdots ,$$

(3.5c) $P_{2n+1}(z) = \sum_{j=0}^{n} p_{2n+1,j}z^j = \beta_0\beta_{2n+1} + \cdots + (\beta_0+\alpha_1)z^n, \quad n = 1,2,3, \cdots,$

(3.5d) $Q_{2n+1}(z) = \sum_{j=0}^{n} q_{2n+1,j}z^j = \beta_{2n+1} + \cdots + z^n, \quad n = 1,2,3, \cdots.$

Using these with the determinant formulas for continued fractions [12, (2.1.9)], one can prove the following

Theorem PC 1. [8, Theorem 2.1]. *Let* (3.3) *be a given PC-fraction. Then there exists a unique pair* (L_0, L_∞) *of formal power series* (fps)

(3.6) $L_0 = c_0^{(0)} + \sum_{k=1}^{\infty} c_k z^k, \quad L_\infty = -c_0^{(\infty)} - \sum_{k=1}^{\infty} c_{-k}z^{-k}, \quad (c_0 := c_0^{(0)} + c_0^{(\infty)})$

such that, for $n = 0,1,2, \cdots,$

(3.7a) $$L_0 - \frac{P_{2n}(z)}{Q_{2n}(z)} = \beta_{2n+2} \prod_{j=0}^{n} \alpha_{2j+1} z^{n+1} + O(z^{n+2}),$$

(3.7b) $$L_\infty - \frac{P_{2n+1}(z)}{Q_{2n+1}(z)} = \frac{-\beta_{2n+3}\prod_{j=1}^{n}\alpha_{2j+1}}{z^{n+1}} + O((\tfrac{1}{z})^{n+2}),$$

(3.8a) $$Q_{2n}L_0 - P_{2n} = \beta_{2n+2} \prod_{j=0}^{n} \alpha_{2j+1} z^{n+1} + O(z^{n+2}),$$

(3.8b) $$Q_{2n}L_\infty - P_{2n} = \prod_{j=0}^{n} \alpha_{2j+1} + O(\tfrac{1}{z}),$$

(3.8c) $$Q_{2n+1}L_0 - P_{2n+1} = -\prod_{j=0}^{n} \alpha_{2j+1} z^n + O(z^{n+1}),$$

(3.8d) $$Q_{2n+1}L_\infty - P_{2n+1} = \frac{-\beta_{2n+3}\prod_{j=0}^{n}\alpha_{2j+1}}{z} + O((\tfrac{1}{z})^2).$$

We note that (3.5) and (3.8) imply that P_{2n}/Q_{2n} and P_{2n+1}/Q_{2n+1} are the weak (n,n) two-point Padé approximants for (L_0, L_∞) of orders $(n+1, n)$ and $(n, n+1)$, respectively. Here the symbol $O(z^r)$ denotes a fps in increasing powers of z, starting with a power not less than r.

In the following it is convenient to introduce the determinants Δ_n, Φ_n and Θ_n corresponding to $\{c_n\}$ as follows:

$$\Delta_n := \begin{vmatrix} c_0 & c_1 & \cdots & c_n \\ c_{-1} & c_0 & \cdots & c_{n-1} \\ \vdots & \vdots & & \vdots \\ c_{-n} & c_{-n+1} & \cdots & c_0 \end{vmatrix}, \quad n = 0,1,2, \cdots \qquad (\Delta_{-1} := 1)$$

and

$$\Phi_n := \begin{vmatrix} c_1 & c_2 & \cdots & c_n \\ c_0 & c_1 & \cdots & c_{n-1} \\ \vdots & \vdots & & \vdots \\ c_{-n+2} & c_{-n+3} & \cdots & c_1 \end{vmatrix}, \quad \Theta_n := \begin{vmatrix} c_{-1} & c_0 & \cdots & c_{n-2} \\ c_{-2} & c_{-1} & \cdots & c_{n-3} \\ \vdots & \vdots & & \vdots \\ c_{-n} & c_{-n+1} & \cdots & c_{-1} \end{vmatrix},$$

for $n = 1,2,3,\cdots$.

Theorem PC 2. [8, Theorem 2.2] (A) *Let* (3.3) *be a given PC-fraction and let* (L_0, L_∞) *denote the pair of corresponding fps* (3.6). *Then*

(3.9) $$\Delta_n \neq 0, \quad n = 0,1,2,\cdots$$

and for $n = 1,2,3,\cdots$

(3.10a) $$\alpha_1 = -\Delta_0, \quad \alpha_{2n+1} = \Delta_n \Delta_{n-2}/\Delta_{n-1}^2,$$

(3.10b) $$\beta_{2n} = (-1)^n \Phi_n/\Delta_{n-1}, \quad \beta_{2n+1} = (-1)^n \Theta_n/\Delta_{n-1},$$

(3.11) $$Q_{2n}(z) = \frac{1}{\Delta_{n-1}} \begin{vmatrix} c_0 & c_1 & \cdots & c_n \\ c_{-1} & c_0 & \cdots & c_{n-1} \\ \vdots & \vdots & & \vdots \\ c_{-n+1} & c_{-n+2} & \cdots & c_1 \\ z^n & z^{n-1} & \cdots & 1 \end{vmatrix}, \quad Q_{2n+1}(z) = \frac{1}{\Delta_{n-1}} \begin{vmatrix} c_0 & c_{-1} & \cdots & c_{-n} \\ c_1 & c_0 & \cdots & c_{-n+1} \\ \vdots & \vdots & & \vdots \\ c_{n-1} & c_{n-2} & \cdots & c_{-1} \\ 1 & z & \cdots & z^n \end{vmatrix}.$$

(B) *Conversely, let* (L_0, L_∞) *be a given pair of fps* (3.6) *such that* (3.9) *holds. Let* $\{\alpha_{2n+1}\}$ *and* $\{\beta_n\}$ *be defined by* (3.10). *Then* (3.3b) *holds and hence* (3.3a) *is a PC-fraction and it corresponds to* (L_0, L_∞) *in the sense of Theorem PC 1.*

We shall now interpret for the Schur fraction (3.2) the PC-fraction properties given by Theorems PC 1 and PC 2. For that purpose we let P_n and Q_n denote the nth numerator and denominator, respectively, of the Schur fraction (3.2). Further we let (for a given Schur fraction (3.2)) for $n = 1,2,3,\cdots$,

(3.12a) $$\beta_0 := \gamma_0, \quad \beta_{2n} = \overline{\gamma}_0 \gamma_n, \quad \beta_{2n+1} = \overline{\gamma}_n / \overline{\gamma}_0,$$

(3.12b) $$\alpha_1 := (1 - |\gamma_0|^2)/\overline{\gamma}_0, \quad \alpha_{2n+1} := 1 - |\gamma_n|^2.$$

It follows that (3.1b) holds and

(3.13) $$\gamma_n = \beta_{2n}/\overline{\beta}_0, \quad \beta_{2n} = |\gamma_0|^2 \overline{\beta}_{2n+1}, \quad n = 1,2,3,\cdots.$$

We define

$$\Gamma_n := \prod_{j=0}^{n} (1 - |\gamma_j|^2)$$

and thus arrive at:

Theorem 3. *Let* (3.2) *be a given Schur fraction satisfying* (3.1b). *Then there exists a unique pair* (L_0, L_∞) *of fps* (3.6) *such that, for* $n = 0,1,2,\cdots$,

(3.14a) $$L_0 - \frac{P_{2n}(z)}{Q_{2n}(z)} = \gamma_{n+1}\Gamma_n z^{n+1} + O(z^{n+2}),$$

(3.14b) $$L_\infty - \frac{P_{2n+1}(z)}{Q_{2n+1}(z)} = -\frac{\overline{\gamma}_{n+1}\Gamma_{n+1}}{(\overline{\gamma}_0)^2 z^{n+1}} + O((\frac{1}{z})^{n+2}),$$

and

(3.15a) $$Q_{2n}L_0 - P_{2n} = \gamma_{n+1}\Gamma_n z^{n+1} + O(z^{n+2}),$$

(3.15b) $$Q_{2n}L_\infty - P_{2n} = \Gamma_n/\overline{\gamma}_0 + O(\frac{1}{z}),$$

(3.15c) $$Q_{2n+1}L_0 - P_{2n+1} = -(\Gamma_n/\overline{\gamma}_0)z^n + O(z^{n+1}),$$

(3.15d) $$Q_{2n+1}L_\infty - P_{2n+1} = \frac{-\overline{\gamma}_{n+1}\Gamma_n}{(\overline{\gamma}_0)^2 z} + O((\frac{1}{z})^2).$$

It follows from Theorem 3 that P_{2n}/Q_{2n} and P_{2n+1}/Q_{2n+1} are the weak (n,n) two-point Padé approximants for (L_0, L_∞) of orders (n+1, n) and (n, n+1), respectively. We also obtain:

Theorem 4. (A) *Let* (3.1a) *be a given Schur fraction satisfying* (3.1b) *and let* (L_0, L_∞) *denote the pair of fps* (3.6) *to which* (3.1a) *corresponds in the sense of Theorem 3. Then:*

(3.16) $$\Delta_n \neq 0, \quad n = 0,1,2,\cdots,$$

(3.17) $$\gamma_0 = c_0^{(0)}, \quad \gamma_n = \frac{(-1)^n\Phi_n}{\overline{c_0^{(0)}}\,\Delta_{n-1}}, \quad n = 1,2,3,\cdots$$

(3.18) $$0 \neq |c_0^{(0)}| \neq 1, \quad \left|\frac{\Phi_n}{c_0^{(0)}\Delta_{n-1}}\right| \neq 1$$

and $Q_{2n}(z)$ *and* $Q_{2n+1}(z)$ *can be expressed by the determinant formulas* (3.11).

(B) *Conversely, let* (L_0, L_∞) *be a given pair of fps* (3.6) *such that* (3.16) *and* (3.18) *hold. Let* $\{\gamma_n\}$ *be defined by* (3.17). *Then* (3.1a) *is a Schur fraction satisfying* (3.1b) *and it corresponds to* (L_0, L_∞) *in the sense of Theorem 3.*

Proof. (A): The Schur fraction (3.1) is equivalent to the PC-fraction (3.2). Hence (3.16) and (3.17) follow from (3.9) and (3.10), respectively. (3.18) then follows from (3.17) and (3.16). The remaining parts of (A) are consequences of Theorem PC 2(A). (B) can be easily deduced from Theorem PC 2(B). □

Although the correspondence of the Schur fraction (3.1) to a pair (L_0, L_∞) of fps has been known [6], the formulas (3.17) and (3.11) and the conditions (3.18) are believed to be new.

For a given Schur fraction (3.1a) we define sequences of *l.f.t.*'s $\{s_n\}$ and $\{S_n\}$ by

(3.19a) $s_0(z,w) := \gamma_0 + w, \ s_{2n}(z,w) := \dfrac{1}{\gamma_n + w}, \quad n = 1,2,3, \cdots,$

(3.19b) $s_{2n+1}(z,w) := \dfrac{(1-|\gamma_n|^2)z}{\overline{\gamma}_n z + w}, \quad n = 0,1,2, \cdots,$

(3.19c) $S_0(z,w) := s_0(z,w), \ S_n(z,w) := S_{n-1}(z, s_n(z,w)), \quad n = 1,2,3, \cdots .$

It is helpful to note that the *l*.f.t.'s in (3.19) are special cases of those given by (2.11). The nth numerator $A_n(z)$ and denominator $B_n(z)$ of (3.1a) are defined by the difference equations

(3.20a) $$A_{-1} = 1, \ B_{-1} = 0. \ A_0 = \gamma_0, \ B_0 = 1,$$

(3.20b) $$\begin{bmatrix} A_{2n+1} \\ B_{2n+1} \end{bmatrix} = \overline{\gamma}_n z \begin{bmatrix} A_{2n} \\ B_{2n} \end{bmatrix} + (1 - |\gamma_n|^2) z \begin{bmatrix} A_{2n-1} \\ b_{2n-1} \end{bmatrix}, \quad n = 1,2,3, \cdots,$$

(3.20c) $$\begin{bmatrix} A_{2n} \\ B_{2n} \end{bmatrix} = \gamma_n \begin{bmatrix} A_{2n-1} \\ B_{2n-1} \end{bmatrix} + \begin{bmatrix} A_{2n-2} \\ B_{2n-2} \end{bmatrix}, \quad n = 1,2,3, \cdots .$$

If $\{t_n\}$ and $\{T_n\}$ are defined by (1.2), then it can be seen that

(3.21) $t_0(z,w) = s_0 \circ s_1(z,w^{-1}), \ t_n(z,w) = \dfrac{1}{s_{2n} \circ s_{2n+1}(z,w^{-1})}, \quad n = 1,2,3, \cdots .$

Therefore

(3.22) $$S_{2n+1}(z,w) = T_n(z,w^{-1}), \quad n = 0,1,2, \cdots$$

and

(3.23) $$S_{2n+1}(z,0) = T_n(z,\infty), \ S_{2n}(z,0) = T_n(z,0), \quad n = 0,1,2, \cdots .$$

By the well known formula [12, (2.1.7)] we have

$$S_n(z,w) = \frac{A_n(z) + wA_{n-1}(z)}{B_n(z) + wB_{n-1}(z)}, \quad n = 0,1,2, \cdots .$$

Hence

(3.24) $$T_n(z,w) = \frac{A_{2n+1}(z)w + A_{2n}(z)}{B_{2n+1}(z)w + B_{2n}(z)}, \quad n = 0,1,2, \cdots .$$

Defining $\{C_n\}, \{D_n\}, \{E_n\}, \{F_n\}$ by

$$C_n(z)z = A_{2n+1}(z), \ D_n(z) = A_{2n}(z), \ E_n(z)z = B_{2n+1}(z), \ F_n(z) = B_{2n}(z),$$

we obtain from (3.24)

(3.25) $$T_n(z,w) = \frac{C_n(z)zw + D_n(z)}{E_n(z)zw + F_n(z)}, \quad n = 0,1,2, \cdots$$

and from (3.20)

(3.26a) $$C_0 = 1, \ D_0 = \gamma_0, \ E_0 = \overline{\gamma}_0, \ F_0 = 1,$$

$$(3.26b)\qquad \begin{bmatrix} C_n \\ E_n \end{bmatrix} = z \begin{bmatrix} C_{n-1} \\ E_{n-1} \end{bmatrix} + \overline{\gamma}_n \begin{bmatrix} D_{n-1} \\ F_{n-1} \end{bmatrix}, \quad n = 1,2,3, \cdots ,$$

$$(3.26c)\qquad \begin{bmatrix} D_n \\ F_n \end{bmatrix} = \gamma_n z \begin{bmatrix} C_{n-1} \\ E_{n-1} \end{bmatrix} + \begin{bmatrix} D_{n-1} \\ F_{n-1} \end{bmatrix}, \quad n = 1,2,3, \cdots .$$

It can be seen from (3.26) that C_n, D_n, E_n, F_n are polynomials in z of degree at most n. To express relationships between these polynomials we introduce the concept of **indexed reciprocal** H_n^x of H_n, where $\{H_n\}$ is a sequence of indexed polynomials with degree $H_n \leq n$:

$$(3.27)\qquad H_n^x(z) := z^n \overline{H(1/\overline{z})} := z^n \overline{H}_n(1/z)$$

(see, [6, (2.8)]). It follows by induction from (3.26) that

$$(3.28)\quad C_n^x(z) = F_n(z),\ D_n^x(z) = E_n(z),\ E_n^x(z) = D_n(z),\ F_n^x(z) = C_n(z) .$$

By using (3.25) and (3.28) it can be shown that

$$(3.29)\qquad \overline{T_n\left(\frac{1}{\overline{z}}, \frac{1}{\overline{w}}\right)} =: \overline{T}_n\left(\frac{1}{z}, \frac{1}{w}\right) = \frac{1}{T_n(z,w)}, \quad n = 0,1,2, \cdots .$$

Therefore

$$(3.30)\quad T_n(z,0) = \frac{1}{\overline{T_n\left(\frac{1}{\overline{z}}, \infty\right)}} =: \frac{1}{\overline{T}_n\left(\frac{1}{z}, \infty\right)}, \quad n = 0,1,2, \cdots$$

and hence by (3.23)

$$(3.31)\qquad S_{2n}(z,0) = \frac{1}{\overline{S_{2n+1}\left(\frac{1}{\overline{z}}, 0\right)}}, \quad n = 0,1,2, \cdots .$$

Since

$$(3.32)\quad S_{2n}(z,0) = \frac{A_{2n}(z)}{B_{2n}(z)} = \frac{P_{2n}(z)}{Q_{2n}(z)}, \quad S_{2n+1}(z,0) = \frac{A_{2n+1}(z)}{B_{2n+1}(z)} = \frac{P_{2n+1}(z)}{Q_{2n+1}(z)}$$

it follows from (3.31) and Theorem 3 that

$$(3.33)\qquad L_\infty(z) = \frac{1}{\overline{L_0\left(\frac{1}{\overline{z}}\right)}} =: \frac{1}{\overline{L}_0\left(\frac{1}{z}\right)}$$

and hence

$$(3.34)\quad \overline{L}_0(z) L_\infty(z^{-1}) = \left(\overline{c_0^{(0)}} + \sum_{k=1}^{\infty} \overline{c}_k z^k\right)\left(-c_0^{(\infty)} - \sum_{k=1}^{\infty} c_{-k} z^k\right) = 1 .$$

Here (L_0, L_∞) is the pair of fps (3.6) to which the Schur fraction (3.1a) corresponds. From (3.33) we see that one of the fps L_0 and L_∞ determines the other.

4. Positive Schur Fractions. Let Σ_0^+ denote the subclass of all positive Schur fractions (3.1a) satisfying

(4.1) $\qquad 0 \neq \gamma_0 \in \mathbb{R}, \ |\gamma_n| < 1, \ n = 0,1,2, \cdots .$

Theorem 5. (A) *Let* $S\{\gamma_n\}$ *be a given positive Schur fraction (3.1a) in* Σ_0^+ *and let* (L_0, L_∞) *denote the pair of fps (3.6) to which it corresponds. Then*

(4.2a) $$c_0^{(0)} \neq 0, \ -1 < c_0^{(0)} < 1$$

(4.2b) $$|\Delta_n| > \left|\frac{\Phi_{n+1}}{c_0^{(0)}}\right| > |\Phi_{n+1}| \geq 0, \ n = 0,1,2, \cdots$$

and (3.17) *holds.* (B) *Conversely, let* (L_0, L_∞) *be a pair of fps* (3.6) *such that* (4.2) *holds. Let* $\{\gamma_n\}$ *be defined by* (3.17). *Then* (3.1a) *is a positive Schur fraction such that* (4.1) *holds and it corresponds to* (L_0, L_∞) *in the sense of Theorem 3.*

Proof. (A): It follows from Theorem 4(A) that (3.17) holds. By (3.17) and (4.1) we obtain

$$|\gamma_{n+1}| = \left|\frac{\Phi_{n+1}}{c_0^{(0)}\Delta_n}\right| < 1, \ n = 0,1,2, \cdots ,$$

from which we have (4.2b). (4.2a) follows from (4.1) and (3.17). (B): Suppose that (4.2) holds and let $\{\gamma_n\}$ be defined by (3.17). Then clearly (4.1) holds. The correspondence of (3.1a) to (L_0, L_∞) follows from Theorem 4(B). □

We turn next to convergence of the even and odd sequences of approximants of a positive Schur fraction. By well known properties of *l*.f.t.'s

(4.3) $\qquad |t_n(z,w)| < 1, \ \text{for} \ |z| < 1, \ |w| \leq 1 ,$

since $|\gamma_n| < 1, \ n \geq 0$. If $U := [w\colon \ |w| < 1]$, then

(4.4) $$t_n(z,U) \subseteq U \ \text{ for } \ |z| < 1 .$$

Hence by (1.2b)

(4.5) $\qquad T_n(z,U) \subseteq T_{n-1}(z,U) \subseteq t_0(z,U) \subseteq U, \ n = 1,2,3, \cdots .$

It follows from this and (3.23) that

$$|S_{2n}(z,0)| < 1 \ \text{ for } \ |z| < 1 ;$$

thus $\{S_{2n}(z,0)\}$ is uniformly bounded on every compact subset of U. Since $\{S_{2n}(z,0)\}$ corresponds to $L_0(z)$ at $z = 0$ by Theorem 3, we can conclude from [12, Theorem 5.13] that $\{S_{2n}(z,0)\}$ converges to a function $f(z) \in \mathscr{S}_0$; the convergence is uniform on compact subsets of U and $L_0(z)$ is the Taylor series expansion of $f(z)$ about $z = 0$. By a similar argument we can show that $\{S_{2n+1}(z,0)\}$ converges to a function $g(z)$ holomorphic in $|z| > 1$. Moreover, by (3.33) we can conclude that $g(z) = 1/\overline{f}\,(1/z)$ for $|z| > 1$. We summarize these results in the following:

Theorem 6. [6, Theorem 2.2(C)]. *Let* $S\{\gamma_n\} \in \Sigma^+$, *let* $S_{2n}(z,0)$ *and* $S_{2n+1}(z,0)$ *denote its* 2nth *and* (2n+1)th *approximants, respectively, and let* (L_0, L_∞) *denote the pair of fps (3.6) to which* $S\{\gamma_n\}$ *corresponds in the sense of Theorem 3. Then:*

(A) $\{S_{2n}(z,0)\}$ *converges uniformly on compact subsets of* U *to a function* $f(z) \in \mathcal{S}$ *and* $L_0(z)$ *is the Taylor series of* f(z) *about* $z = 0$.

(B) *Suppose in addition that* $S\{\gamma_n\} \in \Sigma_0^+$. *Then* $\{S_{2n+1}(z,0)\}$ *converges uniformly on compact subsets of* $|z| > 1$ *to a function* g(z) *holomorphic in* $|z| > 1$ *and* $L_\infty(z)$ *is the Laurent expansion of* g(z) *about* $z = \infty$.

(C) *For* $|z| > 1$

$$g(z) = \frac{1}{\overline{f(1/\bar{z})}} .$$

In the proof of Theorem 6 it remains to show that $\{S_{2n}(z,0)\}$ corresponds to $L_0(z)$ at $z = 0$ even if $S\{\gamma_n\} \in \Sigma^+$ (i.e., even if $\gamma_0 = 0$). But this is a simple consequence of the determinant formulas [12, (2.1.9)].

In Section 1 we described the Schur algorithm that generates a sequence of Schur parameters $\{\gamma_n\}$ starting with a function $f \in \mathcal{J}$. It can be shown that for the resulting positive Schur fraction $S\{\gamma_n\}$ the sequence of even approximants $S_{2n}(z,0)$ converges to f(z) for $|z| < 1$. This follows from the fact that $\{T_n(z, w_n)\}$ corresponds to $L_0(z)$ at $z = 0$ for arbitrary $\{w_n\}$, $w_n \in \mathbb{C}$. Therefore $\{T_n(z, w_n)\}$ converges to a function $h(z) \in \mathcal{J}$ by [12, Theorem 5.13], since $|T_n(z, w_n)| < 1$ for $|z| < 1$, $|w_n| \leq 1$. By setting $w_n := f_{n+1}(z)$ it follows from (1.3) that $h(z) = f(z)$.

To obtain best truncation error bounds for the even approximant of a Schur fraction, we let z be fixed with $|z| < 1$ and define, for $n = 0,1,2, \cdots$,

(4.6) $W_n(z) := [t_{n+1} \circ t_{n+2} \circ \cdots \circ t_{n+m}(z,0)\colon\ |\gamma_k| < 1,\ n+1 \leq k \leq n+m,\ m = 1,2,3, \cdots]$,

where each $t_n(z,w)$ is defined by (1.2a).

Lemma 7. *For* $|z| < 1$,

(4.7) $W_n(z) = U := [w\colon\ |w| < 1],\ n = 0,1,2, \cdots$.

Proof. Let z be fixed, $|z| < 1$. It is readily seen from (1.2a) that

$$[t_{n+1}(z,0)\colon\ |\gamma_{n+1}| < 1] = [\gamma_{n+1}\colon\ |\gamma_{n+1}| < 1] = U,\ n = 0,1,2, \cdots .$$

Therefore

$$U \subseteq W_n(z), \quad n = 0,1,2, \cdots .$$

By (4.4) we also have, for $n \geq 0$, $m \geq 1$,

$$\begin{aligned} t_{n+1} \circ t_{n+2} \circ \cdots \circ t_{n+m}(z,0) &\in t_{n+1} \circ t_{n+2} \circ \cdots \circ t_{n+m}(z,U) \\ &\subseteq t_{n+1} \circ t_{n+2} \circ \cdots \circ t_{n+m-1}(z,U) \\ &\subseteq \cdots \subseteq t_{n+1}(z,U) \subseteq U . \end{aligned}$$

Thus $W_n(z) \subseteq U$. □

Lemma 8. *Let* $S\{\gamma_n\} \in \Sigma^+$ *be given and let* z *be fixed, with* $|z| < 1$. *Then, for* $n \geq 0$ *and* $m \geq 1$

(4.8) $|T_{n+m}(z,0) - T_n(z,0)| \leq \sup [|T_n(z,u) - T_n(z,0)|: \ u \in U]$.

Proof. For each $m \geq 1$ let

$$u_m := t_{n+1} \circ t_{n+2} \circ \cdots \circ t_{n+m}(z,0) .$$

Then by (4.6) and Lemma 7, $u_m \in U$. Hence by (1.2b) $T_{n+m}(z,0) = T_n(z, u_m)$, which proves (4.8). □

Lemma 9. *Let* $S\{\gamma_n\} \in \Sigma^+$ *be given and let* z *be fixed with* $|z| < 1$. *Then, for* $n = 0,1,2, \cdots$,

$$(4.9) \quad \sup [|T_n(z,u) - T_n(z,0)|: \ u \in U] = \frac{\prod_{j=0}^{n} (1-|\gamma_j|^2)|z|^{n+1}}{|B_{2n}(z)| \cdot |B_{2n}(z) - B_{2n+1}(z)|} .$$

Proof. By the determinant formula [12, (2.1.9)] and (3.24)

$$|T_n(z,u) - T_n(z,0)| = \left| \frac{uA_{2n+1}(z)+A_{2n}(z)}{uB_{2n+1}(z)+B_{2n}(z)} - \frac{A_{2n}(z)}{B_{2n}(z)} \right|$$

$$= \left| \frac{u(A_{2n+1}(z)B_{2n}(z) - A_{2n}(z)B_{2n+1}(z))}{B_{2n}(z)(uB_{2n+1}(z)+B_{2n}(z))} \right|$$

$$= \frac{|u| \prod_{j=0}^{n} (1-|\gamma_j|^2)|z|^{n+1}}{|B_{2n}(z)B_{2n+1}(z)| \cdot |u+g_n|} ,$$

where

(4.10) $$g_n := B_{2n}(z)/B_{2n+1}(z) .$$

Since $T_n(z,w)$ is a *l*.f.t. in w, it follows from (4.5) that $\{T_n(z,U)\}$ is a nested sequence of open circular disks. Let $x_n \in U$ be chosen so that $T_n(z, x_n)$ is the center of $T_n(z,U)$. Then since $T_n(z, -g_n) = \infty$ and since inverses of circles are preserved under *l*.f.t., it follows that x_n and $-g_n$ are inverses of each other with respect to the circle $\partial T_n(z,U)$ (see Figure 1). If v_n denotes the point of intersection of the segment $[0, -g_n]$ with the circle ∂U, then

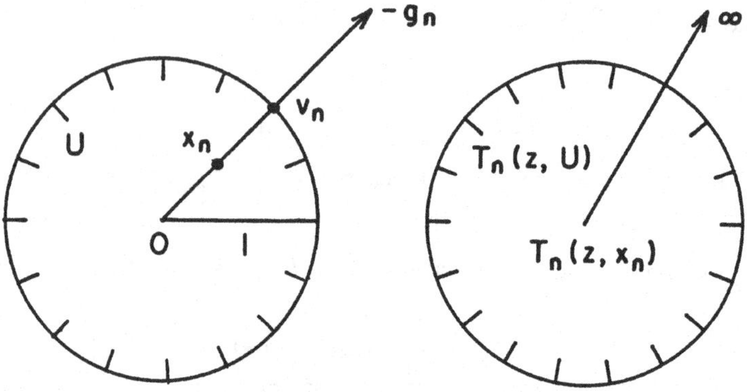

Figure 1.

$$\inf [|u + g_n|: \ u \in U] = |g_n + v_n| = |g_n| - 1 .$$

Therefore, since $\sup [|u|: \ u \in U] = 1$, we obtain

$$\sup [|T_n(z,u) - T_n(z,0)|: \ u \in U] = \frac{\prod_{j=0}^{n} (1-|\gamma_j|^2)|z|^{n+1}}{|B_{2n}(z)B_{2n+1}(z)|\cdot(|g_n| - 1)} .$$

This together with (4.10) implies (4.9). □

We can now prove

Theorem 10. *Let* $S\{\gamma_n\} \in \Sigma^+$ *be given and let* z *be fixed, with* $|z| < 1$. *Let* $S_{2n}(z,0) = T_n(z,0)$ *denote the* 2nth *approximant of the positive Schur fraction* (3.1a), *and let* $f(z) = \lim_{n\to\infty} S_{2n}(z,0)$. *Then:*

(A) *For* $n = 0,1,2, \cdots$,

$$\text{(4.11)} \qquad |f(z) - S_{2n}(z,0)| \leq \frac{\prod_{j=0}^{n} (1-|\gamma_j|^2)|z|^{n+1}}{|B_{2n}(z)|\cdot|B_{2n}(z)-B_{2n+1}(z)|} .$$

(B) *The a posteriori truncation error bound in* (4.11) *is best possible* provided the only Schur parameters that are known are $\gamma_0, \gamma_1, \cdots, \gamma_n$.

Proof. (A) follows immediately from Theorem 6 and Lemmas 8 and 9. (B) follows from (4.6) and Lemmas 7 and 8 since, given $u \in U$ there exist $m \geq 1$ and γ_k, $n+1 \leq k \leq n+m$ such that $|\gamma_k| < 1$ and

$$u = t_{n+1} \circ t_{n+2} \circ \cdots \circ t_{n+m}(z,0) . \qquad \square$$

We conclude with the following remarks on Schur fraction representations of analytic functions. Suppose that $F(\varsigma)$ is a function holomorphic in $|\varsigma - \varsigma_0| < \rho$. Let $G(\varsigma)$

and $H(\varsigma)$ be defined by

$$G(\varsigma) := F(\varsigma) - i\,\mathrm{Im}\,F(\varsigma_0),$$

$$H(\varsigma) := \begin{cases} G(\varsigma) & \text{if } \mathrm{Re}\,F(\varsigma_0) \neq 0 \\ G(\varsigma) + 1 & \text{if } \mathrm{Re}\,F(\varsigma_0) = 0. \end{cases}$$

Then $H(\varsigma)$ is holomorphic in $|\varsigma - \varsigma_0| < \rho$ and

$$0 \neq H(\varsigma_0) \in \mathbb{R}\,.$$

Let ϵ be given so that $0 < \epsilon < \rho$ and let

$$M(\epsilon) := \max\left[|H(\varsigma)|\colon\ |\varsigma - \varsigma_0| \leq \rho - \epsilon\right].$$

Define

$$f(z) := \frac{H((\rho-\epsilon)z+\varsigma_0)}{M}\,, \quad |z| \leq 1\,.$$

Then it follows that $f(z) \in \mathscr{S}_0$ and hence $f(z)$ has a Schur fraction expansion $S\{\gamma_n\} \in \Sigma_0^+$. We obtain an expression for $F(\varsigma)$ by writing

$$F(\varsigma) = G(\varsigma) + i\,\mathrm{Im}\,F(\varsigma_0)$$

$$G(\varsigma) = \begin{cases} H(\varsigma) & \text{if } \mathrm{Re}\,F(\varsigma_0) \neq 0 \\ H(\varsigma) - 1 & \text{if } \mathrm{Re}\,F(\varsigma_0) = 0 \end{cases}$$

$$H(\varsigma) = M f\left(\frac{\varsigma-\varsigma_0}{\rho-\epsilon}\right).$$

References.

1. Bultheel, Adhemar. Algorithms to compute the reflection coefficients of digital filters, *Numerical methods of approximation theory,* vol. 7, (eds. L. Collatz G. Meinardus, H. Werner), Birkhäuser Verlag, Basel (1984), 33-50.

2. Feyh, German, William B. Jones and Clifford T. Mullis. Extension of the Schur algorithm for frequency transformations, Proceedings of the International Symposium on Mathematical Theory of Networks and Systems - 1987, Phoenix, AZ.

3. Frank, Evelyn. On the properties of certain continued fractions, *Proc. Amer. Math. Soc.* 53 (1952), 921-936.

4. Frank, Evelyn and Oskar Perron. Remark on a certain class of continued fractions, *Proc. of the Amer. Math. Soc.* 5, No. 2 (April 1954), 270-283.

5. Gohberg, I., (ed.). *I. Schur Methods in Operator Theory and Signal Processing,* Birkhäuser Verlag, Boston (1986).

6. Jones, William B., Olav Njåstad and W.J. Thron. Schur fractions, Perron-Carathéodory fractions and Szegö polynomials, a survey, in *Analytic Theory of Continued Fractions II* (ed. W.J. Thron), Lecture Notes in Mathematics 1199, Springer-Verlag, New York (1986), 127-158.

7. Jones, William B., Olav Njåstad and W.J. Thron. Continued fractions associated with Wiener's linear prediction method, *Computational and Combinatorial Methods in Systems Theory* (C.I. Byrnes and A. Lindquist, editors), Elsevier Science Publishers B.V. (North Holland) (1986), 327-340.

8. Jones, William B., Olav Njåstad and W.J. Thron. Continued fractions associated with the trigonometric and other strong moment problems, *Constructive Approximation* 2 (1986), 197-211.

9. Jones, William B. and Allan Steinhardt. Digital filters and continued fractions, *Analytic theory of continued fractions*, (eds., W.B. Jones, W.J. Thron and H. Waadeland), Lecture Notes in Mathematics 932, Springer-Verlag, New York (1982), 129-151.

10. Jones, William B. and Allan Steinhardt. Applications of Schur fractions to digital filtering and signal processing, *Rational approximation and interpolation* (eds., P.R. Graves-Morris, E.B. Saff and R.S. Varga), Lecture Notes in Mathematics 1105, Springer-Verlag, New York (1984), 210-226.

11. Jones, William B. and Allan Steinhardt. Finding the poles of the lattice filter, *IEEE Trans. on Acoustics,Speech and Signal Processing*, vol. ASSP-33, No. 5 (October 1985), 1328-1331.

12. Jones, William B. and W.J. Thron. *Continued Fractions: Analytic Theory and Applications*, Encyclopedia of Mathematics and its Applications, 11, Addison-Wesley Publishing Company, Reading, MA (1980), distributed now by Cambridge University Press, New York.

13. Jones, William B. and W.J. Thron. Contraction of the Schur algorithm for functions bounded in the unit circle, *Rocky Mtn. J. Math.*, to appear.

14. Schur, I. Uber Potenzreihen die im Inneren des Einheitskreises beschränkt sind, *J. reine angewandte Math.* 147 (1917), 205-232, 148 (1918/19), 122-145.

15. Thron, W.J. Two-point Padé tables, T-fractions and sequences of Schur, *Padé and Rational Approximation*, (eds. E.B. Saff and R.S. Varga), Academic Press, Inc., New York (1977), 215-226.

16. Wall, H.S. *Analytic Theory of Continued Fractions*, D. Van Nostrand Co., Inc., New York (1948).

SOME RECENT RESULTS IN THE ANALYTIC THEORY OF CONTINUED FRACTIONS

Haakon Waadeland
Department of Mathematics and Statistics
The University of Trondheim
N-7055 Dragvoll
Norway

ABSTRACT. Examples of recent development in the analytic theory of continued fractions are presented, limited to the following: a) In the basic theory: Some new concepts and their role in the theory. b) In the algorithmic part: Methods of computation of values of continued fractions. c) In the applications:A remark on the determination of zero-free regions for polynomials.

The results are largely elementary, and may to a large extent be included in a possible undergraduate course in analytic theory of continued fractions.

1. INTRODUCTION

We shall here discuss continued fractions $K(a_n/b_n)$,

$$(1.1)\qquad \cfrac{a_1}{b_1+\cfrac{a_2}{b_2+\cfrac{a_3}{b_3+\ddots}}} = \frac{a_1}{b_1}+\frac{a_2}{b_2}+\frac{a_3}{b_3}+\dots = \operatorname*{K}_{n=1}^{\infty}\left(\frac{a_n}{b_n}\right),$$

where a_n and b_n are complex numbers, $a_n \neq 0$. Since $a_n \neq 0$, the linear fractional transformations

$$s_n(w) = \frac{a_n}{b_n+w}$$

are all non-singular, and hence also the composition

This research was supported in part by the Alexander von Humboldt-Stiftung, Bonn-Bad Godesberg.

A. Cuyt (ed.), Nonlinear Numerical Methods and Rational Approximation, 299–333.

(1.2) $$S_n(w) = s_1 \circ s_2 \circ \ldots \circ s_n(w) = \frac{a_1}{b_1} + \frac{a_2}{b_2} + \ldots + \frac{a_n}{b_n + w}.$$

Following standard notations and terminology [15], we shall call

(1.3) $$f_n = S_n(0)$$

the nth *approximant*. Sometimes, since we also shall study some $S_n(w_n)$, used as approximants, we shall call (1.3) classical approximants (C-approximants).

Convergence means that (with $\hat{\mathbb{C}}$ = extended complex plane)

(1.4) $$\lim_{n\to\infty} f_n = f$$

exists in $\hat{\mathbb{C}}$. In case of convergence we shall write

(1.5) $$\mathop{K}_{n=1}^{\infty}\left(\frac{a_n}{b_n}\right) = f.$$

We thus use K in the same double meaning as the symbols $\sum$ and Π. The number f is called the *value* of the continued fraction (1.1).

For any fixed N, the continued fraction

(1.6) $$\mathop{K}_{n=1}^{\infty}\left(\frac{a_{N+n}}{b_{N+n}}\right)$$

shall be called the Nth *tail* of the continued fraction (1.1). In case of convergence, its value, $f^{(N)}$, shall be called the Nth *tail value*. Sometimes we need to emphasize that it is the right value, in which case we shall call it the Nth *right tail value*. We have in particular that for $N = 0$ the continued fraction (1.6) (the 0th tail) is the original one (1.1), and in case of convergence we have $f = f^{(0)}$. Obviously the tail values satisfy the relations

(1.7) $$f^{(n)} = \frac{a_{n+1}}{b_{n+1} + f^{(n+1)}}, \quad n = 0,1,2,\ldots .$$

Any sequence $\{g^{(n)}\}_{n=0}^{\infty}$, satisfying

$$(1.8)\qquad g^{(n)} = \frac{a_{n+1}}{b_{n+1} + g^{(n+1)}}, \quad n = 0,1,2,\dots ,$$

and such that $g^{(0)} \neq f^{(0)}$, shall be called a sequence of *wrong tails* (wrong tail values). If in particular the continued fraction (1.1) diverges, any sequence of tail values is a sequence of wrong tail values. Wrong tails are just as important as right tails, perhaps more important.

Often it is of advantage to write the approximants as fractions, for instance $f_1 = a_1/b_1$, $f_2 = a_1b_2/(b_1b_2+a_2)$, or more generally

$$(1.9)\qquad f_n = \frac{A_n}{B_n},$$

where A_n, B_n are given by the recurrence relations

$$(1.10)\qquad \begin{aligned} A_n &= b_nA_{n-1} + a_nA_{n-2} \\ B_n &= b_nB_{n-1} + a_nB_{n-2} \end{aligned}, \quad n = 1,2,3,\dots$$

with initial conditions

$$(1.10')\qquad \begin{aligned} A_{-1} &= 1,\ A_0 = 0, \\ B_{-1} &= 0,\ B_0 = 1. \end{aligned}$$

It is easy to see, that with these notations we have

$$(1.11)\qquad S_n(w) = \frac{A_n+A_{n-1}w}{B_n+B_{n-1}w}.$$

We shall also need the *determinant formula*

$$A_nB_{n-1} - B_nA_{n-1} = (-1)^{n-1} \prod_{k=1}^{n} a_k, \quad n = 1,2,3,\dots .$$

If all $b_n \neq 0$ the continued fraction

$$(1.12)\qquad \frac{(a_1/b_1)}{1} + \frac{(a_2/b_1b_2)}{1} + \frac{(a_3/b_2b_3)}{1} + \dots$$

has exactly the same sequence of approximants as (1.1). In view of this and also because many of the important special continued fractions are of the form $K(a_n/1)$ we shall throughout the paper assume that in (1.1) all $b_n = 1$. Another point in favor of this slight reduction in generality is that the things we want to illustrate through this paper will loose little in idea and generality and gain much in simplicity by this assumption. The transition (1.1)$\rightsquigarrow$(1.12) is an example of an equivalence transformation of a continued fraction. Another equivalence relation leads, *without loss of generality*, to continued fractions of the form $K(1/e_n)$. There are strong reasons for carrying out investigations of the type we are going to see in the present paper also to continued fractions of this type.

Important concepts in the analytic theory of continued fractions are element- and value regions. Following [15] we shall call E an element region and V a value region for a family of continued fractions $K(a_n/1)$ if (and only if)

$$(1.13)\qquad \begin{cases} a_n \in E \quad \text{for all} \quad n, \\ E \subseteq V, \\ \dfrac{a_n}{1+V} \subseteq V \quad \text{for all} \quad a_n \in E. \end{cases}$$

All approximants of $K(a_n/1)$ are in V, and all values and tail values of the convergent continued fractions in the family are in the closure $\mathrm{cl}\,V$. Here in the introduction we shall mention one particular application: Let $K(\tilde{a}_n/1)$ be a particular continued fraction in the family. For any $n \geq 2$ the set

$$J_{n-1} = \frac{\tilde{a}_1}{1} + \frac{\tilde{a}_2}{1} + \dots + \frac{\tilde{a}_{n-1}}{1+\mathrm{cl}\,V}$$

(called inclusion region) contains the approximants

$$\tilde{f}_m = \frac{\tilde{a}_1}{1} + \frac{\tilde{a}_2}{1} + \dots + \frac{\tilde{a}_m}{1}, \quad m \geq n,$$

and hence

$$|\tilde{f}_{n+k} - \tilde{f}_n| \leq \mathrm{diam}\, J_{n-1}.$$

In case of convergence we find the following a priori truncation error bound:

$$(1.14)\qquad |\tilde{f} - \tilde{f}_n| \leq \mathrm{diam}\, J_{n-1}.$$

J_{n-1} is obtained from $\mathrm{cl}\,V$ by successive use of $n-1$ linear fractional transformations. In many cases this process is simplified by replacing $\mathrm{cl}\,V$ by a larger and simpler set (disk, lens-shaped region etc.). Henrici and Pfluger [4] used inclusion regions for S-fractions. Later similar methods have been used by several authors, see the introduction in [22].

Two simple and well known examples are:

$$(1.15) \qquad E = \{w \mid |w| \leq \tfrac{1}{4}\}$$

$$(1.15') \qquad V = \{w \mid |w| \leq \tfrac{1}{2}\}$$

and

$$(1.16) \qquad E = \{w \mid |w| - \mathrm{Re}(we^{-2i\alpha}) \leq \tfrac{1}{2}\cos^2\alpha\}$$

$$(1.16') \qquad V = \{w \mid \mathrm{Re}(we^{-i\alpha}) \geq -\tfrac{1}{2}\cos\alpha\}, \quad -\tfrac{\pi}{2} < \alpha < \tfrac{\pi}{2}.$$

The first is part of Worpitzky's theorem [33], the second one is part of the uniform parabola theorem [15, Thm. 4.40] and [25].

Closely tied to the two element/value-region results above are two of the most well known convergence results: Worpitzky's theorem: The disk

$$|w| \leq \frac{1}{4}$$

is a *convergence region* for $K(a_n/1)$ (i.e. $K(a_n/1)$ converges whenever all a_n are in the disk.)

The uniform parabola theorem: Any *bounded subset* of the parabolic region

$$|w| - \mathrm{Re}(we^{-2i\alpha}) \leq \frac{1}{2}\cos^2\alpha$$

is a convergence region for $K(a_n/1)$.

We conclude this section by drawing the attention to an entity of importance in the analytic theory of continued fractions:

$$(1.17) \qquad h_1 = 1, \; h_n = B_n/B_{n-1} = 1 + \frac{a_n}{1} + \frac{a_{n-1}}{1} + \dots + \frac{a_2}{1}.$$

From the second example of element/value-region results it follows that if all a_n are in the parabolic region (1.16) then all h_n are in the half plane obtained by translating (1.16') one unit in the real positive direction, i.e. the half plane

$$\mathrm{Re}(we^{-i\alpha}) > \frac{1}{2}\cos\alpha.$$

Another observation on the sequence $\{h_n\}$ is important: Since with $-h_0 = \infty$, $-h_1 = -1$, we have

$$-h_n = \frac{a_{n+1}}{1-h_{n+1}},$$

i.e. the sequence $\{-h_n\}$ satisfies the recurrence relation (1.8) for tails, it is a sequence of tail values for $K(a_n/1)$ (wrong tail values unless $f = \infty$).

2. LIMIT REGIONS. MODIFIED TRUNCATION ERRORS. TAILS.

2.1. Limit regions and modified truncation error estimates

An alternative to the concept of value region is the recent concept of limit region. It was used by L. Jacobsen in [6], formally defined by M. Overholt [19] and given name by W. Thron in the Trondheim seminar. There are some minor differences in the definitions one can find in the literature. We shall base it upon the concept of *pre value* region by Jones and Thron [16]. It is defined by removing the condition $E \subseteq V$ from the definition of value region, i.e.:

A pre value region V for $K(a_n/1)$, $a_n \in E$, is a set V which is such that

$$\frac{a_n}{1+V} \subseteq V \quad \text{for all } a_n \in E.$$

Definition: Given an element region E for the continued fraction $K(a_n/1)$, a limit region corresponding to the element region E is a pre value region containing all the possible values of the continued fractions $K(a_n/1)$, $a_n \in E$, and a best limit region if it contains no other points, except possibly 0.

The advantage (and in some instances disadvantage) of limit regions compared to value regions is that they do not take into account approximants of lower order, and limit regions can therefore be regarded as obtained from value regions by removing less interesting parts. (In some cases we need those parts, as will be seen i.e. in the last chapter of this article.) Often limit regions (in particular best ones) are simpler than value regions (best ones). In the paper [22] several examples are given, showing the difference and the advantages. From [22] we shall merely bring one example, a very simple one. First we need to see how limit regions are used in the computation of the value of a continued fraction. It is very similar to the procedure with value regions:

Let $K(\tilde{a}_n/1)$ be a given continued fraction whose elements all are in E. Assume furthermore that the continued fraction converges to an $\tilde{f} \neq \infty$. Let L be a limit region corresponding to the element region E and let, for a certain n, M_n be the set

$$(2.1) \qquad M_n = \frac{\tilde{a}_1}{1} + \frac{\tilde{a}_2}{1} + \ldots + \frac{\tilde{a}_n}{1+L}.$$

Since $\tilde{f}^{(n)} = \frac{\tilde{a}_{n+1}}{1} + \frac{\tilde{a}_{n+2}}{1} + \ldots$ obviously is in L, it follows that $\tilde{f} \in M_n$. Hence for any $x_n \in L$ we have

$$(2.2) \qquad |\tilde{f} - S_n(x_n)| \leq \text{diam } M_n.$$

$S_n(x_n)$ is a *modified approximant*, and (2.2) is a *modified truncation error estimate*. If $\text{diam } M_n \to 0$ we can use $S_n(x_n)$ as approximants, without even knowing the values of $S_n(0)$.

Over to the special example from [22] mentioned above. There the following proposition is proved:

Proposition 2.1.

Let p and q be positive numbers,

$$0 < p < q < p+1,$$

and define

$$E = [p,q]. \quad \text{(Interval.)}$$

For all continued fractions $K(a_n/1)$ with $a_n \in E$ for all n the following holds: The continued fraction converges to a value

$$f \in [X,Y],$$

where

$$X = \frac{p}{1} + \frac{q}{1} + \frac{p}{1} + \frac{q}{1} + \ldots = \frac{1}{2}\left[\sqrt{(1+p+q)^2-4pq} -1-q+p\right],$$

and

$$Y = \frac{q}{1} + \frac{p}{1} + \frac{q}{1} + \frac{p}{1} + \ldots = \frac{1}{2}\left[\sqrt{(1+p+q)^2-4pq} -1+q-p\right].$$

Moreover, for all n and all $x_n \in [X,Y]$ we have

$$|f - S_n(x_n)| \leq (Y-X)\left(\frac{Y}{1+X}\right)^n. \tag{2.3}$$

Remark: Since $Y-X = q-p < 1$, it follows that $\frac{Y}{1+X} < 1$.

The proof is simple, and shall be omitted, except for a remark on how the estimate (2.3) is obtained: For arbitrary $c \in [p,q]$ the mapping $w \to c/(1+w)$ maps an interval $[r,s] \subset [X,Y]$ onto a subinterval of $[X,Y]$. The ratio of the lengths of the intervals (new length/old length) is bounded above by $\frac{Y}{1+X}$.

Some special complex results are proved in [22], but deeper results on element and value regions are proved by Jacobsen and Thron in [11]. The starting idea is related to the one by Lane in [18], see also [15, Thm. 4.3]. We shall restrict ourselves to one result from [11]. Before quoting this result we need to present another new concept, *modified continued fraction*. It grew out of all recent use and emphasis on modifications. It appeared in print at first in [1]. The formal definition of a continued fraction is an ordered pair $\langle\langle \{a_n\},\{b_n\}\rangle,\{f_n\}\rangle$, where $a_n \in \mathbb{C}$, $a_n \neq 0$, $b_n \in \mathbb{C}$, $f_n \in \hat{\mathbb{C}}$.

$$f_n = S_n(0) \tag{2.4}$$

(Henrici, Pfluger [4]). The formal definition of a modified continued fraction is obtained by replacing the ordinary approximants (2.4) by modified approximants

$$g_n = S_n(w_n), \tag{2.5}$$

and is thus formally written

$$\langle\langle \{a_n\},\{b_n\},\{w_n\}\rangle , \{g_n\}\rangle .$$

In discussing modified continued fractions the classical one is sometimes referred to as the *reference continued fraction* . The nth approximant $S_n(w_n)$ of the modified continued fraction is a modified approximant of the reference continued fraction. A basic theory with element and value regions (or sequences of element and value regions) can be established for modified continued fractions [1]. We shall use the abbreviated notation $K(a_n,b_n,w_n)$.

Let $V(\Gamma,\rho)$ be the disk given by

$$|w-\Gamma| \leqslant \rho \tag{2.6}$$

and $E(\Gamma,\rho)$ be the Cartesian oval given by

$$|w(1+\Gamma)-\Gamma(|1+\Gamma|^2-\rho^2)|+\rho|w| \leqslant \rho(|1+\Gamma|^2-\rho^2). \tag{2.7}$$

In terms of modified continued fractions the announced result by Jacobsen and Thron takes the form:

<u>Theorem 2.2.</u>

Let $0 < \rho < \mathrm{Re}(\Gamma + \frac{1}{2})$. Then the modified continued fraction $K(a_n,1,w_n)$ converges for

$$a_n \in E(\Gamma,\rho)\ ,\quad w_n \in V(\Gamma,\rho)\ ,\quad n \geqslant 1\ ,$$

to a value in $V(\Gamma,\rho)$, independent of the choice of $\{w_n\}$.

The transition of results from modified to ordinary continued fractions is made possible (also in the non-trivial case when $0 \notin V$) by a result by Jacobsen [8] saying: If E and V are corresponding element and value regions for $K(a_n,1,w_n)$ such that $K(a_n,1,w_n)$ converges to a limit independent of the choice of $\{w_n\}$, $w_n \in V$, and finally V is bounded and contains more than one point, then $K(a_n/1)$ converges to the same limit.

2.2. Tails

The question about which sequences $\{f_n\}^{\infty}$, are sequences of approximants for some continued fraction $K(a_n/b_n)$ or $K(a_n/1)$ is easily settled (see e.g. [15, Thm. 2.7]. In the latter case the only conditions needed are $f_{n-1} \neq f_n$, $f_{n-1} \neq f_{n+1}$ for all n. In the paper [29] was raised the question which sequences $\{g^{(n)}\}$ are sequences of right tail values for some continued fraction $K(a_n/1)$. The answer is (under the additional condition that all $g^{(n)} \neq 0,\infty$) that with

$$H_n = -\frac{1+g^{(n)}}{g^{(n)}}, \tag{2.8}$$

$$1 + H_1 + H_1H_2 + H_1H_2H_3 + \ldots = \infty \tag{2.9}$$

is necessary and sufficient for $\{g^{(n)}\}$ to be a sequence of right tail values.

In the same paper is also given a result, where the value of a convergent continued fraction $K(a_n/1)$ can be determined by using a sequence of wrong tail values for the continued fraction. [29, Formula 2.1].

Another use (among several) of wrong tail values is for analytic continuation, see for instance the survey articles [26] and [13].

An interesting property of wrong tail values is the following result by L. Jacobsen, here slightly rephrazed.

Theorem 2.3.

Let $\{g^{(k)}\},\{\tilde{g}^{(k)}\}$ be two sequences of wrong tail values for a convergent continued fraction $K(a_n/b_n)$, and if $d(u,v)$ is the chordal distance, we have

(2.10) $$\lim_{n\to\infty} d(g^{(n)},\tilde{g}^{(n)}) = 0.$$ [3] [9]

Remark: This theorem can be interpreted in the following way: Pick two arbitrary, different points $g^{(0)}$ and $\tilde{g}^{(0)}$ on the Riemann sphere. Generate the sequences $\{g^{(n)}\}$ and $\{\tilde{g}^{(n)}\}$ by using the recurrence relations (1.8). Then, unless $g^{(0)}$ or $\tilde{g}^{(0)}$ happens to be the value of the continued fraction, we will have (2.10).

Some of the results on tail values are known earlier, but usually in much more special versions, and with no reference to the tails of continued fractions. See e.g. the reference list in [29].

3. GENERAL CONVERGENCE

We have been discussing convergence of continued fractions and computed it by using the approximants $S_n(0)$. Some times we have used modified approximants $S_n(w_n)$, in which case we would have to find a way to conclude from $S_n(w_n) \to f$ to $S_n(0) \to f$ (if we are discussing convergence of a classical continued fraction). One way of doing this, due to L. Jacobsen, was mentioned in the previous chapter. But we cannot always conclude convergence of $S_n(0)$ from convergence of $S_n(w_n)$, as seen in the following example due to L. Jacobsen [9]:

The 3-periodic continued fraction

$$\frac{2}{1} + \frac{1}{1} - \frac{1}{1} + \frac{2}{1} + \frac{1}{1} - \frac{1}{1} + \dots$$

has the ordinary approximants

$$f_{3n-2} = \frac{2^n}{2^{n+1}-3}, \quad f_{3n-1} = \frac{2^{n-1}}{2^n-1}, \quad f_{3n} = 0.$$

Hence $f_{3n-2} \to \frac{1}{2}$, $f_{3n-1} \to \frac{1}{2}$ and $f_{3n} \to 0$. By using the formula (1.11) with w_n instead of w, it is easily seen that

$$\lim_{n\to\infty} S_n(w_n) = \frac{1}{2}$$

for all sequences $\{w_n\}$ that are bounded away from 0, -1 and ∞. "Since $\{S_n(0)\}$ diverges, the divergence of the continued fraction appears to be an unfortunate result of the definition of convergence" [9, p. 478]. L. Jacobsen is not the first one to suggest something else instead of convergence of $\{S_n(0)\}$ as definition of convergence of continued fraction. In [26] are included historical remarks, and also in [9]. But, on the basis of its simplicity, conceptually as well as in use, the concept *general convergence* by L. Jacobsen seems to be the right one ($d(u,v)$ means the chordal distance):

Definition: $K(a_n/b_n)$ is said to converge generally to $f \in \hat{\mathbb{C}}$ if there exist two sequences $\{v_n\},\{w_n\}$ of extended complex numbers such that

$$\underline{\lim}\, d(w_n,v_n) > 0$$

and

$$\lim_{n\to\infty} S_n(v_n) = \lim_{n\to\infty} S_n(w_n) = f.$$

This definition calls for a proof of the uniqueness of f. This is established in [9]. Furthermore, it follows easily by taking $w_n = 0$, $v_n = \infty$, that convergence implies general convergence. Another important property is that general convergence to f implies convergence to f of any sequence $\{S_n(w_n)\}$ such that

$$\underline{\lim}\, d(w_n,-h_n) > 0.$$

One property of general convergence is that it picks up cases where the continued fraction diverges but "ought to converge". More important, however, is that it is easier to handle in practice, since from convergence of some $S_n(w_n)$ we do not have to prove convergence of $S_n(0)$. Once we have proved $\lim S_n(w_n) = \lim S_n(v_n)$ for two sequences with $\underline{\lim}\, d(w_n,v_n) > 0$, we are through.

4. COMPUTATION OF CONTINUED FRACTIONS

4.1. The problem. An auxiliary function

In the present section we shall discuss some recent results on the computation of the value of a continued fraction

$$(4.1)\qquad \mathop{K}_{n=1}^{\infty} \frac{a_n}{1} = \frac{a_1}{1} + \frac{a_2}{1} + \ldots + \frac{a_n}{1} + \ldots$$

where

$$(4.2)\qquad a_n = a + \delta_n$$

and $|\delta_n| \leq r'$, where r' is such that (4.1) converges to a finite value. In the methods to be described the function

$$(4.3)\qquad F(z) = \mathop{K}_{n=1}^{\infty} \frac{a+\delta_n z}{1}$$

plays a crucial role. We shall therefore start by discussing briefly some properties of the function (4.3):

Let a be a fixed complex number not on the ray $(-\infty, -\frac{1}{4}]$. Let furthermore R be a positive number such that the disk

$$(4.4)\qquad |w-a| \leq R$$

is a convergence region for the continued fraction (4.1), and that the value of any continued fraction (4.1) with all a_n in (4.4) is finite. By the convergence neighborhood theorem [15, Thm. 4.45] (based upon the parabola theorem, [15, Thm. 4.40 and 42], [24], [25]) we find that R can be chosen as

$$(4.5)\qquad \begin{aligned} R = R(a) &= \sqrt{\frac{|a|+\mathrm{Re}(a)}{2}} && \text{for } |a| \geq \frac{1}{4}, \\ R = R(a) &= \left|a+\frac{1}{4}\right| && \text{for } |a| < \frac{1}{4}. \end{aligned}$$

Let, for a fixed a, $M = M(a)$ be such that

$$\left|\mathop{K}_{n=1}^{\infty} \frac{a_n}{1}\right| \leq M(a)$$

if for all n

$$|a_n - a| \leq R(a).$$

By [24] we know that $M(a)$ can be taken to be

$$(4.5') \qquad M(a) = 2(|a|+R(a))\sqrt{\frac{2|a|}{|a|+\text{Re}\,a}} \qquad (\text{for } |a| > \tfrac{1}{4}) .$$

We shall also need to know that under the same condition on a_n, all approximants and hence also the value are in the half plane given by

$$(4.6) \qquad \text{Re}(we^{-i\alpha}) \geq -\frac{1}{2}\cos\alpha ,$$

where $\alpha = \frac{1}{2}\arg a$ if $|a| > \frac{1}{4}$ and $\alpha = \frac{1}{2}\arg a'$ if $|a| \leq \frac{1}{4}$, a' being the point where the line through $-\frac{1}{4}$ and a meets the circle $|w| = \frac{1}{4}$. [15, Cor. 4.16] and [15, p. 112].

Proposition 4.1.

Let a and R be as described above. Let furthermore r be such that

$$0 < r < R$$

and

$$(4.7) \qquad \sup|\delta_n| < r \quad \text{for all } n.$$

Then the function

$$(4.3) \qquad F(z) = \mathop{K}_{n=1}^{\infty} \frac{a+\delta_n z}{1}$$

is holomorphic in $|z| \leq \frac{R}{r}$.

Proof: The approximants are all rational, and bounded in $|z| < \frac{R}{r-\varepsilon}$ for some $\varepsilon > 0$, and hence holomorphic. Since the sequence of approximants is uniformly bounded (all are in (4.4)), it is normal. The pointwise convergence thus implies (by the Stieltjes-Vitali theorem) that the convergence is uniform on compact subsets of $|z| < \frac{R}{r-\varepsilon}$. Hence the limit function is holomorphic there.

Proposition 4.2.

Let F be the function in Proposition 4.1, and let $P_k(z)$ be the kth Taylor polynomial. Then

(4.8) $$|F(1)-P_k(1)| \leq \frac{M}{1-\frac{r}{R}}\left(\frac{r}{k}\right)^{k+1}$$

Proof: Standard use of the Cauchy integral.

Remark: The result (4.8) obviously also holds for the tails of the continued fraction (4.1). In order to apply it to the tails we recall that $f^{(N)}$ denotes the value of the Nth tail, i.e.

$$f^{(N)} = \overset{\infty}{\underset{n=1}{K}} \frac{a_{n+N}}{1} \qquad (f^{(0)} = f)$$

Let furthermore $w_{k,N}$ be the value for $z = 1$ of the kth Taylor polynomial of the function

$$\overset{\infty}{\underset{n=1}{K}} \frac{a+\delta_{n+N}z}{1}.$$

With

(4.7') $$\sup_{n\geq N+1} |\delta_n| = r_N \qquad (r_0 = r)$$

we get from (4.8)

(4.8') $$|f^{(N)} - w_{k,N}| \leq \frac{M}{1-\frac{r_N}{R}}\left(\frac{r_N}{R}\right)^{k+1}.$$

This will be an important estimate, since $w_{k,N}$ shall be used in the computation of the value of (4.1).

The computation of $w_{k,N}$ will depend upon the following lemma.

Lemma 4.3.

Let F be the function in Propositions 4.1 and 2, and let

(4.9) $$\Gamma = \overset{\infty}{\underset{n=1}{K}}\left(\frac{a}{1}\right) = \frac{1}{2}\left[\sqrt{1+4a}-1\right].$$

Then, with

(4.10) $$G_n(z) = \frac{a+\delta_1 z}{1} + \frac{a+\delta_2 z}{1} + \ldots + \frac{a+\delta_n z}{1+\Gamma}$$

we have

(4.11) $$\lim_{n\to\infty} G_n(z) = F(z) \quad \text{for} \quad |z| \leq \frac{R}{r}.$$

Proof: Let $F_n(z)$ denote the nth approximant of (4.3). For a fixed z we have, with the notations from section 1:

$$F_n(z)-G_n(z) = \frac{A_n}{B_n} - \frac{A_n+\Gamma A_{n-1}}{B_n+\Gamma B_{n-1}} = \frac{\Gamma(A_n B_{n-1}-A_{n-1}B_n)}{B_n(B_n+\Gamma B_{n-1})}$$

$$= \frac{\Gamma(F_n(z)-F_{n-1}(z))}{h_n(h_n+\Gamma)}$$

Here, since

$$h_n = \frac{B_n}{B_{n-1}} + 1 + \frac{a_n}{1} + \frac{a_{n-1}}{1} + \ldots + \frac{a_2}{1},$$

h_n is in the halfplane obtained from (4.6) by a translation of 1, and thus

$$|h_n| \geq \mathrm{Re}(h_n e^{-i\alpha}) \geq \frac{1}{2}\cos\alpha$$

Since Γ is an interior point of the halfplane (4.6), we easily find

$$|h_n+\Gamma| \geq \mathrm{Re}(\Gamma e^{-i\alpha}) + \frac{1}{2}\cos\alpha > 0.$$

Hence the denominator is bounded away from 0. Since furthermore $F_n(z) \to F(z)$, we find that

$$F_n(z)-G_n(z) \to 0,$$

and the theorem is thus proved.

Remark: By the uniform boundedness of the sequence $\{C_n(z)\}$ it follows that the convergence is uniform on compact subsets of $|z| < \frac{R}{r-\varepsilon}$ for some $\varepsilon > 0$, and hence that for any p

$$G_n^{(p)}(z) \to F^{(p)}(z) \quad \text{when} \quad n \to \infty.$$

4.2. Taylor modifications and truncation error estimates

We compute the value of the continued fraction (4.1) by computing the numbers of the sequence $\{S_N(u_N)\}$ for some (not any!) sequence $\{u_N\}$. Usually the Backward Recurrence Algorithm is used. For all $u_N = 0$ the limit exists and is the right one, since we use the definition of convergence. $S_N(0)$ is the Nth classical approximant. But rather general sequences work: If all u_N are contained in a closed and bounded subset of the interior of the halfplane (4.6) we find (by a rather similar argument to the one in the proof of Lemma 4.3.) that $\{S_N(u_N)\}$ converges to f, to name but one example. Simple computation, together with $f = S_N(f^{(N)})$, gives

$$(4.12) \qquad f-S_N(u_N) = \frac{h_N}{(h_N+f^{(N)})(h_N+u_N)} (f_{N-1}-f_N)\cdot(f^{(N)}-u_N),$$

where $f_N = S_N(0)$ is the Nth classical approximant. We have in particular

$$(4.13) \qquad f-S_N(0) = \frac{f^{(N)}}{(h_N+f^{(N)})} (f_{N-1}-f_N).$$

It is a general result in the analytic theory of continued fractions that

$$f_{N-1} \neq f_N$$

[15, Thm. 2.8]. Furthermore, for the continued fraction (4.1) from the previous section we know that

$$h_N+f^{(N)} \neq 0.$$

Let us assume in addition that

$$(4.14) \qquad f^{(N)} \neq 0.$$

From (4.12) and (4.13) we find

$$(4.15) \qquad \frac{f-S_N(u_N)}{f-S_N(0)} = \frac{h_N}{h_N+u_N} \cdot \frac{f^{(N)}-u_N}{f^{(N)}}.$$

We shall here in particular use

$$u_N = w_{k,N} ,$$

in which case we have, from (4.8'):

$$|f^{(N)}-u_N| = |f^{(N)}-w_{k,N}| \leq \frac{M}{1-\frac{r_N}{R}} \left(\frac{r_N}{R}\right)^{k+1}$$

From now on (except for a case in the concluding remarks) we shall assume that the continued fraction in question is limit-1-periodic, i.e. that

$$r_N \to 0 \quad \text{when} \quad N \to \infty.$$

From (4.8') we find, when $N \to \infty$:

$$\lim_{N\to\infty} |f^{(N)}-w_{k,N}| = 0$$

Since $f^{(N)} \to \Gamma$, [20, Satz 2.41], this gives

$$\lim_{N\to\infty} w_{k,N} = \Gamma.$$

Since furthermore $h_N \to 1+\Gamma$ [28] we conclude, by using (4.15) and (4.8')

Proposition 4.4.

Let a be a fixed complex number not on the ray $(-\infty, -\frac{1}{4}]$, and let R be such that the disk $|w-a| \leq R$ is a convergence region with finite values for the continued fraction

$$(4.1) \qquad \overset{\infty}{\underset{n=1}{K}} \frac{a_n}{1}.$$

Let furthermore

$$(4.16) \qquad |a_n-a| \leq r_n \to 0$$

when $n \to \infty$. Then for all k

$$(4.15') \qquad \left|\frac{f-S_N(w_{k,N})}{f-S_N(0)}\right| \leq K_N \cdot \left(\frac{r_N}{R}\right)^{k+1},$$

where

$$\lim_{N\to\infty} K_N = \frac{(1+\Gamma)}{(1+2\Gamma)\cdot\Gamma}\cdot M .$$

The right-hand side of (4.15') tells completely at which rate $S_N(w_{k,N})$ tends to f compared to what $S_N(0)$ does. It also tells about the acceleration caused by $k \rightsquigarrow k+1$. Nevertheless, this result is completely useless, unless we can find explicitely $w_{k,N}$. So far this is only done for $k = 0,1,2$.

4.3. Taylor modification of order 0,1,2. Numerical examples.

We obviously have

$$(4.17) \qquad w_{0,N} = \Gamma,$$

see [27] and the references there. Next we have

$$(4.18) \qquad w_{1,N} = \Gamma + \left(\frac{d}{dz}\left(\underset{n=1}{\overset{\infty}{K}} \frac{a+\delta_{n+N}z}{1}\right)\right)_{z=0} \cdot 1 .$$

In order to compute this, we need to be able to compute $F'(0)$, where

$$(4.3) \qquad F(z) = \underset{n=1}{\overset{\infty}{K}} \frac{a+\delta_n z}{1} .$$

From the paper [30] we know the following: Let

$$(4.19) \qquad H(z_1,z_2,\ldots,z_{n+1},\ldots) = \underset{n=1}{\overset{\infty}{K}} \frac{z_n}{1},$$

where we assume continuity. Let $(\ )_0$ denote evaluation at $z_1 = z_2 = \ldots = z_n = \ldots = a$, and let Γ be as defined earlier. Then

$$\left(\frac{\partial H}{\partial z_{n+1}}\right)_0 = \frac{1}{1+\Gamma}\left(\frac{-\Gamma}{1+\Gamma}\right)^n .$$

Keeping in mind that

$$(4.10) \qquad G_n(z) = \frac{a+\delta_1 z}{1} + \frac{a+\delta_2 z}{1} + \ldots + \frac{a+\delta_n z}{1+\Gamma}$$

we find

$$G_n'(z) = \frac{1}{1+\Gamma} \sum_{k=0}^{n-1} \left(\frac{-\Gamma}{1+\Gamma}\right)^k \delta_{k+1} ,$$

and by Lemma 5.3

$$F'(0) = \frac{1}{1+\Gamma} \sum_{k=0}^{\infty} \left(\frac{-\Gamma}{1+\Gamma}\right)^k \delta_{k+1} .$$

The analog holds for the tail functions, i.e. the ones where δ_{k+1} is replaced by δ_{N+k+1}. We thus find:

Proposition 4.5.

The Taylor modification of order 1 is

$$(4.18') \qquad w_{1,N} = \Gamma + \frac{1}{1+\Gamma} \sum_{k=0}^{\infty} \left(\frac{-\Gamma}{1+\Gamma}\right)^k \delta_{k+N+1} .$$

See e.g. [30].

Remark: We know from Proposition 4.4 the rate at which $S_N(w_{1,N})$ converges to f compared to $S_N(0)$. But the practical use of the modification (4.18') is highly dependent upon how easily we can find the values of $w_{1,N}$, $N = 0,1,2,3,\ldots$. The sum is a series in powers of $\left(\frac{-\Gamma}{1+\Gamma}\right)$. In simple cases it turns out to be a tail of a known series, like the log- or arctan-series [31], or even a geometric series.

Example.

If $\delta_k = C \cdot t^k$ we find

$$w_{1,N} = \Gamma + \frac{Ct^{N+1}}{1+\Gamma+t\Gamma}$$

In Table I C means 'classical approximation", J means $w_{0,N}$-

approximation (Γ-appr.) and D1 means $w_{1,N}$-approximation.

TABLE I. APPROXIMANTS OF $\underset{n=1}{\overset{\infty}{\mathrm{K}}} \frac{30+2(-0.7)^n}{1}$

1C=	2.86000E+01	J= 4.76667E+00	D1= 4.47434E+00
2C=	8.94309E-01	J= 4.64035E+00	D1= 4.46123E+00
3C=	1.41446E+01	J= 4.56604E+00	D1= 4.46683E+00
4C=	1.67821E+00	J= 4.52404E+00	D1= 4.46462E+00
5C=	9.61952E+00	J= 4.49937E+00	D1= 4.46554E+00
6C=	2.34138E+00	J= 4.48517E+00	D1= 4.46517E+00
7C=	7.53498E+00	J= 4.47686E+00	D1= 4.46532E+00
8C=	2.88095E+00	J= 4.47204E+00	D1= 4.46526E+00
9C=	6.40217E+00	J= 4.46922E+00	D1= 4.46528E+00
10C=	3.30438E+00	J= 4.46758E+00	D1= 4.46527E+00
11C=	5.72915E+00	J= 4.46662E+00	D1= 4.46528E+00
12C=	3.62666E+00	J= 4.46606E+00	D1= 4.46528E+00
13C=	5.30737E+00	J= 4.46573E+00	D1= 4.46528E+00
14C=	3.86604E+00	J= 4.46554E+00	D1= 4.46528E+00
15C=	5.03395E+00	J= 4.46543E+00	D1= 4.46529E+00
16C=	4.01056E+00	J= 4.46537E+00	D1= 4.46528E+00
17C=	4.85274E+00	J= 4.46533E+00	D1= 4.46528E+00
18C=	4.16604E+00	J= 4.46531E+00	D1= 4.46528E+00
19C=	4.73086E+00	J= 4.46530E+00	D1= 4.46528E+00
20C=	4.25534E+00	J= 4.46529E+00	D1= 4.46528E+00
21C=	4.64806E+00	J= 4.46528E+00	D1= 4.46528E+00
22C=	4.31344E+00	J= 4.46528E+00	D1= 4.46528E+00
23C=	4.59143E+00	J= 4.46528E+00	D1= 4.46528E+00
24C=	4.36279E+00	J= 4.46528E+00	D1= 4.46528E+00
25C=	4.55251E+00	J= 4.46528E+00	D1= 4.46528E+00
26C=	4.39366E+00	J= 4.46528E+00	D1= 4.46528E+00
27C=	4.52567E+00	J= 4.46528E+00	D1= 4.46528E+00
28C=	4.41556E+00	J= 4.46528E+00	D1= 4.46528E+00
29C=	4.50713E+00	J= 4.46528E+00	D1= 4.46528E+00
30C=	4.43069E+00	J= 4.46528E+00	D1= 4.46528E+00

In order to obtain the same accuracy in C as in J and D1, we need to take $N \geqslant 83$.

In the numerical example below the continued fraction is

$$\underset{n=1}{\overset{\infty}{\mathrm{K}}} \frac{2+n^{-1}}{1}$$

The value is 2.2474, correctly rounded in the 4th decimal place. The smallest N-values for which the approximants, rounded in the 4th place coincide with this value for this and all greater N are in the three cases

28 (for C) 15 (for J) 5 (for D1)

In order to find $w_{2,N}$ we need to determine

$$G_{m,n} = \left(\frac{\partial^2 H}{\partial z_{n+1}^2}\right)_0 \quad \text{and} \quad G_{m,n} = \left(\frac{\partial^2 H}{\partial z_{m+1}\partial z_{n+1}}\right)_0, \quad 0 \leqslant m < n.$$

With A_n, B_n as defined in Chapter 1, but here related to the continued fraction (4.19), and with $H^{(n)}$ as the nth tail

$$H^{(n)}(z_1, z_2, \ldots) = H(z_{n+1}, z_{n+2}, \ldots)$$

we have

$$H = \frac{A_n + A_{n-1}H^{(n)}}{B_n + B_{n-1}H^{(n)}}, \tag{4.20}$$

and we find, since

$$H^{(n)} = \frac{z_{n+1}}{1+H^{(n+1)}}, \tag{4.21}$$

that

$$\frac{\partial^2 H}{\partial z_{n+1}^2} = \frac{-2(A_{n-1}B_n - A_nB_{n-1})B_{n-1}}{(B_n + B_{n-1}H^{(n)})^3(1+H^{(n+1)})^2} \tag{4.22}$$

By using the formulas

$$-(A_{n-1}B_n - A_nB_{n-1}) = (-1)^{n-1}\prod_{k=1}^{n} z_k$$

(determinant formula, see e.g. [15, formula (2.1.9)]),

$$(B_n + B_{n-1}H^{(n)}) = \prod_{m=1}^{n}(1+H^{(m)})$$

[12, formula (3)],

$$B_{n-1} = \sum_{p=1}^{n} (-1)^{n-p} \left(\prod_{\nu=p}^{n-1} H^{(\nu)}\right)\left(\prod_{\nu=1}^{p-1} (1+H^{(\nu)})\right)$$

[12, formula (6)], and (4.21), we find, after simple computations

$$G_{m,n} = \left(\frac{\partial^2 H}{\partial x_{n+1}^2}\right)_0 = \frac{-2}{(2\Gamma+1)(\Gamma+1)^2}\left(\frac{-\Gamma}{1+\Gamma}\right)^n \left[1 - \left(\frac{-\Gamma}{1+\Gamma}\right)^n\right] . \tag{4.22'}$$

A similar, but slightly more complicated procedure gives for $0 \leqslant m < n$:

$$G_{m,n} = \left(\frac{\partial^2 H}{\partial x_{m+1}\partial x_{n+1}}\right)_0 = \frac{1}{\Gamma(\Gamma+1)(2\Gamma+1)}\left(\frac{-\Gamma}{\Gamma+1}\right)^n \left[1+2\Gamma\left(\frac{-\Gamma}{1+\Gamma}\right)^m\right] . \tag{4.23}$$

We omit the proof. We also omit the proof of the second order Taylor approximant:

$$f \approx w_{2,0} = \Gamma + \frac{1}{\Gamma+1}\sum_{n=0}^{\infty}\left(\frac{-\Gamma}{1+\Gamma}\right)^n \delta_{n+1} + \frac{1}{2}\sum_{n=0}^{\infty} G_{n,m}\,\delta_{n+1}^2 +$$

$$+ \sum_{n=m+1}^{\infty}\sum_{m=0}^{\infty} G_{m,n}\cdot\delta_{m+1}\cdot\delta_{n+1}$$

The modifying factor $w_{2,N}$ is obtained by the transition $\delta_k \rightsquigarrow \delta_{k+N}$, and we have:

<u>Proposition 4.6.</u>

The Taylor modification of order 2 is

$$\begin{aligned} w_{2,N} &= \Gamma + \frac{1}{\Gamma+1}\sum_{k=0}^{\infty}\left(\frac{-\Gamma}{1+\Gamma}\right)^k \delta_{k+N+1} - \\ &- \frac{1}{(2\Gamma+1)(\Gamma+1)^2}\sum_{k=0}^{\infty}\left(\frac{-\Gamma}{1+\Gamma}\right)^k\left[1-\left(\frac{-\Gamma}{1+\Gamma}\right)^k\right]\delta_{k+N+1}^2 + \\ &+ \frac{1}{\Gamma(\Gamma+1)^2(2\Gamma+1)}\sum_{n=k+1}^{\infty}\sum_{k=0}^{\infty}\left(\frac{-\Gamma}{\Gamma+1}\right)^n\left[1+2\left(\frac{-\Gamma}{\Gamma+1}\right)^k\right]\delta_{k+N+1}\delta_{n+N+1} \end{aligned} \tag{4.24}$$

Remark: Also here the practical use of the result depends upon the accessability of a closed form or a simple and good approximate value. One particular case, when we easily can find a closed form is when

$$\delta_{k+1} = C\cdot T^{k+1}, \quad 0 < |T| < 1. \tag{4.25}$$

In this particular case (4.24) takes the form

$$\begin{aligned} w_{2,N} = \Gamma + \frac{C\cdot T^{N+1}}{1+(T+1)\Gamma} + {} \\ + \frac{C^2 T^{2N+4}\cdot\Gamma}{(1+\Gamma)(1+(1+T^2)\Gamma)((1+\Gamma)^2-T^2\Gamma^2)} - {} \\ - \frac{C^2 T^{2N+3}}{(1+\Gamma)(2\Gamma+1)(1+(1+T)\Gamma)}\left[\frac{1}{1+(1+T^2)\Gamma} + \frac{2\Gamma(1+\Gamma)}{(1+\Gamma)^2-T^2\Gamma^2}\right]. \end{aligned} \tag{4.24'}$$

In Table II the $w_{2,N}$-approximation is denoted by D2.

TABLE II. APPROXIMANTS OF $\overset{\infty}{\underset{n=1}{\mathrm{K}}}\left(\frac{30+0.9^n}{1}\right)$

1C= 3.09000E+01	J= 5.15000E+00	D1= 5.08463E+00	D2= 5.08507E+00
2C= 9.71393E-01	J= 5.03667E+00	D1= 5.08536E+00	D2= 5.08506E+00
3C= 1.56770E+01	J= 5.12176E+00	D1= 5.08486E+00	D2= 5.08507E+00
4C= 1.85765E+00	J= 5.05762E+00	D1= 5.08520E+00	D2= 5.08506E+00
5C= 1.08172E+01	J= 5.10580E+00	D1= 5.08497E+00	D2= 5.08507E+00
6C= 2.62101E+00	J= 5.06953E+00	D1= 5.08513E+00	D2= 5.08507E+00
7C= 8.53842E+00	J= 5.09677E+00	D1= 5.08502E+00	D2= 5.08507E+00
8C= 3.24649E+00	J= 5.07628E+00	D1= 5.08509E+00	D2= 5.08507E+00
9C= 7.28204E+00	J= 5.09167E+00	D1= 5.08505E+00	D2= 5.08507E+00
10C= 3.73832E+00	J= 5.08011E+00	D1= 5.08508E+00	D2= 5.08507E+00
11C= 6.52716E+00	J= 5.08879E+00	D1= 5.08506E+00	D2= 5.08507E+00
12C= 4.11263E+00	J= 5.08227E+00	D1= 5.08507E+00	D2= 5.08507E+00
13C= 6.05004E+00	J= 5.08717E+00	D1= 5.08506E+00	D2= 5.08507E+00
14C= 4.39046E+00	J= 5.08349E+00	D1= 5.08507E+00	D2= 5.08507E+00
15C= 5.73877E+00	J= 5.08625E+00	D1= 5.08506E+00	D2= 5.08507E+00
16C= 4.59286E+00	J= 5.08418E+00	D1= 5.08507E+00	D2= 5.08507E+00
17C= 5.53149E+00	J= 5.08573E+00	D1= 5.08507E+00	D2= 5.08507E+00
18C= 4.73830E+00	J= 5.08457E+00	D1= 5.08507E+00	D2= 5.08507E+00
19C= 5.39159E+00	J= 5.08544E+00	D1= 5.08507E+00	D2= 5.08507E+00
20C= 4.84179E+00	J= 5.08478E+00	D1= 5.08507E+00	D2= 5.08507E+00
21C= 5.29629E+00	J= 5.08528E+00	D1= 5.08507E+00	D2= 5.08507E+00
22C= 4.91490E+00	J= 5.08491E+00	D1= 5.08507E+00	D2= 5.08507E+00
23C= 5.23098E+00	J= 5.08519E+00	D1= 5.08507E+00	D2= 5.08507E+00
24C= 4.96629E+00	J= 5.08498E+00	D1= 5.08507E+00	D2= 5.08507E+00
25C= 5.18604E+00	J= 5.08513E+00	D1= 5.08507E+00	D2= 5.08507E+00
26C= 5.00228E+00	J= 5.08502E+00	D1= 5.08507E+00	D2= 5.08507E+00
27C= 5.15507E+00	J= 5.08510E+00	D1= 5.08507E+00	D2= 5.08507E+00
28C= 5.02743E+00	J= 5.08504E+00	D1= 5.08507E+00	D2= 5.08507E+00
29C= 5.13356E+00	J= 5.08509E+00	D1= 5.08507E+00	D2= 5.08507E+00
30C= 5.04497E+00	J= 5.08505E+00	D1= 5.08507E+00	D2= 5.08507E+00

For the same accuracy in C we need $N \geqslant 87$ (J: $N \geqslant 39$).

TABLE III. APPROXIMANTS OF $\overset{\infty}{\underset{n=1}{K}} \frac{90-2(-0.7)^n}{1}$

1C= 9.14000E+01	J= 9.14000E+00	D1= 9.38867E+00	D2= 9.38539E+00
2C= 1.01533E+00	J= 9.23046E+00	D1= 9.38403E+00	D2= 9.38541E+00
3C= 4.63742E+01	J= 9.28734E+00	D1= 9.38604E+00	D2= 9.38541E+00
4C= 2.01015E+00	J= 9.32359E+00	D1= 9.38514E+00	D2= 9.38541E+00
5C= 3.14933E+01	J= 9.34639E+00	D1= 9.38554E+00	D2= 9.38541E+00
6C= 2.95788E+00	J= 9.36082E+00	D1= 9.38536E+00	D2= 9.38541E+00
7C= 2.41734E+01	J= 9.36991E+00	D1= 9.38544E+00	D2= 9.38542E+00
8C= 3.83975E+00	J= 9.37565E+00	D1= 9.38540E+00	D2= 9.38541E+00
9C= 1.98875E+01	J= 9.37926E+00	D1= 9.38542E+00	D2= 9.38541E+00
10C= 4.64383E+00	J= 9.38154E+00	D1= 9.38541E+00	D2= 9.38541E+00
11C= 1.71217E+01	J= 9.38297E+00	D1= 9.38542E+00	D2= 9.38541E+00
12C= 5.36408E+00	J= 9.38387E+00	D1= 9.38541E+00	D2= 9.38541E+00
13C= 1.52241E+01	J= 9.38444E+00	D1= 9.38541E+00	D2= 9.38541E+00
14C= 5.99927E+00	J= 9.38480E+00	D1= 9.38541E+00	D2= 9.38541E+00
15C= 1.38672E+01	J= 9.38503E+00	D1= 9.38541E+00	D2= 9.38541E+00
16C= 6.55189E+00	J= 9.38517E+00	D1= 9.38541E+00	D2= 9.38541E+00
17C= 1.28681E+01	J= 9.38526E+00	D1= 9.38541E+00	D2= 9.38541E+00
18C= 7.02709E+00	J= 9.38532E+00	D1= 9.38541E+00	D2= 9.38541E+00
19C= 1.21165E+01	J= 9.38535E+00	D1= 9.38541E+00	D2= 9.38541E+00
20C= 7.43162E+00	J= 9.38538E+00	D1= 9.38541E+00	D2= 9.38541E+00
21C= 1.15419E+01	J= 9.38539E+00	D1= 9.38541E+00	D2= 9.38541E+00
22C= 7.77308E+00	J= 9.38540E+00	D1= 9.38541E+00	D2= 9.38541E+00
23C= 1.10973E+01	J= 9.38540E+00	D1= 9.38541E+00	D2= 9.38541E+00
24C= 8.05921E+00	J= 9.38541E+00	D1= 9.38541E+00	D2= 9.38541E+00
25C= 1.07500E+01	J= 9.38541E+00	D1= 9.38541E+00	D2= 9.38541E+00
26C= 8.29756E+00	J= 9.38541E+00	D1= 9.38541E+00	D2= 9.38541E+00
27C= 1.04766E+01	J= 9.38541E+00	D1= 9.38541E+00	D2= 9.38541E+00
28C= 8.49510E+00	J= 9.38541E+00	D1= 9.38541E+00	D2= 9.38541E+00
29C= 1.02603E+01	J= 9.38541E+00	D1= 9.38541E+00	D2= 9.38541E+00
30C= 8.65814E+00	J= 9.38541E+00	D1= 9.38541E+00	D2= 9.38541E+00

For the same accuracy in C we need $N \geqslant 158$.

Some of the results in the present chapter are extended to the more general case when we study continued fractions

$$\overset{\infty}{\underset{n=1}{K}} \frac{a_n+\delta_n}{1}$$

near ($\delta_n \to 0$) a very well known one

$$\overset{\infty}{\underset{n=1}{K}} \frac{a_n}{1},$$

i.e. the value $f^{(0)} = f$ and all the tail values $f^{(n)}$ are known. This is done in the $w_{0,N}$-case [5], [7] and to some extent in the $w_{1,N}$-case [32]. Also some results for $a_n \to \infty$ are established. The main difference in the formulas is that products

$$\prod_{k=1}^{n} \left(\frac{-f^{(k)}}{1+f^{(k)}}\right)$$

replace powers $\left(\frac{-\Gamma}{1+\Gamma}\right)^n$. It is not yet done in the $w_{2,N}$-case, at least the computational part can be worked out, but it will be rather complicated expressions.

It is likely that most of the results in the present chapter can be obtained for continued fractions of the form

$$\mathop{K}_{n=1}^{\infty} \frac{1}{b_n}$$

(and it ought to be done). Some results on

$$\mathop{K}_{n=1}^{\infty} \frac{a_n}{b_n}$$

are also within reach, and should be of some interest.

4.4. Computation of continued fractions with asymptotic side conditions

As remarked modification with $w_{1,N}$ or $w_{2,N}$ is often complicated. In some cases we have additional asymptotic information, which suggests a simpler modification at a low cost.

We have already used one type of asymptotic result: With

$$(4.26) \qquad \varepsilon_N = \mathop{K}_{n=1}^{\infty} \frac{a+\delta_{N+n}}{1} - \Gamma$$

we know from Perron [20, Satz 2.41] that

$$\delta_N \to 0 \Rightarrow \varepsilon_N \to 0.$$

This is generalized in [10]. In many important cases we know more than $\delta_N \to 0$, we even know that

$$(4.27) \qquad \frac{\delta_N}{\delta_{N-1}} \to t, \quad 0 \leq |t| \leq 1.$$

This is for instance the case for C-fraction expansions of certain hypergeometric functions or ratios of hypergeometric functions, where $t = -1$.

Assume that (4.27) holds, and let us also assume that $|\delta_N| \downarrow 0$. Then, from (4.18'), (4.26) and (4.8') we find

(4.28) $$\varepsilon_N = \frac{1}{1+\Gamma}\sum_{k=0}^{\infty}\left(\frac{-\Gamma}{1+\Gamma}\right)^k \delta_{N+k+1} + O(r_N^2),$$

Due to the monotonicity we have

$$\frac{\varepsilon_N}{\delta_{N+1}} = \frac{1}{1+\Gamma}\sum_{k=0}^{\infty}\left(\frac{-\Gamma}{1+\Gamma}\right)^k \frac{\delta_{N+k+1}}{\delta_{N+1}} + O(r_N),$$

and hence, by taking the limit when $N \to \infty$:

(4.29) $$\lim_{N\to\infty}\frac{\varepsilon_N}{\delta_{N+1}} = \frac{1}{1+\Gamma(1+t)} .$$

This strongly suggests the use of

(4.30) $$\mu_N = \Gamma + \frac{\delta_{N+1}}{1+\Gamma(1+t)} .$$

So far no proper investigation of the possible gain by the transition $\Gamma \rightsquigarrow u_N$ has been undertaken. It is likely that not much will be lost compared to the Taylor modification of order 1 (and for $\delta_N = Ct^N$ it coincides with it). Observe also that in the particular case when $t = -1$ the suggested modification is

$$\mu_N = \Gamma + \delta_{N+1}$$

The modifying factor (4.30) is independently suggested by A. Lembarki in his thesis [17], even without requirement of monotonicity, and with a quite different proof. L. Jacobsen's proof in [14] is also without requirement of monotonicity.

In the example below approximants for the continued fraction

(4.31) $$\mathop{K}_{N=1}^{\infty} \frac{20+(1+\frac{1}{N})0.9^N}{1}$$

are computed. C and J have the meaning defined earlier, Ny-app means approximation obtained by using the modifying factor μ_N in (4.30). See Table IV.

TABLE IV. APPROXIMANTS OF (4.31).

1	C-APP= 2.18000E+01	J-APP= 4.36000E+00	NY-APP= 4.24019E+00
2	C-APP= 9.81319E-01	J-APP= 4.15792E+00	NY-APP= 4.23366E+00
3	C-APP= 1.10911E+01	J-APP= 4.28779E+00	NY-APP= 4.23538E+00
4	C-APP= 1.84468E+00	J-APP= 4.19900E+00	NY-APP= 4.23496E+00
5	C-APP= 7.75121E+00	J-APP= 4.26061E+00	NY-APP= 4.23515E+00
6	C-APP= 2.54384E+00	J-APP= 4.21715E+00	NY-APP= 4.23510E+00
7	C-APP= 6.23396E+00	J-APP= 4.24792E+00	NY-APP= 4.23510E+00
8	C-APP= 3.07339E+00	J-APP= 4.22600E+00	NY-APP= 4.23512E+00
9	C-APP= 5.43012E+00	J-APP= 4.24163E+00	NY-APP= 4.23510E+00
10	C-APP= 3.45448E+00	J-APP= 4.23044E+00	NY-APP= 4.23511E+00
11	C-APP= 4.96954E+00	J-APP= 4.23845E+00	NY-APP= 4.23511E+00
12	C-APP= 3.71870E+00	J-APP= 4.23271E+00	NY-APP= 4.23511E+00
13	C-APP= 4.69376E+00	J-APP= 4.23683E+00	NY-APP= 4.23511E+00
14	C-APP= 3.89717E+00	J-APP= 4.23387E+00	NY-APP= 4.23511E+00
15	C-APP= 4.52431E+00	J-APP= 4.23600E+00	NY-APP= 4.23511E+00
16	C-APP= 4.01558E+00	J-APP= 4.23447E+00	NY-APP= 4.23511E+00
17	C-APP= 4.41853E+00	J-APP= 4.23557E+00	NY-APP= 4.23511E+00
18	C-APP= 4.09320E+00	J-APP= 4.23478E+00	NY-APP= 4.23511E+00
19	C-APP= 4.35186E+00	J-APP= 4.23535E+00	NY-APP= 4.23511E+00
20	C-APP= 4.14367E+00	J-APP= 4.23494E+00	NY-APP= 4.23511E+00
21	C-APP= 4.30959E+00	J-APP= 4.23523E+00	NY-APP= 4.23511E+00
22	C-APP= 4.17632E+00	J-APP= 4.23502E+00	NY-APP= 4.23511E+00
23	C-APP= 4.28269E+00	J-APP= 4.23517E+00	NY-APP= 4.23511E+00
24	C-APP= 4.19737E+00	J-APP= 4.23506E+00	NY-APP= 4.23511E+00
25	C-APP= 4.26553E+00	J-APP= 4.23514E+00	NY-APP= 4.23511E+00
26	C-APP= 4.21090E+00	J-APP= 4.23509E+00	NY-APP= 4.23511E+00
27	C-APP= 4.25457E+00	J-APP= 4.23513E+00	NY-APP= 4.23511E+00
28	C-APP= 4.21959E+00	J-APP= 4.23510E+00	NY-APP= 4.23511E+00
29	C-APP= 4.24756E+00	J-APP= 4.23512E+00	NY-APP= 4.23511E+00
30	C-APP= 4.22517E+00	J-APP= 4.23510E+00	NY-APP= 4.23511E+00
31	C-APP= 4.24308E+00	J-APP= 4.23511E+00	NY-APP= 4.23511E+00
32	C-APP= 4.22874E+00	J-APP= 4.23511E+00	NY-APP= 4.23511E+00
33	C-APP= 4.24021E+00	J-APP= 4.23511E+00	NY-APP= 4.23511E+00
34	C-APP= 4.23103E+00	J-APP= 4.23511E+00	NY-APP= 4.23511E+00
35	C-APP= 4.23837E+00	J-APP= 4.23511E+00	NY-APP- 4.23511E+00

For C-approximants the same accuracy needs $N \geqslant 65$.

It is a rather natural question to ask for consequences of (4.27). Is it true, that (4.27) implies

$$\frac{\varepsilon_{N+1}}{\varepsilon_N} \to t \ ?$$

It follows directly from (4.29) that the answer is <u>yes</u>. (We have no trouble in (4.29), since $|t| \leqslant 1$ and $\mathrm{Re}\,\Gamma > -\frac{1}{2}$.) But this proof depends upon monotonicity of the sequence $\{|\delta_N|\}$, since that is used in (4.28). L. Jacobsen [14] has a proof which is much better in two respects: 1) It is essentially a pure "continued fraction type" of proof, and uses nothing about derivatives of continued fractions. 2) It does not require monotonicity of $\{|\delta_N|\}$, only that $|\delta_N| \to 0$ (and (4.27)).

Also Lembarki [17] has independently proved this result.

L. Jacobsen even proved a more general version: For limit k-periodic continued fractions (convergent) with limit k-periodic sequence $\left\{\frac{\delta_{N+1}}{\delta_N}\right\}$

she proved limit k-periodic behavior of $\left\{\frac{\varepsilon_{N+1}}{\varepsilon_N}\right\}$. Like in the limit 1-periodic case this also suggests modifying factors. We shall here restrict ourselves to one single numerical example, just to give the flavor of it. The continued fractions is

$$\frac{3+\frac{C}{1^k}}{1}+\frac{4+\frac{3C}{2^k}}{1}+\dots+\frac{3+\frac{C}{(2n+1)^k}}{1}+\frac{4+\frac{3C}{(2n+2)^k}}{1}+\dots \tag{4.32}$$

The modifying factors to be used turned out to be

$$z_N = 1 - \frac{C/4}{(N+1)^k} \quad \text{for even } N$$

$$z_N = 2 + \frac{7C/4}{(N+1)^k} \quad \text{for odd } N$$

In Table V we have $C = -0.01$, $K = 0.1$. Observe also: $z_0 = 1.0025$.

TABLE V. APPROXIMANTS OF (4.32) WITH C = -0.01, K = 0.1.

1	C-APP= 2.99000E+00	J-APP= 9.96667E-01	NY-APP= 1.00212E+00
2	C-APP= 6.01367E-01	J-APP= 1.00134E+00	NY-APP= 1.00208E+00
3	C-APP= 1.49857E+00	J-APP= 1.00034E+00	NY-APP= 1.00204E+00
4	C-APP= 8.59093E-01	J-APP= 1.00179E+00	NY-APP= 1.00203E+00
5	C-APP= 1.15464E+00	J-APP= 1.00148E+00	NY-APP= 1.00202E+00
6	C-APP= 9.53281E-01	J-APP= 1.00194E+00	NY-APP= 1.00202E+00
7	C-APP= 1.05156E+00	J-APP= 1.00184E+00	NY-APP= 1.00202E+00
8	C-APP= 9.85662E-01	J-APP= 1.00199E+00	NY-APP= 1.00202E+00
9	C-APP= 1.01838E+00	J-APP= 1.00196E+00	NY-APP= 1.00202E+00
10	C-APP= 9.96558E-01	J-APP= 1.00201E+00	NY-APP= 1.00202E+00
11	C-APP= 1.00745E+00	J-APP= 1.00200E+00	NY-APP= 1.00202E+00
12	C-APP= 1.00020E+00	J-APP= 1.00201E+00	NY-APP= 1.00202E+00
13	C-APP= 1.00382E+00	J-APP= 1.00201E+00	NY-APP= 1.00202E+00
14	C-APP= 1.00141E+00	J-APP= 1.00202E+00	NY-APP= 1.00202E+00
15	C-APP= 1.00262E+00	J-APP= 1.00201E+00	NY-APP= 1.00202E+00
16	C-APP= 1.00182E+00	J-APP= 1.00202E+00	NY-APP= 1.00202E+00
17	C-APP= 1.00222E+00	J-APP= 1.00202E+00	NY-APP= 1.00202E+00
18	C-APP= 1.00195E+00	J-APP= 1.00202E+00	NY-APP= 1.00202E+00
19	C-APP= 1.00208E+00	J-APP= 1.00202E+00	NY-APP= 1.00202E+00
20	C-APP= 1.00199E+00	J-APP= 1.00202E+00	NY-APP= 1.00202E+00
21	C-APP= 1.00204E+00	J-APP= 1.00202E+00	NY-APP= 1.00202E+00
22	C-APP= 1.00201E+00	J-APP= 1.00202E+00	NY-APP= 1.00202E+00
23	C-APP= 1.00202E+00	J-APP= 1.00202E+00	NY-APP= 1.00202E+00
24	C-APP= 1.00201E+00	J-APP= 1.00202E+00	NY-APP= 1.00202E+00
25	C-APP= 1.00202E+00	J-APP= 1.00202E+00	NY-APP= 1.00202E+00
26	C-APP= 1.00202E+00	J-APP= 1.00202E+00	NY-APP= 1.00202E+00
27	C-APP= 1.00202E+00	J-APP= 1.00202E+00	NY-APP= 1.00202E+00
28	C-APP= 1.00202E+00	J-APP= 1.00202E+00	NY-APP= 1.00202E+00
29	C-APP= 1.00202E+00	J-APP= 1.00202E+00	NY-APP= 1.00202E+00
30	C-APP= 1.00202E+00	J-APP= 1.00202E+00	NY-APP= 1.00202E+00

5. ZERO-FREE REGIONS. AN EXAMPLE OF APPLICATION

There are several results on zero-free regions for polynomials determined by three-term recurrence relations, and many of them are based in the analytic theory of continued fractions. See e.g. the recent survey article by de Bruin, Gilewicz and Runckel [2]. Most of the results are proved by using methods developed for this particular purpose, for instance the parabola theorem by Saff and Varga [23] or the generalization to the complex case by Runckel [21], to name but two. In some cases, however, results can be obtained by using directly basic element- and value-region results in the established analytic theory of continued fractions. There are several examples, but we shall here, in this last chapter of the article, show only one example.

Let $B_n(z)$ be a sequence of polynomials given by the following recurrence relations and initial conditions:

$$(5.1)\qquad \begin{aligned} &B_n(z) = B_{n-1}(z) + a_n z^2 B_{n-2}(z), \; n \geq 2 \\ &B_0(z) = 1, \; B_1(z) = 1. \end{aligned}$$

In addition we shall also put certain restrictions on the numbers a_n, but that will be done later. We shall be aiming at a statement on zero-free regions for the polynomials $B_n(z)$.

Remark: The reason for the apparently un-natural choice of z^2 instead of z in the recurrence relations is that it makes the computation, as well as the geometry of the problem simpler and more appealing. To transpose the statement to a statement for polynomials determined by (5.1) with z^2 replaced by z is simple.

$B_n(z)$ is the normalized denominator of the nth approximant of the continued fraction

$$(5.2)\qquad \frac{1}{1} + \frac{a_2 z^2}{1} + \frac{a_3 z^2}{1} + \ldots + \frac{a_n z^2}{1} + \ldots$$

From the "value region part" of the theorem in the analytic theory of continued fraction called the parabola theorem [24] and [15, Thm. 4.42] it follows that $B_n(z) \neq 0$ for all n if, for all n, $a_n z^2$ is contained in some parabolic region

$$(5.3)\qquad |w| \leq \operatorname{Re}(we^{-2i\theta}) + \frac{1}{2}\cos^2\theta, \quad -\frac{\pi}{2} < \theta < \frac{\pi}{2} .$$

(Here we use the fact that $A_n(z)$ and $B_n(z)$ cannot have any non-trivial factor in common.) With

$$(5.4)\qquad a_n = |a_n|e^{2i\psi_n}, \quad -\frac{\pi}{2} < \theta < \frac{\pi}{2}$$

we find by simple computation that this set is the parallel strip

$$(5.3')\qquad |\mathrm{Im}(ze^{i(\psi_n-\theta)})| \leqslant \frac{\cos\theta}{2\sqrt{|a_n|}} .$$

We shall call this $S_n(\theta)$. Obviously all $B_n(z) \neq 0$ for all

$$(5.5)\qquad z \in \bigcap_{n=2}^{\infty} S_n(\theta) =: S(\theta)$$

Since this holds for all $\theta \in \left(-\frac{\pi}{2}, \frac{\pi}{2}\right)$, we find that $B_n(z) \neq 0$ for all n when

$$(5.6)\qquad z \in \bigcup_{\theta\in\left(-\frac{\pi}{2},\frac{\pi}{2}\right)} S(\theta) =: S$$

We shall here restrict ourselves to the two very simplest cases:

1. *All* $a_n > 0$

If $A = \sup a_n$, $S(\theta)$ is the parallel strip

$$(5.7)\qquad |\mathrm{Im}(ze^{-i\theta})| \leqslant \frac{\cos\theta}{2\sqrt{A}} .$$

(For $A = \infty$ the strip degenerates to a straight line.) Since all intersect the imaginary axis in the interval between $\pm \frac{i}{2\sqrt{A}}$, we find that S is the whole plane minus the two cuts from $\frac{i}{2\sqrt{A}}$ to $i\infty$ and from $\frac{-i}{2\sqrt{A}}$ to $-i\infty$ along the positive and the negative axis.

Remark: If in the recurrence relation z^2 is replaced by z, i.e. we study polynomials $Q_n(z)$, where

$$(5.8)\qquad \begin{aligned} &Q_n(z) = Q_{n-1}(z) + a_n z Q_{n-2}(z), \; n \geqslant 2 \\ &Q_0(z) = Q_1(z) = 1 \end{aligned}$$

and $a_n > 0$, then all $Q_n(z) \neq 0$ in the cut plane

$$\left\{ z \in \mathbf{C} \;\middle|\; \arg(z + \tfrac{1}{4A}) < \pi \right\},$$

where $A = \sup a_n$ $(0 < A \leqslant \infty)$. This is a well known result for *Stieltjes fractions*, i.e. continued fractions

$$(5.9) \qquad \frac{a_1 z}{1} + \frac{a_2 z}{1} + \dots + \frac{a_n z}{1} + \dots ,$$

where $a_n > 0$ for all n.

The situation when ψ_n has a fixed value, not necessarily 0, is equally easy to handle.

We now go back to the recurrence relations (5.1) (with z^2). We shall look at the simplest case when ψ_n only takes a finite number > 1 of values:

2. $a_n \in \boldsymbol{R}$.

Let $\sup a_n = A_+$, $\sup -a_n = A_-$, $0 < A_\pm \leqslant \infty$. From (5.3') and (5.5) we find that $S(\theta)$ is the intersection of the two parallel strips

$$(5.10a) \qquad |\mathrm{Im}(ze^{-i\theta})| \leqslant \frac{\cos\theta}{2\sqrt{A_+}} ,$$

$$(5.10b) \qquad |\mathrm{Im}(zie^{-i\theta})| = |\mathrm{Re}(ze^{-i\theta})| \leqslant \frac{\cos\theta}{2\sqrt{A_-}} ,$$

i.e. a rectangle. Observe that the sides (or extension of sides) of the rectangle for all values of θ pass through 4 fixed points:

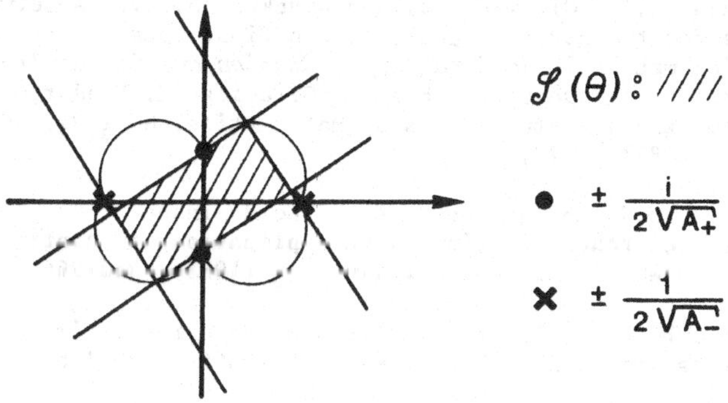

The boundary lines of (5.10a) pass through $\pm \frac{i}{2\sqrt{A_+}}$, and of (5.10b) through $\pm \frac{1}{2\sqrt{A_-}}$. When θ varies from $-\frac{\pi}{2}$ to $\frac{\pi}{2}$, the corners of the rectangle describe four half circles as shown on the illustration, and S , a zero-free region for all $B_n(z)$, is the region bounded by those half circles. The result is more simply expressed by switching from z to $\frac{1}{z}$.

Theorem 5.1.

Let $\{B_n(z)\}$ be the sequence of polynomials, given by the recurrence relations

$$(5.1)\qquad \begin{aligned} B_n(z) &= B_{n-1}(z) + a_n z^2 B_{n-2}(z),\ n \geqslant 2,\\ B_0(z) &= B_1(z) = 1, \end{aligned}$$

where $a_n \in \mathbf{R}$ for all n, and where $\sup a_n = A_+ > 0$, $\sup(-a_n) = A_- > 0$. Then the zeros of all $B_n(\frac{1}{z})$ are contained in the closed rhombus region with corners

$$\pm\ 2i\sqrt{A_+} \quad \text{and} \quad \pm\ 2\sqrt{A_-}\ .$$

REFERENCES

1. Baltus, C. and Jones, W.B., 'A family of best value regions for modified continued fractions'. *Analytic theory of continued fractions II*. (Ed. W.J. Thron.) Lecture Notes in Mathematics, No 1199, Springer-Verlag (1986), 1-20.

2. De Bruin, M.G., Gilewicz, J. and Runckel, H.-J., 'A survey of bounds for the zeros of analytic functions obtained by continued fraction methods'. *Rational approximation and its applications in mathematics and physics*. (Eds. J. Gilewicz, M. Pindor, W. Siemaszko.) Lecture Notes in Mathematics, No 1237, Springer-Verlag (1987), 1-23.

3. De Bruin, M.G. and Jacobsen, L., 'The dominance concept for linear recurrence relations with applications to continued fractions'. Nieuw Archief voor wiskunde 3 (1985), 253-266.

4. Henrici, P. and Pfluger, P., 'Truncation error estimates for Stieltjes fractions'. Numer. Math. 9 (1966), 120-138.

5. Jacobsen, L., 'Convergence acceleration for continued fractions $K(a_n/1)$'. Trans. Amer. Math. Soc. **275** (1983), 265-285.

6. Jacobsen, L., 'Some periodic sequences of circular convergence regions'. *Analytic theory of continued fractions*. (Eds. W.B. Jones, W.J. Thron, H. Waadeland.) Lecture Notes in Mathematics, No **932**. Springer-Verlag (1982), 74-86.

7. Jacobsen, L., 'Further results on convergence acceleration for continued fractions $K(a_n/1)$'. Trans. Amer. Math. Soc. **281** (1984), 129-146.

8. Jacobsen, L., 'A theorem on simple convergence regions for continued fractions'. *Analytic theory of continued fractions II*. (Ed. W.J. Thron.) Lecture Notes in Mathematics, No **1199**, Springer-Verlag (1986), 59-66.

9. Jacobsen, L., 'General convergence for continued fractions'. Trans. Amer. Math. Soc. **294** (1986), 477-485.

10. Jacobsen, L., 'Convergence of limit k-periodic continued fractions $K(a_n/b_n)$, and of subsequences of their tails'. Proc. London Math. Soc. (3) **51** (1985), 563-576.

11. Jacobsen, L. and Thron, W.J., 'Oval convergence regions and circular limit regions for continued fractions $K(a_n/1)$'. *Analytic theory of continued fractions II*. (Ed. W.J. Thron.) Lecture Notes in Mathematics, No **1199**, Springer-Verlag (1986), 59-66.

12. Jacobsen, L. and Waadeland, H., 'Some useful formulas involving tails of continued fractions'. *Analytic theory of continued fractions*. (Eds. W.B. Jones, W.J. Thron, H. Waadeland.) Lecture Notes in Mathematics, No **932**, Springer-Verlag (1982), 99-105.

13. Jacobsen, L. and Waadeland, H., 'Modification of continued fractions'. *Padé approximation and its applications Bad Honnef 1983*. (Eds. H. Werner, H.J. Bünger.) Lecture Notes in Mathematics, No **1071**, Springer-Verlag (1984) 176-196.

14. Jacobsen, L. and Waadeland, H., 'An asymptotic property for tails of limit periodic continued fractions'. Submitted.

15. Jones, W.B. and Thron, W.J., *Continued fractions: Analytic theory and applications*. Encyclopedia of Mathematics and its Applications **11**, Addison-Wesley Publishing Company, Reading. Mass. (1980), distributed now by Cambridge University Press, New York.

16. Jones, W.B. and Thron, W.J., 'Twin convergence regions for continued fractions $K(a_n/1)$'. Trans. Amer. Math. Soc. **150** (1970), 93-119.

17. Lembarki, A., 'Accélération des fractions continues'. Thèse pour obtenir de titre de Docteur D'Etat, L'Universite des Sciences et Techniques de Lille Flandres Artois.

18. Lane, R.E., 'The value region problem for continued fractions'. Duke Math. J. 12 (1945), 207-216.

19. Overholt, M., 'The values of continued fractions with complex elements'. *Padé approximants and continued fractions*. (Eds. H. Waadeland, H. Wallin.) Det Kongelige Norske Videnskabers Selskab. Skrifter No. 1, Universitetsforlaget, (1983), 109-116.

20. Perron, O., *Die Lehre von den Kettenbrüchen*. Dritte, verbesserte und erweiterte Auflage (1957), Band 2, B.G. Teubner Verlags-gesellschaft, Stuttgart.

21. Runckel, H.J., 'Zero-free parabolic regions for polynomials with complex coefficients'. Proc. Amer. Math. Soc. 88 (1983), 299-304.

22. Rye, E. and Waadeland, H., 'Reflections on value regions, limit regions and truncation errors for continued fractions'. Numer. Math. 47 (1985) 191-215.

23. Saff, E.B. and Varga, R.S., 'Zero-free parabolic regions for sequences of polynomials'. SIAM J. Math. Anal. 7 (1976), 344-357.

24. Thron, W.J., 'On parabolic convergence regions for continued fractions'. Math. Zeitschr. 69 (1958), 173-182.

25. Thron, W.J. and Leighton, W., 'Continued fractions with complex elements'. Duke Math. J. 9 (1942), 763-775.

26. Thron, W.J. and Waadeland, H., 'Modification of continued fractions, a survey'. *Analytic theory of continued fractions*. (Eds. W.B. Jones, W.J. Thron, H. Waadeland.) Lecture Notes in Mathematics, No 932, Springer-Verlag (1982) 38-66.

27. Thron, W.J. and Waadeland, H., 'Accelerating convergence of limit periodic continued fractions $K(a_n/1)$'. Numer. Math. 34 (1980), 155-170.

28. Thron, W.J. and Waadeland, H., 'Truncation error bounds for limit periodic continued fractions'. Math. of Comp. 40, (1983), 583-597.

29. Waadeland, H., 'Tales about tails'. Proc. Amer. Math. Soc. 90 (1984), 54-64.

30. Waadeland, H., 'Local properties of continued fractions'. *Rational approximation and its applications in mathematics and physics*. (Eds. J. Gilewicz, M. Pindor, W. Siemaszko.) Lecture Notes in Mathematics, No 1237, Springer-Verlag (1987), 239-250.

31. Waadeland, H., 'Derivatives of continued fractions with applications to hypergeometric functions'. Journal of computational and Applied Mathematics **19** (1987).

32. Waadeland, H., 'Linear approximations to continued fractions'. Journal of Computational and Applied Mathematics. To appear.

33. Worpitzky, J.D.T., 'Untersuchungen über die Entwickelung der monodromen und monogenen Funktionen durch Kettenbrüche'. *Friedrichs-Gymnasium und Realschule Jahresbericht*. Berlin (1865), 3-39.

BEST A POSTERIORI TRUNCATION ERROR ESTIMATES FOR CONTINUED FRACTIONS $K(a_n/1)$ WITH TWIN ELEMENT REGIONS

William B. Jones*
Department of Mathematics
University of Colorado - Boulder
Boulder, Colorado 80309, U.S.A.

Walter M. Reid
Department of Mathematics
University of Wisconsin - Eau Claire
Eau Claire, Wisconsin 54702, U.S.A.

ABSTRACT. Best truncation error estimates are determined for continued fractions $K(a_n/1)$ where the elements, a_n, are drawn from the twin-convergence regions of Lange (1966). These error estimates are established by proving that the twin-value regions of Lange, which contain the tails $f^{(n)}$ of $K(a_n/1)$, satisfy sufficient conditions to be best twin-value regions. The principal result, stated in Theorem 2, is that

$$|f - f_n| \leq \frac{\prod_{j=1}^{n} |a_j|}{|B_n B_{n-1}|} \cdot \frac{|\rho^2 - |\Gamma_n|^2|}{\rho|h_n| - |\rho^2 - \overline{\Gamma}_n(\Gamma_n + (-1)^n h_n)|}$$

where f and f_n are the limit and nth approximant, respectively, of the continued fraction. The estimates are a posteriori since the elements a_j, $1 \leq j \leq n$ are known, fixed, and used in determining the value of h_n, the ratio of partial denominators of the continued fraction.

1. Introduction.

A continued fraction

$$K = K(a_n/1) := \frac{a_1}{1} + \frac{a_2}{1} + \frac{a_3}{1} + \cdots \tag{1.1}$$

with nth approximant

$$f_n := \frac{a_1}{1} + \frac{a_2}{1} + \cdots + \frac{a_n}{1}, \quad n = 1,2,3, \cdots \tag{1.2}$$

*Research supported in part by the U.S. National Science Foundation under grant No. DMS-8401717.

A. Cuyt (ed.), Nonlinear Numerical Methods and Rational Approximation, 335–347.

is said to **converge** to a finite value $f \in \mathbb{C}$ provided that

$$\lim_{n\to\infty} f_n = f \in \mathbb{C} . \tag{1.3}$$

Subsets (E_1, E_2) of $\mathbb{C}$ are said to be **twin-convergence regions for** $K(a_n/1)$ provided that the continued fraction converges whenever the elements, a_n, satisfy

$$0 \neq a_{2n-1} \in E_1 \text{ and } 0 \neq a_{2n} \in E_2 \tag{1.4}$$

for $n = 1,2,3, \cdots$. Let $\mathcal{K}(E_1, E_2)$ be the set of all continued fractions of the form (1.1) whose elements satisfy (1.4).

Some early examples of twin-convergence regions for $K(a_n/1)$ are those of Leighton and Wall [6] where $|a_{2n-1}| \leq 1/4$ and $|a_{2n}| \geq 25/4$. Later Thron [8] showed that $|a_{2n-1}| \leq \rho$ and $|a_{2n}| \geq 2(\rho - \cos(\arg a_{2n}))$ constitute twin-convergence regions for $\rho > 1$.

In this paper the twin-convergence regions of Lange and Thron [4,5] are studied. These are defined by

$$E_1 = [w: |w\bar{\Gamma} - \Gamma(\rho^2 - |\Gamma|^2)| + \rho|w| \leq \rho(\rho^2 - |\Gamma|^2)] \tag{1.5a}$$

and

$$E_2 = [w: |w(1 + \bar{\Gamma}) - (1 + \Gamma)(\rho^2 - |1+\Gamma|^2)| - \rho|w| \geq \rho|\rho^2 - |1 + \Gamma|^2|] \tag{1.5b}$$

where

$$|\Gamma| < \rho < |1 + \Gamma| , \quad \Gamma \in \mathbb{C} . \tag{1.6}$$

Explicit parameterizations for these regions have been determined by one of the authors [7] to be:

$$E_1 = [w = |w|e^{i\theta}: 0 \leq |w| \leq t_1(\theta) - \sqrt{t_1^2(\theta) - k_1}\,]$$

where

$$t_1(\theta) = \rho^2 - |\Gamma|^2\cos(\theta - 2\arg\Gamma), \text{ and}$$

$$k_1 = (\rho^2 - |\Gamma|^2)^2 ,$$

and

$$E_2 = \mathbb{C} \;\; [w = |w|e^{i\theta}: B_{-1}(\theta) < |w| < B_{+1}(\theta), \;\; \theta \in \Omega]$$

where

$$B_{\pm 1}(\theta) = t_2(\theta) \pm \sqrt{t_2^2(\theta) - k_2} ,$$

$$t_2(\theta) = \rho^2 - |1 + \Gamma|^2\cos(\theta - 2\arg(1 + \Gamma)) ,$$

$$k_2 = (\rho^2 - |1 + \Gamma|^2)^2 , \text{ and}$$

$$\Omega = [\theta: |\theta - 2\arg(1 + \Gamma)| \geq \alpha, \; -\pi \leq \theta \leq \pi]$$

where

$$\alpha = \cos^{-1}(2(\frac{\rho}{|1+\Gamma|})^2 - 1) .$$

Examples of such sets are shown in Figure 1. (The scales are different.) We shall establish truncation error estimates for $K(a_n/1) \in \mathcal{K}(E_1, E_2)$; that is, we determine bounds for the **truncation error** $|f - f_n|$.

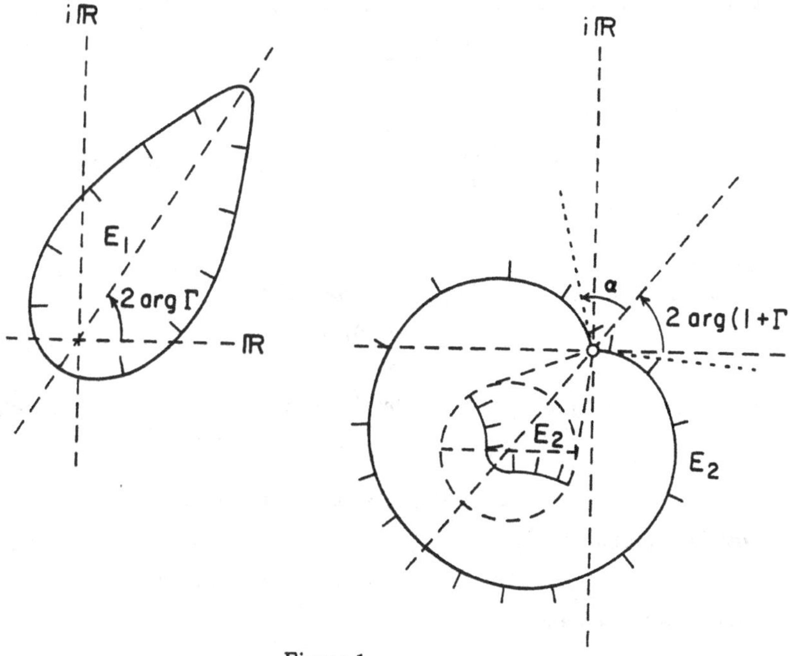

Figure 1.

To introduce the linear fractional transformation (l.f.t.) structure upon which the analysis rests, we define

$$s_n(w) := \frac{a_n}{1+w} , \quad n = 1,2,3, \cdots , \tag{1.7}$$

and, for $m = 0,1,2,3, \cdots$,

$$S_0^{(m)}(w) := w, \text{ and} \tag{1.8a}$$

$$S_n^{(m)}(w) := s_{m+1} \circ s_{m+2} \circ \cdots \circ s_{m+n}(w) = S_{n-1}^{(m)}(s_{m+n}(w)) , \quad n = 1,2,3, \cdots . \tag{1.8b}$$

Further let

$$S_n(w) := S_n^{(0)}(w), \quad n = 0,1,2,3, \cdots . \tag{1.9}$$

Then

$$f_n = \frac{a_1}{1} \, \frac{a_2}{+ \, 1} \, + \, \cdots \, + \frac{a_n}{1} = S_n(0), \tag{1.10}$$

and

$$f = S_m(f^{(m)}), \quad m = 0,1,2, \cdots \tag{1.11}$$

where $f^{(m)}$, the mth **tail** of $K(a_n/1)$, is the continued fraction

$$f^{(m)} := \frac{a_{m+1}}{1} \, \frac{a_{m+2}}{+ \, 1} \, + \, \cdots \, , \tag{1.12}$$

which has as its nth approximant

$$f_n^{(m)} = \frac{a_{m+1}}{1} \, \frac{a_{m+2}}{+ \, 1} \, + \, \cdots \, + \frac{a_{m+n}}{1} = S_n^{(m)}(0) \, . \tag{1.13}$$

With this notation it is clear that

$$f = f^{(0)} \, ,$$

when these limits exist. It can be shown (see [3]) that

$$S_n(w) = \frac{A_n + wA_{n-1}}{B_n + wB_{n-1}} \, , \quad n = 0,1,2, \cdots , \tag{1.14}$$

where the A_n and B_n are defined by the second order linear **difference equations**

$$A_n := A_{n-1} + a_n A_{n-2} \, , \quad n = 1,2,3, \cdots \tag{1.15a}$$

$$B_n := B_{n-1} + a_n B_{n-2} \, , \quad n = 1,2,3, \cdots \tag{1.15b}$$

with initial conditions

$$A_{-1} = 1, \;\; A_0 = 0, \;\; B_{-1} = 0, \;\; B_0 = 1 \, . \tag{1.15c}$$

From (1.10) we obtain

$$f_n = S_n(0) = \frac{A_n}{B_n} \, , \quad n = 1,2,3, \cdots \, . \tag{1.16}$$

Accordingly A_n and B_n are called the nth **numerator** and **denominator**, respectively, for $K(a_n/1)$. It can also be shown that the numerators and denominators satisfy the **determinant formulas**:

$$A_n B_{n-1} - A_{n-1} B_n = (-1)^{n-1} \prod_{j=1}^{n} a_j \, , \quad n = 1,2,3, \cdots \, . \tag{1.17}$$

From (1.4) and (1.17) it follows that $A_n B_{n-1} - A_{n-1} B_n \neq 0$, which implies that S_n is invertible. Of particular importance in our analysis is the quantity h_n defined for $n = 0,1,2, \cdots$, by:

$$h_0 := \infty, \quad h_1 := 1, \tag{1.18a}$$

$$h_n := \frac{B_n}{B_{n-1}} = -S_n^{-1}(\infty), \quad n = 2,3, \cdots \, . \tag{1.18b}$$

From (1.15b) it is clear that

$$h_n = 1 + \frac{a_n}{h_{n-1}}$$

which by iteration leads to the identity

$$h_n = 1 + \frac{a_n}{1} + \frac{a_{n-1}}{1} + \cdots + \frac{a_2}{1}. \tag{1.19}$$

From this basic structure we can now derive the identity that leads to the truncation error estimates. Indeed we have from (1.10), (1.11), (1.15), and (1.16),

$$\begin{aligned} |f-f_n| &= |S_n(f^{(n)}) - S_n(0)| \\ &= \left|\frac{A_n+f^{(n)}A_{n-1}}{B_n+f^{(n)}B_{n-1}} - \frac{A_n}{B_n}\right| \\ &= \frac{|A_nB_{n-1}-A_{n-1}B_n||f^{(n)}|}{|B_n||B_n+f^{(n)}B_{n-1}|} \\ &= \frac{\prod_{j=1}^{n}|a_j|}{|B_nB_{n-1}|} \cdot \frac{|f^{(n)}|}{|h_n+f^{(n)}|}, \end{aligned} \tag{1.20}$$

where the last equality follows from (1.17) and (1.18).

To obtain sharp upper bounds on the right hand expression in (1.20), we determine tight estimates on the locations of the tails, $f^{(n)}$. It is in this light that the importance of *l.f.t.*'s appears since they have the property of mapping the family of circular regions into itself. This mapping property is exploited in the notion of value regions which, in the cases of Lange and Thron, are chosen as disks and complements of disks.

2. Twin-Value Regions.

Subsets (V_1, V_2) of the extended complex plane $\hat{\mathbb{C}} := \mathbb{C} \cup [\infty]$ which satisfy

$$0 \in V_1 \cap V_2, \tag{2.1a}$$

$$-1 \notin V_1 \cup V_2, \tag{2.1b}$$

$$\frac{E_1}{1+V_1} \subseteq V_2, \tag{2.2a}$$

and

$$\frac{E_2}{1+V_2} \subseteq V_1 \tag{2.2b}$$

are called **twin-value regions** corresponding to the twin-convergence regions (E_1, E_2). Here, of course,

$$\frac{E}{1+V} := \left[\frac{a}{1+z} :\ a \in E,\ z \in V\right].$$

We denote by $\mathcal{V}(E_1, E_2)$ the set of all pairs (V_1, V_2) corresponding to (E_1, E_2). The expression "twin-value regions" is appropriate since, as will soon be shown, for $n = 1,2,3, \cdots$

$$f_n = \frac{a_1}{1} + \frac{a_2}{1} + \cdots + \frac{a_n}{1} \in V_2 \tag{2.3}$$

and

$$f_n^{(1)} = \frac{a_2}{1} + \frac{a_3}{1} + \cdots + \frac{a_{n+1}}{1} \in V_1 . \tag{2.4}$$

Thus V_2 *contains* all approximants f_n of $K(a_n/1)$ and V_1 *contains* all partial tails $f_n^{(1)}$ of $K(a_n/1)$.

To reveal the advantages of this twin-value region framework in combination with the *l.f.t.* structure of $K(a_n/1)$, we define for $(V_1, V_2) \in \mathcal{V}(E_1, E_2)$ the sequences $\{E_n\}_{n=1}^{\infty}$ and $\{V_n\}_{n=0}^{\infty}$ as follows:

$$E_{2n-1} := E_1 , \quad E_{2n} := E_2 \tag{2.5a}$$

and

$$V_{2n+1} := V_1 , \quad V_{2n} := V_2 . \tag{2.5b}$$

Relations (2.2) now combine in the form

$$\frac{E_n}{1+V_n} \subseteq V_{n-1} , \quad n = 1,2,3, \cdots . \tag{2.6}$$

It follows that for $m = 0,1,2,3, \cdots , n = 1,2,3, \cdots ,$

$$S_n^{(m)}(V_{n+m}) \subseteq S_{n-1}^{(m)}(V_{n+m-1}) \subseteq V_m . \tag{2.7}$$

The left subset relation can be seen since

$$S_n^{(m)}(V_{n+m}) = S_{n-1}^{(m)}(s_{m+n}(V_{n+m})) = S_{n-1}^{(m)}\left(\frac{a_{m+n}}{1+V_{n+m}}\right)$$

$$\subseteq S_{n-1}^{(m)}\left(\frac{E_{n+m}}{1+V_{n+m}}\right) \subseteq S_{n-1}^{(m)}(V_{n+m-1}) .$$

Repeated application of the left subset relation in (2.7) yields the right subset relation in (2.7) since

$$S_{n-1}^{(m)}(V_{n+m-1}) \subseteq \cdots \subseteq S_0^{(m)}(V_m) = V_m .$$

Specifically, in view of (2.1a), for $m = 0$ we find from (1.10) that

$$f_n = S_n^{(0)}(0) = S_n(0) \in V_0 = V_2$$

and for $m = 1$, from (1.13),

$$f_n^{(1)} = S_n^{(1)}(0) \in V_1 .$$

Thus the claims in (2.3) and (2.4) are proved. In summary $\{S_n^{(m)}(V_{n+m})\}_{n=0}^{\infty}$ is a nested sequence of sets. Furthermore $f_n^{(m)} = S_n^{(m)}(0) \in S_n^{(m)}(V_{n+m}) \subseteq V_m$. For $m = 0$, it follows that $\{S_n(V_n)\}_{n=0}^{\infty}$ is nested and that

$$f_m \in S_n(V_n), \quad m = n, n+1, n+2, \cdots .$$

Consequently

$$f = \lim_{m\to\infty} f_m \in C(S_n(V_n)) = S_n(C(V_n)) \subseteq C(V_2) . \tag{2.8}$$

Here $C(A)$ denotes the closure of the set A. Note that if V_2 is bounded then, since $-h_n = -S_n^{-1}(\infty)$, it is clear that,

$$-h_n \notin V_n . \tag{2.9}$$

If the pair $(W_1, W_2) \in \mathcal{V}(E_1, E_2)$ satisfies the conditions that

$$W_1 \subseteq V_1 \quad \text{and} \quad W_2 \subseteq V_2$$

for all $(V_1, V_2) \in \mathcal{V}(E_1, E_2)$, then (W_1, W_2) are called **best** twin-value regions in $\mathcal{V}(E_1, E_2)$. We are interested, then, in determining $C(W_1)$ and $C(W_2)$ for best twin-value regions (W_1, W_2).

Theorem 1. *The pair* (V_1^*, V_2^*) *defined by*

$$V_1^* := [z:\ |z + (1+\Gamma)| \geq \rho] \tag{2.10a}$$

and

$$V_2^* := [z:\ |z - \Gamma| \leq \rho] \tag{2.10b}$$

for $|\Gamma| < \rho < |1+\Gamma|$, $\Gamma \in \mathbb{C}$, *satisfies*

a) $(V_1^*, V_2^*) \in \mathcal{V}(E_1, E_2)$, and (2.11)

b) $V_1^* = C(W_1)$ and $V_2^* = C(W_2)$. (2.12)

Proof of a). Conditions (2.1) follow since $|\Gamma| < \rho < |1+\Gamma|$. Conditions (2.2) are satisfied since these are precisely the conditions used in [2] to define the twin-convergence regions of (1.5).

Proof of b). In [1, Lemma 5.2] Baltus and Jones proved a result which when adapted to the case of twin-convergence regions states that if $(V_1, V_2) \in \mathcal{V}(E_1, E_2)$ are closed and satisfy

$$\frac{E_1}{1+V_1} = V_2 , \tag{2.10}$$

$$\frac{E_2}{1+V_2} = V_1 , \tag{2.14}$$

and

$$\lim_{n\to\infty} \text{diam } S_n(V_n) = 0 \text{ for all } K(a_n/1) \in K(E_1, E_2)\,, \tag{2.15}$$

then $V_1 = C(W_1)$ and $V_2 = C(W_2)$.

We shall show that (V_1^*, V_2^*) defined in (2.10) satisfy (2.13), (2.14), and (2.15); hence (2.12) is a consequence. In fact (2.15) was proved by Lange [5] in establishing (E_1, E_2) as twin-convergence regions. Only (2.13) and (2.14) remain to be shown.

To verify (2.13) we know by (2.1b) that $E_1/(1 + V_1^*) \subseteq V_2^*$. To establish set equality we merely note that for $w \in \mathbb{C}$,

$$\frac{w}{1+V_1^*} = \left[z\colon\ |z - \frac{w\bar{\Gamma}}{\rho^2-|\Gamma|^2}| \le \frac{|w|\rho}{\rho^2-|\Gamma|^2}\right], \text{ a disk.}$$

Hence, in fact, $\dfrac{w}{1+V_1^*} = V_2^*$

$$\text{for } w = \rho^2 \text{ if } \Gamma = 0 \text{ and for } w = e^{i2\arg\Gamma}(\rho^2-|\Gamma|^2) \text{ if } \Gamma \neq 0\,.$$

In both cases it can be seen that $w \in E_1$. Therefore (2.13) is proved for (V_1^*, V_2^*).

To show that (2.14) holds for (V_1^*, V_2^*) is more involved. While, as before, $E_2/(1 + V_2^*) \subseteq V_1^*$ by (2.1c), it remains to show that set equality holds. But in this case we have for $0 \neq w \in \mathbb{C}$, $w/(1 + V_2^*)$ is the disk

$$\frac{w}{1+V_2^*} = \left[z\colon\ |z - \frac{w(1+\bar{\Gamma})}{|1+\Gamma|^2-\rho^2}| \le \frac{|w|\rho}{|1+\Gamma|^2-\rho^2}\right]$$

while V_1^* is the complement of a disk. It must be shown, then, that for each $z \in V_1^*$ there is a $w(z) \in E_2$ such that

$$z \in \frac{w(z)}{1+V_2^*} \subseteq V_1^*\,. \tag{2.16}$$

(2.16) will place $w(z) \in E_2$ and will yield

$$V_1^* \subseteq \bigcup_{z\in V_1^*} \frac{w(z)}{1+V_1^*} = \frac{E_2}{1+V_2^*}\,.$$

To this end, $z \in V_1^*$ can be expressed in the form

$$z = -(1+\Gamma) + r\rho e^{i\theta}\,, \quad r \ge 1\,. \tag{2.17}$$

We select $m \ge 1$ to determine $w(z)$ of the form

$$w(z) = \frac{(|1+\Gamma|^2-\rho^2)[-(1+\Gamma)+mr\rho e^{i\theta}]}{(1+\bar{\Gamma})}\,. \tag{2.18}$$

This choice will place $w(z)$ on the ray from $-(1+\Gamma)$ through z. The additional requirement

$$|z - \frac{w(z)(1+\bar{\Gamma})}{|1+\Gamma|^2-\rho^2}| = \frac{|w(z)|\rho}{|1+\Gamma|^2-\rho^2} \tag{2.19}$$

will place $z \in \partial(\frac{w(z)}{1+V_2^*})$ and allow a single argument to serve for $z \in \mathrm{Int}(V_1^*)$ and $z \in \partial(V_1^*)$. (See Figure 2.) From (2.10a) and (2.18) it follows that $z \in \partial(\frac{w(z)}{1+V_2^*})$ if and only if

$$|z - \frac{w(z)(1+\overline{\Gamma})}{|1+\Gamma|^2-\rho^2}| = r\rho|m-1| . \tag{2.20}$$

From (2.18) and (2.5) we find that (2.4) holds if and only if m satisfies the quadratic equation

$$m^2[r^2(|1+\Gamma|^2-\rho^2)] - 2m[r|1+\Gamma|(r|1+\Gamma| - \rho\cos(\theta-\gamma))] + |1+\Gamma|^2(r^2-1) = 0$$

where $\gamma = \arg(1+\Gamma)$. The roots of this equation are

$$m_\pm = \frac{|1+\Gamma|}{r} \cdot \frac{(r|1+\Gamma|-\rho\cos(\theta-\gamma)) \pm \sqrt{D}}{(|1+\Gamma|^2-\rho^2)} \tag{2.21}$$

where

$$D := (r|1+\Gamma|-\rho\cos(\theta-\gamma))^2 - (|1+\Gamma|^2-\rho^2)(r^2-1) . \tag{2.22}$$

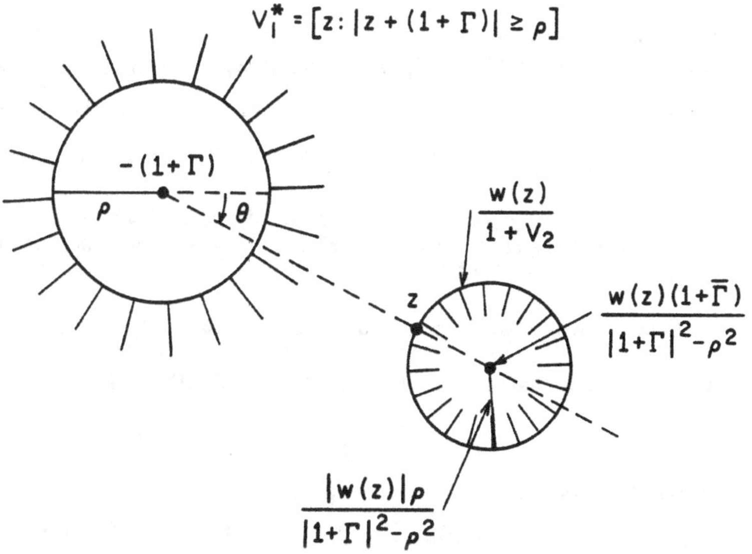

Figure 2.

That $D \geq 0$ follows since

$$D = \rho(1-\cos(\theta-\gamma))(2r|1+\Gamma| - \rho(1+\cos(\theta-\gamma))) + (r\rho-|1+\Gamma|)^2$$

$$\geq \rho(1-\cos(\theta-\gamma))\cdot 2\cdot(r|1+\Gamma|-\rho) + (r\rho-|1+\Gamma|)^2 \geq 0 .$$

Note next that $m_+ \geq 1$ is equivalent to the inequality

$$|1+\Gamma| \sqrt{D} \geq \rho[|1+\Gamma|\cos(\theta-\gamma)-r\rho] . \tag{2.23}$$

This inequality clearly holds for $\cos(\theta-\gamma) \leq \frac{r\rho}{|1+\Gamma|}$ in which case the right hand side is negative. The inequality also holds for $\cos(\theta-\gamma) > \frac{r\rho}{|1+\Gamma|}$ since (2.23), upon squaring both sides, simplifying, and dividing throughout by the positive quantity $(|1+\Gamma|^2-\rho^2)$, is equivalent to

$$r^2\rho^2 - 2r\rho|1+\Gamma|\cos(\theta-\gamma) + |1+\Gamma|^2 \geq 0 . \tag{2.24}$$

But (2.24) holds since the left hand side of (2.24) is bounded below by

$$[r\rho - |1+\Gamma|]^2$$

which is nonnegative.

Hence $D \geq 0$ and $m_+ \geq 1$. We may now choose $w(z)$ by selecting m in (2.18) to be m_+ in (2.21). With this choice we now have $z \in \frac{w(z)}{1+V_2^*} \subseteq V_1^*$ as desired. Finally notice, as a matter of interest, that in (2.21) $m_+ = m_-$ when $D = 0$. This case occurs when $r = \frac{|1+\Gamma|}{\rho}$ and $\theta = \gamma = \arg(1+\Gamma)$. Consequently $m = m_+ = m_- = 1$, $z = w(z) = 0$ and we see the degenerate case where, recalling that $-1 \notin V_2^*$,

$$z = 0 \in \frac{w(z)}{1+V_2^*} = [0] \subseteq V_1^* \text{ as required.}$$

This completes the proof of (2.14) and hence the proof of Theorem 1. Consequently the twin-value regions of (2.10) are the closures of best twin-value regions for the twin-convergence regions of (1.5).

3. Best A Posteriori Truncation Error Estimates.

We now assume that the first n elements of $K \in \mathcal{K}(E_1, E_2)$ are *fixed* and denote them by

$$a_1^*, a_2^*, \cdots, a_n^* . \tag{3.1}$$

Define the subset $\mathcal{K}_n^*(E_1, E_2)$ of $\mathcal{K}(E_1, E_2)$ by

$$\mathcal{K}_n^* = \mathcal{K}_n^*(E_1, E_2) := [K \in \mathcal{K}(E_1, E_2)\colon\ a_i = a_i^*, 1 \leq i \leq n]. \tag{3.2}$$

For $K \in \mathcal{K}_n^*$ it follows from (1.2) and (1.19) that

$$f_n = \frac{a_1^*}{1} + \frac{a_2^*}{1} + \cdots + \frac{a_n^*}{1}, \text{ and} \tag{3.3a}$$

$$h_n = 1 + \frac{a_n^*}{1} + \frac{a_{n-1}^*}{1} + \cdots + \frac{a_2^*}{1} \tag{3.3b}$$

are completely determined. From (1.20) it follows that

$$|f-f_n| = \frac{\prod_{j=1}^{n} |a_j^*|}{|B_n B_{n-1}|} \cdot \frac{|f^{(n)}|}{|h_n + f^{(n)}|} \tag{3.4}$$

and hence $|f-f_n|$ depends only on the nth tail $f^{(n)} = \frac{a_{n+1}}{1} + \frac{a_{n+2}}{1} + \cdots$. Bounds on $|f-f_n|$ determined by $f^{(n)}$ with known values in (3.1) and (3.3b) constitute *a posteriori* truncation error estimates for $K \in \mathcal{K}_n^*$.

By exploiting knowledge from Theorem 1 that V_1^* and V_2^* of (2.10) are the closures of best twin-value regions for (E_1, E_2), we will obtain in Theorem 2 below *best a posteriori* truncation error estimates for $K \in \mathcal{K}_n^*$.

Theorem 2. *If* $K \in \mathcal{K}_n^*$, *then*

$$|f-f_n| \le \frac{\prod_{j=1}^{n} |a_j^*|}{|B_n B_{n-1}|} \cdot \frac{|\rho^2 - |\Gamma_n|^2|}{\rho |h_n| - |\rho^2 - \overline{\Gamma}_n(\Gamma_n + (-1)^n h_n)|}, \tag{3.5}$$

constitutes a best upper bound for $|f-f_n|$. *Here*

$$\Gamma_n := \begin{cases} \Gamma, & n \text{ even} \\ -(1+\Gamma), & n \text{ odd} \end{cases} \tag{3.6}$$

are elements of a sequence $\{\Gamma_n\}_{n=0}^{\infty}$ *corresponding to* $\{V_n\}_{n=0}^{\infty}$.

Proof. If (W_1, W_2) are the best twin-value regions in $\mathcal{K}(E_1, E_2)$ and the sequence $\{W_n\}_{n=1}^{\infty}$ is defined by

$$W_{2n+1} := W_1, \quad W_{2n} := W_2, \quad n = 0,1,2, \cdots,$$

then by [3, Theorem 4.1], for $k = 0,1,2, \cdots$,

$$W_k = \left[\frac{a_{k+1}}{1} + \frac{a_{k+2}}{1} + \cdots + \frac{a_{k+m}}{1} ; \; a_j \in E_j, \; k+1 \le j \le k+m, \; m = 1,2,3, \cdots \right]. \tag{3.7}$$

Here $E_{2n-1} := E_1$, $E_{2n} := E_2$, $n = 1,2,3, \cdots$. By Theorem 1

$$V_n^* = C(W_n), \quad n = 0,1,2, \cdots$$

where $V_{2n+1}^* = V_1^*$ and $V_{2n}^* = V_2^*$, $n = 0,1,2, \cdots$. It follows from this and (3.7) that V_n^* contains all tails $f^{(n)}$ of continued fractions in $\mathcal{K}_n^*$ and consists only of such tails (terminating or non-terminating). From this and (3.4) it follows that

$$-|f-f_n| = \frac{\prod_{j=1}^{n} |a_j^*|}{|B_n B_{n-1}|} \cdot \frac{|f^{(n)}|}{|h_n+f^{(n)}|} = \frac{\prod_{j=1}^{n} |a_j^*|}{|B_n B_{n-1}|} \cdot \frac{1}{|\frac{h_n}{f^{(n)}}+1|}$$

$$\leq \frac{\prod_{j=1}^{n} |a_j^*|}{|B_n B_{n-1}|} \cdot \frac{1}{\min|\frac{h_n}{V_n^*}+1|} . \tag{3.8}$$

Now it is easily shown that, for V_1^* and V_2^* as in (2.10),

$$\frac{1}{V_1^*} = \left[z: \; |z + \frac{(1+\overline{\Gamma})}{|1+\Gamma|^2-\rho^2}| \geq \frac{\rho}{|1+\Gamma|^2-\rho^2}\right]$$

and

$$\frac{1}{V_2^*} = \left[z: \; |z + \frac{\overline{\Gamma}}{\rho^2-|\Gamma|^2}| \geq \frac{\rho}{\rho^2-|\Gamma|^2}\right].$$

From these we obtain the general expression

$$\frac{h_n}{V_n^*} + 1 = [z: \; |z-C_n| \geq R_n] \tag{3.9}$$

where

$$C_n := 1 - \frac{h_n\overline{\Gamma}_n}{|\rho^2-|\Gamma_n|^2|} , \tag{3.10a}$$

and

$$R_n := \frac{|h_n|\rho}{|\rho^2-|\Gamma_n|^2|} \tag{3.10b}$$

for $n = 1,2,3, \cdots$ and Γ_n as defined in (3.6). Furthermore from (2.10b), (2.7) with $m = 0$, and (2.5b) we know that $S_n(V_n^*)$ is bounded. Hence by (2.9), $h_n \notin V_n^*$. It follows that

$$0 \notin \frac{h_n}{V_n^*} + 1$$

which in turn implies that $R_n > |C_n|$. Consequently, noting that $|\rho^2-|\Gamma_n|^2| = (-1)^n(\rho^2-|\Gamma_n|^2)$, we see that

$$\min|\frac{h_n}{V_n^*} + 1| = R_n - |C_n| = \frac{\rho|h_n|-|\rho^2-\overline{\Gamma}_n(\Gamma_n+(-1)^n h_n)|}{|\rho^2-|\Gamma_n|^2|} .$$

This bound combined with (3.8) yields the result (3.5). This completes the proof of Theorem 2.

References:

1. Baltus, C., Jones, W.B.: Truncation error bounds for limit-periodic continued fractions $K(a_n/1)$ with $\lim a_n = 0$, *Numer. Math.* 46 (1985), 541-569.
2. Jones, W.B., Thron, W.J.: Twin-convergence regions for continued fractions $K(a_n/1)$, *Trans. AMS* 150 (1970), 93-119.
3. Jones, W.B., Thron, W.J.: *Continued Fractions: Analytic Theory and Applications*, Encyclopedia of Mathematics and Its Applications, Reading, MA, Addison Wesley (1980), distributed now by Cambridge University Press.
4. Lange, L.J.: On a family of twin convergence regions for continued fractions, *Illinois J. Math.* 10 (1966),, 97-108. MR32 #4253.
5. Lange, L.J., Thron, W.J.: A two-parameter family of best twin convergence regions for continued fractions, *Math. Z.* 73 (1960), 295-311. MR22 #6887.
6. Leighton, W., Wall, H.S.: On the transformation and convergence of continued fractions, *Amer. J. Math.* 58 (1936), 267-281.
7. Reid, W.M.: Parameterizations and factorizations of element regions for continued fractions $K(a_n/1)$, *Lecture Notes in Mathematics* No. 932, Springer-Verlag, Berlin, 1982.
8. Thron, W.J.: Zwillingskonvergenzgebiete für Kettenbrüche $1 + K(a_n/1)$, deren eines die Kreisscheibe $|a_{2n-1}| \leq \rho^2$ ist, *Math. Zeitschr.* 70 (1959), 310-344. MR21 #4229.

CONVERGENCE ACCELERATION FOR MILLER'S ALGORITHM

Paul Levrie and Robert Piessens
Department of Computer Science
K.U.Leuven
Celestijnenlaan 200A
B-3030 Heverlee
Belgium

ABSTRACT. In this paper Miller's recurrence algorithm for calculating a minimal solution of a p-th order linear homogeneous recurrence relation is modified with the intention of avoiding the occurrence of overflow and underflow. This algorithm is a generalisation of Gautschi's continued fraction-algorithm for second-order recurrence relations. It uses a generalisation of a continued fraction which will be called a G-continued fraction. Convergence of this G-continued fraction is defined and some convergence results are given. The concept of modification of a G-continued fraction is introduced. The main result in this paper is the proof of convergence acceleration for a suitable modification in the case of a recurrence relation of Perron-Kreuser type. It is assumed that the characteristic equations for this recurrence relation have only simple roots with differing absolute values.

1. INTRODUCTION AND NOTATION

In his paper of 1977 R.V.M. Zahar [13] discusses convergence and stability of the Miller algorithm for the computation of solutions of p-th-order linear homogeneous recurrence relations

$$\sum_{i=0}^{p} a_i(n) y_{n+p-i} = 0\,, \quad a_0(n) a_p(n) \neq 0\,, \quad n = 0, 1, \ldots . \tag{1.1}$$

One of his main results is that the Miller algorithm converges if the equation adjoint to (1.1) possesses a maximal solution. A problem that still remains however is that when we use the Miller algorithm overflow or underflow is very likely to occur. In this paper we present a method of rescaling for the Miller algorithm which avoids overflow and underflow, and which reduces to the continued fraction-method of Gautschi [2] in the case of second-order recurrence relations. The method was first proposed by Gautschi (Zahar [14]).

A. Cuyt (ed.), Nonlinear Numerical Methods and Rational Approximation, 349–370.

In the first paragraph we introduce some notations and definitions. In the second paragraph we give some convergence results. In the third paragraph finally we shall apply L. Jacobsen's method of convergence acceleration for ordinary continued fractions [3, 9] to our rescaled version of the Miller algorithm.

The adjoint equation of (1.1) is given by

$$\sum_{i=0}^{p} a_{p-i}(n+p-i)y_{n+p-i} = 0\,,\quad n=0,1,\ldots\,. \tag{1.2}$$

We denote by $A_n^{(h)}$ $(h=1,\ldots,p)$ the solution of (1.1) with initial values

$$A_n^{(h)} = \delta_{n,h-1}\,,\quad n=0,1,\ldots p-1\,. \tag{1.3}$$

We define for arbitrary solutions $f_n^{(1)}, f_n^{(2)}, \ldots, f_n^{(m)}$ of (1.1)

$$E_N(f_n^{(1)},\ldots,f_n^{(m)}) = \det \begin{pmatrix} f_{N+p-m+1}^{(1)} & \cdots & f_{N+p-m+1}^{(m)} \\ \vdots & & \vdots \\ f_{N+p}^{(1)} & \cdots & f_{N+p}^{(m)} \end{pmatrix}$$

for $N=0,1,\ldots$.

We now assume that we want to calculate a solution f_n of (1.1). Let N be an integer, and define

$$\begin{aligned} y_{N+i}^N &= 0\,, & i&=2,\ldots,p \\ y_{N+1}^N &= 1\,, \\ y_n^N &= -\sum_{i=1}^{p} \frac{a_{p-i}(n)}{a_p(n)} y_{n+i}^N\,, & n&=N,\ldots,0\,. \end{aligned} \tag{1.4}$$

(y_n^N is a solution of (1.1).) We may assume that one of the initial values f_k with $k \in \{0, 1,\ldots, p-1\}$ is different from zero. We then take as an approximation to f_n:

$$f_n^N = f_k \frac{y_n^N}{y_k^N}\,. \tag{1.5}$$

This algorithm is called the Miller algorithm [12]. The Miller algorithm (1.4-5) is said to converge if

$$\lim_{N\to\infty} f_n^N = f_n\,,\quad n=0,1,\ldots\,. \tag{1.6}$$

From now on we assume that $f_0 \neq 0$ and we take $k=0$ in (1.5). In many cases the calculation of the quantities y_n^N will lead to overflow or underflow. To avoid this we calculate

$$\begin{aligned} b^N(n) &= \frac{y_{n+1}^N}{y_n^N} & &\text{if } n=0,\ldots,N \\ &= 0 & &\text{if } n=N+1,\ldots,N+p-1 \end{aligned} \tag{1.7}$$

instead of the y_n^N. The Miller algorithm then reduces to

$$
\begin{aligned}
b^N(N+i) &= 0\,, && i = 1,\ldots, p-1 \\
b^N(n) &= -\frac{a_p(n)}{\sum_{i=1}^{p} a_{p-i}(n)\left(\prod_{h=1}^{i-1} b^N(n+h)\right)}\,, && n = N,\ldots, 0 \qquad (1.8)\\
f_n^N &= \left(\prod_{h=0}^{n-1} b^N(h)\right) f_0\,.
\end{aligned}
$$

The relationship between y_n^N and $b^N(n)$ is given by

$$
y_n^N = \frac{\prod_{h=0}^{n-1} b^N(h)}{\prod_{h=0}^{N} b^N(h)}\,. \tag{1.9}
$$

If $p = 2$ and $a_0(n) \equiv 1$, then $b^N(0)$ is the $N+1$-th approximant of the continued fraction:

$$
-\frac{a_2(0)|}{|a_1(0)-}\,\frac{a_2(1)|}{|a_1(1)-}\,\frac{a_2(2)|}{|a_1(2)-}\cdots \tag{1.10}
$$

and furthermore we have

$$
b^N(0) = -\frac{A_{N+2}^{(1)}}{A_{N+2}^{(2)}}\,,\quad N = 0, 1,\ldots \tag{1.11}
$$

if both members of this equation are well-defined.
We shall now prove the corresponding result for the continued fraction-like algorithm defined in (1.8). To do this we define transformations $\{s_n\}$ en $\{S_n\}$ ($n = 0, 1,\ldots$)

$$
\begin{aligned}
s_n(w^{(1)},\ldots,w^{(p-1)}) &= -\frac{a_p(n)}{\sum_{i=1}^{p} a_{p-i}(n)\left(\prod_{h=1}^{i-1} w^{(h)}\right)}\,,\\
S_0(w^{(1)},\ldots,w^{(p-1)}) &= s_0(w^{(1)},\ldots,w^{(p-1)}) \qquad (1.12)\\
S_n(w^{(1)},\ldots,w^{(p-1)}) &= S_{n-1}(s_n(w^{(1)},\ldots,w^{(p-1)}),w^{(1)},\ldots,w^{(p-2)})
\end{aligned}
$$

for $n = 1, 2,\ldots$ and with $w^{(1)},\ldots, w^{(p-1)} \in \mathbb{C}$.
As an immediate consequence of these definitions we have

$$
S_N(0,\ldots,0) = b^N(0)\,. \tag{1.13}
$$

The connection between the transformations defined in (1.12) and the solutions $A_n^{(h)}$ ($h = 1,\ldots, p$) of (1.1) is given by

THEOREM 1. *For $n=0,1,\ldots$ we have*

$$S_n(w^{(1)},\ldots,w^{(p-1)}) = -\frac{\Gamma_{n+1}^{(1,2)} + \sum_{i=2}^{p}(-1)^{i-1}\Gamma_{n+1}^{(i,2)}(\prod_{h=1}^{i-1} w^{(h)})}{\Gamma_{n+1}^{(1,1)} + \sum_{i=2}^{p}(-1)^{i-1}\Gamma_{n+1}^{(i,1)}(\prod_{h=1}^{i-1} w^{(h)})} \tag{1.14}$$

where $\Gamma_n^{(i,j)}$ $(n=0,1,\ldots,\ 1\leqslant i,j\leqslant p)$ is defined by

$$\Gamma_n^{(i,j)} = \det\begin{vmatrix} A_n^{(1)} & \ldots & A_n^{(j-1)} & A_n^{(j+1)} & \ldots & A_n^{(p)} \\ \vdots & & \vdots & \vdots & & \vdots \\ A_{n+i-2}^{(1)} & \ldots & A_{n+i-2}^{(j-1)} & A_{n+i-2}^{(j+1)} & \ldots & A_{n+i-2}^{(p)} \\ A_{n+i}^{(1)} & \ldots & A_{n+i}^{(j-1)} & A_{n+i}^{(j+1)} & \ldots & A_{n+i}^{(p)} \\ \vdots & & \vdots & \vdots & & \vdots \\ A_{n+p-1}^{(1)} & \ldots & A_{n+p-1}^{(j-1)} & A_{n+p-1}^{(j+1)} & \ldots & A_{n+p-1}^{(p)} \end{vmatrix}. \tag{1.15}$$

PROOF. Using the recurrence relation (1.1) and some elementary properties of determinants it is easy to show that for $n=0,1,\ldots$ and $m=1,\ldots,p$

$$\begin{aligned} \Gamma_{n+1}^{(p,m)} &= \Gamma_n^{(1,m)} \\ \Gamma_{n+1}^{(i,m)} &= (-1)^{p-i}\frac{a_{p-i}(n)}{a_0(n)}\Gamma_n^{(1,m)} + (-1)^{p-1}\frac{a_p(n)}{a_0(n)}\Gamma_n^{(i+1,m)} \end{aligned} \tag{1.16}$$

for $i=1,\ldots,p-1$.

We now prove (1.14) using induction. Since

$$\Gamma_0^{(i,1)} = \delta_{i,1}$$
$$\Gamma_0^{(i,2)} = \delta_{i,2}$$

we may rewrite the right member of (1.14) for $n=0$ as :

$$-\frac{(-1)^{p-1}\dfrac{a_p(0)}{a_0(0)}}{(-1)^{p-1}\dfrac{a_{p-1}(0)}{a_0(0)} + \sum_{i=2}^{p}(-1)^{i-1}(-1)^{p-i}\dfrac{a_{p-i}(0)}{a_0(0)}(\prod_{h=1}^{i-1} w^{(h)})}.$$

Multiplying the numerator and the denominator in this expression by $(-1)^{p-1}a_0(0)$ we get $s_0(w^{(1)},\ldots,w^{(p-1)})$ as defined in (1.12).

We now assume that (1.14) is true for $n=m-1$, then we have for $n=m$:

$$\begin{aligned} &S_m(w^{(1)},\ldots,w^{(p-1)}) \\ &= S_{m-1}(s_m(w^{(1)},\ldots,w^{(p-1)}),w^{(1)},\ldots,w^{(p-2)}) \end{aligned}$$

$$= -\frac{\Gamma_m^{(1,2)} - s_m(w^{(1)},\ldots,w^{(p-1)})\sum_{i=2}^{p}(-1)^{i-2}\Gamma_m^{(i,2)}(\prod_{h=1}^{i-2}w^{(h)})}{\Gamma_m^{(1,1)} - s_m(w^{(1)},\ldots,w^{(p-1)})\sum_{i=2}^{p}(-1)^{i-2}\Gamma_m^{(i,1)}(\prod_{h=1}^{i-2}w^{(h)})}.$$

We multiply the numerator and the denominator of this expression with the denominator of $s_m(w^{(1)},\ldots,w^{(p-1)})$ in (1.12) and then we find:

$$S_m(w^{(1)},\ldots,w^{(p-1)})$$

$$= -\frac{\Gamma_m^{(1,2)}\sum_{i=1}^{p}a_{p-i}(m)(\prod_{h=1}^{i-1}w^{(h)}) + a_p(m)\sum_{i=2}^{p}(-1)^{i-2}\Gamma_m^{(i,2)}(\prod_{h=1}^{i-2}w^{(h)})}{\Gamma_m^{(1,1)}\sum_{i=1}^{p}a_{p-i}(m)(\prod_{h=1}^{i-1}w^{(h)}) + a_p(m)\sum_{i=2}^{p}(-1)^{i-2}\Gamma_m^{(i,1)}(\prod_{h=1}^{i-2}w^{(h)})}.$$

If we divide the numerator and the denominator by $a_0(m)$, then we get after some rearrangements and using (1.16):

$$= -\frac{\Gamma_{m+1}^{(p,2)}\prod_{h=1}^{p-1}w^{(h)} + (-1)^{p-1}\sum_{i=1}^{p-1}(-1)^{i-1}\Gamma_{m+1}^{(i,2)}\prod_{h=1}^{i-1}w^{(h)}}{\Gamma_{m+1}^{(p,1)}\prod_{h=1}^{p-1}w^{(h)} + (-1)^{p-1}\sum_{i=1}^{p-1}(-1)^{i-1}\Gamma_{m+1}^{(i,1)}\prod_{h=1}^{i-1}w^{(h)}}$$

$$= -\frac{(-1)^{p-1}[(-1)^{p-1}\Gamma_{m+1}^{(p,2)}\prod_{h=1}^{p-1}w^{(h)} + \sum_{i=1}^{p-1}(-1)^{i-1}\Gamma_{m+1}^{(i,2)}\prod_{h=1}^{i-1}w^{(h)}]}{(-1)^{p-1}[(-1)^{p-1}\Gamma_{m+1}^{(p,1)}\prod_{h=1}^{p-1}w^{(h)} + \sum_{i=1}^{p-1}(-1)^{i-1}\Gamma_{m+1}^{(i,1)}\prod_{h=1}^{i-1}w^{(h)}]}$$

$$= -\frac{\Gamma_{m+1}^{(1,2)} + \sum_{i=2}^{p}(-1)^{i-1}\Gamma_{m+1}^{(i,2)}(\prod_{h=1}^{i-1}w^{(h)})}{\Gamma_{m+1}^{(1,1)} + \sum_{i=2}^{p}(-1)^{i-1}\Gamma_{m+1}^{(i,1)}(\prod_{h=1}^{i-1}w^{(h)})}.$$

□

COROLLARY. *For* $N = 0, 1, \ldots$ *we have*

$$b^N(0) = -\frac{\Gamma_{N+1}^{(1,2)}}{\Gamma_{N+1}^{(1,1)}}. \tag{1.17}$$

If $p = 2$ (1.14) is equal to

$$S_N(w^{(1)}) = -\frac{A_{N+2}^{(1)} - A_{N+1}^{(1)}w^{(1)}}{A_{N+2}^{(2)} - A_{N+1}^{(2)}w^{(1)}}. \tag{1.18}$$

This is the well-known connection formula from the theory of ordinary continued fractions [11]. If we take $w^{(1)} = 0$ we get (1.11).

From now on we assume that $a_0(n) \equiv 1$.

If we let N tend to ∞ in $S_N(0,\ldots,0)$ we call the infinite expression which is generated a *G-continued fraction*. For the G-continued fraction defined above we adopt the notation:

$$\mathop{\mathrm{K}}_{n=0}^{\infty}\left(\frac{-a_p(n)}{a_{p-1}(n)\,;\cdots;a_1(n)}\right).$$

The expression

$$t^{(m)} = \mathop{\mathrm{K}}_{n=m}^{\infty}\left(\frac{-a_p(n)}{a_{p-1}(n)\,;\cdots;a_1(n)}\right)$$

is called the *m-th tail* of the G-continued fraction.
The quantity

$$b^{N-1}(0) = -\frac{\Gamma_N^{(1,2)}}{\Gamma_N^{(1,1)}}$$

is called the *N-th approximant* of the G-continued fraction.
In definition 1 we generalise the concept of *equivalence of continued fractions* to G-continued fractions:

DEFINITION 1. *The G-continued fractions*

$$\mathop{\mathrm{K}}_{n=0}^{\infty}\left(\frac{-a_p(n)}{a_{p-1}(n)\,;\cdots;a_1(n)}\right) \quad and \quad \mathop{\mathrm{K}}_{n=0}^{\infty}\left(\frac{-\hat{a}_p(n)}{\hat{a}_{p-1}(n)\,;\cdots;\hat{a}_1(n)}\right)$$

are said to be equivalent if their N-th approximants $b^{N-1}(0)$ and $\hat{b}^{N-1}(0)$ satisfy

$$b^{N-1}(0) = \hat{b}^{N-1}(0) \tag{1.19}$$

for $N = 1, 2, \ldots$.

THEOREM 2. *If there exists a sequence of non-zero constants $\{r_n\}_{n=-p+1}^{\infty}$ with $r_{-p+1} = 1$ such that*

$$\hat{a}_i(n) = r_n r_{n-1} \cdots r_{n-i+1} a_i(n) \ , \ i = 1, \ldots, p \ , \tag{1.20}$$

for $n = 0, 1, \ldots$, then the G-continued fractions

$$\mathop{\mathrm{K}}_{n=0}^{\infty}\left(\frac{-a_p(n)}{a_{p-1}(n)\,;\cdots;a_1(n)}\right) \quad and \quad \mathop{\mathrm{K}}_{n=0}^{\infty}\left(\frac{-\hat{a}_p(n)}{\hat{a}_{p-1}(n)\,;\cdots;\hat{a}_1(n)}\right)$$

are equivalent.

PROOF. Let $A_n^{(h)}$ $(h = 1, \ldots, p)$ be the solutions of (1.1) defined by (1.3), and let $\hat{A}_n^{(h)}$ be the corresponding solutions of

$$\sum_{i=0}^{p} \hat{a}_i(n) y_{n+p-i} = 0\,, \quad \hat{a}_0(n) = 1\,, \quad n = 0, 1, \ldots . \tag{1.21}$$

Then it is easy to prove using (1.20) that

$$\hat{A}_n^{(h)} = r_{-p+h} \cdots r_{-p+n} A_n^{(h)} \tag{1.22}$$

for $n = p, p+1, \ldots$. From (1.17) and (1.22) we get

$$\hat{b}^N(0) = -\frac{E_N(\hat{A}_n^{(1)}, \hat{A}_n^{(3)}, \ldots, \hat{A}_n^{(p)})}{E_N(\hat{A}_n^{(2)}, \hat{A}_n^{(3)}, \ldots, \hat{A}_n^{(p)})}$$

$$= -\frac{r_{-p+1} \cdots r_{N-p+2}}{r_{-p+2} \cdots r_{N-p+2}} \frac{E_N(A_n^{(1)}, A_n^{(3)}, \ldots, A_n^{(p)})}{E_N(A_n^{(2)}, A_n^{(3)}, \ldots, A_n^{(p)})}$$

$$= r_{-p+1} b^N(0)$$

$$= b^N(0)$$

for $N = 0, 1, \ldots$.

□

2. CONVERGENCE OF G-CONTINUED FRACTIONS

Next we introduce what is meant by a convergent G-continued fraction:

DEFINITION 2. *The G-continued fraction*

$$\mathop{\mathrm{K}}_{n=0}^{\infty} \left| \frac{-a_p(n)}{a_{p-1}(n)\,; \cdots ; a_1(n)} \right| \tag{2.1}$$

is said to converge if and only if

$$\lim_{N \to \infty} \frac{\Gamma_N^{(1,2)}}{\Gamma_N^{(1,1)}} \in \mathbb{C}\,.$$

The value of the G-continued fraction (2.1) is then given by

$$\xi_0 = -\lim_{N \to \infty} \frac{\Gamma_N^{(1,2)}}{\Gamma_N^{(1,1)}}\,.$$

If the m th tail $t^{(m)}$ of the G-continued fraction (2.1) converges, we shall denote its value by ξ_m.

We now have the following convergence criterion, which is a generalisation of Pincherle's theorem [2].

THEOREM 3 (Zahar [14]). *If the recurrence relation (1.1) has a fundamental system of solutions* $\{\{f_n^{(1)}\}, \{f_n^{(2)}\}, \ldots, \{f_n^{(p)}\}\}$ *with* $f_0^{(p)} \neq 0$ *and for which*

$$\lim_{N\to\infty} \frac{E_N(f_n^{(1)}, \ldots, f_n^{(i-1)}, f_n^{(p)}, f_n^{(i+1)}, \ldots, f_n^{(p-1)})}{E_N(f_n^{(1)}, \ldots, f_n^{(p-1)})} = 0 \tag{2.2}$$

for $i = 1, \ldots, p-1$, *then the G-continued fraction (2.1) converges to* $f_1^{(p)} / f_0^{(p)}$.

PROOF. It is easy to prove using (2.2) and $f_0^{(p)} \neq 0$ that $\{\{f_n^{(p)}\}, \{A_n^{(2)}\}, \ldots, \{A_n^{(p)}\}\}$ is also a fundamental system of solutions for (1.1). Furthermore we have:

$$\lim_{N\to\infty} \frac{E_N(A_n^{(2)}, \ldots, A_n^{(i-1)}, f_n^{(p)}, A_n^{(i+1)}, \ldots, A_n^{(p)})}{E_N(A_n^{(2)}, \ldots, A_n^{(p)})} = 0 \tag{2.3}$$

for $i = 2, \ldots, p$. Since

$$f_n^{(p)} = \sum_{i=1}^{p} f_{i-1}^{(p)} A_n^{(i)}$$

we get from (2.3) with $i = 2$:

$$0 = \lim_{N\to\infty} \frac{E_N(\sum_{i=1}^{p} f_{i-1}^{(p)} A_n^{(i)}, A_n^{(3)}, \ldots, A_n^{(p)})}{E_N(A_n^{(2)}, \ldots, A_n^{(p)})}$$

$$= \lim_{N\to\infty} \frac{E_N(f_0^{(p)} A_n^{(1)}, A_n^{(3)}, \ldots, A_n^{(p)}) + E_N(f_1^{(p)} A_n^{(2)}, A_n^{(3)}, \ldots, A_n^{(p)})}{E_N(A_n^{(2)}, \ldots, A_n^{(p)})}$$

$$= f_0^{(p)} \lim_{N\to\infty} \frac{E_N(A_n^{(1)}, A_n^{(3)}, \ldots, A_n^{(p)})}{E_N(A_n^{(2)}, \ldots, A_n^{(p)})} + f_1^{(p)} .$$

$$= f_0^{(p)} \lim_{N\to\infty} \frac{\Gamma_N^{(1,2)}}{\Gamma_N^{(1,1)}} + f_1^{(p)} .$$

This means that the G-continued fraction converges, and its value is given by

$$\xi_0 = \frac{f_1^{(p)}}{f_0^{(p)}} .$$

□

As a consequence of this theorem we have:

COROLLARY (Zahar [13]). *The Miller algorithm (1.4-5) converges if the adjoint equation of (1.1) possesses a maximal solution, i.e. (1.2) has a fundamental system of solutions* $\{\{g_n^{(1)}\}, \{g_n^{(2)}\}, \ldots, \{g_n^{(p)}\}\}$ *with*

$$\lim_{n\to\infty} \frac{g_n^{(i)}}{g_n^{(1)}} = 0 \tag{2.4}$$

for $i = 2, \ldots, p$. *(*$g_n^{(1)}$ *is called a maximal solution of (1.2).)*

REMARK. The following converse of theorem 3 can be proved: if the G-continued fraction (2.1) and its tails $t^{(1)}, \ldots, t^{(p-2)}$ converge, then the recurrence relation (1.1) has a fundamental system of solutions $\{\{f_n^{(1)}\}, \{f_n^{(2)}\}, \ldots, \{f_n^{(p)}\}\}$ for which (2.2) is satisfied.

Another convergence theorem is given by the following generalization of Pringsheim's theorem [11] to G-continued fractions:

THEOREM 4. *If the coefficients of the recurrence relation (1.1) satisfy the inequality*

$$|a_{p-1}(n)| \geqslant 1 + \sum_{i=1}^{p-2} |a_i(n)| + |a_p(n)| \tag{2.5}$$

for $n = 0, 1, \ldots$, *then the G-continued fraction (2.1) converges.*

PROOF. see [5].

3. CONVERGENCE ACCELERATION FOR G-CONTINUED FRACTIONS

The speed of convergence of the Miller algorithm (1.4) is closely related to the speed with which the limits in (2.2) approach zero as N tends to infinity : with y_n^N as in (1.4) and $f_n^{(i)}$ ($i = 1, \ldots, p$) as in theorem 2 we have

$$\frac{y_n^N}{y_0^N} = \frac{f_n^{(p)} - \sum_{i=1}^{p-1} f_n^{(i)} \dfrac{E_N(f_n^{(1)}, \ldots, f_n^{(i-1)}, f_n^{(p)}, f_n^{(i+1)}, \ldots, f_n^{(p-1)})}{E_N(f_n^{(1)}, \ldots, f_n^{(p-1)})}}{f_0^{(p)} - \sum_{i=1}^{p-1} f_0^{(i)} \dfrac{E_N(f_n^{(1)}, \ldots, f_n^{(i-1)}, f_n^{(p)}, f_n^{(i+1)}, \ldots, f_n^{(p-1)})}{E_N(f_n^{(1)}, \ldots, f_n^{(p-1)})}} .$$

Sometimes the convergence of the Miller algorithm is slow. Scraton [9] has shown however that when good asymptotic estimates are known for the wanted solution f_n, convergence may be accelerated. For second-order recurrence relations this is done by replacing the starting values $y_{N+1}^N = 1$, $y_{N+2}^N = 0$ by $y_{N+1}^N = 1$, $y_{N+2}^N = \gamma_N$ where γ_N is an asymptotic estimate for f_{N+2}/f_{N+1}. For higher-order recurrences the same method may be applied, but it is not always easy to get asymptotic estimates of the solution you want to calculate. In many cases it is possible to deduce good starting values from the coefficients of the recurrence relation. In the theory of ordinary continued fractions the same technique is applied to accelerate the convergence of slowly converging continued fractions [3,10].

It is an immediate consequence of the definition of a G-continued fraction that

$$\xi_0 = S_n(\xi_{n+1}, \xi_{n+2}, \ldots, \xi_{n+p-1}) \tag{3.1}$$

for all $n = 0, 1, \ldots$, if all ξ_n exist. From this we see that in our modification of Miller's algorithm (1.8) we calculate the value of the G-continued fraction by replacing its tails by zero. If approximations to the tails are known, the convergence of the G-continued fraction will be accelerated if we replace its tails by their approximation instead of by zero. Hence the following definition:

DEFINITION 3. *For a given sequence* $\{w_n\}_{n=0}^{\infty}$ *of numbers from* $\mathbb{C}$, *we define a modification of (2.1) to be the sequence of numbers*

$$\{S_n(w_n, w_{n+1}, \ldots, w_{n+p-2})\}_{n=0}^{\infty} .$$

Let us now assume that the recurrence relation (1.1) is of Perron-Kreuser type (see appendix 1), and furthermore that for every λ the roots of the characteristic equation (formula A4) belonging to the λ–th class are all different in modulus. Let u_i $(i = e_{\lambda-1}+1, \ldots, e_\lambda)$ be the roots of the characteristic equation for the λ–th class, with

$$|u_{e_{\lambda-1}+1}| > \cdots > |u_{e_\lambda}| . \tag{3.2}$$

Since $r^{(\lambda)}_{e_\lambda - e_{\lambda-1}} = d_{e_\lambda} \neq 0$ we also have $|u_i| > 0$ for all i.
It is easy to prove that in this case the recurrence relation (1.1) has a basis ordered by domination : using (A6) we are able to choose for each $i = 1, \ldots, p$ a solution $f_j^{(i)}$ of (1.1) which satisfies :

$$\lim_{n\to\infty} \frac{f_{n+1}^{(i)}}{n^{\alpha_\lambda} f_n^{(i)}} = u_i \quad \text{for } i = e_{\lambda-1}+1, \ldots, e_\lambda \tag{3.3}$$

and using (3.2) we then have

$$\lim_{n\to\infty} \frac{f_n^{(i+1)}}{f_n^{(i)}} = 0 \quad \text{for } i = 1, \ldots, p-1. \tag{3.4}$$

Hence the $\{f_n^{(i)}\}$ form a basis which is ordered by domination. From (3.3) we have :

$$\frac{f_{n+1}^{(i)}}{f_n^{(i)}} = n^{\alpha_\lambda} u_i(1 + o(1)) \;\; (n\to\infty)$$

$$= v_i(n)(1 + o(1)) \;\; (n\to\infty) \tag{3.5}$$

where we have put $v_i(n) = n^{\alpha_\lambda} u_i$. As a consequence of the conditions imposed above the $f_n^{(i)}$ $(i = 1, \ldots, p)$ satisfy (2.2), and we have the following theorem:

THEOREM 5. *Let us assume that the recurrence relation (1.1) is of Perron-Kreuser type and that the above conditions are satisfied. If the G-continued fraction (2.1) and its tails $t^{(1)}, \ldots, t^{(p-2)}$ converge, then we have*

$$\lim_{n\to\infty} \frac{\xi_0 - S_n(w_n, w_{n+1}, \ldots, w_{n+p-2})}{\xi_0 - S_n(0, 0, \ldots, 0)} = 0 \tag{3.6}$$

where ξ_0 is the value of the G-continued fraction, and w_n is defined by

$$\begin{aligned} w_n &= u_p \quad \text{for } n = 0, 1, \ldots, p-2 \\ &= (-p+n+2)^{\alpha_\sigma} u_p \quad \text{for } n = p-1, p, \ldots . \end{aligned} \tag{3.7}$$

PROOF. First we note that it is always possible using an equivalence transformation to obtain that the linear segment $[e_{\sigma-1}e_\sigma]$ of the Newton-Puiseux polygon is horizontal. To do this we define

$$\begin{aligned} r_n &= 1 \quad \text{for } n = -p+1, \ldots, 0 \\ &= n^\tau \quad \text{for } n = 1, 2, \ldots \end{aligned}$$

with

$$\tau = -\frac{m_{e_\sigma} - m_{e_{\sigma-1}}}{e_\sigma - e_{\sigma-1}} = -\alpha_\sigma .$$

If we define $\hat{a}_i(n)$ as in (1.20) we have

$$\hat{a}_i(n) = d_i n^{m_i + i\tau}(1+o(1)) \quad (n\to\infty)$$

and since

$$m_{e_{\sigma-1}} + e_{\sigma-1}\tau = m_{e_\sigma} + e_\sigma \tau$$

we get the desired result. Now using the same method as in the proof of theorem 2 it is easy to show that

$$\hat{S}_n(u_p, u_p, \ldots, u_p) = S_n(w_n, w_{n+1}, \ldots, w_{n+p-2})$$

where by $\hat{S}_n$ we denote the transformations (1.12) for the recurrence relation with coefficients $\hat{a}_i(n)$, w_n is defined by (3.7). Hence we may now assume that for the given recurrence relation (1.1) the right-hand segment of the Newton-Puiseux diagram is horizontal, and we shall show that under the conditions of the theorem we have

$$\lim_{n\to\infty} \frac{\xi_0 - S_n(u_p, u_p, \ldots, u_p)}{\xi_0 - S_n(0, 0, \ldots, 0)} = 0 . \tag{3.8}$$

We note that u_p may be calculated using the adjoint equation (1.2): it is a consequence of the Perron-Kreuser theorem that

$$\lim_{n\to\infty} \frac{g_n^{(1)}}{n^{\alpha_\sigma} g_{n+1}^{(1)}} = u_p$$

for every maximal solution $g_n^{(1)}$ of (1.2).
Later on we shall also need the following lemma:

LEMMA. *Let* $\Delta_n^{(i,j)}$ *(*$n=0, 1,\ldots, 1 \leqslant i,j \leqslant p$*) be the determinant we get by deleting the i-th row and j-th column of*

$$\Delta_n = \det \begin{vmatrix} f_n^{(1)} & \cdots & f_n^{(p)} \\ \vdots & & \vdots \\ f_{n+p-1}^{(1)} & \cdots & f_{n+p-1}^{(p)} \end{vmatrix}. \tag{3.9}$$

then we have

$$a) \lim_{n\to\infty} \frac{\Delta_n^{(p,i)}}{\Delta_n^{(p,j)}} = 0 \quad if \quad 1 \leqslant i < j \leqslant p, \tag{3.10}$$

$$b) \frac{\Delta_n^{(j,i)}}{\Delta_n^{(p,i)}} = \phi_{p-j}(v_1,\ldots,v_{i-1},v_{i+1},\ldots,v_p)(1+o(1)) \quad (n\to\infty) \tag{3.11}$$

if $1 \leqslant i \leqslant p$ and $1 \leqslant j \leqslant p-1$,

$$c) \frac{\Delta_n^{(p,i)}}{\Delta_{n-1}^{(p,i)}} = \phi_{p-1}(v_1,\ldots,v_{i-1},v_{i+1},\ldots,v_p)(1+o(1)) \quad (n\to\infty) \tag{3.12}$$

if $1 \leqslant i \leqslant p$.

The ϕ_i *are the elementary symmetric functions defined in appendix 2.*

PROOF. This follows immediately from the Perron-Kreuser theorem.

□

Since $\{\{A_n^{(1)}\},\ldots,\{A_n^{(p)}\}\}$ is a fundamental system of solutions for the recurrence relation (1.1), it follows that the sequences $\{\Gamma_n^{(i,1)}\},\ldots,\{\Gamma_n^{(i,p)}\}$ $(1 \leqslant i \leqslant p)$ are linearly independent (Wimp [12]). The solutions $f_n^{(1)},\ldots,f_n^{(p)}$ also form a fundamental system, so the same thing is true for the sequences $\{\Delta_n^{(i,1)}\},\ldots,\{\Delta_n^{(i,p)}\}$, $(1 \leqslant i \leqslant p)$. Furthermore we have

$$\text{vect}\,\{\{\Gamma_n^{(i,1)}\},\ldots,\{\Gamma_n^{(i,p)}\}\} = \text{vect}\,\{\{\Delta_n^{(i,1)}\},\ldots,\{\Delta_n^{(i,p)}\}\} \tag{3.13}$$

where for a set U vect U means the vectorspace spanned by the elements of U. This means that for $i=1,\ldots,p$

$$\Gamma_n^{(i,j)} = \sum_{h=1}^{p} \gamma_h^{(j)} \Delta_n^{(i,h)} \tag{3.14}$$

for some $\gamma_1^{(j)},\ldots,\gamma_p^{(j)}$. Note that the coefficients $\gamma_h^{(j)}$ do not depend on i. From the convergence of the G-continued fraction and its tails $t^{(1)},\ldots,t^{(p-2)}$ we may deduce that for $i=1,\ldots,p-1$ we have

$$\prod_{h=0}^{i-1} \xi_h = (-1)^i \lim_{n\to\infty} \frac{\Gamma_n^{(p,i+1)}}{\Gamma_n^{(p,1)}} \tag{3.15}$$

with

$$\frac{\Gamma_n^{(p,i+1)}}{\Gamma_n^{(p,1)}} = \frac{\gamma_p^{(i+1)}\Delta_n^{(p,p)} + \sum_{h=1}^{p-1}\gamma_h^{(i+1)}\Delta_n^{(p,h)}}{\gamma_p^{(1)}\Delta_n^{(p,p)} + \sum_{h=1}^{p-1}\gamma_h^{(1)}\Delta_n^{(p,h)}}\,. \tag{3.16}$$

As a consequence of (3.10) and (3.15) it follows that $\gamma_p^{(1)}\neq 0$ and hence

$$\text{vect}\,\{\{\Gamma_n^{(i,1)}\},\{\Delta_n^{(i,1)}\},\ldots,\{\Delta_n^{(i,p-1)}\}\}=\text{vect}\,\{\{\Delta_n^{(i,1)}\},\ldots,\{\Delta_n^{(i,p)}\}\}\,. \tag{3.17}$$

Hence for $j=2,\ldots,p$ there exist numbers $\zeta_1^{(j)},\ldots,\zeta_p^{(j)}$ for which

$$\Gamma_n^{(i,j)} = \zeta_1^{(j)}\Gamma_n^{(i,1)} + \sum_{h=2}^{p}\zeta_h^{(j)}\Delta_n^{(i,p-h+1)},\quad i=1,\ldots,p \tag{3.18}$$

and with

$$\forall j\in\{2,\ldots,p\}\,\exists t\in\{2,\ldots,p\}(\zeta_h^{(j)}=0\ (h=2,\ldots,t-1)\wedge\zeta_t^{(j)}\neq 0)\,. \tag{3.19}$$

Furthermore we have ($i=1,\ldots,p$ and $j=1,\ldots,p-1$).

$$\lim_{n\to\infty}\frac{\Delta_n^{(i,j)}}{\Gamma_n^{(i,1)}} = 0\ ,\quad \frac{\Gamma_n^{(p,1)}}{\Gamma_{n-1}^{(p,1)}} = \phi_{p-1}(\nu_1,\ldots,\nu_{p-1})(1+o(1)) \tag{3.20}$$

for $n\to\infty$.

The proof is similar to the proof of the main theorem in [1]. It is in three parts:

(a) $S_{m+j}(0,\ldots,0)-S_j(0,\ldots,0)\neq 0$ for $j\geqslant j_0$, $m\geqslant m_0$.

(b) $\displaystyle\lim_{m\to\infty}\frac{S_{m+j}(0,\ldots,0)-S_j(u_p,\ldots,u_p)}{S_{m+j}(0,\ldots,0)-S_j(0,\ldots,0)}\in\mathbb{C}$.

(c) the proof of (3.8).

Using (1.14) we may rewrite the expression in (a) as

$$\begin{aligned} S_j(0,\ldots,0)-S_{m+j}(0,\ldots,0) &= \frac{\Gamma_{m+j+1}^{(1,2)}}{\Gamma_{m+j+1}^{(1,1)}} - \frac{\Gamma_{j+1}^{(1,2)}}{\Gamma_{j+1}^{(1,1)}} \\ &= \frac{\Gamma_{m+j+2}^{(p,2)}}{\Gamma_{m+j+2}^{(p,1)}} - \frac{\Gamma_{j+2}^{(p,2)}}{\Gamma_{j+2}^{(p,1)}} \\ &= \frac{\Gamma_{m+j+2}^{(p,2)}\Gamma_{j+2}^{(p,1)} - \Gamma_{j+2}^{(p,2)}\Gamma_{m+j+2}^{(p,1)}}{\Gamma_{m+j+2}^{(p,1)}\Gamma_{j+2}^{(p,1)}}\,. \end{aligned}$$

Equation (3.20) implies that the denominator is different from zero for $j\geqslant j_1$ and $m\geqslant 0$. For the numerator we get, using (3.18):

$$\begin{aligned} &\Gamma_{m+j+2}^{(p,2)}\Gamma_{j+2}^{(p,1)} - \Gamma_{j+2}^{(p,2)}\Gamma_{m+j+2}^{(p,1)} \\ &= \sum_{h=t}^{p}\zeta_h^{(2)}(\Delta_{m+j+2}^{(p,p-h+1)}\Gamma_{j+2}^{(p,1)} - \Delta_{j+2}^{(p,p-h+1)}\Gamma_{m+j+2}^{(p,1)}) \end{aligned}$$

$$= \Gamma^{(p,1)}_{m+j+2}\Delta^{(p,p-t+1)}_{j+2}\sum_{h=t}^{p}\zeta_h^{(2)}\left(\frac{\Delta^{(p,p-h+1)}_{m+j+2}}{\Gamma^{(p,1)}_{m+j+2}}\frac{\Gamma^{(p,1)}_{j+2}}{\Delta^{(p,p-t+1)}_{j+2}} - \frac{\Delta^{(p,p-h+1)}_{j+2}}{\Delta^{(p,p-t+1)}_{j+2}}\right).$$

From (3.13) en (3.20) it follows that the factor before the summation sign is different from zero for $j \geqslant j_2 \geqslant j_1$ and $m \geqslant 0$. For the sum we have

$$\left|\sum_{h=t}^{p}\zeta_h^{(2)}\left(\frac{\Delta^{(p,p-h+1)}_{m+j+2}}{\Gamma^{(p,1)}_{m+j+2}}\frac{\Gamma^{(p,1)}_{j+2}}{\Delta^{(p,p-t+1)}_{j+2}} - \frac{\Delta^{(p,p-h+1)}_{j+2}}{\Delta^{(p,p-t+1)}_{j+2}}\right)\right|$$

$$= \left|\sum_{h=t}^{p}\zeta_h^{(2)}\frac{\Delta^{(p,p-h+1)}_{j+2}}{\Delta^{(p,p-t+1)}_{j+2}}\left(\frac{\Delta^{(p,p-h+1)}_{m+j+2}}{\Gamma^{(p,1)}_{m+j+2}}\frac{\Gamma^{(p,1)}_{j+2}}{\Delta^{(p,p-h+1)}_{j+2}} - 1\right)\right|$$

$$\geqslant \frac{1}{2}\left|\zeta_t^{(2)}\left(\frac{\Delta^{(p,p-t+1)}_{m+j+2}}{\Gamma^{(p,1)}_{m+j+2}}\frac{\Gamma^{(p,1)}_{j+2}}{\Delta^{(p,p-t+1)}_{j+2}} - 1\right)\right|$$

$$\geqslant \frac{1}{4}\left|\zeta_t^{(2)}\right| > 0$$

from some m and j on. The inequalities are consequences of (3.10) and (3.20). This proves (a).

We rewrite the expression in (b) using (1.14):

$$\frac{S_{m+j}(0,\ldots,0) - S_j(u_p,\ldots,u_p)}{S_{m+j}(0,\ldots,0) - S_j(0,\ldots,0)}$$

$$= \left(\sum_{h=1}^{p}(-u_p)^{h-1}\frac{\Gamma^{(h,1)}_{j+1}}{\Gamma^{(1,1)}_{j+1}}\right)^{-1}\sum_{h=1}^{p}(-u_p)^{h-1}z_{m,j,h} \tag{3.21}$$

where

$$z_{m,j,h} = \frac{\Gamma^{(1,2)}_{m+j+1}\Gamma^{(h,1)}_{j+1} - \Gamma^{(1,1)}_{m+j+1}\Gamma^{(h,2)}_{j+1}}{\Gamma^{(1,2)}_{m+j+1}\Gamma^{(1,1)}_{j+1} - \Gamma^{(1,1)}_{m+j+1}\Gamma^{(1,2)}_{j+1}}$$

$$= \frac{\sum_{i=t}^{p}\zeta_i^{(2)}\left(\frac{\Delta^{(1,p-i+1)}_{m+j+1}}{\Gamma^{(1,1)}_{m+j+1}}\Gamma^{(h,1)}_{j+1} - \Delta^{(h,p-i+1)}_{j+1}\right)}{\sum_{i=t}^{p}\zeta_i^{(2)}\left(\frac{\Delta^{(1,p-i+1)}_{m+j+1}}{\Gamma^{(1,1)}_{m+j+1}}\Gamma^{(1,1)}_{j+1} - \Delta^{(1,p-i+1)}_{j+1}\right)}.$$

For the factor before the summation sign in (3.21) it is easy to prove that

$$\frac{\Gamma^{(h,1)}_{j+1}}{\Gamma^{(1,1)}_{j+1}} = \frac{\Delta^{(h,p)}_{j+1}}{\Delta^{(1,p)}_{j+1}}(1+o(1))$$

$$= \frac{\Delta^{(h,p)}_{j+1}}{\Delta^{(p,p)}_{j+1}}\frac{\Delta^{(p,p)}_{j+1}}{\Delta^{(1,p)}_{j+1}}(1+o(1))$$

$$= \frac{\phi_{p-h}(v_1(j),\ldots,v_{p-1}(j))}{\phi_{p-1}(v_1(j),\ldots,v_{p-1}(j))}(1+o(1))\quad (j\to\infty)$$

(here we use (3.10–12)).

Using (3.5) and the special form of the elementary symmetric functions we find:

$$\lim_{j\to\infty}\left(\sum_{h=1}^{p}(-u_p)^{h-1}\frac{\Gamma_{j+1}^{(h,1)}}{\Gamma_{j+1}^{(1,1)}}\right)$$

$$=\sum_{h=1}^{p-e_{\sigma-1}}(-u_p)^{h-1}\frac{\phi_{p-e_{\sigma-1}-h}(u_{e_{\sigma-1}+1},\ldots,u_{p-1})}{\phi_{p-e_{\sigma-1}-1}(u_{e_{\sigma-1}+1},\ldots,u_{p-1})}=Q\neq 0$$

since $u_p\neq u_i$ for $i=e_{\sigma-1}+1,\ldots,p-1$. For the second factor in (3.21) we have

$$\lim_{m\to\infty} z_{m,j,h}=\frac{\sum_{i=t}^{p}\zeta_i^{(2)}\Delta_{j+1}^{(h,p-i+1)}}{\sum_{i=t}^{p}\zeta_i^{(2)}\Delta_{j+1}^{(1,p-i+1)}}$$

$$=\frac{\sum_{i=t}^{p}\zeta_i^{(2)}\Delta_{j+1}^{(h,p-i+1)}/\Delta_{j+1}^{(1,p-t+1)}}{\sum_{i=t}^{p}\zeta_i^{(2)}\Delta_{j+1}^{(1,p-i+1)}/\Delta_{j+1}^{(1,p-t+1)}}$$

with

$$\left|\sum_{i=t}^{p}\zeta_i^{(2)}\Delta_{j+1}^{(1,p-i+1)}/\Delta_{j+1}^{(1,p-t+1)}\right|\geqslant\frac{1}{2}\left|\zeta_t^{(2)}\right|>0$$

for $j\geqslant j_0$. This proves (b). Furthermore we have

$$\lim_{j\to\infty}\lim_{m\to\infty} z_{m,j,h}=\lim_{j\to\infty}\frac{\sum_{i=t}^{p}\zeta_i^{(2)}\Delta_{j+1}^{(h,p-i+1)}/\Delta_{j+1}^{(1,p-t+1)}}{\sum_{i=t}^{p}\zeta_i^{(2)}\Delta_{j+1}^{(1,p-i+1)}/\Delta_{j+1}^{(1,p-t+1)}}$$

$$=\lim_{j\to\infty}\frac{\zeta_t^{(2)}\Delta_{j+1}^{(h,p-t+1)}/\Delta_{j+1}^{(1,p-t+1)}}{\zeta_t^{(2)}}$$

since

$$\frac{\Delta_{j+1}^{(h,p-i+1)}}{\Delta_{j+1}^{(1,p-t+1)}}=\frac{\Delta_{j+1}^{(h,p-i+1)}}{\Delta_{j+1}^{(1,p-i+1)}}\frac{\Delta_{j+1}^{(1,p-i+1)}}{\Delta_{j+1}^{(1,p-t+1)}}$$

$$=\frac{\Delta_{j+1}^{(h,p-i+1)}}{\Delta_{j+1}^{(p,p-i+1)}}\frac{\Delta_{j+1}^{(p,p-i+1)}}{\Delta_{j+1}^{(1,p-i+1)}}\frac{\Delta_{j+1}^{(1,p-i+1)}}{\Delta_{j+1}^{(1,p-t+1)}} \tag{3.22}$$

$$=\frac{\phi_{p-h}(v_1,\ldots,v_{p-i},v_{p-i+2},\ldots,v_p)}{\phi_{p-1}(v_1,\ldots,v_{p-i},v_{p-i+2},\ldots,v_p)}\frac{\Delta_{j+1}^{(1,p-i+1)}}{\Delta_{j+1}^{(1,p-t+1)}}(1+o(1))\quad(j\to\infty)$$

where we have used (3.14). For $i>t$ and $h=1,\ldots,p$ this gives

$$\lim_{j\to\infty}\frac{\Delta_{j+1}^{(h,p-t+1)}}{\Delta_{j+1}^{(1,p-t+1)}}=0\,.$$

From (3.22) we get

$$\lim_{j\to\infty}\lim_{m\to\infty} z_{m,j,h}$$

$$=\lim_{j\to\infty}\frac{\phi_{p-h}(v_1,\ldots,v_{p-t},v_{p-t+2},\ldots,v_p)}{\phi_{p-1}(v_1,\ldots,v_{p-t},v_{p-t+2},\ldots,v_p)}$$

$$=\frac{\phi_{p-e_{\sigma-1}+1-h}(u_{e_{\sigma-1}+1},\ldots,u_p)}{\phi_{p-e_{\sigma-1}}(u_{e_{\sigma-1}+1},\ldots,u_p)} \quad \text{if } p-t+1\leqslant e_{\sigma-1} \text{ and } h=1,\ldots,p-e_{\sigma-1}+1$$

$$=0 \quad \text{if } p-t+1\leqslant e_{\sigma-1} \text{ and } h=p-e_{\sigma-1}+2,\ldots,p$$

$$=\frac{\phi_{p-e_{\sigma-1}-h}(u_{e_{\sigma-1}+1},\ldots,u_{p-t},u_{p-t+2},\ldots,u_p)}{\phi_{p-e_{\sigma-1}-1}(u_{e_{\sigma-1}+1},\ldots,u_{p-t},u_{p-t+2},\ldots,u_p)}$$

$$\text{if } p-t+1>e_{\sigma-1} \text{ and } h=1,\ldots,p-e_{\sigma-1}$$

$$=0 \quad \text{if } p-t+1>e_{\sigma-1} \text{ and } h=p-e_{\sigma-1}+1,\ldots,p$$

Now we prove (c). From (1.14) we deduce

$$\lim_{j\to\infty}\frac{\xi_0-S_j(u_p,\ldots,u_p)}{\xi_0-S_j(0,\ldots,0)}$$

$$=\lim_{j\to\infty}\lim_{m\to\infty}\frac{S_{m+j}(0,\ldots,0)-S_j(u_p,\ldots,u_p)}{S_{m+j}(0,\ldots,0)-S_j(0,\ldots,0)}$$

$$=\lim_{j\to\infty}\left\{\left[\sum_{h=1}^{p}(-u_p)^{h-1}\frac{\Gamma_{j+1}^{(h,1)}}{\Gamma_{j+1}^{(1,1)}}\right]^{-1}\sum_{h=1}^{p}(-u_p)^{h-1}z_{m,j,h}\right\}$$

$$\begin{cases} =Q^{-1}\displaystyle\sum_{h=1}^{p-e_{\sigma-1}+1}(-u_p)^{h-1}\frac{\phi_{p-e_{\sigma-1}+1-h}(u_{e_{\sigma-1}+1},\ldots,u_p)}{\phi_{p-e_{\sigma-1}}(u_{e_{\sigma-1}+1},\ldots,u_p)} & \text{if } p-t+1\leqslant e_{\sigma-1}\\ =Q^{-1}\displaystyle\sum_{h=1}^{p-e_{\sigma-1}}(-u_p)^{h-1}\frac{\phi_{p-e_{\sigma-1}-h}(u_{e_{\sigma-1}+1},\ldots,u_{p-t},u_{p-t+2},\ldots,u_p)}{\phi_{p-e_{\sigma-1}-1}(u_{e_{\sigma-1}+1},\ldots,u_{p-t},u_{p-t+2},\ldots,u_p)} & \\ \qquad\qquad \text{if } p-t+1>e_{\sigma-1}\,. \end{cases}$$

Both expressions are zero since $t>1$. □

REMARK. If the conditions of theorem 5 are satisfied, we get the following expression for the rate of convergence of the modified Miller algorithm (1.8):

$$\frac{S_{n+1}(0,\ldots,0)-\xi_0}{S_n(0,\ldots,0)-\xi_0}=\frac{v_p(n)}{v_{p-t+1}(n)}(1+o(1))\ \ (n\to\infty)\,. \tag{3.23}$$

where t is the smallest index for which $\zeta_t^{(2)}$ is different from zero in (3.18).

This follows immediately from:

$$\frac{S_{n+1}(0,...,0) - S_n(0,...,0)}{S_n(0,...,0) - S_{n-1}(0,...,0)} = \frac{v_p(n)}{v_{p-t+1}(n)}(1+o(1)) \quad (n \to \infty)$$

(see [6]). In most cases $t = 2$.
The convergence will be slow if $\lim_{n\to\infty} |v_p(n)/v_{p-t+1}(n)|$ is close to one. In this case the modification (3.7) will be very effective (see example 1 and 2).

It is a consequence of theorem 5 that the Miller algorithm (1.4) may be improved by using

$$y_{N+i}^N = w_{N+i-2} \ , i = 2,..., p$$
$$y_{N+1}^N = 1$$

as starting values instead of (1.4), or, better still, by using

$$b^N(N+i) = w_{N+i-1} \ , i = 1,..., p-1$$

in (1.8) to calculate the value of the G-continued fraction.

EXAMPLE 1. Let $p = 3$ and let the coefficients of the recurrence relation (1.1) be given by

$$\begin{cases} a_0(n) = 1 \\ a_1(n) = -3 - (\frac{1}{2})^n \\ a_2(n) = 2.99 + (\frac{2}{3})^n \\ a_3(n) = -0.99 \end{cases} \qquad (3.24)$$

for $n = 0, 1, \ldots$. This recurrence relation is of Poincaré-type, and the characteristic equation has the roots 0.9, 1.0, 1.1. In table 1 we compare the values of $S_n(0,0)$ and $S_n(0.9,0.9)$. In table 2 we have calculated approximations to f_n with
(first column) N=60 in (1.8) and $b^N(N+1) = b^N(N+2) = 0.9 : \tilde{f}_n^{60}$;
(second column) N=60 in (1.8) and $b^N(N+1) = b^N(N+2) = 0.0 : f_n^{60}$;
The third column contains the exact values of f_n ($n = 0,..., 60$).

REMARK. Another modification which leads to convergence acceleration (if the conditions of theorem 5 are satisfied) is for instance

$$\{S_n(\tilde{w}_n, \tilde{w}_{n+1}, \ldots, \tilde{w}_{n+p-2})\}_{n=0}^{\infty}$$

with

$$\tilde{w}_n = \frac{a_p(n+1)}{a_p(n+2)} \cdot \frac{g_{n+1}}{n^{\alpha_\sigma} g_{n+2}} \qquad (3.25)$$

where g_n is a maximal solution of the adjoint equation (1.2). This choice is based on the remark following equation (3.8).

n	$S_n(0.0, 0.0)$	$S_n(0.9, 0.9)$
10	0.412422676037	0.412303308250
40	0.412342767116	0.412340090669
50	0.412341002014	0.412340090688
100	0.412340095336	0.412340090688
150	0.412340090712	0.412340090688
180	0.412340090689	0.412340090688
190	0.412340090688	0.412340090688
200	0.412340090688	0.412340090688

Table 1. Approximants of the G-continued fraction associated with (3.24) using different tail-values.

n	$\tilde{f}_n^{60}$	f_n^{60}	f_n
0	0.100000000000d+01	0.100000000000d+01	0.100000000000d+01
1	0.412340090688d+00	0.412340406101d+00	0.412340090688d+00
5	0.307922668624d-01	0.307928697631d-01	0.307922668624d-01
10	0.647194076603d-02	0.646724645155d-02	0.647194076601d-02
20	0.178779151982d-02	0.176172137206d-02	0.178779151976d-02
30	0.620614150743d-03	0.584484660572d-03	0.620614150662d-03
40	0.216378141870d-03	0.178049747611d-03	0.216378141783d-03
50	0.754462924576d-04	0.399944952965d-04	0.754462923761d-04
60	0.263064949898d-04	0.131242019188d-05	0.263064949289d-04

Table 2. Approximations to the solution f_n of (3.24) calculated with the G-continued fraction (2.1) using different values for the tails.

EXAMPLE 2. $p = 3$ and the coefficients of the recurrence relation are given by:

$$\begin{cases} a_0(n) &= 1 \\ a_1(n) &= -((n+1)^2 - \dfrac{1}{(10n+11)^2}) \\ a_2(n) &= 1.9(n+1)^2 + \dfrac{2}{(10n+11)^3} \\ a_3(n) &= -0.9(n+1)^2 \end{cases} \tag{3.26}$$

for $n = 0, 1, \ldots$. In this example $u_p = 0.9$. In table 3 we compare the values of $S_n(0,0)$, $S_n(0.9,0.9)$ and $S_n(\tilde{w}_n, \tilde{w}_{n+1})$, with $\tilde{w}_n$ as defined in (3.25), and g_n the solution of (1.2) with initial values $g_0 = g_1 = 0$, $g_2 = 1$.

n	$S_n(0.0, 0.0)$	$S_n(0.9, 0.9)$	$S_n(\tilde{w}_n, \tilde{w}_{n+1})$
1	0.628602892889	0.491031523661	0.571542043041
50	0.515131923352	0.515123928266	0.515124282368
100	0.515124129245	0.515124095910	0.515124096211
133	0.515124097105	0.515124096124	0.515124096129
134	0.515124097007	0.515124096124	0.515124096129
135	0.515124096918	0.515124096125	0.515124096128
153	0.515124096245	0.515124096127	0.515124096128
154	0.515124096233	0.515124096127	0.515124096128
155	0.515124096222	0.515124096128	0.515124096128
200	0.515124096129	0.515124096128	0.515124096128
201	0.515124096129	0.515124096128	0.515124096128
202	0.515124096128	0.515124096128	0.515124096128

Table 3. Approximants of the G-continued fraction associated with (3.26) using different tail-values.

APPENDIX 1: THE PERRON-KREUSER THEOREM

The recurrence relation (1.1) is said to be of Perron-Kreuser type if its coefficients satisfy

$$a_i(n) = d_i n^{m_i}(1 + o(1)) \quad (n \to \infty), \quad i = 1,\dots, p \tag{A1}$$

where the d_i are real or complex numbers. The m_i are real or $-\infty$ with $m_p > -\infty$. With such a recurrence relation we can associate a uniquely defined Newton-Puiseux polygon in a rectangular (x,y)-coordinate system :

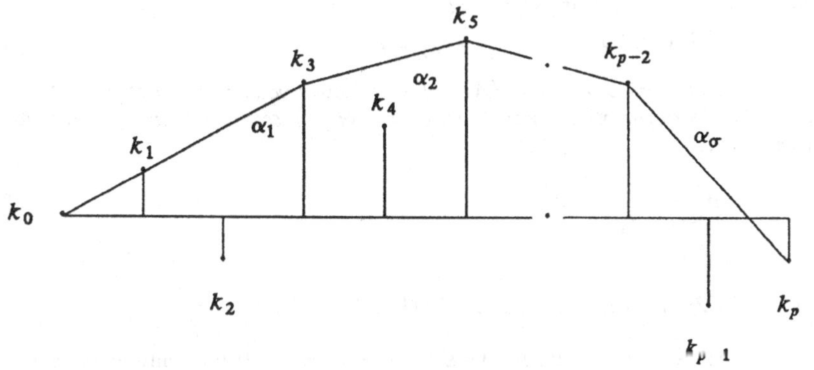

Let the points $k_0, k_1, \ldots, k_p$ be defined by $x = i$, $y = m_i$ $(i= 0,\ldots, p)$ with $m_0 = 0$. Then some of the points $k_0, \ldots, k_p$ are connected with linear segments in such a way that the resulting polygon is concave downwards, each point k_i $(i= 1,\ldots, p-1)$ being on or below the resulting figure, k_0 and k_p being the endpoints of the figure . If the polygon so constructed has σ distinct linear segments, their respective slopes are denoted by $\alpha_1, \ldots, \alpha_\sigma$ with $\alpha_1 > \alpha_2 \cdots > \alpha_\sigma$, and their abscissas are denoted by $0 = e_0 < e_1 < \cdots < e_\sigma = p$. It follows that

$$\alpha_\lambda = \frac{m_{e_\lambda} - m_{e_{\lambda-1}}}{e_\lambda - e_{\lambda-1}}, \quad \lambda = 1,\ldots, \sigma \tag{A2}$$

and the Perron-Kreuser theorem states :

THEOREM (Perron [7,8], Kreuser [4]). *A linear recurrence relation with the above Newton-Puiseux polygon has a fundamental system of p solutions which fall into σ classes. Each of these classes is further broken into subclasses, the λ–th class $(\lambda= 1,\ldots, \sigma)$ containing β_λ subclasses. Let $c_\gamma^{(\lambda)}$ $(\gamma= 1,\ldots, \beta_\lambda)$ denote the number of linearly independent solutions in the γ–th subclass of the λ–th class. Then each of the $c_\gamma^{(\lambda)}$ solutions f_n and their nonzero linear combinations satisfy*

$$\limsup_{n\to\infty} \left(\frac{|f_n|}{(n!)^{\alpha_\lambda}} \right)^{\frac{1}{n}} = \rho_\gamma^{(\lambda)}. \tag{A3}$$

Here the $\rho_\gamma^{(\lambda)}$ are distinct positive numbers which are the moduli of the roots of the following characteristic equation corresponding to the λ–th class :

$$r_0^{(\lambda)} x^{e_\lambda - e_{\lambda-1}} + r_1^{(\lambda)} x^{e_\lambda - e_{\lambda-1}-1} + \cdots + r_{e_\lambda - e_{\lambda-1}}^{(\lambda)} = 0, \tag{A4}$$

where $r_i^{(\lambda)} = d_{e_{\lambda-1}+i}$ or 0 depending on whether the point $(e_{\lambda-1}+i, m_{e_{\lambda-1}+i})$ falls, respectively, on or below the λ–th side of the Newton-Puiseux polygon. The number $c_\gamma^{(\lambda)}$ is equal to the number of roots, counting their multiplicities, of (A4) with absolute value $\rho_\gamma^{(\lambda)}$. Thus, it follows that

$$c_1^{(\lambda)} + c_2^{(\lambda)} + \cdots + c_{\beta_\gamma}^{(\lambda)} = e_\lambda - e_{\lambda-1}. \tag{A5}$$

Further, to each simple root u of (A4) whose absolute value is distinct from the absolute values of the other roots, there corresponds a solution f_n in the λ–th class which satisfies

$$\lim_{n\to\infty} \frac{f_{n+1}}{n^{\alpha_\lambda} f_n} = u. \tag{A6}$$

APPENDIX 2: ELEMENTARY SYMMETRIC FUNCTIONS

The elementary symmetric functions for a set of m arbitrary complex numbers $x_1, \ldots, x_m$ are defined as

$$\phi_0(x_1,\ldots,x_m) = 1$$
$$\phi_1(x_1,\ldots,x_m) = x_1 + \cdots + x_m$$
$$\phi_2(x_1,\ldots,x_m) = x_1x_2 + x_1x_3 + \cdots + x_{m-1}x_m$$
$$\ldots$$
$$\phi_m(x_1,\ldots,x_m) = x_1x_2 \cdots x_m.$$

If we construct from these ϕ_i the polynomial

$$P_{x_1,\ldots,x_m}(x) = x^m - \phi_1 x^{m-1} + \phi_2 x^{m-2} + \cdots + (-1)^m \phi_m$$

then this polynomial has the roots $x_1,\ldots,x_m$, and so we have for $i = 1,\ldots,m$:

$$P_{x_1,\ldots,x_m}(x_i) = x_i^m - \phi_1 x_i^{m-1} + \phi_2 x_i^{m-2} + \cdots + (-1)^m \phi_m = 0\,.$$

This can be seen as a system of m equations in the m unknown ϕ_i ($i = 1,\ldots,m$). If all the x_i are different we can solve this system using Cramer's rule:

$$\phi_{m-i+1} = \frac{\det V^{(i)}}{\det V^{(m+1)}} \qquad i=1,\ldots,m$$

where $V^{(i)}$ is the square matrix arising from the matrix

$$V = \begin{pmatrix} 1 & 1 & . & 1 \\ x_1 & x_2 & . & x_m \\ . & . & . & . \\ x_1^m & x_2^m & . & x_m^m \end{pmatrix}$$

by removing the i-th row.

REFERENCES

[1] de Bruin, M. G. & Jacobsen L., 'Modification of generalised continued fractions I', *Lect. Notes Math.* 1237, Springer-Verlag, pp 161-176, 1987.

[2] Gautschi W., 'Computational Aspects of Three-Term Recurrence Relations', *SIAM Review* 9, pp 24-82, 1967.

[3] Jacobsen, L., 'Modified approximants for continued fractions, construction and applications', *Skr., K. Nor. Vidensk. Selsk. no.* 3, 1983.

[4] Kreuser, P., *Uber das Verhalten der Integrale homogener linearer Differenzengleichungen im Unendlichen*, Thesis (Tubingen), Borna-Leipzig, 1914.

[5] Levrie P., *Het numeriek oplossen van lineaire recursiebetrekkingen: een veralgemening van de kettingbreukmethode van Gautschi*, Thesis, K. U. Leuven, 1987.

[6] Niethammer, W. & Wietschorke H., 'On the acceleration of limit periodic continued fractions', *Numer. Math.* **44**, pp 129-137, 1984.

[7] Perron O., 'Uber Summengleichungen und Poincarésche Differenzengleichungen', *Math. Ann.* **84**, pp 1-15, 1921.

[8] Perron O., 'Uber lineare Differenzengleichungen und eine Anwendung auf lineare Differentialgleichungen mit Polynomkoeffizienten', *Math. Z.* **72**, pp 16-24, 1959.

[9] Scraton R.E., 'A Modification of Miller's Recurrence Algorithm', *BIT* **12**, pp 242-251, 1972.

[10] Thron, W. J. & Waadeland H., 'Accelerating convergence of limit-periodic continued fractions $K(a_n/1)$', *Numer. Math.* **34**, pp 155-170, 1980.

[11] Wall H.S., *Analytic Theory of Continued Fractions*, Van Nostrand, New York, 1948.

[12] Wimp J., *Computation with Recurrence Relations*, Pitman Advanced Publishing Program, Boston-London-Melbourne, 1984.

[13] Zahar R.V.M., 'A Mathematical Analysis of Miller's Algorithm', *Numer. Math.* **27**, pp 427-447, 1977.

[14] Zahar R.V.M., *Computational Algorithms for Linear Difference Equations*, Thesis, Purdue University, 1968.

CONVERGENCE ACCELERATION

Chairmen:

H. Waadeland

Invited communications:

C. Brezinski

A new approach to convergence acceleration methods.

Short communications:

F. Cordellier*

L' accélération de la convergence des suites par des procédés δ^2 d'Aitken et W de Lubkin.

A. Draux*

On composite sequence transformations.

A. Sidi*

Recent developments in vector extrapolation methods.

---

*Lecture notes are not included.

A NEW APPROACH TO CONVERGENCE ACCELERATION METHODS

Claude BREZINSKI
Laboratoire d'Analyse Numérique et d'Optimisation
UFR IEEA M3
Université de Lille 1
59655 Villeneuve d'Ascq - Cedex
FRANCE

ABSTRACT. A new notion, the perfect estimation of the error of a sequence, is introduced. This approach explains and relates many concepts, ideas and algorithmic procedures used for accelerating the convergence of sequences, which were indepently developed. It thus provides a more synthetic and profound view of the entire field.

1 - INTRODUCTION.

Let (S_n) be a sequence converging to S. All the methods to accelerate the convergence of (S_n) consist in transforming (S_n) into another sequence (T_n) with the hope that :

$$T_n - S = o(S_n - S) \qquad (n \to \infty).$$

Such a method is called a sequence transformation. There are two ways for obtaining sequence transformations :

1°) Let N be a set of sequences. The sequence transformation T : $(S_n) \to (T_n)$ is built such that $\forall (S_n) \in N$, then $\forall n$, $T_n = S$ where S is the limit of (S_n). The set N is called the kernel of the transformation T. Many transformations have been obtained by this procedure and quite often the numbers T_n are defined as ratios of determinants. Thus the first duty of a numerical analyst is to derive a recursive algorithm to calculate the T_n's without computing the determinants involved in their definition since a numerical analyst dont't know how to compute determinants. Then one has to look for the convergence and acceleration properties of the transformation T that is to find classes of sequences for which the transformed sequence (T_n) converges to the same limit as (S_n) and faster. See [7] for an exposition of this kind of approach and [10, 29] for a very general algorithm of this type.

2°) The second way consists in giving directly an algorithm for computing the new sequence (T_n). This was, for example, the method followed for the

A. Cuyt (ed.), Nonlinear Numerical Methods and Rational Approximation, 373–405.

θ-algorithm [5]. This approach is more difficult than the first one since one has to find the theoretical properties (and, in particular, the kernel) of the transformation from the rules of the algorithm and then, the convergence and acceleration properties. A good idea of these difficulties is given by the θ_2-algorithm [18].

The aim of this paper is to give a third approach which is very much more synthetic and thus illuminating. It is based on the obvious observation that accelerating the convergence is equivalent to estimating the error. Moreover this approach will explain and relate many concepts, ideas and algorithmic procedures which were developed independently without any link. Thus we shall arrive at a better and more profound understanting of the field of convergence acceleration of sequences.

2 - BASIC DEFINITIONS AND RESULTS.

Let us begin with some definitions and examples.

Definition 1. Let (S_n) be a sequence converging to S and let (D_n) be a sequence converging to zero.

If the ratio $(S-S_n)/D_n$ has no limit when n tends to infinity or if it tends to o or infinity then (D_n) is said to be a bad estimation of the error of (S_n). If it has a limit $a \neq 1$ then (D_n) is said to be a good estimation the error of (S_n) and if $a = 1$ it is said to be a perfect estimation of the error of (S_n). If $\forall n$, $(S-S_n)/D_n=1$ then (D_n) is said to be an exact estimation of the error of (S_n).

Definition 2. Let T be the transformation : $(S_n) \rightarrow (T_n)$. T is said to accelerate the convergence of (S_n) if $\lim_{n\to\infty} (T_n-S)/(S_n-S) = 0$. In that case (T_n) is said to converge faster than (S_n).

The problem of finding a perfect estimation of the error of (S_n) is equivalent to the problem of finding a transformation which accelerates its convergence as we shall see now.

Indeed let (T_n) be a sequence converging faster than (S_n). We set $D_n = T_n - S_n$. We have :

$$\lim_{n\to\infty} \frac{D_n}{S-S_n} = 1 + \lim_{n\to\infty} \frac{T_n-S}{S-S_n} = 1$$

which shows that (D_n) is a perfect estimation of the error of (S_n). Conversely let (D_n) be a perfect estimation of the error of (S_n). We set $T_n = S_n+D_n$. We have

$$\lim_{n\to\infty} \frac{T_n-S}{S_n-S} = 1 - \lim_{n\to\infty} \frac{D_n}{S-S_n} = 0$$

which shows that (T_n) converges faster than (S_n). This is nothing else than the well known result that if $D_n \sim (S-S_n)$ then $D_n-(S-S_n) = o(D_n) = o(S-S_n)$, (see [19] for example). More generally, if (D_n) is a perfect or a good estimation of the error of (S_n), then

$$S - S_n = a\,D_n + o(D_n)$$

which means that $a\,D_n$ is the first term in an asymptotic expansion of the error $S-S_n$.

Most of the convergence acceleration methods are in fact more complicated. They transform the sequence (S_n) to be accelerated into a set of sequences $(T_1^{(n)}), (T_2^{(n)}),\ldots$ by an algorithm of the form

$$T_o^{(n)} = S_n \qquad n = 0,1,\ldots$$

$$T_{k+1}^{(n)} = T_k^{(n)} + D_{k+1}^{(n)} \qquad k,n = 0,1,\ldots$$

where the $D_{k+1}^{(n)}$'s are obtained by some auxiliary rule [7] (It must be noticed that, in some algorithms, $T_k^{(n+1)}$ appears instead of $T_k^{(n)}$ in the above rule but this remark is without any importance for the understanding of our ideas).

The use of such an algorithm consists in assuming that, $\forall n$

$$S = S_n + D_1^{(n)} + D_2^{(n)} + \ldots$$

and to take

$$T_k^{(n)} = S_n + D_1^{(n)} + \ldots + D_k^{(n)}$$

as an approximation of S.

If each transformation converges faster than the preceding one, that is if for $k = 0,1,\ldots$

$$\lim_{n\to\infty} (T_{k+1}^{(n)} - S)/(T_k^{(n)} - S) = 0$$

then $(D_{k+1}^{(n)})$ is a perfect estimation of the error of $(T_k^{(n)})$ and $\forall i \geq 2$

$$\lim_{n\to\infty} D_{k+i}^{(n)} \,/\, D_{k+1}^{(n)} = 0.$$

This means that $\forall k$, $D_{k+1}^{(n)} = o\,(D_k^{(n)})$ when n tends to infinity or, in other words that the set $\{(D_1^{(n)}), (D_2^{(n)}),\ldots\}$ forms a scale of comparison and that $S-S_n$ has an asymptotic expansion with respect to this scale of comparison (with all the coefficients equal to one). Moreover

$$S - T_k^{(n)} = D_{k+1}^{(n)} + o(D_{k+1}^{(n)}).$$

Reciprocally if the $(D_k^{(n)})$'s form a scale of comparison and if all the coefficients equal one, each transformation converges faster than the

preceding one.
All these ideas were already given, in the case of the θ-algorithm, in [6, §. 3.5, pp. 87-88] and the above estimation of the error was used in the subroutines which can be found in [8].

Let us give a more sophisticated example showing that convergence acceleration is equivalent to the knowledge of a perfect estimation of the error. It deals with continued fractions and was, in fact, the catalyser of the ideas developed in this paper. Let C be a converging continued fraction with convergents $C_n = A_n/B_n$, n = 0,1,... and let (D_n) be a perfect estimation of the error of (C_n). We have $C = b_o + \frac{a_1|}{|b_1} + \frac{a_2|}{|b_2} + \ldots$ and

$$C - C_n = \frac{A_n + R_n A_{n-1}}{B_n + R_n B_{n-1}} - \frac{A_n}{B_n}$$

where R_n is the nth tail defined by

$$R_n = \frac{a_{n+1}|}{|b_{n+1}} + \frac{a_{n+2}|}{|b_{n+2}} + \ldots$$

Let (r_n) be the sequence defined by

$$D_n = \frac{A_n + r_n A_{n-1}}{B_n + r_n B_{n-1}} - \frac{A_n}{B_n}$$

that is $r_n = -D_n h_n/(D_n + C_n - C_{n-1})$ with $h_n = B_n/B_{n-1}$.

It is easy to see that [14]

$$\frac{S_n(r_n) - C}{C_n - C} = \frac{1 - r_n/R_n}{1 - \frac{r_n}{R_n}\frac{C_n - C}{C_{n-1} - C}} = 1 - \frac{D_n}{C - C_n}$$

with $S_n(w) = \frac{A_n + wA_{n-1}}{B_n + wB_{n-1}}$.

Thus $S_n(r_n) - C = o\,(C_n - C)$ and we see that $S_n(r_n) = C_n + D_n$. Our choice of (r_n) always accelerated the convergence. In fact it corresponds to the transformation $(C_n) \to (C_n + D_n)$

Conversely let us assume the knowledge of a sequence (r_n) such that $S_n(r_n) - C = o\,(C_n - C)$. We have

$$\frac{S_n(r_n) - C}{C_n - C} = 1 - \frac{D_n}{C - C_n}$$

with $D_n = S_n(r_n) - C_n$ and $\lim_{n\to\infty} D_n/(C-C_n) = 1$. It is easy to see that

$$D_n = - r_n \frac{C_n - C_{n-1}}{h_n + r_n} \text{ that is } S_n(r_n) = \frac{h_n C_n + r_n C_{n-1}}{h_n + r_n}$$

and that, since $h_n^{-1} = - R_n^{-1}(C_n - C)/(C_{n-1} - C)$

$$\frac{D_n}{C-C_n} = \frac{\dfrac{C_n - C}{C_{n-1} - C} - 1}{\dfrac{C_n - C}{C_{n-1} - C} - \dfrac{R_n}{r_n}}$$

which tends to one when n goes to infinity.

Let us now consider the case of the exponential series. We set

$$S_n = 1 + \frac{x}{1!} + \ldots + \frac{x^n}{n!} .$$

It has been proved by Gautschi [22] that $R_n = \frac{x^{n+1}}{(n+1)!} e^{\theta_n x}$ with $\lim_{n\to\infty} \theta_n = 0$ for all $x \neq 0$.

Thus $(x^{n+1}/(n+1)!)$ is a perfect estimation of the error of (S_n) and $T_n = S_{n+1}$. For $x = 1$ we obtain :

n	$S_n - e$	$T_n - e$	$(T_n - e)/(S_n - e)$
1	-0.71828	-0.21828	0.30389
5	$-0.16152\ 10^{-2}$	$-0.22627\ 10^{-3}$	0.14009
10	$-0.27313\ 10^{-7}$	$-0.22606\ 10^{-8}$	$0.82766\ 10^{-1}$
15	$-0.50848\ 10^{-13}$	$-0.31086\ 10^{-14}$	$0.61135\ 10^{-1}$

As a consequence we showed that (S_n) converges superlinearly that is $\lim_{n\to\infty} (e^x - S_{n+1})/(e^x - S_n) = 0$. Moreover $\forall \varepsilon > 0$, $\lim_{n\to\infty} (e^x - S_{n+1})/(e^x - S_n)^{1+\varepsilon} = \infty$.

Any (D_n) of the form $D_n = a_n x^{n+1}/(n+1)!$ with $\lim_{n\to\infty} a_n = 1$ is a perfect estimation of the error of (S_n). However since the faster (a_n) tends to 1 the faster (T_n) converges, then the best choice is $\forall n$, $a_n = 1$.

The same idea can be applied to other MacLaurin series inside the disc of convergence. We set, for a fixed value of x

$$S_n = f(0) + \frac{x}{1!} f(o) + \ldots + \frac{x^n}{n!} f^{(n)}(0) , \qquad S = f(x)$$

and

$$R_n = S - S_n = \frac{x^{n+1}}{(n+1)!} f^{(n+1)}(\theta_n x) \text{ with } 0 < \theta_n < 1.$$

Thus $(D_n = \frac{x^{n+1}}{(n+1)!} f^{(n+1)}(0))$ is a perfect estimation of the error of (S_n) if and only if $\lim_{n\to\infty} f^{(n+1)}(\theta_n x)/f^{(n+1)}(0) = 1$ or, in other words, if and only if (S_n) converges superlinearly.

If $(m_n)_n$ and $(M_n)_n$ are known such that, $\forall n$

$$m_n \le f^{(n+1)}(\theta_n x) \le M_n$$

then $(D_n = m_n x^{n+1}/(n+1)!)$ and $(D'_n = M_n x^{n+1}/(n+1)!)$ are perfect estimations of the error of (S_n) if $\lim_{n\to\infty} M_n/m_n = 1$.

The estimations (D_n) and (D'_n) obtained with $m_n = f^{(n+1)}(0)$ and $M_n = f^{(n+1)}(x)$ are not, in general, perfect but usually only good ones. For example if

$$f(x) = c_o + c_1 x + c_2 x^2 + \ldots$$

$$S_n = c_o + c_1 + \ldots + c_n x^n$$

$$R_n = c_{n+1} x^{n+1} + c_{n+2} x^{n+2} + \ldots$$

We have $D_n = c_{n+1} x^{n+1}$ if we take $m_n = f^{(n+1)}(0)$. Thus if we assume that $\exists \xi$, $\lim_{n\to\infty} c_{n+1}/c_n = \xi$ then

$$\frac{R_n}{D_n} = 1 + \frac{c_{n+2}}{c_{n+1}} x + \frac{c_{n+3}}{c_{n+1}} x^2 + \ldots, \quad \lim_n R_n/D_n = (1-\xi x)^{-1}$$

if $|x| < |\xi^{-1}|$ (which means that x belongs to the disc of convergence of f). Thus (D_n) is a good estimation of the error of (S_n). As we shall see in the next section $(D_n^1 = -\frac{\Delta S_n}{\Delta D_n} D_n)$ is a perfect estimation of the error of (S_n) and we get

$$T_n = S_n - \frac{\Delta S_n}{\Delta D_n} D_n = c_o + c_1 x + \ldots + c_n x^n + x^{n+1} \frac{c_{n+1}}{1 - c_{n+2} x / c_{n+1}} = [n+1/1]_f(x)$$

which is the classical Padé approximant of f with a numerator of degree n+1 and a denominator of degree one.

The ideas developed in this section also apply to Padé-type approximation. The definitions and notations are those of [11]. We have

$$(n+k/k)_f(t) = \sum_{i=o}^{n} c_i t^i + t^{n+1} (k-1/k)_{f_n}(t)$$

with $f_n(t) = c_{n+1} + c_{n+2} t + \ldots$, $v_k^{(n+1)}$ the generating polynomial of $(k-1/k)_{f_n}$ and $w_k^{(n+1)}$ its associated polynomial.

When k is fixed and n tends to infinity the sequence $((n+k/k)_f)_n$ converges faster than the sequence of the partial sums $S_n(t) = \sum_{i=o}^{n} c_i t^i$ if and only if $(t^{n+1} (k-1/k)_{f_n}(t))_n$ is a perfect estimation of the error of $(S_n(t))$ that is, since $f(t) - S_n(t) = t^{n+1} f_n(t)$

$$\lim_{n\to\infty} (k-1/k)_{f_n}(t)/f_n(t) = 1.$$

We set $(k-1/k)_{f_n}(t) = e_{n+1}^{(n)} + e_{n+2}^{(n)} t + \ldots$ where the $e_i^{(n)}$ also depend on the index k. We have $e_i^{(n)} = c_i$ for $i = n+1, \ldots, n+k$, and

$$\frac{(k-1/k)_{f_n}(t)}{f_n(t)} = \frac{1 + \dfrac{c_{n+2}}{c_{n+1}} + \ldots + \dfrac{c_{n+k}}{c_{n+1}} t^{k-1} + \dfrac{e_{n+k+1}^{(n)}}{c_{n+1}} t^k + \ldots}{1 + \dfrac{c_{n+2}}{c_{n+1}} t + \dfrac{c_{n+3}}{c_{n+1}} t^2 + \ldots}$$

Let us assume that $\exists \xi$ such that $\lim_{n\to\infty} c_{n+1}/c_n = \xi$ and that $\lim_n e_{n+1}^{(n)}/c_n = \xi$.

Since $\dfrac{e_{n+k+i}^{(n)}}{c_{n+1}} = \dfrac{e_{n+k+i}^{(n)}}{c_{n+k+i-1}} \dfrac{c_{n+k+i-1}}{c_{n+k+i-2}} \cdots \dfrac{c_{n+2}}{c_{n+1}}$ then $\lim_{n\to\infty} \dfrac{e_{n+k+i}^{(n)}}{c_{n+1}} =$

$\lim_{n\to\infty} \dfrac{c_{n+k+i}}{c_{n+1}} = \xi^{k+i-1}$ and $\lim_{n\to\infty} \dfrac{(k-1/k)_{f_n}(t)}{f_n(t)} = \dfrac{1+\xi t+\xi^2 t^2 + \ldots}{1+\xi t+\xi^2 t^2 + \ldots} = 1$ if

$|t| < R = |\xi^{-1}|$ the radius of convergence of the series f. The same results hold for Padé approximants the only difference being that $e_i^{(n)} = c_i$ for $i = n+1, \ldots, n+2k$. The main and most difficult point is to check whether the condition $\lim_{n\to\infty} e_{n+1}^{(n)}/c_n = \xi$ is satisfied or not.

3 - QUASI-LINEAR TRANSFORMATIONS.

Let us consider the sequence transformation $T : (S_n) \to (T_n = S_n + D_n)$ and let us assume that (D_n) is a perfect estimation of the error of (S_n).

We consider the sequence $(S'_n = S_n + b)$. Applying T to (S'_n), (D_n) becomes (D'_n). But since $S'-S'_n = S-S_n$ it is, of course, highly desirable that $D'_n = D_n$ in order that (D'_n) be a perfect estimation of the errors of (S'_n).

Let us now consider the sequence $(S'_n = aS_n)$. We have $S'-S'_n = a(S-S_n)$. Thus it is desirable that $D'_n = aD_n$ for (D'_n) to be a perfect estimation of the error of (S'_n).

A sequence transformation having these two properties is said to be quasi-linear. Among such transformations are those considerer by Germain-Bonne [23] :

$$T_n = G_k(S_n, \ldots, S_{n+k})$$

The functions G_k are assumed to be defined and continuous with respect to each of their k+1 variables and to satisfy

H_1 : $G_k(ay_o, \ldots, ay_k) = a\, G_k(y_o, \ldots, y_k)$.

H_2 : $G_k(y_o+b, \ldots, y_k+b) = G_k(y_o, \ldots, y_k)+b$.

From H_1, $G_k(0, \ldots, 0) = 0$ and from H_2, $G_k(b, \ldots, b) = G_k(0, \ldots, 0)+b$.
Thus $\forall y$, $G_k(y, \ldots, y) = y$.

Let A_{k+1} be the subset of vectors $(y_o, \ldots, y_k) \in \mathbb{R}^{k+1}$ such that $\Delta y_i \neq 0$ for $i = 0, \ldots, k-1$. Then G_k has the form [23] :

$$G(y_o, \ldots, y_k) = y_o + \Delta y_o\, g\left(\frac{\Delta y_1}{\Delta y_o}, \ldots, \frac{\Delta y_{k-1}}{\Delta y_{k-2}}\right)$$

Thus

$$T_n = S_n + D_n \text{ with } D_n = \Delta S_n\, g\left(\frac{\Delta S_{n+1}}{\Delta S_n}, \ldots, \frac{\Delta S_{n+k-1}}{\Delta S_{n+k-2}}\right)$$

which shows that (D_n) is a perfect estimation of the error of (S_n) if and only if

$$\lim_{n\to\infty} \left(1 - \frac{S_{n+1}-S}{S_n-S}\right) g\left(\frac{\Delta S_{n+1}}{\Delta S_n}, \ldots, \frac{\Delta S_{n+k-1}}{\Delta S_{n+k-2}}\right) = 1.$$

If (S_n) converges linearly, that is if $\exists \xi \neq 1$ such that

$$\lim_{n\to\infty} (S_{n+1}-S)/(S_n-S) = \lim_{n\to\infty} \Delta S_{n+1}/\Delta S_n = \xi$$

then this condition writes

$$\forall \xi \neq 1 \quad g(\xi, \ldots, \xi) = 1/(1-\xi).$$

As stated by Smith and Ford [47] this condition is equivalent to saying that the transformation is exact on geometric sequences that is if $S_n = S + \alpha\xi^n$, $\xi \neq 1$ then $\forall n$, $T_n = S$.

This result can be considered as an extension to non-linear sequence transformations of Okada's theorem [43] for linear summation processes [20, 46] which roughly says that if a summation process sums geometric series then a domain exists in which an arbitrary power series can be analytically continued. The connection between these two results deserves further study.

Let us mention that Germain-Bonne extended the above results to sequence transformations of the form

$$T_n = G_k(S_n, \ldots, S_{n+k} \; ; \; x_n, \ldots, x_{n+k})$$

where (x_n) is an auxiliary sequence tending to zero, and that the same conclusions hold [24].

4 - The θ-procedure.

Let us now assume that (D_n) is a good estimation of the error of (S_n). We set

$$D'_n = -\frac{\Delta S_n}{\Delta D_n} D_n \, .$$

We have

$$\frac{D'_n}{S-S_n} = \frac{D_n}{S-S_n} \frac{-\Delta S_n}{\Delta D_n}$$

Thus if $\exists \alpha < 1 < \beta$, $\exists N$ such that $\forall n \geq N$, $(S_{n+1}-S)/(S_n-S) \notin [\alpha, \beta]$ then $\lim_{n\to\infty} -\Delta S_n/\Delta D_n = a$ and it follows that

$$\lim_{n\to\infty} D'_n/(S-S_n) = 1$$

which shows that (D'_n) is a perfect estimation of the error of (S_n) and that

$$S_n + D'_n = S_n - \frac{\Delta S_n}{\Delta D_n} D_n$$

converges faster than (S_n).

We have thus replaced a sequence transformation of the form $T_n = S_n + D_n$ by another one of the form $S_n - \frac{\Delta S_n}{\Delta D_n} D_n$. This is the so-called θ-procedure studied in [12]. The new transformation, called the θ-type algorithm associated with the transformation T, is denoted by $\theta(T_n)$. It is identical with the second standard process of Germain-Bonne [25, p. 6].

It must be noticed that if (D_n) is a perfect estimation of the error of (S_n) then, under the above conditions so is (D'_n). If the above conditions on (S_n) are not satisfied, in particular if (S_n) converges logarithmically, then (D'_n) is not necessarily a perfect estimation of the error of (S_n) which explains why logarithmic sequences are difficult to accelerate.

It is easy to see that the condition $(S_{n+1}-S)/(S_n-S) \notin [\alpha,\beta]$ can be replaced by $D_{n+1}/D_n \notin [\alpha,\beta]$.

Let us now give an example to illustrate our purpose. The simplest choice for an estimation of the error is

$$D_n = S_{n+1} - S_n.$$

We have

$$\frac{D_n}{S-S_n} = 1 - \frac{S_{n+1}-S}{S_n - S}.$$

Thus (D_n) is a perfect estimation of the error of (S_n) if and only if (S_n) converges superlinearly. This is an uninteresting case for our purpose since such sequences converge fast enough and don't need to be accelerated. If (S_n) converges linearly, but not logarithmically, that is when the ratio $(S_{n+1}-S)/(S_n-S)$ has a limit different from 0 and 1 then (D_n) is a good estimation of the error of (S_n). Applying the θ-procedure, we get a perfect estimation of the error and

$$\theta(T_n) = S_n - \frac{(\Delta S_n)^2}{\Delta^2 S_n}$$

which is the well-known Aitken's Δ^2 process.

As mentionned above this explains why logarithmic sequences are difficult to accelerate. For such sequences there is no universal method for constructing a perfect estimation of the error from a good one. The same is true for sequences for which the ratio $(S_{n+1}-S)/(S_n-S)$ has no limit when n tends to infinity.

As proved in [27] since the set of logarithmic sequences is remanent, a perfect estimation of the error for all logarithmic sequences cannot exist. Such estimations exist for subsets of this set.

In fact, by applying the θ-procedure to $D_n = \Delta S_n$ we obtain a perfect estimation of the error for the sequences satisfying

$$\lim_{n\to\infty} \frac{1 - (S_{n+1}-S)/(S_n-S)}{1 - \Delta S_{n+1}/ S_n} = 1.$$

This is the necessary and sufficient condition for Aitken's Δ^2 process to accelerate the convergence of (S_n).

This result can also be brought together with L'Hospital's rule for sequences and series proved in [9] stating that if (S_n) is such that

$$\lim_{n\to\infty} \left(\frac{\Delta S_{n+1}}{\Delta^2 S_{n+1}} - \frac{\Delta S_n}{\Delta^2 S_n}\right) = 0$$

then $S-S_n \sim -(\Delta S_n)^2/\Delta^2 S_n$ and thus $T_n = S_n-(\Delta S_n)^2/\Delta^2 S_n$ converges faster than (S_n).

More generally the θ-procedure accelerates the convergence of (S_n) if and only if

$$\lim_{n\to\infty} \frac{D_{n+1}/D_n - 1}{(S_{n+1}-S)/(S_n-S)-1} = \lim_{n\to\infty} \frac{\dfrac{\beta_n}{\beta_{n+1}}\dfrac{S_{n+1}-S}{S_n - S} - 1}{\dfrac{S_{n+1}-S}{S_n - S} - 1} = 1$$

with $\beta_n = (S - S_n)/D_n$.

Thus if $\exists\alpha < 1 < \beta$, $\exists N$, $\forall n \geq N$, $(S_{n+1}-S)/(S_n-S) \notin [\alpha,\beta]$ then a necessary and sufficient condition for the θ-procedure to accelerate the convergence of (S_n) is that

$$\lim_{n\to\infty} \beta_{n+1}/\beta_n = 1.$$

This condition is more general than $\lim_{n\to\infty} \beta_n = a \neq 0$ which shows that the θ-procedure can accelerate the convergence of sequences even if only a bad estimation of the error is known provided the ratio β_{n+1}/β_n tends to one.

If the above condition on (S_n) is not satisfied we set

$$\frac{S_{n+1}-S}{S_n-S} = 1 - \lambda_n$$

where (λ_n) is a sequence with zero as an accumulation point. A necessary and sufficient condition for the θ-procedure to accelerate the convergence of (S_n) is that

$$\lim_{n\to\infty} (1 - \beta_n/\beta_{n+1}(1-\lambda_n))/\lambda_n = 1.$$

Let us give an application to the so-called Kaluza sequences that is sequences such that, $\forall n$

$$0 < S_n \leq S_0 = 1 \text{ and } S_{n+1}^2 \leq S_n S_{n+2}.$$

It has been proved by Kaluza [31] that these sequences satisfy

$$S_o = 1$$
$$S_n = \sum_{i=1}^{n} c_i S_{n-i} \quad n = 1,2,\ldots$$

with $\forall n$, $c_n \geq 0$ and $\sum_{n=1}^{\infty} c_n \leq 1$. A sequence satisfying such relations is called a renewal sequence. Kendall [32] proved that (S_n) is a Kaluza sequence if and only if $\forall \alpha > 0$, (S_n^{α}) is a revewal sequence. Renewal sequences have applications in numerical analysis ; in particular they arise in some algorithms for the computation of moment integrals [49] and, thus, it is important to accelerate them. Let us remark that totally monotonic sequences are Kaluza sequences.

Let (S_n) be a sequence converging to S. We have the

Theorem : If $(S_n\text{-}S)$ is a Kaluza sequence then $\forall n$

$$0 \leq S \leq T_n \leq S_{n+2}$$

where $T : (S_n) \to (T_n)$ is Aitken's Δ^2 process. Moreover if (S_n) is non logarithmic then $(-(\Delta S_{n+1})^2/\Delta^2 S_n)$ is a perfect estimation of the error of (S_{n+2}) and thus (T_n) converges faster than (S_{n+2}).

Proof : as used by Kingman [33] $\forall n$, $(S_{n+1}-S)^2 \leq (S_n-S)(S_{n+2}-S)$ if and only if $\forall t$, $\forall n$, $(S_{n+2}-S)-2t(S_{n+1}-S)+t^2(S_n-S) \geq 0$. Thus if (S_n-S) is a Kaluza sequence then $\Delta^2(S_n-S) = \Delta^2 S_n \geq 0$ and we have

$$T_n-S = S_{n+2} - S - \frac{(\Delta S_{n+1})^2}{\Delta^2 S_n}$$

which proves the first part of the theorem. Moreover, $\forall n$

$$\frac{S_{n+1}-S}{S_n - S} \leq \frac{S_{n+2}-S}{S_{n+1}-S}$$

Since (S_n) converges to S these ratios are bounded from above by 1 and they converge. If the limit is different from 1 then $\Delta S_{n+1}/\Delta S_n$ has also the same limit. It follows immediately that $(-(\Delta S_{n+1})^2/\Delta^2 S_n)$ is a perfect estimation of the error of (S_{n+2}) and that $T_n-S = o(S_{n+2}-S)$. □

5 - ERROR ESTIMATES.

In order to prove the convergence of a given sequence (S_n) one must have some asymptotic information on it. For example the classical convergence tests for series are of this type. They usually take the form of an inequality which must be satisfied $\forall n \geq N$ as in d'Alembert's test : $|\Delta S_{n+1}/\Delta S_n| \leq K < 1$, $\forall n \geq N$.

Now if we want to accelerate the convergence of our given sequence (S_n) we must have a more precise asymptotic information on it. For example instead of an inequality satisfied $\forall n \geq N$ one of the members of the inequality must have a limit : $\exists K \neq 1$ such that $\lim_{n\to\infty} \Delta S_{n+1}/\Delta S_n = K$.

In fact, as we saw in the second section, one must know a perfect estimation of the error of (S_n). The aim of this section is to show that such estimations can be obtained from the usual convergence tests for sequences and series. I don't pretend to give here an extensive study of these tests but just to show how to use the most powerful of them which are based on comparison with a known sequence.

Let (x_n) be an auxiliary increasing sequence with a known limit x. (S_n) is also assumed to be increasing. We set $R_n = S - S_n$, $r_n = x - x_n$, $A_n = \Delta x_n/\Delta S_n = \Delta r_n/\Delta R_n$ and we assume that $\exists A$ such that $A = \lim_{n\to\infty} A_n$ is known. Then (S_n) converges and [37], $\forall n$

$$r_n/A \leq R_n \leq r_n/A_n \qquad \text{if } (A_n) \text{ is increasing}$$

$$r_n/A \geq R_n \geq r_n/A_n \qquad \text{if } (A_n) \text{ is decreasing.}$$

Thus in both cases, $(D_n = r_n/A)$ and $(D'_n = r_n/A_n)$ are perfect estimations of the error of (S_n), and the corresponding sequence transformations are

$$T_n = S_n + (x - x_n)/A \qquad n = 0,1,\dots$$

which is Germain-Bonne's first standard process [25] and

$$T'_n = S_n - \frac{\Delta S_n}{\Delta x_n}(x_n - x) \qquad n = 0,1,\dots$$

which is Germain-Bonne's second standard process [25] or, equivalently, the application of the θ-procedure to (T_n).

If $x_n = \Delta S_n$ the preceding convergence test reduces to d'Alembert's and (T'_n) is Aitken's Δ^2 process. When $\Delta x_n = \xi^n$ with $0 < \xi < 1$ the test is Cauchy's. For $x_n = n\,\Delta S_n$ it corresponds to the Raabe-Duhamel's criterion and (T'_n) is a transformation studied by Kowalewski [34, p. 96] for accelerating the convergence of some logarithmic sequences.

More general forms of this test and their connections to series transformations can be found in [37]. Further extensions of sequence transformations of this type were studied by Levin [36]. The choice $x_n = \Delta S_n$ corresponds to its t-transform, $x_n = (n+1)\Delta S_n$ gives its u-transform while $x_n = \Delta S_n \Delta S_{n-1}/\Delta^2 S_{n-1}$ leads to its v-transform. As shown by the numerical examples worked out by Smith and Ford [48] these three transformations, along with the θ-algorithm, are among the most powerful acceleration methods.

The case $x_n = \Delta S_n$ was studied in more details by Pennacchi [45] and Hillion [30].

We set $\xi_n = \Delta S_n/\Delta S_{n-1}$ and $\xi = \lim_{n\to\infty} \xi_n$. If $\xi \neq 1$ and if $\forall n$

$$1 > \frac{\Delta S_{n+1}}{\Delta S_n} > \frac{\Delta S_n}{\Delta S_{n-1}} > 0, \text{ then } \forall n$$

$$\frac{\Delta S_n}{1-\xi_{n+1}} < S-S_n < \frac{\Delta S_n}{1-\xi} .$$

If $\forall n \quad 0 < \dfrac{\Delta S_{n+1}}{\Delta S_n} < \dfrac{\Delta S_n}{\Delta S_{n-1}} < 1$, then $\forall n$

$$\frac{\Delta S_n}{1-\xi_{n+1}} > S-S_n > \frac{\Delta S_n}{1-\xi} .$$

Thus, in both cases, $(D_n = \Delta S_n/(1-\xi))$ and $(D'_n = \Delta S_n/(1-\xi_{n+1}))$ are perfect estimations of the error of (S_n). The choice (D'_n) leads again to Aitken's Δ^2 process while the choice (D_n) corresponds to the optimal value of the parameter appearing in Baranger's transformation [1, 4]. Moreover we get a method for controlling the error (other methods will be described in section 10).

Let us now consider Kummer's test. Let (a_n) be an auxiliary sequence such that

$$\lim_{n\to\infty} a_n \Delta S_{n-1} = 0.$$

We set

$$D_n = a_n \Delta S_{n-1} \Delta S_n/(a_n \Delta S_{n-1} - a_{n+1}\Delta S_n) \qquad n = 1,2,...$$

Under the assumptions that $\forall n$, $\Delta S_n > 0$ and $\exists \mu > 0$ such that

$$\mu = \lim_{n\to\infty} (a_n \Delta S_{n-1} - a_{n+1}\Delta S_n)/\Delta S_n$$

then the sequence (S_n) converges. Moreover $\lim_{n\to\infty} D_n/(S-S_n) = 1$.

This last result given in [40] (see also [41]) can be proved under weaker assumptions and we have the

Theorem. Let (S_n) be a sequence converging to S and let (a_n) be an auxiliary sequence such that

i) $\lim_{n\to\infty} a_n \Delta S_{n-1} = 0$

ii) $\exists\mu \neq 0$ finite $\lim_{n\to\infty} (a_n \Delta S_{n-1} - a_{n+1}\Delta S_n)/\Delta S_n = \mu$

iii) $\exists\lambda \neq 0$ finite $\lim_{n\to\infty} a_n \Delta S_{n-1}/(S-S_n) = \lambda$.

Let $D_n = a_n \Delta S_{n-1}\Delta S_n/(a_n \Delta S_{n-1} - a_{n+1}\Delta S_n)$ $\quad n = 1,2,\ldots$

If $\lambda = \mu$ the sequence $(T_n = S_n + D_n)$ converges faster than (S_n).

If $\lambda \neq \mu$ the sequence $(T_n = S_n - \frac{\Delta S_n}{\Delta D_n} D_n)$ converges faster than (S_n).

Remarks. If ii) is satisfied and if (S_n) is monotonic then it follows [1, theorem 15, p. 13] that iii) holds with $\lambda = \mu$.

If iii) is satisfied and if $\exists\alpha < 1 < \beta$, $\exists N$, $\forall n \geq N$, $(S_{n+1}-S)/(S_n-S) \notin [\alpha,\beta]$ or $a_{n+1}\Delta S_n/a_n \Delta S_{n-1} \notin [\alpha,\beta]$ it follows [7, theorem 12, p. 12] that ii) holds with $\mu = \lambda$.

If $(a_n = 1)$ Kummer's test is d'Alembert's while for $(a_n = n)$ it reduces to the Raabe-Duhamel's criterion. The existence of a sequence (a_n) of positive terms satisfying ii) is, in the case of a strictlyincreasing sequence (S_n), a necessary and sufficient condition for the convergence of (S_n) (see [38] for this result and other forms of it). It is also known [39] that if $\forall n$, $0 < \Delta S_n < \Delta S_{n-1}$ then $\lim_{n\to\infty} n\,\Delta S_{n-1} = 0$ and that if $\forall n$, $a_n > 0$ and $\Delta S_n > 0$ then $\lim_{n\to\infty} \Delta S_{n+1}/\Delta S_n = 0$ if and only if $\lim_{n\to\infty} a_n = 0$ [42]. Particular cases of Kummer's test were studied in [39], see also Bromwich [17, p. 37].

Perfect estimations can, in some cases, be obtained from integral tests for the convergence of series, an idea already exploited in [2].

If $S_n = \sum_{i=1}^{n} f(i)$ with $f(x) \geq 0$ decreasing for $x \geq 0$. Then it is well known (Cauchy) that, $\forall n$

$$D_{n+1} \leq R_n = S-S_n \leq D_n = \int_n^\infty f(x)dx.$$

If $f(x) = 1/x^2$ then $D_n = 1/n$ and from the inequality

$$1/(n+1) < R_n \leq 1/n$$

we have $\lim_{n\to\infty} R_n/D_n = 1$ which shows that $(1/n)$ is a perfect estimation of the error of (S_n). We obtain the following numerical results with $T_n = S_n + D_n$

n	$S_n - \pi^2/6$	$T_n - \pi^2/6$	$(T_n - \pi^2/6)/(S_n - \pi^2/6)$
2	-0.39493	0.10506	-0.26603
5	-0.18132	$0.18677\ 10^{-1}$	-0.10300
10	$-0.95166\ 10^{-1}$	$0.48337\ 10^{-2}$	$-0.50792\ 10^{-1}$
15	$-0.64494\ 10^{-1}$	$0.21729\ 10^{-2}$	$-0.33691\ 10^{-1}$
20	$-0.48771\ 10^{-1}$	$0.12292\ 10^{-2}$	$-0.25203\ 10^{-1}$

In the case of series with positive terms each convergence test provides an upper bound of R_n. For obtaining a perfect estimation of the error this is, in general, not sufficient. A lower bound is also needed and one must prove that the ratio of these two bounds tends to one. Let us give some examples of this situation. They are taken from [21] and both bounds are perfect estimations of the error.

For the series $\sum_{n=1}^{\infty} 1/n^{1+a}$ we have for $a > 0$

$$1/a(n+1)^a < R_n < 1/an^a.$$

For $\sum_{n=1}^{\infty} 1/(n+h)^{1+a}$ with $h > 0$ we have

$$(n+k-2a-1)/a(n+h)^{1+a} < R_n < (n+h-1)/a(n+k)^{1+a}$$

where k is any real number strictly greater than one.

Consider now the series $\sum_{n=1}^{\infty} u_n$ with

$$\frac{u_{n+1}}{u_n} = \frac{n^p + a_1 n^{p-1} + a_2 n^{p-2} + \ldots}{n^p + b_1 n^{p-1} + b_2 n^{p-2} + \ldots} \quad \text{and } a = b_1 - a_1 - 1 > 0.$$

Let h and k be real numbers such that $h > b_1 + \frac{a_2 - b_2}{a+1} > k$, then

$$(n+k-2a-1)u_n/a < R_n < (n+h-1)u_n/a.$$

The last example concerns the series

$$\sum_{n=1}^{\infty} \frac{(1-b)(2-b)\ldots(n-b)}{(n+1)(p+a)(p+a+1)\ldots(p+a+n)} = \sum_{n=1}^{\infty} u_n$$

with $b \geq 0$ and $p+a > 0$. We have

$$(n-p-a-2b)u_n/(p+a+b) < R_n < (n+p+a+1)\ u_n/(p+a+b).$$

For example if b = 0 and p+a = 1 we have

$$\sum_{n=1}^{\infty} u_n = \sum_{n=1}^{\infty} 1/(n+1)^2 = \pi^2/6 - 1 = S$$

and we obtain the inequalities

$$\frac{n-1}{(n+1)^2} < R_n < \frac{n+2}{(n+1)^2}$$

We set

$$T_n = S_n + (n-1)/(n+1)^2$$

$$T'_n = S_n + (n+2)/(n+1)^2$$

$$S_n = \sum_{i=1}^{n} 1/(i+1)^2$$

n	$S-T_n$	$S-S_n$	$S-T'_n$
1	0.39493	0.39493	- 0.35506
2	0.17271	0.28382	- 0.16062
3	0.09632	0.22132	- 0.09118
4	0.06132	0.18132	- 0.05868
5	0.04243	0.15354	- 0.04090

6 - EXTRACTION PROCESSES.

If (S_n) is a logarithmic sequence, that is if $(S_{n+1}-S)/(S_n-S)$ tends to 1 then $D_n = \Delta S_n$ is not a good estimation of the error of (S_n) since $\lim_{n\to\infty} D_n/(S-S_n) = 0$. However the θ-procedure can be applied and interesting results can be obtained for some subsets of logarithmic sequences.

Let $\mathcal{L}$ be the subset of logarithmic sequences satisfying

$$\exists a \neq 0 \text{ such that } \lim_{n\to\infty} \frac{1-(S_{n+1}-S)/(S_n-S)}{1 - \Delta S_{n+1}/\Delta S_n} = a$$

Theorem. $\mathcal{L}$ is accelerable.

Proof.

$$-\frac{(\Delta S_n)^2}{\Delta^2 S_n (S-S_n)} = \frac{1 - (S_{n+1}-S)/(S_n-S)}{1 - \Delta S_{n+1}/\Delta S_n}$$

Thus $(-(\Delta S_n)^2/\Delta^2 S_n)$ is a good (or a perfect if a = 1) estimation of the error of (S_n). Let

$$t_n = S_n - (\Delta S_n)^2/\Delta^2 S_n .$$

We have :

$$\lim_{n\to\infty} \frac{t_n - S}{S_n - S} = 1-a.$$

Since $1-a \neq 1$, Ƅ is of synchronous type and thus [26] it is accelerable.

Remarks.

1) The set Ƅ which was considered in [34] is a subset of the set called LOGSF which is not accelerable [35].
The set L studied in [35] is a subset of Ƅ. Thus L is accelerable which gives a positive answer to a question raised in [35].

2) For the definition of a set of synchronous type and for the result used in the proof, see [26]. The complete proof is given in [28]. The algorithm for accelerating the convergence consists in extracting a subsequence of (S_n) converging linearly and then accelering it by, for example, Aitken's Δ^2 process.

3) Aitken's Δ^2 process accelerates the subset of Ƅ of sequences for which a = 1 (perfect estimation). In that case the extracted subsequences is the whole sequence.

The preceding results can be generalized to any synchronous subset of logarithmic sequences such that $\exists N, \forall n \geq N, \Delta S_n$ has a constant sign or $\forall n \geq 0, S_n - S$ has a constant sign.

Let A be such a subset. Then there exists a transformation

$$T : (S_n) \to (T_n) \text{ such that } \forall (S_n) \in A, \exists b \neq 1$$

$$\lim_{n\to\infty} (T_n - S)/(S_n - S) = b.$$

Since T_n can always be written $T_n = S_n + D_n$ it means that $\forall (S_n) \in A$, (D_n) is a good estimation of the error of (S_n). But, by Germain-Bonne's result [24], A is accelerable which means that $\exists T' : (S_n) \to (T'_n = S_n + D'_n)$ such that $\forall (S_n) \in A$, $\lim_{n\to\infty} (T'_n - S)/(S_n - S) = 0$ or equivalently, $\forall (S_n) \in A$, (D'_n) is a perfect estimation of the error of (S_n). However it is more or less a matter of chance (most of the time more) to find the estimation (D'_n) since the θ-procedure usually does not work.

Another way of proceeding is to extract a subsequence converging linearly as described in [16]. Another method is given in [28]. Since (S_n) is logarithmic and $D_n/(S-S_n)$ tends to a limit different from zero then (D_n) is also logarithmic and monotone. Such an algorithm will be given farther.

Then from $(D_{b(n)} = D'_n)$ we can extract a linear subsequence $(D'_{a(n)})$ as follows : let $\xi \in \,]0,1[$. We define the sequence $(a(n))$ as

$$a(0) = 0$$

$$a(n) = \min \{i > a(n-1) \mid |D'_i| < \xi^n |D_o|\} \quad n = 1,2,\ldots$$

It can be proved [41, theorems 2 and 4] that

$$\lim_{n\to\infty} \frac{S_{a(n+1)}-S}{S_{a(n)} - S} = \lim_{n\to\infty} \frac{D'_{a(n+1)}}{D'_{a(n)}} = \xi.$$

Thus the sequence $(S_{a(n)})$ can be accelerated by, for example, Aitken's Δ^2 process :

$$T_i = S_{a(i)} - \frac{(S_{a(i+1)}-S_{a(i)})^2}{S_{a(i+2)}-2S_{a(i+1)}+S_{a(i)}} \qquad i = 0,1,\ldots$$

or by the θ-procedure

$$T_i = S_{a(i)} - \frac{S_{a(i+1)}-S_{a(i)}}{D'_{a(i+1)}-D'_{a(i)}} D'_{a(i)} \qquad i = 0,1,\ldots$$

Let (t_n) be defined by

$$t_n = T_i \quad \text{for } a(i) \le n < a(i+1).$$

Then it has been proved in [16, theorem 1] that

$$\lim_{n\to\infty} (t_n-S)/(S_n-S) = 0.$$

Let us now show how to extract a monotone logarithmic sequence $(D_{b(n)})$. The extraction procedure is based on the following remarks. Since the ratio D_{n+1}/D_n tends to 1, then $\exists m$ such that $\forall n \ge m$, $D_{n+1}/D_n > 0$, which means that $\forall n \ge m$, D_n has a constant sign. If $\forall n \ge m$, $D_n > 0$ since it is convergent to zero, a decreasing subsequence can be extracted. If $\forall n \ge m$, $D_n < 0$ then an increasing subsequence can be extracted. It must be remarked that, in practice, we cannot be sure that this index m has been atteined. The extraction algorithm is as follows.

1 - Set m = 0

2 - Set i = 0 and b(0) = m
If $D_{b(o)} > 0$ set k = 0. If $D_{b(o)} < 0$ set k = 1

3 - Set j = 1. If k = 0 go to 4 and if k = 1 go to 5

4 - If $D_{b(i)+j} < 0$ set m = b(i)+j, k = 1 and go to 2.
If $D_{b(i)+j} > 0$ and if $D_{b(i)+j} < D_{b(i)}$ set b(i+1) = b(i)+j, replace i by i+1 and go to 3.
If $D_{b(i)+j} > 0$ and if $D_{b(i)+j} \geq D_{b(i)}$ replace j by j+1 and go to 4.

5 - If $D_{b(i)+j} > 0$ set m = b(i)+j, k = 0 and go to 2.
If $D_{b(i)+j} < 0$ and if $D_{b(i)} < D_{b(i)+j}$ set b(i+1)=b(i)+j, replace i by i+1 and go to 3.
If $D_{b(i)+j} < 0$ and if $D_{b(i)} \geq D_{b(i)+j}$ replace j by j+1 and go to 5.

Point 4 corresponds to the case $\forall n \geq m$, $D_n > 0$ while point 5 treats the other case.

7 - THE E-ALGORITHM.

Let us now generalize the ideas developed in section 2.
We assume that

$$S_n = S + a_1 g_1(n) + a_2 g_2(n) + \ldots$$

where the $(g_i(n))$'s are known sequences but where the coefficients a_i's are unknown. If the g_i's are assumed to form a scale of comparison then, since the a_i's are unknown, $(g_1(n))$ is not, in general, a perfect estimation of the error of (S_n) but only a good one, $(g_2(n))$ is, in general, only a good estimation of the error of $(S_n - a_1 g_1(n))$, and so on.

Thus, in order to obtain perfect estimations of these errors, let us apply the θ-procedure. We get a first sequence transformation

$$E_1^{(n)} = E_o^{(n)} - \frac{\Delta E_o^{(n)}}{\Delta g_{o,1}^{(n)}} g_{o,1}^{(n)} \quad , n = 0,1,\ldots$$

Replacing S_n by its expression we find

$$E_1^{(n)} = S + a_2 g_{1,2}^{(n)} + a_3 g_{1,3}^{(n)} + \ldots \qquad n = 0,1,\ldots$$

with

$$g_{1,i}^{(n)} = g_{o,i}^{(n)} - \frac{\Delta g_{o,i}^{(n)}}{\Delta g_{o,1}^{(n)}} g_{o,1}^{(n)} \qquad n = 0,1,\ldots \;;\; i \geq 2$$

and $g_{o,i}^{(n)} = g_i(n)$, n = 0,1,...

Under the above assumptions $\{(g_{1,2}^{(n)}), (g_{1,3}^{(n)}), \ldots\}$ is again a scale of comparison (if some additional assumptions are satisfied, see [10]) but, in general, $(g_{1,2}^{(n)})$ is not a perfect estimation of the error of $(E_1^{(n)})$ but only a good one. Thus the θ-procedure can be applied a second time and we get a second sequence transformation

$$E_2^{(n)} = E_1^{(n)} - \frac{\Delta E_1^{(n)}}{\Delta g_{1,2}^{(n)}} g_{1,2}^{(n)} \qquad n = 0,1,\ldots$$

Replacing $E_1^{(n)}$ by its above expression we obtain

$$E_2^{(n)} = S + a_3 g_{2,3}^{(n)} + a_4 g_{2,4}^{(n)} + \ldots \qquad n = 0,1,\ldots$$

with

$$g_{2,i}^{(n)} = g_{1,i}^{(n)} - \frac{\Delta g_{1,i}^{(n)}}{\Delta g_{1,2}^{(n)}} g_{1,2}^{(n)} \qquad n = 0,1,\ldots ; \ i \geq 3.$$

Such a procedure can be continued and we finally obtain the algorithm

$$E_o^{(n)} = S_n, \ g_{o,i}^{(n)} = g_i(n) \qquad n = 0,1,\ldots ; \ i=1,2,\ldots$$

$$E_k^{(n)} = E_{k-1}^{(n)} - \frac{\Delta E_{k-1}^{(n)}}{\Delta g_{k-1,k}^{(n)}} g_{k-1,k}^{(n)} \qquad n = 0,1,\ldots ; \ k=1,2,\ldots$$

$$g_{k,i}^{(n)} = g_{k-1\ i}^{(n)} - \frac{\Delta g_{k-1,i}^{(n)}}{\Delta g_{k-1,k}^{(n)}} g_{k-1,k}^{(n)} \qquad n = 0,1,\ldots ; \ k=1,2,\ldots ; \ i \geq k.$$

which is the so-called E-algorithm [10, 29] which includes most of the convergence acceleration algorithms actually known. This algorithm was previously derived by writing

$$S_i = E_k^{(n)} + a_1 g_1(i) + \ldots + a_k g_k(i) \qquad i = n,\ldots,n+k$$

and solving this system either by Gaussian elimination [29] or, equivalently, by using Sylvester's determinantal identity. It has now been issued from a procedure for obtaining a perfect estimation of the error of the sequence to accelerate and the acceleration result given in [10, theorem 7] derived from its very construction.

The case where the g_i's form a scale of comparison is an ideal one since, under the additional assumptions stated in the acceleration result mentioned above, each transformation converges faster than the preceding one. This is not always the case for example if

$$S_n = S + a_1(n+b_1)^{-1} + a_2(n+b_2)^{-1} + a_3g_3(n) + \dots \quad n = 0,1,\dots$$

However in this case, if $g_i(n) = o((n+b_1)^{-1})$ for $i \geq 3$, then

$$\lim_{n\to\infty} (S_n-S)/(n+b_1)^{-1} = a_1+a_2$$

which shows that, unless $a_1+a_2 = 0$, $(n+b_1)^{-1}$ is a good estimation of the error of (S_n). Thus the θ-procedure, that is the E-algorithm, can be applied and we get

$$E_1^{(n)} = S + a_2(b_2-b_1)(n+1+b_2)^{-1}(n+b_2)^{-1}+a_3g_{1,3}^{(n)}+\dots \qquad n = 0,1,\dots$$

which shows that $(E_1^{(n)})$ converges faster than (S_n).

More generally if, $\forall n$

$$S_n - S = a_1g_1(n) + a_2R_n$$

where $\lim_{n\to\infty} R_n/g_1(n) = b \neq 0$ then however

$$\lim_{n\to\infty} (S_n-S)/g_1(n) = a_1+a_2b.$$

Thus $(g_1(n))$ is a good estimation of the error of (S_n) and $(E_1^{(n)})$ will converge faster than (S_n).

The importance and usefulness of the E-algorithm arises from the fact that it is usually easier to have informations on $(S-S_n)$ (for example its asymptotic expansion) than on the successive $E_k^{(n)}$ directly. A particular case of the E-algorithm is the well known Richardson extrapolation process with corresponds to the assumption that

$$S_n = S + a_1x_n + a_2x_n^2 + \dots$$

where (x_n) is a given auxiliary sequence with limit zero such that $\exists\alpha < 1 < \beta$, $\exists N$, $\forall n \geq N$, $x_{n+1}/x_n \notin [\alpha,\beta]$.

In that case the E-algorithm reduces to

$$E_k^{(n)} = E_{k-1}^{(n)} - \frac{\Delta E_{k-1}^{(n)}}{x_{n+k}-x_n} x_n \qquad n = 0,1,\dots ; k = 1,2,\dots$$

It is easy to check that $- \Delta E_{k-1}^{(n)} x_n/(x_{n+k}-x_n)$ is a perfect estimation of the error of $E_{k-1}^{(n)}$ if and only if

$$\lim_{n\to\infty} \frac{E_{k-1}^{(n+1)}-S}{E_{k-1}^{(n)}-S} = \lim_{n\to\infty} \frac{x_{n+k}}{x_n}$$

which is the necessary and sufficient condition that $(E_k^{(n)})$ converges faster than $(E_{k-1}^{(n)})$ [7, theorem 26, p. 29].

In particular if $S_n = S+a_1x_n+a_2x_n^2 + \dots$ with $x_{n+1} = ax_n$, $a < 1$ then $E_k^{(n)} = S + a_{k,k}x_n^{k+1} + a_{k,k+1}x_n^{k+2} + \dots$ and

$$\lim_{n\to\infty} \frac{E_{k-1}^{(n+1)}-S}{E_{k-1}^{(n)}-S} = \lim_{n\to\infty} \frac{x_{n+k}}{x_n} = a^k .$$

Romberg's method for accelerating the trapezoidal rule enters in this case.

8 - OVERHOLT'S PROCESS.

Overholt's process [44] is the only convergence acceleration method with does not fit exactly into the framework of the E-algorithm. It consists in assuming that, $\forall n$

$$S_n = S+a_1(S_{n-1}-S) + a_2(S_{n-1}-S)^2 + \dots \text{ with } a_1 \neq 1$$

and then to construct higher and higher approximations of S by the scheme

$$E_o^{(n)} = S_n \qquad n = 0,1,\dots$$

$$E_k^{(n)} = E_{k-1}^{(n)} - \frac{\Delta E_{k-1}^{(n)}}{(\Delta S_{n+k})^k - (\Delta S_{n+k-1})^k} (\Delta S_{n+k-1})^k$$

$$n=0,1,\dots \; ; \; k=1,2,\dots$$

Of course this algorithm looks very much like the E-algorithm. In fact it is exactly the principal rule of the E-algorithm with $g_{k-1,k}^{(n)} = (\Delta S_{n+k-1})^k$ but such quantities cannot be obtained from the auxiliary rule of the algorithm with $g_i(n) = (S_{n-1}-S)^i$ since S is unknown. Thus Overholt's process does not fit perfectly into this algorithm. The reason is that the E-algorithm computes the exact values of the coefficients a_i's in the expansion of S_n-S while Overholt's process computes only approximate values of them. In particular

$$a_1^{(n)} = \Delta S_{n+1}/\Delta S_n = a_1 + (1+a_1)a_2d_n + O(d_n^2)$$

with $d_n = S_n - S$. This approximation is sufficient to ensure that $E_1^{(n)}$ is a second order approximation of S. More generally

$$E_k^{(n)} = S + a_{kk} d_n^{k+1} + a_{k,k+1} d_n^{k+2} + \ldots$$

Thus $\lim_{n\to\infty} (E_{k-1}^{(n+1)} - S)/(E_{k-1}^{(n)} - S) = a_1^k = \lim_{n\to\infty} (\Delta S_{n+k}/\Delta S_{n+k-1})^k$ which shows that $-\Delta E_{k-1}^{(n)} (\Delta S_{n+k-1})^k / [(\Delta S_{n+k})^k - (\Delta S_{n+k-1})^k]$ is a perfect estimation of the error of (S_n). Therefore Overholt's process arises from the construction of a perfect estimation of the error of sequences of the form $d_n = a_1 d_{n-1} + a_2 d_{n-1}^2 + \ldots$.

9 - COMPOSITE SEQUENCE TRANSFORMATIONS.

Such transformations were introduced in [15] . Let $t_1 : (S_n) \to (t_1^{(n)})$ and $t_2 : (S_n) \to (t_2^{(n)})$ be two sequence transformations. The rank two composite transformation $T : (S_n) \to (T_n)$ is defined by

$$T_n = (1-a_n) t_1^{(n)} + a_n t_2^{(n)}$$

where (a_n) is chosen such that the kernel of T (that is the set of sequences transformed by T into a constant sequence) contains the kernels of t_1 and t_2. This property is satisfied by the choice

$$a_n = -\Delta t_1^{(n)} / (\Delta t_2^{(n)} - \Delta t_1^{(n)})$$

where the denominator is assumed to be different from zero $\forall n$. We have

$$T_n = t_1^{(n)} - \frac{\Delta t_1^{(n)}}{\Delta t_2^{(n)} - \Delta t_1^{(n)}} (t_2^{(n)} - t_1^{(n)}).$$

Since

$$t_2^{(n)} = t_1^{(n)} + (t_2^{(n)} - t_1^{(n)})$$

then T_n is obtained by applying the θ-procedure to that expression.

Thus, as shown in section 4, if $D_n = t_2^{(n)} - t_1^{(n)}$ is a good estimation of the error of $t_1^{(n)}$ and if $(t_1^{(n+1)} - S)/(t_1^{(n)} - S)$ does not approach one then $-\frac{\Delta t_1^{(n)}}{\Delta (t_2^{(n)} - t_1^{(n)})} (t_2^{(n)} - t_1^{(n)})$ will be a perfect estimation of the error of $t_1^{(n)}$. Since we have

$$\frac{D_n}{S-t_1^{(n)}} = \frac{t_2^{(n)}-S}{S-t_1^{(n)}} - \frac{t_1^{(n)}-S}{S-t_1^{(n)}} = 1 - \frac{t_2^{(n)}-S}{t_1^{(n)}-S}$$

Then $(t_2^{(n)}-t_1^{(n)})$ is a good estimation of the error of $(t_1^{(n)})$ iff $\exists a \neq 1,0$ such that $\lim_{n\to\infty} (t_2^{(n)}-S)/(t_1^{(n)}-S) = a$. This is exactly the result of [15, theorem 3, p. 315]. In that case we shall have $T_n-S=o(t_1^{(n)}-S)$ and, moreover, $T_n-S=o(t_2^{(n)}-S)$.

Since composite transformations of higher rank (that is those combining more than 2 transformations) can be implemented by the E-algorithm, they can be included in our framework.

Let us consider the particular case where $t_1^{(n)} = S_n$. For simplicity we shall write t_n instead of $t_2^{(n)}$ and $t_n = S_n+D_n$ as usual. We have

$$T_n = S_n - \frac{\Delta S_n}{\Delta(t_n-S_n)}(t_n-S_n) = S_n - \frac{\Delta S_n}{\Delta D_n} D_n$$

which is the θ-procedure. In [34] this procedure has been proved to accelerate the convergence of some subsets of logarithmic sequences.

Let us look at this point more carefully.
We have

$$\frac{T_n-S}{S-S_n} = \frac{t_n-S_n}{S-S_n}\,\frac{\Delta S_n}{\Delta(S_n-t_n)} - 1$$

$$\frac{t_n-S_n}{S-S_n} = 1 - \frac{t_n-S}{S_n-S} \qquad \text{and} \qquad \frac{\Delta(S_n-t_n)}{\Delta S_n} = 1 - \frac{\Delta t_n}{\Delta S_n}$$

Thus if $\exists b \neq 1$ such that $\lim_{n\to\infty} (t_n-S)/(S_n-S) = b$ (that is if the transformation $t = t_2$ is synchronous) and if $\lim_{n\to\infty} \Delta t_n/\Delta S_n = b$ then $T_n-S=o(S_n-S)$.

In particular when t is Aitken's Δ^2 process then T is the θ_2-algorithm which was proved to accelerate some subsets of logarithmic sequences. It is a recursive use of the procedure θ as described in [12].

If the condition $\lim_{n\to\infty}\Delta t_n/\Delta S_n = b$ is not satisfied, or if it is not possible to check if it is satisfied then, as shown in section 6, an extraction procedure can be applied to the sequence $(D_n=t_n-S_n)$ and the subsequence $(S_{a(n)})$ thus obtained is linear and can be easily accelerated.

If $\lim_{n\to\infty} \Delta t_n/\Delta S_n = a \neq b$ then it means that $-\frac{\Delta S_n}{\Delta D_n} D_n$ was not a perfect estimation of the error of (S_n) but only a good one. Thus (at least in the case of a non-logarithmic sequence) it means that (D_n) was a bad estimation of the error of (S_n). Therefore (again for non-logarithmic sequences) a perfect estimation of the error of (S_n) can be obtained by applying again the θ-procedure.

Let us give two examples to illustrate this point. We take

$$S-S_n = (-1)^n \lambda^n \quad \text{and} \quad D_n = \lambda^n, \ \lambda \in]-1,1[.$$

We have $(S-S_n)/D_n = (-1)^n$. Thus (D_n) is a bad estimation of the error of (S_n). Applying the θ-procedure once we get

$$D'_n = -\frac{\Delta S_n}{\Delta D_n} D_n = (-1)^{n+1} \lambda^n \frac{\lambda+1}{\lambda-1}$$

$$(S-S_n)/D'_n = -(\lambda-1)/(\lambda+1)$$

which shows that (D'_n) is a good estimation of the error of (S_n) but not a perfect one. Let us apply the θ-procedure a second time. We get

$$D''_n = -\frac{\Delta S_n}{\Delta D'_n} D'_n = (-1)^n \lambda^n$$

which shows that (D''_n) is a perfect estimation (even an exact one) of the error of (S_n). We must remark that, although the ratio $(S-S_n)/D_n$ has no limit, the ratios $(S_{n+1}-S)/(S_n-S)$ and D_{n+1}/D_n are both convergent.

Let us give a second example. We consider the choice $D_n = 1/(n+1)$ and we assume that $\exists a \neq 1$ such that

$$\lim_{n\to\infty} (S_{n+1} - S)/(S_n - S) = a.$$

(D_n) is a bad estimation of the error of (S_n) since $(S-S_n)/D_n$ tends to zero. We have

$$D'_n = -\frac{\Delta S_n}{\Delta D_n} D'_n = -\frac{\Delta S_n}{\frac{n+3}{n+2}\,\frac{\Delta S_{n+1}}{\Delta S_n} - 1}$$

and $(S-S_n)/D''_n$ tends to 1 since $\Delta S_{n+1}/\Delta S_n$ tends to a.

The transformation $T : (S_n) \to (S_n+D_n)$ accelerates the set of logarithmic sequences such that $(S-S_n)(n+1)$ tends to 1. The transformation

$T' : (S_n) \to (S_n + D'_n)$ accelerated the set of logarithmic sequences such that $(n+2)\,[1-(S_{n+1}-S)/(S_n-S)]$ tends to 1. But neither T nor T' accelerates the set of linearly converging sequences, that is those corresponding to the above property with $a \neq 0,1$.

10 - ERROR CONTROL.

A method for controlling the error in convergence acceleration methods was proposed in [13]. We consider a perfect estimation (D_n) of the error of (S_n) and the transformation $T_n = S_n + D_n$. We then define

$$T_n(b) = S_n + (1-b)D_n = T_n - bD_n$$

$$e_n = 1 - (S-S_n)/D_n$$

and let $I_n(b)$ be the interval with bounds $T_n(b)$ and $T_n(-b)$. A necessary and sufficient condition that $\forall n \geq N$, $S \in I_n(b)$ is that $\forall n \geq N$, $|e_n| \leq |b|$.

If (D_n) is a perfect estimation of the error of (S_n) then e_n tends to zero. Thus it always exists an index N such that $|e_n| \leq |b|$ and so $\forall n \geq N$, $S \in I_n(b)$. This method allows to control the error since $\forall n \geq N$

$$|T_n - S| \leq |T_n(b) - T_n(-b)|/2.$$

However the benefit of accelerating the convergence has been lost since the bounds $(T_n(b))$ and $(T_n(-b))$ do not converge faster than (S_n). To have simultaneously both properties we have to use a sequence (b_n) instead of a constant value of b.

Of course, a necessary and sufficient condition that $\forall n \geq N$, $S \in I_n(b_n)$ is that $\forall n \geq N$, $|e_n| \leq |b_n|$. Moreover, a necessary and sufficient condition that $T_n(b_n) - S = o(S_n - S)$ is that $b_n = o(1)$. We have

$$T_n(b_n) - S = (b_n - e_n)(S_n - S)/(1-e_n).$$

Thus the best choice of b_n is $b_n = e_n$ since it leads to $T_n(e_n) = S$, $\forall n$. Of course this choice is impossible in practice since S is unknown and we shall choose a sequence (b_n) such that $\forall n \geq N$, $|e_n| \leq |b_n|$ and such that

$$\lim_{n\to\infty} b_n/e_n = 1.$$

This condition is equivalent to the condition for $(-b_n D_n)$ to be a perfect estimation of the error of (T_n). Indeed we have

$$\frac{S-T_n}{-b_n D_n} = \frac{1-(S-S_n)/D_n}{b_n} = \frac{e_n}{b_n} .$$

Thus $(T_n(b_n))$ will converge faster than (T_n) while, as remarked in [13], $(T_n(-b_n))$ will not. A method for avoiding this drawback is proposed in [3].

Another method for controlling the error with an interval whose bounds converge faster than (S_n) is to use two sequence transformations accelerating (S_n) in cascade, that is

$$T_n = S_n + D_n \qquad T_n - S = o(S_n - S)$$

$$V_n = T_n + D'_n \qquad V_n - S = o(T_n - S) .$$

Then we consider

$$V_n(b) = T_n + (1-b) D'_n$$

and the interval $J_n(b)$ with bounds $V_n(b)$ and $V_n(-b)$. It is easy to see that $\forall b \geq 0$, $\exists N$ such that $\forall n \geq N$, $S \in J_n(b) \subsetneq I_m(b)$. Moreover $V_n(\pm b) - S = o(S_n - S)$.

Since the two transformations accelerate (S_n) in cascade then $D'_n = o(S_n - S)$. We have

$$V_n(\pm b) = S_n + D_n + (1 \mp b) D'_n$$

and

$$\frac{D_n + (1 \mp b) D'_n}{S-S_n} = \frac{D_n}{S-S_n} + (1 \mp b) \frac{D'_n}{S-S_n} .$$

The first ratio in the right hand side converges to 1 since (D_n) is a perfect estimation of the error of (S_n), while the second tends to zero. Thus $(D_n + (1 \mp b) D'_n)$ is also a perfect estimation of the error of (S_n) and so $(V_n(\pm b))$ converges faster than (S_n).

Therefore the methods proposed in [13] for controlling the error in convergence acceleration processes can be explained in our framework.

REFERENCES.

[1] J. BARANGER
Approximation optimale de la somme d'une série.
C.R. Acad. Sci. Paris, 271A (1970), 149-152.

[2] A. BECKMAN, B. FORNBERG, A. TENGVALD
A method of acceleration of the convergence of infinite series.
BIT, 9 (1969), 78-80.

[3] M. BELLALIJ
Procédés de contrôle de l'erreur et accélération de la convergence.
Thèse de 3e cycle, Université de Lille 1, 1985.

[4] C. BREZINSKI
Résultats sur les procédés de sommation et l'ε-algorithme.
RIRO, R3 (1970), 147-153.

[5] C. BREZINSKI
Etudes sur les ε et ρ-algorithmes.
Numer. Math., 17 (1971), 153-162.

[6] C. BREZINSKI
Méthodes d'accélération de la convergence en analyse numérique.
Thèse d'Etat, Université de Grenoble, 1971.

[7] C. BREZINSKI
Accélération de la convergence en analyse numérique.
LNM 584, Springer-Verlag, Heidelberg, 1977.

[8] C. BREZINSKI
Algorithmes d'accélération de la convergence. Etude numérique.
Editions Technip, Paris, 1978.

[9] C. BREZINSKI
Limiting relationships and comparison theorems for sequences.
Rend. Circ. Mat. Palermo, serie 2, 28 (1979), 273-280.

[10] C. BREZINSKI
A general extrapolation algorithm.
Numer. Math., 35 (1980), 175-187.

[11] C. BREZINSKI
Padé-type approximation and general orthogonal polynomials.
ISNM vol. 50, Birkhäuser Verlag, Basel, 1980.

[12] C. BREZINSKI
Some new convergence acceleration methods.
Math. Comp., 39 (1982), 133-145.

[13] C. BREZINSKI
Error control in convergence acceleration processes.
IMA J. Numer. Anal., 3 (1983), 65-80.

[14] C. BREZINSKI
How to accelerate continued fractions.
In "Informatique et calcul", P. Chenin et al. eds, Masson, Paris, 1985.

[15] C. BREZINSKI
Composite sequence transformation.
Numer. Math., 46 (1985), 311-321.

[16] C. BREZINSKI, J.P. DELAHAYE, B. GERMAIN-BONNE
Convergence acceleration by extraction of linear subsequences.
SIAM J. Numer. Anal., 20 (1983), 1099-1105.

[17] T.J. BROMWICH
An introduction to the theory of infinite series.
Macmillan, London, 2d ed., 1949.

[18] F. CORDELLIER
Caractérisation des suites que la première étape du θ-algorithme transforme en suites constantes.
C.R. Acad. Sci. Paris, 284 A (1977), 389-392.

[19] J. DIEUDONNE
Calcul infinitésimal.
Hermann, Paris, 1968.

[20] M. EIERMANN
On the convergence of Padé-type approximants to analytic functions.
J. Comp. Appl. Math., 10 (1984), 219-227.

[21] E. FABRY
Théorie des séries à termes constants.
Hermann, Paris, 1910.

[22] W. GAUTSCHI
A note on the successive remainders of the exponential series.
El. Math., 37 (1982), 46-49.

[23] B. GERMAIN-BONNE
Transformations de suites.
RAIRO, R1 (1973), 84-90.

[24] B. GERMAIN-BONNE
Transformations non linéaires de suites.
Publication 40, Laboratoire de Calcul, Université de Lille 1, 1973.

[25] B. GERMAIN-BONNE
Estimation de la limite de suites et formalisation des procédés d'accélération de la convergence.
Thèse, Université de Lille 1, 1978.

[26] B. GERMAIN-BONNE
Conditions suffisantes d'accélération de la convergence.
in "*Padé approximation and its applications*", LNM 1071,
H. Werner and J. Bünger eds, Springer-Verlag, Heidelberg, 1984.

[27] B. GERMAIN-BONNE, J.P. DELAHAYE.
The set of logarithmically convergent sequences cannot be accelerated.
SIAM J. Numer. Anal., 19 (1982), 840-844.

[28] B. GERMAIN-BONNE, C. KOWALEWSKI
Accélération de la convergence par utilisation d'une transformation auxiliaire.
Publication ANO 77, Université de Lille 1, 1982.

[29] T. HÅVIE
Generalized Neville type extrapolation schemes.
BIT, 19 (1979), 204-213.

[30] P. HILLION
Intervalle d'approximation pour les suites à quotients de différences successives monotones.
C.R. Acad. Sci. Paris, 277 A (1973), 853-856.

[31] T. KALUZA
Über die Koeffizienten reziproker Potenreihen.
Math. Z., 28 (1928), 161-170.

[32] D.G. KENDALL
Unitary dilatations of Markov transition operators.
in "*Probability and statistics*", U. Grenander ed., Almquist and Wiksell, Stockholm, 1959.

[33] J.F.C. KINGMAN
Regenerative phenomena.
Wiley, London, 1972.

[34] C. KOWALEWSKI
Possibilités d'accélération de la convergence logarithmique.
Thèse de 3e cycle, Université de Lille 1, 1981.

[35] C. KOWALEWSKI
Accélération de la convergence pour certaines suites à convergence logarithmique.
in "*Padé approximation and its applications*", M.G. de Bruin and H. Van Rossum eds., LNM 888, Springer-Verlag, Heidelberg, 1981.

[36] D. LEVIN
Development of non-linear transformations for improving convergence of sequences.
Internat. J. Comput. Math., ser. B, 3 (1973), 371-388.

[37] L.A. LYUSTERNIK, A.R. YANPOL'SKII, eds.
Mathematical analysis.
Pergamon Press, London, 1965.

[38] A. NEY
Contribution à l'étude de la rapidité de convergence des séries à termes positifs.
Mathematica, 4 (1962), 77-105.

[39] A. NEY
Un procédé général pour l'étude de la convergence et pour le calcul des séries à termes positifs.
Mathematica, 5 (1963), 97-108.

[40] A. NEY
Observations concernant la formule d'extrapolation d'Aitken.
Rev. Anal. Numer. Th. Approx., 5 (1976), 59-62.

[41] A. NEY
Nouvelle étude sur l'accélération de la convergence des séries numériques.
Rev. Anal. Numer. Th. Approx., 7 (1978), 81-93.

[42] A. NEY
On the improvement of Kummer transformation.
Report, Dept. of Mathematical Sciences, Tel-Aviv University.

[43] Y. OKADA
Über die Annäherung analytischer Funktionen.
Math. Z., 23 (1925), 62-71.

[44] K.J. OVERHOLT
Extended Aitken acceleration.
BIT, 5 (1965), 122-132.

[45] R. PENNACCHI
Somma di serie numeriche mediante la trasformazione quadratica $T_{2,2}$.
Calcolo, 5 (1968), 51.

[46] R.E. POWELL, S.M. SHAH
Summability theory and its applications.
Van Nostrand, London, 1972.

[47] D.A. SMITH, W. F. FORD
Acceleration of linear and logarithmic convergence.
SIAM J. Numer., Anal., 16 (1979), 223-240.

[48] D.A. SMITH, W.F. FORD
Numerical comparisons of nonlinear convergence accelerators.
Math. Comp., 38 (1982), 481-499.

[49] J. WIMP, B. KLINE, A. GALARDI, D. COLTON
Some preliminary observations on an algorithm for the computation of moment integrals.
J. Comp. Appl. Math., to appear.

APPLICATIONS

Chairmen:

J. Meinguet

M. Gutknecht

Short communications:

S. Cooper
General T-fraction solutions to Riccati differential equations.

A. K. Common*
Continued fraction solutions to the Riccati equation and related lattice systems.

S. Paszkowski
Evaluation of Fermi-Dirac integral.

M. Pindor
An application of operator Padé approximants to multireggeon processes.

M. J. Rodrigues
A simple alternative principle for rational τ-method approximation.

---

* Lecture notes are not included.

GENERAL T-FRACTION SOLUTIONS TO RICCATI DIFFERENTIAL EQUATIONS

S. Clement Cooper*
Arne Magnus
Department of Mathematics
Colorado State University
Fort Collins, CO 80523 USA

William B. Jones*
Department of Mathematics
University of Colorado
Boulder, CO 80309 USA

ABSTRACT. We construct a general T-fraction solution to a Riccati differential equation. The general T-fraction corresponds to a formal power series solution of the Riccati equation at $z = 0$ and to a formal Laurent series solution at $z = \infty$. If the T-fraction converges uniformly in a neighborhood of $z = 0$, then it converges to the unique analytic solution of the Riccati equation that vanishes at $z = 0$. A similar result holds at $z = \infty$. Finally an example is given.

1. INTRODUCTION

Riccati differential equations arise in many areas of science including general relativity [9, 10], systems theory [8], and acoustics [6]. For certain Riccati equations

$$R_0[W_0(z)]:=zA_0(z) + B_0(z)W_0(z) + C_0(z)[W_0(z)]^2 -zW_0'(z)=0 \quad (1.1a)$$

where

$$W_0(0) = 0 \quad (1.1b)$$

we obtain (Section 2) general T-fractions

$$T_1[R_0(z)]:=G_0z + \mathop{K}_{n=1}^{\infty}\left[\frac{F_nz}{1+G_nz}\right], \quad F_n,G_{n-1} \in \mathbb{C}\setminus\{0\}, \ n=1,2,\ldots \quad (1.2)$$

*Research supported in part by the National Science Foundation under Grant No. DMS-84-01717.

A. Cuyt (ed.), Nonlinear Numerical Methods and Rational Approximation, 409–425.

which are formal solutions of (1.1a) at $z = 0$ and at $z = \infty$. The treatment of the case $z = \infty$ is facilitated by transforming equation (1.1a) into

$$R_0^*[V_0(\zeta)] = \zeta A_0^*(\zeta) + B_0^*(\zeta)V_0(\zeta) + C_0^*(\zeta)[V_0(\zeta)]^2 - \zeta V_0'(\zeta) = 0 \qquad (1.3a)$$

where

$$z = \frac{1}{\zeta}, \quad V_0(\zeta) = \frac{1}{\zeta} W_0\left[\frac{1}{\zeta}\right] = \frac{1}{\zeta} W_0(z), \quad V_0(0) = \frac{-B_0^*(0)}{C_0^*(0)}. \qquad (1.3b)$$

A continued fraction with n^{th} approximant f_n is said to be a <u>formal solution of a differential equation</u> $D[W(z)] = 0$ <u>at</u> $z = 0$ <u>or</u> $z = \infty$ if

$$\Lambda_0(D[f_n(z)]) = O(z^{k_n}) \quad \text{where } \lim_{n\to\infty} k_n = \infty \qquad (1.4)$$

or

$$\Lambda_\infty(D[f_n(z)]) = O_-(z^{-\ell_n}) \quad \text{where } \lim_{n\to\infty} l_n = \infty, \qquad (1.5)$$

respectively. Here $\Lambda_0(F(z))$ ($\Lambda_\infty(F(z))$) denotes the Taylor series (Laurent series) of $F(z)$ about $z = 0$ ($z = \infty$). The symbol $O(z^{k_n})$ denotes a power series (possibly divergent) whose first non-vanishing term has degree k_n or greater. $O_-(z^{-\ell_n})$ denotes a Laurent series (possibly divergent) whose nonzero term of highest degree has degree $-\ell_n$ or less.

From the theory of continued fractions [5] it is known that the general T-fraction (1.2) corresponds to a pair (L_0, L_∞) of formal Laurent series

$$L_0(z) = \sum_{k=1}^{\infty} \hat{p}_k z^k, \quad L_\infty(z) = \sum_{k=0}^{\infty} \hat{p}_k^* z^{-k} \qquad (1.6)$$

at $z = 0$ and $z = \infty$, respectively, in the sense that

$$\Lambda_0(f_n(z)) = \hat{p}_1 z + \hat{p}_2 z^2 + \ldots + \hat{p}_n z^n + O(z^{n+1}) \qquad (1.7a)$$

and

$$\Lambda_\infty(f_n(z)) = \hat{p}_0^* + \hat{p}_{-1}^* z^{-1} + \ldots + \hat{p}_{-n+1} z^{-n+1} + O_-(z^{-n}). \qquad (1.7b)$$

Here $f_n(z)$ denotes the nth approximant of $T_1[R_0(z)]$. We show (Theorem 5.1) that $L_0(z)$ and $L_\infty(z)$ are the formal power series solutions of $R_0[W_0(z)] = 0$ at $z = 0$ and $z = \infty$, respectively. Finally, it is shown (Section 6) that under mild restrictions on $A_0(z)$, $B_0(z)$ and $C_0(z)$, the general T-fraction $T_1[R_0(z)]$ and power series solutions, $P_0(z)$ and $P_\infty(z)$, may converge point-wise to solutions of (1.1). The method described in

Section 2 for obtaining the continued fraction solution is algorithmic in character and thus may yield a useful computational procedure. The theory described in Section 6 can be used to establish the convergence of the general T-fraction (1.2) to a solution of (1.1) in specific cases. An example of this is given in Section 7.

Continued fraction solutions of the Riccati equation (1.1) have been studied in a number of earlier papers. Perhaps the chief motivation for considering continued fractions and the main reason for their success is the fact that under a linear fractional transformation

$$y = \frac{\alpha(z)w + \beta(z)}{\gamma(z)w + \delta(z)} \tag{1.8}$$

a Ricatti equation

$$y'(z) = a(z) + b(z)y(z) + c(z)[y(z)]^2 \tag{1.9}$$

is transformed into another Riccati equation

$$w'(z) = \tilde{a}(z) + \tilde{b}(z)w(z) + \tilde{c}(z)[w(z)]^2. \tag{1.10}$$

Merkes and Scott exploited this property in [11] by considering

$$y = \frac{az^{\alpha}}{1 + w}. \tag{1.11}$$

Their resulting solutions were C-fractions

$$\frac{a_1 z^{\alpha_1}}{1} + \frac{a_2 z^{\alpha_2}}{1} + \frac{a_3 z^{\alpha_3}}{1} + \cdots, \quad a_n \in \mathbb{C}, \tag{1.12}$$

α_n a positive integer. Stokes [12] and Khovanskii [7] considered regular C-fractions ((1.12) with $\alpha_n = 1$ for all n); Fair [4] used Padé approximants and associated continued fractions; Ellis [3] employed J-fractions and Chisholm [1] other continued fractions. The approximants of general T-fractions employed in this paper are two-point Padé approximants and hence provide approximate solutions in neighborhoods of both $z = 0$ and $z = \infty$. For continued fraction notation used in this paper we refer to [5].

2. THE CONSTRUCTION OF $T_1[R_0(z)]$

We start with a Riccati differential equation

$$R_0[W_0(z)]=zA_0(z) + B_0(z)W_0(z) + C_0(z)W_0^2(z) - zW_0'(z)=0 \tag{2.1}$$

at $z = 0$. If certain conditions (to be stated later) are satisfied we are able to construct an associated general T-fraction

$$T_1[R_0(z)] = G_0 z + \frac{F_1 z}{1 + G_1 z} + \frac{F_2 z}{1 + G_2 z} + \cdots + \frac{F_n z}{1 + G_n z} + \cdots \tag{2.2}$$

where G_{n-1} and F_n are nonzero complex constants for n = 1, 2, ... The construction depends on the differential equation at both z = 0 and z = ∞. We obtain an equation at z = ∞ by setting

$$z = \frac{1}{\varsigma}, \quad W_0(z) = W_0\left[\frac{1}{\varsigma}\right] = U_0(\varsigma) = \frac{1}{\varsigma} V_0(\varsigma). \tag{2.3}$$

This yields the Riccati equation

$$R_0^*[V_0(\varsigma)] = -A_0\left[\frac{1}{\varsigma}\right] + (1 - B_0\left[\frac{1}{\varsigma}\right])V_0(\varsigma) - \frac{1}{\varsigma} C_0\left[\frac{1}{\varsigma}\right]V_0^2(\varsigma) - \varsigma V_0'(\varsigma) = 0 \tag{2.4}$$

which can be rewritten as

$$R_0^*[V_0(\varsigma)] = \varsigma A_0^*(\varsigma) + B_0^*(\varsigma)V_0(\varsigma) + C_0^*(\varsigma)V_0^2(\varsigma) - \varsigma V_0^1(\varsigma) = 0 \tag{2.5}$$

by letting

$$\begin{cases} \varsigma A_0^*(\varsigma) = -A_0\left[\frac{1}{\varsigma}\right] \\ B_0^*(\varsigma) = 1 - B_0\left[\frac{1}{\varsigma}\right] \\ C_0^*(\varsigma) = -\frac{1}{\varsigma} C_0\left[\frac{1}{\varsigma}\right]. \end{cases} \tag{2.6}$$

In a similar manner we transform $T_1[R_0(z)]$ to a general T-fraction in $\varsigma = \frac{1}{z}$. Substituting $\frac{1}{\varsigma}$ for z, using equivalence transformations, and finally multiplying through by ς, we obtain

$$T_1^*[R_0^*(\varsigma)] = G_0 + \frac{F_1^*\varsigma}{1 + G_1^*\varsigma} + \frac{F_2^*\varsigma}{1 + G_2^*\varsigma} + \cdots + \frac{F_n^*\varsigma}{1 + G_n^*\varsigma} + \cdots \tag{2.7}$$

where

$$F_1^* = \frac{F_1}{G_1}, \quad F_n^* = \frac{F_n}{G_{n-1}G_n}, \; n=2,3,\ldots, \text{ and } G_n^* = \frac{1}{G_n}, \; n=1,2,\ldots . \tag{2.8}$$

The strategy is to use transformations suggested by $T_1[R_0(z)]$ and $T_1^*[R_0^*(\varsigma)]$ to generate two sequences of Riccati equations

$$\{R_n[W_n(z)] = 0\}_{n=0}^{\infty} \quad \text{and} \quad \{R_n^*[V_n(\varsigma)] = 0\}_{n=0}^{\infty} \tag{2.9}$$

in which all of the Riccati equations have the same basic form. In order for the process to succeed (and make sense) it suffices to place the following restrictions on the coefficient functions

$$\begin{cases} \text{i) } A_n(z),\ B_n(z) \text{ and } C_n(z) \text{ are analytic at } z=0,\ n=0,1,\ldots \\ \text{ii) } A_n^*(\varsigma),\ B_n^*(\varsigma) \text{ and } C_n^*(\varsigma) \text{ are analytic at } \varsigma=0,\ n=0,1,\ldots \\ \text{iii) } B_0(0) \notin \mathbf{Z}^+,\ B_0^*(0) \notin \mathbf{Z}_0^-,\ C_0^*(0) \neq 0 \\ \text{iv) } A_n(0) \neq 0,\ \ A_n^*(0) \neq 0,\ n = 0,1,2,\ldots . \end{cases} \tag{2.10}$$

Notice that $A_0^*(\varsigma)$ is analytic at $\varsigma = 0$ if and only if $A_0(z)$ is analytic at $z = \infty$ and $A_0(\infty) = 0$ and similarly for $C_0^*(\varsigma)$. $B_0^*(\varsigma)$ is analytic at $\varsigma = 0$ if and only if $B_0(z)$ is analytic at $z = \infty$.

We now assume the conditions in (2.10) are satisfied for the Riccati equations listed in (2.9) and proceed with the construction of $T_1[R_0(z)]$. Starting with (2.1) we use the substitution

$$W_0(z) = G_0 z + W_1(z) \tag{2.11}$$

to obtain

$$R_1[W_1(z)] = zA_1(z)+B_1(z)W_1(z)+C_1(z)W_1^2(z)-zW_1'(z)=0 \tag{2.12}$$

where

$$\begin{cases} A_1(z) = -G_0 + A_0(z) + G_0B_0(z) + G_0^2zC_0(z) \\ B_1(z) = B_0(z) + 2G_0zC_0(z) \\ C_1(z) = C_0(z). \end{cases} \tag{2.13}$$

From (2.13), it is clear that $A_1(z)$, $B_1(z)$ and $C_1(z)$ are analytic at $z = 0$ for any choice of G_0.

From the substitution

$$V_0(\varsigma) = G_0 + V_1(\varsigma) \tag{2.14}$$

we obtain $R_1^*[V_1(\varsigma)] = 0$ where

$$\begin{cases} \varsigma A_1^*(\varsigma) = G_0B_0^*(\varsigma) + G_0^2C_0^*(\varsigma) + \varsigma A_0^*(\varsigma) \\ B_1^*(\varsigma) = B_0^*(\varsigma) + 2G_0C_0^*(\varsigma) \\ C_1^*(\varsigma) = C_0^*(\varsigma). \end{cases} \tag{2.15}$$

Clearly $B_1^*(\varsigma)$ and $C_1^*(\varsigma)$ are analytic at $\varsigma = 0$ for any choice of G_0. However, $A_1^*(\varsigma)$ is analytic at $\varsigma = 0$ if and only if

$$G_0B_0^*(0) + G_0^2C_0^*(0) = 0 \tag{2.16}$$

or since $G_0 \neq 0$, in case

$$G_0 = \frac{-B_0^*(0)}{C_0^*(0)} . \tag{2.17}$$

Condition iii) in (2.10) assures us that G_0 is a nonzero complex constant.

We use the substitutions

$$W_n(z) = \frac{F_n z}{1+G_n z + W_{n+1}(z)} \text{ and } V_n(\varsigma) = \frac{F_n^*\varsigma}{1 + G_n^*\varsigma + V_{n+1}(\varsigma)} \tag{2.18}$$

to obtain $R_{n+1}[W_{n+1}(z)] = 0$ and $R^*_{n+1}[V_{n+1}(\varsigma)] = 0$, respectively, for $n = 1,2,\dots$. The first one gives us

$$\begin{cases} zA_{n+1}(z)=1-\dfrac{A_n(z)}{F_n} - B_n(z)-\left[2\dfrac{G_n}{F_n}A_n(z)+G_nB_n(z)+F_nC_n(z)\right]z \\ \qquad\qquad -\dfrac{G_n^2}{F_n}z^2A_n(z) \\ B_{n+1}(z)=1-\dfrac{2A_n(z)}{F_n} - B_n(z) - \dfrac{2G_n}{F_n}zA_n(z) \\ C_{n+1}(z) = \dfrac{-A_n(z)}{F_n} . \end{cases} \tag{2.19}$$

For $F_n \neq 0$, $B_{n+1}(z)$ and $C_{n+1}(z)$ are analytic at $z = 0$, while $A_{n+1}(z)$ is analytic at $z = 0$ if and only if

$$F_n = \frac{A_n(0)}{1 - B_n(0)}. \tag{2.20}$$

Condition iv) in (2.10) assures us that $A_n(0) \neq 0$. It is easy to show that $B_n(0) = B_0(0)-n+1$ and hence condition iii) assures us that $B_n(0)\neq 1$. Therefore, F_n, determined by (2.20), is a nonzero complex constant.

We get expressions for $A^*_{n+1}(\varsigma)$, $B^*_{n+1}(\varsigma)$ and $C^*_{n+1}(\varsigma)$ identical to those in (2.19) except that they are in $A^*_n(\varsigma)$, $B^*_n(\varsigma)$, $C^*_n(\varsigma)$, F^*_n and G^*_n. These imply that

$$F^*_n = \frac{A^*_n(0)}{1 - B^*_n(0)} . \tag{2.21}$$

Again condition iv) in (2.10) assures us that $A^*_n(0) \neq 0$. Also, it is easy to show that $B^*_n(0) = -B^*_0(0) - n + 1$ and hence from condition iii), $B^*_n(0) \neq 1$. Thus, F^*_n is also a nonzero complex constant.

Once we have F_n and F^*_n we may compute G_n and G^*_n using (2.8),

$$G_n = \frac{F_n}{F^*_nG_{n-1}} \quad\text{and}\quad G^*_n = \frac{1}{G_n} . \tag{2.22}$$

One can verify that

$$G_{2n}=\prod_{k=1}^{n} \frac{A_{2k}(0)A^*_{2k-1}(0)}{A^*_{2k}(0)A_{2k-1}(0)}\cdot\frac{(2k+B^*_0(0))(2k-1-B_0(0))}{(2k-B_0(0))(2k-1+B^*_0(0))}, \tag{2.23}$$

$n = 1, 2, \dots,$ and

$$G_{2n+1} = \prod_{k=0}^{n} \frac{A_{2k+1}(0)}{A^*_{2k+1}(0)} \cdot \frac{(2k+1 + B^*_0(0))}{(2k+1 - B_0(0))}$$
$$\cdot \prod_{k=1}^{n} \frac{A^*_{2k}(0)(2k - B_0(0))}{A_{2k}(0)(2k + B^*_0(0))}, \quad n=0,1,\ldots . \tag{2.24}$$

We summarize the above construction as follows: Given the Riccati equation $R_0[W_0(z)] = 0$,

1. Construct $R^*_0[V_0(\varsigma)] = 0$ with coefficient functions $A^*_0(\varsigma)$, $B^*_0(\varsigma)$ and $C^*_0(\varsigma)$ given by (2.6).
2. Compute G_0 using (2.17).
3. For $n \geq 1$ compute F_n and F^*_n by means of (2.20) and (2.21).
4. For $n \geq 1$ compute G_n and G^*_n by means of (2.22).
5. For $n \geq 1$ compute $A_{n+1}(z)$, $B_{n+1}(z)$ and $C_{n+1}(z)$ by means of (2.19).
6. For $n \geq 1$ compute $A^*_{n+1}(\varsigma)$, $B^*_{n+1}(\varsigma)$ and $C^*_{n+1}(\varsigma)$ by means of equations similar to those found in (2.19).
7. Return to step 3 with the next value of n.

This algorithm will succeed provided the conditions given in (2.10) are met by the coefficient functions.

3. FORMAL POWER SERIES SOLUTIONS

We now prove two theorems about initial value problems (IVP) involving the Riccati equation (2.1).

Theorem 3.1. The initial value problem

$$\begin{cases} R_0[W_0(z)]=zA_0(z)+B_0(z)W_0(z)+C_0(z)W_0^2(z)-zW_0'(z)=0 \\ W_0(0) = 0 \end{cases} \tag{3.1}$$

has a unique formal power series (fps) solution

$$P_0(z) = \sum_{n=1}^{\infty} p_n z^n \tag{3.2}$$

at $z = 0$ provided

$$A_0(z),\ B_0(z) \text{ and } C_0(z) \text{ are analytic at } z = 0, \tag{3.3}$$

and

$$B_0(0) \text{ is not a positive integer.} \tag{3.4}$$

Proof: Since $A_0(z)$, $B_0(z)$ and $C_0(z)$ are analytic at $z = 0$ they can be expressed as

$$A_0(z) = \sum_{n=0}^{\infty} a_{n,0}z^n, \quad B_0(z) = \sum_{n=0}^{\infty} b_{n,0}z^n \text{ and } C_0(z) = \sum_{n=0}^{\infty} c_{n,0}z^n.$$

We substitute into the Riccati equation to obtain

$$\sum_{n=0}^{\infty} a_{n,0}z^{n+1}+\left[\sum_{n=0}^{\infty} b_{n,0}z^n\right]\left[\sum_{n=1}^{\infty} p_n z^n\right]+\left[\sum_{n=0}^{\infty} c_{n,0}z^n\right]\left[\sum_{n=1}^{\infty} p_n z^n\right]^2 = \sum_{n=1}^{\infty} np_n z^n \tag{3.5}$$

which can be rewritten as

$$(a_{0,0}+b_{0,0}p_1-p_1)z+\sum_{n=2}^{\infty}\left[a_{n-1,0}+\sum_{k=0}^{n-1} b_{k,0}p_{n-k}+\sum_{k=0}^{n-2} c_{k,0}\sum_{j=1}^{n-k-1} p_jp_{n-k-j}-np_n\right]z^n = 0. \tag{3.6}$$

Thus,

$$p_1 = \frac{a_{0,0}}{1-b_{0,0}} \tag{3.7}$$

which is well defined since $b_{0,0} \neq 1$, and

$$p_n = \frac{a_{n-1,0}+f_n(p_1,p_2,\ldots,p_{n-1})}{n-b_{0,0}}, \quad n = 2, 3, \ldots \tag{3.8}$$

where $f_n(p_1, p_2, \ldots, p_{n-1}) = \sum_{k=1}^{n-1} b_{k,0}p_{n-k} + \sum_{k=0}^{n-2} c_{k,0}\left[\sum_{j=1}^{n-k-1} p_jp_{n-k-j}\right]$ which is also well defined for n = 2, 3, ... since $b_{0,0} = B_0(0) \notin \mathbb{Z}^+$. ∎

In Section 2, we saw that the substitutions $z = \frac{1}{\varsigma}$ and $W_0(z) = U_0(\varsigma) = \frac{1}{\varsigma}V_0(\varsigma)$ in the Riccati equation (2.1) led to the Riccati equation (2.4). We now give a theorem about an IVP involving (2.4).

Theorem 3.2. The initial value problem

$$\begin{cases} R_0^*[V_0(\varsigma)] = \varsigma A_0^*(\varsigma) + B_0^*(\varsigma)V_0(\varsigma) + C_0^*(\varsigma)V_0^2(\varsigma) - \varsigma V_0'(\varsigma) = 0 \\ V_0(0) = \dfrac{-B_0^*(0)}{C_0^*(0)} \end{cases} \tag{3.9}$$

has a unique fps solution

$$P_\infty^*(\varsigma) = \sum_{n=0}^{\infty} p_n^* \varsigma^n \tag{3.10}$$

at $\varsigma = 0$ provided

$$A_0^*(\varsigma),\ B_0^*(\varsigma) \text{ and } C_0^*(\varsigma) \text{ are analytic at } \varsigma = 0 \text{ and} \tag{3.11}$$

$$C_0^*(0) \neq 0,\ B_0^*(0) \notin Z_0^-. \tag{3.12}$$

Proof: Since $A_0^*(\varsigma)$, $B_0^*(\varsigma)$ and $C_0^*(\varsigma)$ are analytic at $\varsigma = 0$ they can be expressed as

$$A_0^*(\varsigma) = \sum_{n=0}^{\infty} a_{n,0}^* \varsigma^n,\quad B_0^*(\varsigma) = \sum_{n=0}^{\infty} b_{n,0}^* \varsigma^n \text{ and } C_0^*(\varsigma) = \sum_{n=0}^{\infty} c_{n,0}^* \varsigma^n.$$

We substitute into the Riccati equation and simplify to obtain

$$b_{0,0}^* p_0^* + c_{0,0}^* (p_0^*)^2 + \sum_{n=1}^{\infty} \left[a_{n-1,0}^* + \sum_{k=0}^{n} b_{k,0}^* p_{n-k}^* + \sum_{k=0}^{n} c_{k,0}^* \sum_{j=0}^{n-k} p_j^* p_{n-k-j}^* - np_n \right] \varsigma^n = 0. \tag{3.13}$$

In order to satisfy the initial condition we must choose

$$p_0^* = \frac{-b_{0,0}^*}{c_{0,0}^*}, \tag{3.14}$$

which is well defined and nonzero by (3.12). We also conclude from (3.13) that

$$p_n^* = \frac{a_{n-1,0}^* + f_n^*(p_0^*, p_1^*, \ldots, p_{n-1}^*)}{b_{0,0}^* + n}, \quad n = 1, 2, \ldots \tag{3.15}$$

where

$$f_n^*(p_0^*, p_1^*, \dots, p_{n-1}^*) = \sum_{k=1}^{n} b_{k,0}^* p_{n-k}^* + c_{0,0}^* \sum_{j=1}^{n-1} p_j^* p_{n-j}^* + \sum_{k=1}^{n} c_{k,0}^* \sum_{j=0}^{n-k} p_j^* p_{n-k-j}^*$$

which is finite for $n = 1, 2, \dots$ since $B_0^*(0) \notin \mathbb{Z}_0^+$. ■

An immediate consequence of Theorem 3.2 is given in the following corollary.

Corollary 3.3. The Riccati equation

$$R_0[W_0(z)]=zA_0(z) + B_0(z)W_0(z) + C_0(z)W_0^2(z) - zW_0'(z)=0 \tag{3.16}$$

has a fLs solution at $z = \infty$

$$P_\infty(z) = p_0^* z + p_1^* + \frac{p_2^*}{z} + \frac{p_3^*}{z^2} + \dots . \tag{3.17}$$

It is uniquely determined by the condition $p_0^* \neq 0$, provided

$$A_0(z),\ B_0(z) \text{ and } C_0(z) \text{ are analytic at } z = \infty \text{ and} \tag{3.18a}$$

$$A_0(\infty) = 0 = C_0(\infty), \text{ and} \tag{3.18b}$$

$$\lim_{z\to\infty} -zC_0(z) \neq 0,\ \lim_{z\to\infty} (1 - B_0(z)) \notin \mathbb{Z}_0^-. \tag{3.19}$$

4. FORMAL CONTINUED FRACTION SOLUTIONS

In showing that $T_1[R_0(z)]$ is a formal solution to the Riccati equation (2.1) we make use of some preliminary results. We work with the sequences

$$\{R_n[W_n(z)] = 0\}_{n=0}^{\infty} \quad \text{and} \quad \{R_n^*[V_n(\zeta)] = 0\}_{n=0}^{\infty} \tag{4.1}$$

that appeared in Section 2 in the construction of $T_1[R_0(z)]$.

It is easy to show that

$$R_0^*[f(\zeta)] = -\zeta R_0[\tfrac{1}{\zeta}f(\zeta)] = -\tfrac{1}{z}R_0[zf\left[\tfrac{1}{z}\right]] \tag{4.2}$$

for functions f such that f is analytic at $\zeta = 0$ and $zf\left[\frac{1}{z}\right]$ is analytic at $z = 0$.

In Section 2 we saw that

$$T_1[R_0(z)]=G_0 z+ \mathop{K}_{n=1}^{\infty} \frac{F_n z}{1+ G_n z} \text{ and } T_1^*[R_0^*(\zeta)]=G_0+ \mathop{K}_{n=1}^{\infty} \frac{F_n^*\zeta}{1+ G_n^*\zeta} \tag{4.3}$$

where $F^*_1 = \frac{F_1}{G_1}$, $F^*_n = \frac{F_n}{G_{n-1}G_n}$, $n = 2, 3, \ldots$, and $G^*_n = \frac{1}{G_n}$, $n = 1, 2, \ldots$.

Since $T^*_1[R^*_0(\zeta)]$ was obtained from $T_1[R_0(z)]$ by replacing z by $\frac{1}{\zeta}$, an equivalence transformation, and multiplying by ζ, the approximants $f_n(z)$ of $T_1[R_0(z)]$ and the approximants $f^*_n(\zeta)$ of $T^*_1[R^*_0(\zeta)]$ are related by

$$f_n(z) = \frac{1}{\zeta} f^*_n(\zeta) \quad n = 1, 2, \ldots . \tag{4.4}$$

In generating the two sequences of Riccati equations (4.1) we use the transformations

$$W_0(z) = G_0 z + W_1(z) \text{ and } W_n(z) = \frac{F_n z}{1 + G_n z + W_{n+1}(z)}, \; n = 1, 2, \ldots \tag{4.5}$$

and

$$V_0(\zeta) = G_0 + V_1(\zeta) \text{ and } V_n(\zeta) = \frac{F^*_n \zeta}{1 + G^*_n \zeta + V_{n+1}(\zeta)}, \; n = 1, 2, \ldots . \tag{4.6}$$

Direct substitutions yield the following relations,

$$R_0[W_0(z)] = R_1[W_1(z)]$$

and

$$R_n[W_n(z)] = \frac{-F_n z}{(1 + G_n z + W_{n+1}(z))^2} R_{n+1}[W_{n+1}(z)]. \tag{4.7}$$

Thus, we have

$$R_0[W_0(z)] = \frac{(-1)^n F_1 \cdots F_n z^n \cdot R_{n+1}[W_{n+1}(z)]}{[(1+G_1 z + W_2(z)) \cdots (1+G_n z + W_{n+1}(z))]^2}, \; n = 1, 2, \ldots . \tag{4.8}$$

In a similar manner we see that

$$R^*_0[V_0(\zeta)] = \frac{(-1)^n F^*_1 \cdots F^*_n \zeta^n \cdot R^*_{n+1}[V_{n+1}(\zeta)]}{[(1+G^*_1 \zeta + V_2(\zeta)) \cdots (1+G^*_n \zeta + V_{n+1}(\zeta))]^2}, \; n = 1, 2, \ldots . \tag{4.9}$$

Let M_n and N_n denote the nth numerator and n^{th} denominator, respectively, of $T_1[R_0(z)]$ and M^*_n and N^*_n denote the nth numerator and n^{th} denominator, respectively, of $T^*_1[R^*_0(\zeta)]$. Setting $W_{n+1}(z) = 0$ we obtain

$$W_0(z) = \frac{M_n}{N_n} \tag{4.10}$$

and

$$W_k(z) = \frac{F_k z}{1+G_k z} + \frac{F_{k+1} z}{1+G_{k+1} z} + \cdots + \frac{F_n z}{1+G_n z}, \; k = 1, \ldots, n \tag{4.11}$$

and therefore by (4.8)

$$R_0\left[\frac{M_n}{N_n}\right]=\frac{(-1)^n F_1 \ldots F_n z^n \cdot zA_{n+1}(z)}{[(1+G_1z+W_2(z))\ldots(1+G_{n-1}z+W_n(z))(1+G_nz)]^2}, \quad n=1,2,\ldots \quad (4.12)$$

Since $W_k(z) = O(z)$, we have $\frac{1}{1 + G_{k-1}z + W_k(z)} = O(1)$ and hence

$$R_0\left[\frac{M_n}{N_n}\right] = O(z^{n+1}), \; n = 1, 2, \ldots . \quad (4.13)$$

By a similar argument

$$R_0^*\left[\frac{M_n^*}{N_n^*}\right] = O(\varsigma^{n+1}), \; n = 1, 2, \ldots . \quad (4.14)$$

We are now ready to prove the following theorem.

Theorem 4.1. The general T-fraction

$$T_1[R_0(z)] = G_0z + \frac{F_1z}{1+G_1z} + \cdots + \frac{F_nz}{1+G_nz} + \cdots \quad (4.15)$$

constructed from the Riccati equation

$$R_0[W_0(z)]=zA_0(z) + B_0(z)W_0(z) + C_0(z)W_0^2(z) - zW_0'(z)=0 \quad (4.16)$$

(as described in Section 2) is a formal solution to $R_0[W_0(z)]=0$ at $z = 0$ and $z = \infty$.

Proof: From (4.13) we have

$$R_0[f_n(z)] = O(z^{n+1}), \; n = 1, 2, \ldots . \quad (4.17)$$

Therefore $T_1[R_0(z)]$ is a formal solution to $R_0[W_0(z)] = 0$ at $z = 0$.

By (4.14)

$$R_0^*[f_n^*(\varsigma)] = O(\varsigma^{n+1}), \; n = 1, 2, \ldots \quad (4.18)$$

and from (4.2) and (4.4) we have

$$R_0^*[f_n^*(\varsigma)]=-\varsigma R_0[\tfrac{1}{\varsigma}f_n^*(\varsigma)]=-\varsigma R_n[f_n(z)]=-\tfrac{1}{z}R_0[f_n(z)], \quad n=1,2,\ldots \quad (4.19)$$

Therefore,

$$R_0[f_n(z)] = O_-(z^{-n}), \; n = 1, 2, \ldots \quad (4.20)$$

and hence $T_1[R_0(z)]$ is also a formal solution to the Riccati equation $R_0[W_0(z)] = 0$ at $z = \infty$. ∎

5. $T_1[R_0(z)]$ AND THE FORMAL POWER SERIES SOLUTIONS

In the introduction we mentioned that a general T-fraction

$$G_0z + \frac{F_1z}{1+G_1z} + \cdots + \frac{F_nz}{1+G_nz} + \cdots \tag{5.1}$$

where $F_n \neq 0$ and $G_n \neq 0$ for n = 1, 2, ... corresponds to a fps at z = 0 and to a fLs in $\varsigma = \frac{1}{z}$ at $\varsigma = 0$ ($z = \infty$). The next theorem establishes the connection between these two series, the fps solution in Theorem 3.1 and the fLs solution mentioned in Corollary 3.3.

Theorem 5.1. The general T-fraction $T_1[R_0(z)]$, as constructed in Section 2, corresponds to the pair of series (P_0, P_∞) defined in Theorem 3.1 and Corollary 3.3.

Proof: The relationship between a general T-fraction and its corresponding power series, $L_0(z)$, at z = 0 is characterized by

$$L_0(z) - f_n(z) = O(z^{n+1}),\ n = 1, 2, \ldots \tag{5.2}$$

where $f_n(z)$ is the nth approximant of the general T-fraction in question. Let $f_n(z)$ be the nth approximant of $T_1[R_0(z)]$ and let $L_0(z)$ be its corresponding power series at z = 0. Then by (5.2) we see that $L_0(0)=0$. Now compute

$$\begin{aligned} R_0[L_0(z)] &= R_0[f_n(z) + (L_0(z) - f_n(z))] \\ &= zA_0(z) + B_0(z)[f_n(z) + (L_0(z) - f_n(z))] \\ &\quad + C_0(z)[f_n(z) + (L_0(z) - f_n(z))]^2 \\ &\quad - z[f_n(z) + (L_0 - f_n(z))]' \\ &= R_0[f_n(z)] + B_0(z)(L_0(z) - f_n(z)) \\ &\quad + C_0(z)(L_0^2(z) - f_n^2(z)) - z(L_0'(z) - (f_n(z))') \end{aligned}$$

(the derivative of $L_0(z)$ is taken formally)

$$= O(z^{n+1}) + O(z^{n+1}) + O(z^{n+1}) + O(z^{n+1})$$

since $R_0[f_n(z)] = O(z^{n+1})$ (see (4.13)). Thus, $R_0[L_0(z)] = O(z^{n+1})$ for all n and hence $R_0[L_0(z)] = 0$. The fps solution that vanishes at z = 0 is unique and hence $L_0(z) = P_0(z)$.

Let $L_\infty(z)$ be the Laurent series corresponding to $T_1[R_0(z)]$ at $z = \infty$. The relationship between a general T-fraction and its corresponding Laurent series at $z = \infty$ is characterized by

$$L_\infty(z) - f_n(z) = O_-(z^{-n}), \; n = 1, 2, \ldots . \tag{5.3}$$

Thus,

$$\frac{1}{z}L_\infty(z) - \frac{1}{z}f_n(z) = O_-(z^{-n-1}), \; n = 1, 2, \ldots . \tag{5.4}$$

Letting $\zeta = \frac{1}{z}$ and $L^*_\infty(\zeta) = \zeta L_\infty\left[\frac{1}{\zeta}\right]$ and recalling that $f_n(z) = \frac{1}{\zeta}f^*_n(\zeta)$ (see (4.4)) we see that (5.4) can be rewritten as

$$L^*_\infty(\zeta) - f^*_n(\zeta) = O(\zeta^{n+1}), \; n = 1, 2, \ldots . \tag{5.5}$$

It follows that $L^*_\infty(0) = G_0 \neq 0$ and $L_\infty(z)$ has a simple pole at $z = \infty$.

By an argument completely analogous to that in the first half of the proof we have

$$R^*_0[L^*_\infty(\zeta)] = O(\zeta^{n+1}), \; n = 1, 2, \ldots \tag{5.6}$$

and hence $R^*_0[L^*_\infty(\zeta)] = 0$. Therefore,

$$L^*_\infty(\zeta) = P^*_\infty(\zeta) \tag{5.7}$$

where $P^*_\infty(\zeta)$ is the fps appearing in Theorem 3.2. Therefore, $L_\infty(z)$ is the fLs solution to $R_0[W_0(z)] = 0$ at $z = \infty$ mentioned in Corollary 3.3. ∎

6. SOLUTIONS TO $R_0[W_0(z)] = 0$

Solutions to the Riccati equation $R_0[W_0(z)] = 0$ may be obtained from $T_1[R_0(z)]$.

Theorem 6.1. Suppose the Riccati equation

$$R_0[W_0(z)] = zA_0(z)+B_0(z)W_0(z)+C_0(z)W_0^2(z)-zW_0'(z) = 0 \tag{6.1}$$

admits the formal general T-fraction solution $T_1[R_0(z)]$. Moreover, suppose $T_1[R_0(z)]$ converges uniformly in a neighborhood of $z = 0$ to a function $W(z)$. Then $W(z)$ is the unique solution of $R_0[W_0(z)] = 0$, analytic in a region containing $z = 0$ with $W(0) = 0$.

Proof: Let $f_n(z)$ be the nth approximant of $T_1[R_0(z)]$. By Theorem 5.13 [5], $W(z) = \lim_{n\to\infty} f_n(z)$ is analytic in a neighborhood of $z = 0$ and the Taylor series expansion of $W(z)$ is the power series $L_0(z)$ to which $T_1[R_0(z)]$ corresponds at $z = 0$. By Theorem 5.1 $L_0(z)$ is a fps solution of $R_0[W_0(z)] = 0$ at $z = 0$. It is therefore a solution of $R_0[W_0(z)] = 0$ in the neighborhod of $z = 0$ in which it converges. The assertion follows from the fact that $W(z) = L_0(z)$ for z in this neighborhood of $z = 0$. ∎

Meromorphic solutions to the Riccati equation $R_0[W_0(z)] = 0$ at $z = \infty$ may also be obtained from $T_1[R_0(z)]$.

Theorem 6.2. Suppose the Riccati equation

$$R_0[W_0(z)] = zA_0(z)+B_0(z)W_0(z)+C_0(z)W_0^2(z)-zW_0'(z) = 0 \qquad (6.2)$$

admits the formal general T-fraction solution $T_1[R_0(z)]$. Moreover, suppose $T_1^*[R_0^*(\varsigma)] = \varsigma T_1[R_0(\varsigma)]$ (where $\varsigma = \frac{1}{z}$) converges uniformly in a neighborhood of $\varsigma = 0$ to a function $V(\varsigma)$, then $W(z) = \frac{1}{\varsigma}V(\varsigma)$ is the unique solution of $R_0[W_0(z)] = 0$ at $z = \infty$ which has a simple pole at $z = \infty$.

Proof: Very similar to that of Theorem 6.1.

7. AN EXAMPLE

Merkes and Scott [11] found a C-fraction solution to the Riccati equation

$$R_0[W_0(z)]=\frac{\frac{a(b-c)}{c}z}{1-z} + \frac{(-c+(b-a)z)}{1-z}W_0(z)- \frac{cW_0^2(z)}{1-z} -zW_0'(z)=0 \qquad (7.1)$$

at $z = 0$. This differential equation has the analytic solution

$$W(z) = \frac{F(a,\ b;\ c;\ z)}{F(a,\ b+1;\ c+1;\ z)} - 1 \qquad (7.2)$$

which vanishes at $z = 0$. We use our method to find a general T-fraction that converges to (7.2) in a neighborhood of $z = 0$. It also converges to a similar function at $z = \infty$.

At $z = \infty$ we have

$$R_0^*[V_0(\varsigma)]= \frac{\frac{a(b-c)}{c}\varsigma}{1-\varsigma} + \frac{((b-a+1)-(c+1)\varsigma)}{1-\varsigma}V_0((\varsigma) - \frac{c[V_0(\varsigma)]^2}{1-\varsigma} -\varsigma V_0'(\varsigma)=0. \qquad (7.3)$$

Following the algorithm for computing $T_1[R_0(z)]$ we find that

$$\left.\begin{aligned}
G_0 &= \frac{b-a+1}{c} \\
A_n(z) &= \frac{(b+n)(a-c-n)}{(c+n-1)(1-z)},\quad n=1,2,\ldots \\
B_n(z) &= \frac{-(c+n-1)+(a-b-(n+1))z}{1-z},\quad n=1,2,\ldots \\
C_n(z) &= \frac{-(c+n-1)}{1-z},\quad n=1,2,\ldots \\
F_n &= \frac{(b+n)(a-c-n)}{(c+n-1)(c+n)},\quad n=1,2,\ldots \\
G_n &= \frac{b-a+n+1}{c+n},\quad n=1,2,\ldots \\
A_1^*(\zeta) &= \frac{(b+1)(a-c-1)}{c},\quad A_n^*(\zeta)=\frac{(b+n)(a-c-n)}{(b-a+n)(1-\zeta)},\quad n=2,3,\ldots \\
B_n^*(\zeta) &= \frac{(a-b-n)-(c+n)\zeta}{1-\zeta},\quad n=1,2,\ldots \\
C_1^*(\zeta) &= \frac{-c}{1-\zeta},\quad C_n^*(\zeta)=\frac{a-b-n}{1-\zeta},\quad n=2,3,\ldots \\
F_1^* &= \frac{(b+1)(a-c-1)}{c(b-a+2)},\quad F_n^*=\frac{(b+n)(a-c-n)}{(b-a+n)(b-a+n+1)},\quad n=2,3,\ldots \\
G_n^* &= \frac{c+n}{b-a+n+1},\quad n=1,2,\ldots\,.
\end{aligned}\right\} \tag{7.4}$$

Therefore,

$$T_1[R_0(z)] = \frac{b-a+1}{c}z + \mathop{K}_{n=1}^{\infty}\left[\frac{\frac{(b+n)(a-c-n)}{(c+n-1)(c+n)}z}{1+\frac{(b-a+n+1)}{c+n}z}\right] \tag{7.5}$$

and

$$T_1^*[R_0^*(\zeta)] = \frac{b-a+1}{c} + \frac{\frac{(b+1)(a-c-1)}{c(b-a+2)}\zeta}{1+\frac{c+1}{b-a+2}\zeta} + \mathop{K}_{n=2}^{\infty}\left[\frac{\frac{(b+n)(a-c-n)}{(b-a+n)(b-a+n+1)}\zeta}{1+\frac{c+n}{b-a+n+1}\zeta}\right]. \tag{7.6}$$

In [2] the authors showed that $T_1[R_0(z)]$ converges to

$$W(z) = \frac{F(a,\ b;\ c;\ z)}{F(a,\ b+1;\ c+1;\ z)} - 1 \tag{7.7}$$

at $z = 0$ and to

$$W(z) = \frac{b-a+1}{c}z\,\frac{F(1-a,\ c-a;\ b-a+1;\ \frac{1}{z})}{F(1-a,\ c-a+1;\ b-a+2;\ \frac{1}{z})} - 1 \tag{7.8}$$

at $z = \infty$.

REFERENCES

1. Chisolm, J.S.R., 'Continued fraction solution of the general Riccati equation', Rational Approximation and Interpolation, Proc. of the U.K.-U.S. Conf., Tampa, FL, 1983, Lecture Notes in Mathematics 1105 (Springer-Verlag, Berlin, 1984), 109-116.

2. Cooper, S.Clement, William B. Jones, and Arne Magnus, 'General T-fraction expansions for ratios of hypergeometric functions', to appear in Appl. Numer. Math.

3. Ellis, Homer G., 'Continued fraction solutions of the general Riccati differential equation', Rocky Mountain J. Math., V. 4, no. 2 (1974), 353-6.

4. Fair, Wyman, 'Padé approximation to the solution of the Riccati equation', Math. of Comp. 18 (1964), 627-634.

5. Jones, William B. and W. J. Thron, Continued Fractions: Analytic Theory and Applications, Encyclopedia of Mathematics and Its Applications 11, Addison-Wesley Publ. Co., Reading, MA (1980), distributed now by Cambridge Univ. Press, NY.

6. Kergomard, J., 'Continued fraction solution of the Riccati equation: Applications to acoustic horns and layered-inhomogeneous media, with equivalent electrical circuits', to appear in Wave Motion.

7. Khovanskii, A. N., The Application of Continued Fractions and Their Generalizations to Problems in Approximation Theory, (translated by Peter Wynn), Noordhoff, Groningen, 1963.

8. Lamb, Alan J., 'Algebraic aspects of generalized eigenvalue problems for solving Riccati equations', Computational and Combinatorial Methods in Systems Theory, (C. F. Byrnes and A. Lindquist, eds.) Elsevier Science Publishers B. V. (North Holland) (1986), 213-227.

9. McVittie, G. C., 'The mass-particle in an expanding universe', Mon.Not.Roy.Ast.Soc. 93, 325 (1933).

10. McVittie, G. C., 'Elliptic functions in spherically symmetric solutions of Einstein's equations', Ann. Inst. Henri Poincaré 40, 3, 231 (1984).

11. Merkes, E. P. and W. T. Scott, 'Continued fraction solutions of the Riccati equation', J. Math. Anal. Appl., V. 4 (1962), 309-327.

12. Stokes, A. N., 'Continued fraction solutions of the Riccati equation', Bull. Austral. Math. Soc., V. 25 (1982), 207-214.

A SIMPLE ALTERNATIVE PRINCIPLE FOR RATIONAL τ - METHOD APPROXIMATION

Manuel R. da Silva
Maria João Rodrigues
Grupo de Matemática Aplicada
Faculty of Sciences, University of Porto
4000 Porto - Portugal

ABSTRACT. Given an equation of the form $Dy = f$, where f is an algebraic polynomial and D a linear (algebraic, differential or integral) operator mapping polynomials into polynomials, together with $\nu \geq 0$ supplementary conditions on y, the basic idea of Lanczos' τ - method is to perturb the given equation through the addition to its r.h.s. of an algebraic polynomial, H_n, so chosen that the perturbed problem, $Dy_n = f + H_n$, has a unique polynomial solution, y_n, satisfying the given supplementary conditions.

To circumvent the difficulty lying in the choice of H_n, the first author has recently proposed the following alternative principle for polynomial τ - method approximation of y: take a basis (preferably orthogonal for rapid convergence) for the space of algebraic polynomials of degree $\leq n$, express y_n in it, and determine its coefficients by making y_n satisfy the given supplementary conditions and Dy_n agree with Dy as far as possible or desired.

Our aim here is to extend that alternative principle to construct rational τ - method approximants to functions defined by operator equations of the kind referred to above and examples will be given by way of illustration.

1. INTRODUCTION

Given a problem of the form

$$Dy(x) = f(x) \quad , \quad a \leq x \leq b \quad , \quad |a|, \ |b| < \infty \tag{1.1}$$

$$g_j(y) = \sigma_j \quad , \quad j = 1(1)\nu \quad , \tag{1.2}$$

where D is a νth order linear ordinary differential operator with polynomial coefficients,

A. Cuyt (ed.), Nonlinear Numerical Methods and Rational Approximation, 427–434.

$$D \equiv \sum_{r=0}^{\nu} p_r(x) \frac{d^r}{dx^r} ,$$

$f(x) = f_0 + f_1 x + \ldots + f_N x^N$, and the g_j's are given linear functionals representing the supplementary (initial, boundary or mixed) conditions in (1.2) through linear combinations of function and derivative values of y , the basic idea of the τ-method, as conceived by Lanczos in [8], [9] and [10] for the construction of a polynomial approximation y_n of y , is to perturb the given ODE (1.1) through the addition to its r.h.s. of a polynomial H_n , usually chosen to be a linear combination of Chebyshev or Legendre polynomials [11] with free coefficients, called the τ-parameters, which are to be determined so that y_n is the unique polynomial solution of the following τ-problem

$$Dy_n(x) = f(x) + H_n(x) \quad , \quad a \leq x \leq b \tag{1.3}$$

$$g_j(y_n) = \sigma_j \quad , \quad j = 1(1)\nu . \tag{1.4}$$

The choice of H_n , however, is not a simple matter [14], as it depends essentially on the very nature of D , and so, as an alternative to the Lanczos' original perturbation idea, the following approximation principle has evolved.

Take a basis $v = \{v_k(x)\}_{k=0(1)n}$ (preferably orthogonal for rapid convergence) for the space of algebraic polynomials of degree $\leq n$, express y_n in it,

$$y_n = \sum_{k=0}^{n} \alpha_k^{(n)} v_k ,$$

and determine the coefficients $\alpha_k^{(n)}$ by making y_n satisfy the supplementary conditions and Dy_n agree with Dy as far as possible.

Emerging from [7], [12], and [13], this approximation principle has recently been shown in [15] to lead naturally to good polynomial approximants y_n of y in the sense of the τ-method.

It is our aim here to use this principle to construct rational τ-method approximants to functions defined in terms of a linear operator mapping polynomials into polynomials, such as D as in (1.1) or its integrated forms.

2. OPERATIONAL CONSTRUCTION OF POLYNOMIAL τ-APPROXIMANTS

Here, we will follow the notation in [15]. By furnishing zero coefficients, if need be, all vectors in the sequel are infinite-dimensional. Column-vectors are underlined once and row-vectors twice.

If we let $y = \underline{\underline{a}}\,\underline{x}$ be the formal power series solution of (1.1) - -(1.2), then $\underline{\underline{a}}$ is such that

$$\underline{\underline{a}}\,\Gamma_x = \underline{\underline{b}} \tag{2.1}$$

where

$$\underline{\underline{b}} = (\sigma_1,\ldots,\sigma_\nu\,;\,f_0,\ldots,f_N,\,0,\,0,\ldots)$$

$$\Gamma_x = (B_x\,;\,\Pi_x)$$

$$B_x = (\beta_{ij})\ ,\quad \beta_{ij} = g_j(x^i)\ ,\quad j = 1(1)\nu\ ;\quad i = 0,\,1,\ldots$$

$$\Pi_x = \sum_{r=0}^{\nu} \eta^r\, p_r(\mu)\ ,$$

$$\eta = \begin{bmatrix} 0 & & & & \\ 1 & 0 & & & \\ & 2 & 0 & & \\ & & 3 & 0 & \\ & & & & \cdots \end{bmatrix}\ ,\qquad \mu = \begin{bmatrix} 0 & 1 & & & & \\ & 0 & 1 & & & \\ & & 0 & 1 & & \\ & & & 0 & 1 & \\ & & & & & \cdots \end{bmatrix}\ .$$

Truncation of the system (2.1) to its first $n+1$ equations,

$$\underline{\underline{a}}^{(n)}\,\Gamma_x = \underline{\underline{b}}\ ,$$

leads to the coefficient vector $\underline{\underline{a}}^{(n)}$ of the polynomial solution of the τ-problem (1.3) - (1.4),

$$y_n(x) = \underline{\underline{a}}^{(n)}\,\underline{x}\ .$$

If we let $y = \underline{\underline{\alpha}}\,\underline{v}$ be the v-series expansion of the solution of (1.1) - (1.2), $\underline{v} = V\underline{x}$, then $\underline{\underline{\alpha}}$ is such that

$$\underline{\underline{\alpha}}\,\Gamma_v = \underline{\underline{\beta}}\ , \tag{2.2}$$

where

$$\underline{\underline{\beta}} = (\sigma_1,\ldots,\sigma_\nu\,;\,F_0,\ldots,F_N,\,0,\,0,\ldots)\ ,\quad \underline{\underline{F}} = \underline{\underline{f}}\,V^{-1}\ ,$$

$$\Gamma_v = (B_v\,;\,\Pi_v)\ ,\quad B_v = V\,B_x\ ,\quad \Pi_v = V\,\Pi_x\,V^{-1}\ .$$

Truncation of the system (2.2) to its first $n+1$ equations,

$$\underline{\underline{\alpha}}^{(n)}\,\Gamma_v = \underline{\underline{\beta}}\ ,\qquad n \geq N+\nu\ ,$$

leads to the coefficient vector $\underline{\underline{a}}^{(n)}$ of

$$y_n = \underline{\underline{a}}^{(n)} \underline{v} ,$$

which solves the τ - problem (1.3) - (1.4) with the perturbation

$$H_n = \sum_{i=1}^{\nu+h} \tau_i^{(n)} v_{n-\nu+i} , \quad \tau_i^{(n)} = \underline{\underline{a}}^{(n)} \Pi_\nu \underline{e}_{n-\nu+i} , \quad i = 1(1)\nu+h ,$$

$\underline{e}_j$ being the jth column of the identity matrix and h the height of D (see [15] for details).

3. RATIONAL τ - METHOD APPROXIMATION

3.1. In what follows we will be mainly interested in endpoint approximation [11], and so we will take $v = \{P_k^*(x)\}_{k=0,1,\ldots}$, the Legendre polynomials shifted to $[0, 1]$.

Assuming that a τ - approximation is required for the solution y(x) of (1.1) - (1.2) on the interval $[0, \gamma]$, where γ is a nonzero parameter, we take

$$v_k(x) = P_k^*(x/\gamma) , \quad k = 0, 1, \ldots$$

and the solution of the τ - problem (1.3) - (1.4) will appear in the form

$$y_n(x) = \sum_{k=0}^{n} \alpha_k^{(n)} P_k^*(x/\gamma) , \quad \alpha_k^{(n)} = \frac{p_k(1/\gamma)}{q_n(1/\gamma)} ,$$

where $p_k(1/\gamma)$ and $q_n(1/\gamma)$ are algebraic polynomials in $1/\gamma$, and thus $y_n(x)$ will be a polynomial whenever $1/\gamma$ is a number, and a ratio of two polynomials when $x/\gamma = 1$.

<u>Example</u> 1.

$$Dy \equiv y' + y = 0 , \quad 0 \le x \le 1$$
$$y(0) = 1 .$$

Legendre polynomials on $0 \le x \le \gamma$:

$$P_0^* = 1 , \quad P_1^* = 2x/\gamma - 1 ,$$

$$P_{n+1}^* = \frac{2n+1}{n+1} P_1^* P_n^* - \frac{n}{n+1} P_{n-1}^* , \quad n = 1, 2, \ldots$$

$$DP_0^* = P_0^*$$
$$DP_1^* = 2/\gamma\, P_0^* + P_1^*$$
$$DP_2^* = \qquad 6/\gamma\, P_1^* + P_2^*$$
$$\vdots$$

$$\Pi_v = \begin{bmatrix} 1 & & & \\ 2/\gamma & 1 & & \\ & 6/\gamma & 1 & \\ & & \ddots & \ddots \end{bmatrix}$$

$$y_2 = \alpha_0^{(2)} P_0^* + \alpha_1^{(2)} P_1^* + \alpha_2^{(2)} P_2^* \quad , \quad \Gamma_v = \left[\begin{array}{c|c} 1 & \\ -1 & \Pi_v \\ 1 & \\ \vdots & \end{array}\right]$$

$$\begin{cases} \alpha_0^{(2)} - \alpha_1^{(2)} + \alpha_2^{(2)} = 1 \\ \alpha_0^{(2)} + 2/\gamma\, \alpha_1^{(2)} = 0 \\ \alpha_1^{(2)} + 6/\gamma\, \alpha_2^{(2)} = 0 \end{cases} , \tag{3.1}$$

$$\alpha_0^{(2)} = 12/(\gamma^2 q_2(1/\gamma)) \quad , \quad \alpha_1^{(2)} = -6/(\gamma q_2(1/\gamma)) \quad ,$$
$$\alpha_2^{(2)} = 1/q_2(1/\gamma)$$
$$q_2(1/\gamma) = 1 + 6/\gamma + 12/\gamma^2$$

$$y_2 = (12/\gamma^2 P_0^*(x/\gamma) - 6/\gamma\, P_1^*(x/\gamma) + P_2^*(x/\gamma)) \,/\, q_2(1/\gamma) \tag{3.2}$$

$$x/\gamma = 1 \;\rightarrow\; y_2 = \frac{12 - 6x + x^2}{12 + 6x + x^2} \; .$$

It is worth noticing that

i) $y_2(-x) \cdot y_2(x) = 1$, in correspondence with $e^{-x} \cdot e^{x} = 1$

ii) $y_2(x) = [2/2]_{e^{-x}}$, diagonal Padé approximant

iii) $\| e^{-x} - y_2(x) \|_\infty = 5.4 \times 10^{-4}$

iv) The approximant y_2 in (3.2) will be a polynomial when $1/\gamma$ is a fixed number, a rational function when $x/\gamma = 1$. In either case, however, the system (3.1) for the coefficients of y_2 will be the same.

3.2. Turning to segmented rational τ-method approximation, let $y_2^{[i]}(x)$ be a τ-approximant of e^{-x} on $[x_i , x_{i+1}]$, $x_i = i / 10$, $i = 0(1)9$,

$$y_2^{[0]}(x) = \frac{12 - 6x + x^2}{12 + 6x + x^2}, \quad 0 \le x \le 0.1,$$

$y_i = y_2^{[i-1]}(x_i)$, $i = 1(1)10$, and let

$$y_2^{[i]}(x) = \alpha_0 P_0^* + \alpha_1 P_1^* + \alpha_2 P_2^*, \quad P_j^* = p_j^*((x - x_i)/\gamma)), \quad \gamma = 1/10$$

be the τ-solution of the IVP

$$Dy \equiv y' + y = 0, \quad x_i \le x \le x_{i+1},$$

$$y(x_i) = y_i .$$

The linear algebraic equations for the α_j's are those in (3.1) with the first changed to

$$\alpha_0 - \alpha_1 + \alpha_2 = y_i ,$$

and thus

$$\alpha_0 = y_i \frac{12}{\gamma^2 q_2(1/\gamma)}, \quad \alpha_1 = -y_i \frac{6}{\gamma q_2(1/\gamma)}, \quad \alpha_2 = y_i \frac{1}{q_2(1/\gamma)}$$

$$y_2^{[i]}(x) = y_i \frac{12 - 6(x - x_i) + (x - x_i)^2}{12 + 6(x - x_i) + (x - x_i)^2}, \quad i = 0(1)9,$$

$$y_o = 1$$

$$y_{i+1} = y_2^{[i]}(x_{i+1}) = \frac{1141}{1261} y_i, \quad i = 0(1)9,$$

$$\max_{1 \le i \le 10} |e^{-x_i} - y_i| \le 5.2 \times 10^{-8} .$$

There is no difficulty in the extension of this to the segmented rational τ-method approximation of the solution $y(x)$ of (1.1) - (1.2). The linear systems for the coefficients of $y_n^{[i]}(x)$ will be all the same, except for the first ν equations.

3.3. Let us suppose that the exact solution of a given problem is a rational function,

$$y = P/Q, \quad P = p_0 + p_1 x + \dots, \quad Q = 1 + q_1 x + \dots$$

It might be interesting to understand under what conditions the τ-method will reproduce it.

The rational function y is obviously the exact solution of the IVP

$$(Qy)' = P' , \quad y(0) = p_0$$

and the corresponding τ-solution y_n is such that

$$(Qy_n)' = P' + H_n(x/\gamma) , \quad 0 \leq x \leq \gamma$$

$$y_n(0) = p_0 .$$

The τ-error function, $\eta_n(x, \gamma) = y_n(x, \gamma) - y(x)$, is such that

$$(Q\eta_n)' = H_n(x/\gamma) , \quad \eta_n(0, \gamma) = 0 ,$$

therefore

$$Q(x)\, \eta_n(x, \gamma) = \int_0^x H_n(\xi/\gamma)\, d\xi$$

$$Q(x)\, \eta_n(x, x) = \int_0^x H_n(\xi/x)\, d\xi$$

$$= x \int_0^1 H_n(t)\, dt ,$$

so $\eta_n(x, x) \equiv 0$ for $n \geq 1$ iff $\int_0^1 H_n(t)\, dt = 0$ and this will be the case provided that $H_n(t) = \tau P_n^*(t)$.

4. τ - APPROXIMATION IN THE COMPLEX PLANE

To extend our rational τ-approximation to a segment $[0, z]$ of the complex plane, it suffices to take $\gamma = z$, the only condition being that $[0, \gamma]$ should not contain any singularity of the function $y(z)$ to be approximated.

If we want to approximate $y(z)$ throughout a given region of the complex plane, we take a polynomial basis which is closely associated with that region, namely the Faber polynomials (see [1] - [6] and references given therein) if they are available, and use the approximation principle in §1 to generate polynomial and rational τ-method approximations in terms of those polynomials.

ACKNOWLEDGMENT

We thank John Coleman, Durham University, for having sent us some comments on rational approximation by the Lanczos' τ-method.

REFERENCES

1. J.P. COLEMAN and R.A. SMITH, 'The Faber Polynomials for Circular Sectors', to appear in Math. Comp., 1987.

2. J.P. COLEMAN, 'Complex Polynomial Approximation by the Lanczos' τ - Method : Dawson's Integral', presented at the ICCAM, Leuven, 1986.

3. ————, 'Polynomial Approximations in the Complex Plane', to appear in JCAM.

4. S.W. ELLACOTT, 'Computation of Faber Series with Application to Numerical Polynomial Approximation', Math. Comp. **40** (1983), 575 - 587.

5. ———— and M.H. GUTKNECHT, 'The Polynomial Carathéodory - Féjer Approximation Method for Jordan Regions', IMA J. Numer. Anal. **3** (1983), 207 - 220.

6. ———— and E.B. SAFF, 'On Clenshaw's Method and a Generalization to Faber Series', presented at the ICCAM, Leuven, 1986.

7. L. FOX and I.B. PARKER , Chebyshev Polynomials in Numerical Analysis Oxford Univ. Press, 1968.

8. C. LANCZOS, 'Trigonometric Interpolation of Empirical and Analytical Functions', J. Math. Phys. **17**(1938), 123 - 199.

9. ————, 'Tables of Chebyshev Polynomials $S_n(x)$ and $C_n(x)$, Introduction', Nat. Bur. Standards Appl. Math. Ser. **9**, U.S. Govt. Printing Of., Whashington, 1952.

10. ————, Applied Analysis, Pitman, London, 1957.

11. ————,'Legendre Versus Chebyshev Polynomials', in Topics in Numerical Analysis, ed. by J.J.H. Miller, AP, London, 1973, 191 - 201.

12. E.L. ORTIZ and H. SAMARA, 'An Operational Approach to the Tau - Method for the Numerical Solution of Nonlinear Differential Equations, Computing **27**(1981), 15 - 25.

13. M.R. da SILVA, 'LACALGEBRA Versions of Lanczos' Tau - Method for the Numerical Solution of Differential Equations', Port. Math. **41** (1982), 295 - 316.

14. ————, 'A Quick Survey of Recent Developments and Applications of the τ - Method', in Numerical Approximation of Partial Differential Equations, ed. by E.L. Ortiz, North Holland, 1987, 297 - 308.

15. ————, 'Numerical Treatment of Differential Equations with the τ - Method, to appear in the JCAM, 1987.

EVALUATION OF FERMI–DIRAC INTEGRAL

S. Paszkowski
Polish Academy of Sciences
Institute of Low Temperature and Structure Research
Pl. Katedralny 1, P.O.Box 937, 50-950 Wrocław, Poland

ABSTRACT. The Fermi–Dirac integral is defined by

$$F_\mu(z) := \int_0^\infty \frac{x^\mu\,dx}{1+e^{x-z}} \qquad (\mu > -1).$$

Some exact expressions for F_μ are well known. Basing on them one obtains some new approximate expressions. Three cases are distinguished:

Case a ($\mu = 1, 2, \ldots, z \geq 0$). The non-polynomial part of $F_\mu(z)$ is expanded, after a suitable variable transformation, into Chebyshev series (Section 2).

Case b ($\mu = -\frac{1}{2}, \frac{1}{2}, \ldots,$ $z \geq 0$ is sufficiently small). $F_\mu(z)$ is expanded in powers of $x = 1 - (1+e^z)^{\frac{1}{2}}$. Then Padé approximation is used (Section 3).

Case c ($\mu = -\frac{1}{2}, \frac{1}{2}, \ldots, z \geq u^2$, with a sufficiently large u). $F_\mu(z)$ is expanded, at first, into a series containing the functions Erfi and Erfc and, after that, into Chebyshev series with the variable $u/\sqrt{z}$ (Section 4).

These methods seem to be more general and, from some points of view, better than the earlier ones.

1. INTRODUCTION

The *Fermi–Dirac integral* (FDI) with a real parameter $\mu > -1$ is defined by

$$F_\mu(z) := \int_0^\infty \frac{x^\mu\,dx}{1+e^{x-z}} \qquad (\mu > -1).$$

It appears in many applications associated with the Fermi–Dirac statistics, e.g. in the theory of semiconductors. FDI is used mainly for integer and half-integer values of μ. Only for $\mu = 0$ FDI is expressed in a finite form by elementary functions: $F_0(z) = \log(1+e^z)$. For all others μ some approximate formulas are necessary.

A. Cuyt (ed.), Nonlinear Numerical Methods and Rational Approximation, 435–444.

In the early period of applications many tables of FDI were published; cf. the references in [2]. Recently new approximation techniques are developed and it seems useful to apply them to FDI. We present here new results concerning the approximation of FDI for integer and half-integer μ and real nonnegative z.

2. FDI WITH AN INTEGER PARAMETER

For any integer $n \geq 0$ and for $z \geq 0$ the following formula holds [4]:

$$\begin{aligned} F_n(z) = &\frac{1}{n+1} z^{n+1} + \\ &+ \sum_{i=0}^{[(n-1)/2]} (-1)^i \binom{n}{2i+1} \frac{2^{2i+1}-1}{i+1} \pi^{2i+2} B_{2i+2} z^{n-2i-1} + \\ &+ (-1)^n n! R_n(e^{-z}), \end{aligned} \tag{2.1}$$

where

$$R_n(y) := \sum_{k=1}^{\infty} \frac{(-1)^{k-1}}{k^{n+1}} y^k, \tag{2.2}$$

and where B_k is the kth Bernoulli number. In particular,

$$\begin{aligned} F_1(z) &= \frac{1}{2} z^2 + \frac{1}{6}\pi^2 - R_1(e^{-z}), \\ F_2(z) &= \frac{1}{3} z^3 + \frac{1}{3}\pi^2 z + 2R_2(e^{-z}), \\ F_3(z) &= \frac{1}{4} z^4 + \frac{1}{2}\pi^2 z^2 + \frac{7}{60}\pi^4 - 6R_3(e^{-z}). \end{aligned}$$

The series (2.2) converges for any $y \in [-1, 1]$, i.e. (2.1) is applicable for every $z \geq 0$ (then $y \in [0, 1]$). If, however, $z \approx 0$ then the series $R_n(e^{-z})$ converges slowly, especially for small n. Therefore one should transform it as follows. As

$$R_1'(y) = \frac{1}{y} \log(1+y),$$

we introduce a new variable $x := \log(1+y)$ (then in (2.1) $x = \log(1+e^{-z})$) and a new function

$$r_1(x) := R_1(y).$$

One can verify that

$$r_1'(x) = \frac{x}{1-e^{-x}} = \sum_{k=0}^{\infty} \frac{(-1)^k B_k}{k!} x^k,$$

$$r_1(x) = \sum_{k=1}^{\infty} \frac{(-1)^{k-1} B_{k-1}}{k!} x^k$$
$$= x + \frac{1}{4}x^2 + \frac{1}{36}x^3 - \frac{1}{3600}x^5 + \frac{1}{211680}x^7 - \frac{1}{10886400}x^9 + \dots$$

This series converges for $|x| < 2\pi$. If $y \in [0,1]$, then $x \in [0, \log 2] \approx [0, 0.693]$. For these x the convergence is very fast.

For $n > 1$

$$R_n'(y) = \frac{1}{y} R_{n-1}(y).$$

Taking $r_n(x) := R_n(y)$ we obtain

$$r_n'(x) = \left(\sum_{k=0}^{\infty} \frac{(-1)^k B_k}{k!} x^k \right) \frac{r_{n-1}(x)}{x}, \qquad r_n(0) = 0,$$

which permits us to expand recursively $r_2, r_3, \dots$ in a power series. In particular,

$$r_2(x) = x + \frac{3}{8}x^2 + \frac{17}{216}x^3 + \frac{5}{576}x^4 + \frac{7}{54000}x^5 - \frac{7}{86400}x^6 + \dots,$$
$$r_3(x) = x + \frac{7}{16}x^2 + \frac{151}{1296}x^3 + \frac{137}{6912}x^4 + \frac{12493}{6480000}x^5 + \frac{161}{5184000}x^6 + \dots$$

For all series $r_1, r_2, \dots, r_5$ the sequence of coefficients converges to 0 with the same speed.

Table 1

k	r_{1k}	r_{2k}	r_{3k}
1	0.999999999998	1.000000000014	0.999999999998
2	0.250000000298	0.374999998087	0.437500000266
3	0.027777771070	0.078703747843	0.116512338967
4	-0.000277994092	0.000131199153	0.001927620665
5	0.000000408219	-0.000084323185	0.000031866381
6	0.000004401257	0.000000001710	-0.000016783199

k	r_{4k}	r_{5k}
0	0.999999999976	0.999999999997
1	0.468750003422	0.484375000432
2	0.139531814089	0.152413399516
3	0.028634459124	0.034365627481
4	0.004028936709	0.005717237503
5	0.000345754261	0.000682131020
6	-0.000001571204	0.000050005710

In order to speed up the evaluation of approximate values of R_n we introduce a new variable $u := x/\log 2$ which ranges the interval $[0,1]$ and we transform the power series for $r_n(x)/x$ into Chebyshev series in shifted polynomials $T_k^*(u)$. Its convergence is much faster than for the power series. In fact, e.g., the coefficient of u^{24} in the power series for $r_2(x)/x$ equals $1.6_{10}-24$ and the coefficient of $T_{15}^*(u)$ in the corresponding Chebyshev series equals $2.3_{10}-24$.

In order to obtain an expression approximating FDI with required accuracy it suffices to take a sufficiently large partial sum of Chebyshev series. In particular, for $n = 1,2,\ldots,5$ we can approximate the auxiliary functions $R_n(e^{-s})$ from (2.1), with an error not exceeding 10^{-11}, by the formula

$$R_n(e^{-s}) \approx x\sum_{k=0}^{6} r_{nk}x^k \qquad (x := \log(1+e^{-s}))\,.$$

The coefficients r_{nk} are given in Table 1.

3. FDI WITH A HALF-INTEGER PARAMETER FOR SMALL x

The evaluation of $F_\mu(x)$ for a non-integer μ is more difficult than for an integer μ. In particular, it is rather necessary to divide the interval $[0,\infty)$ of x into two parts at least and to use in each of them a separate approximation method. In this section we consider the case of half-integer μ and of nonnegative and positive, sufficiently small x.

We start from a formula [6] which is true only for $x \le 0$:

$$F_\mu(x) = \Gamma(\mu+1)R_\mu(e^x). \tag{3.1}$$

In fact, it contains the series

$$\sum_{k=1}^{\infty}\frac{(-1)^{k-1}}{k^{\mu+1}}e^{kx}$$

(cf. (2.2)) which diverges for $x>0$. One can, however, transform it so that its new form be useful even for small positive x. The transformation method is suitable for every half-integer μ. It suffices to explain it for $\mu=\frac{1}{2}$.

The formulae (3.1), (2.2) imply

$$F_{1/2}(x) = \frac{1}{2}\sqrt{\pi}e^x\sum_{k=0}^{\infty}\frac{(-1)^k}{(k+1)^{3/2}}y^k \quad \text{where} \quad y := e^x. \tag{3.2}$$

Taking into account asymptotic behaviour of the series coefficients for $k\to\infty$, the same as for the series $\sqrt{1+y}$, we introduce the new variable

$$z := 1-\sqrt{1+y}.$$

Hence $y = z^2 - 2z$ and

$$F_{1/2}(x) = \frac{1}{2}\sqrt{\pi}e^x\varphi(z), \tag{3.3}$$

where an auxiliary function φ is defined by

$$\varphi(z) := \sum_{k=0}^{\infty}(k+1)^{-3/2}(2z - z^2)^k = \sum_{k=0}^{\infty} f_k z^k, \tag{3.4}$$

Obviously,

$$f_k = \sum_{j=[(k+1)/2]}^{k} (-1)^{k-j}2^{2j-k}\binom{j}{k-j}(j+1)^{-3/2} \qquad (k = 0, 1, \ldots).$$

The coefficients of the series (3.2) tend to 0 very slowly, whereas for (3.4) the relation $f_k/f_{k+1} \to 2$ can be proved. Then the series (3.4) converges at least for $z \in (-2, 2)$. Since for y growing from -1 to 8 the variable z decreases from 1 to -2, one can expect that the series (3.4) allows us to express, through (3.3), the function $F_{1/2}(x)$ for all $y \in (-1, 8)$, i.e. for $x \in (-\infty, \log 8) \approx (-\infty, 2.08)$.

Table 2

k	f_k	k	f_k
0	1	10	0.002375007720362467
1	0.7071067811865475	11	0.001210915382154220
2	0.4162469683262273	12	0.0006161226781448234
3	0.2301996410804990	13	0.0003129543525702463
4	0.1235335953297407	14	0.0001587358022046199
5	0.06515720460753861	15	0.00008041603204965210
6	0.03398990068027349	16	0.00004069667501215044
7	0.01759749081467424	17	0.00002057709772421416
8	0.009060989343989115	18	0.00001039603268983052
9	0.004646366914321636	19	0.000005248683683361831

One should remark, however, that the f_k's are linear combinations, with great coefficients, of the numbers $(j+1)^{-3/2}$. Then it is evident that the evaluation of the f_k's causes a very considerable loss of accuracy. Therefore, in order to obtain the f_k's with a fixed relative accuracy (e.g., with 16 significant decimal digits) one should use much more accurate values of the roots $(j+1)^{-3/2}$ (even up to 40 digits after the decimal point).

The values of the f_k's, evaluated up to 16 significant digits, are given in Table 2.

To the series (3.4) with these f_k's Padé approximation was applied. In order to examine its utility, the values of some Padé approximants were explicitly calculated. One can expect that all the digits given in Table 3 are exact.

Table 3

z	Approximant	Approx. value of $F_{1/2}(z)$
0.0	[7/7]	0.678093895153101
0.5	[7/7]	0.99020924871280
1.0	[9/9]	1.396375280666564
1.5	[9/9]	1.90083346106439
2.0	[9/9]	2.50245782601
2.5	[9/9]	3.1965986994
3.0	[9/9]	3.97698535
3.5	[9/9]	4.8370639
4.0	[9/9]	5.770727

The approximate values of $F_{1/2}(z)$ were additionally compared with the tables of [1]. The numerical results seem to prove that Padé approximants are useful even beyond the convergence interval of (3.4), up to, say, $z = 3$ or (if the required accuracy is not very high) $z = 4$.

4. FDI WITH A HALF-INTEGER PARAMETER FOR GREAT z

Let m be natural. We transform $F_{m-1/2}(z)$ as follows:

$$\begin{aligned} F_{m-1/2}(z) &= \int_0^z \frac{x^{m-1/2}dx}{1+e^{x-z}} + \int_z^\infty \frac{x^{m-1/2}dx}{1+e^{x-z}} \\ &= \int_0^z x^{m-1/2}dx + \\ &\quad + \sum_{k=1}^{\infty}(-1)^k \left[e^{-kz}\int_0^z x^{m-1/2}e^{kx}dx - e^{kz}\int_z^\infty x^{m-1/2}e^{-kx}dx\right]. \end{aligned}$$

Substitute $y = \sqrt{kx}$ under the integral sign:

$$F_{m-1/2}(z) = \frac{2}{2m+1}z^{m+1/2} + 2\sum_{k=1}^{\infty}\frac{(-1)^k}{k^{m+1/2}}G_{m,-1}(\sqrt{kz}),$$

where

$$G_{ms}(t) := e^{-t^2}\int_0^t y^{2m}e^{y^2}dy + se^{t^2}\int_t^\infty y^{2m}e^{-y^2}dy \qquad (m = 0,1,\ldots;\ \ s = \pm 1).$$

Using the recurrence formula

$$G_{ms}(t) = \frac{s+1}{2} t^{2m-1} - \frac{2m-1}{2} G_{m-1,-s}(t)$$

one can verify that

$$G_{m,-1}(t) = - \sum_{j=0}^{[m/2-1]} (m - \frac{1}{2} - 2j)_{2j+1} t^{2m-4j-3} + (-1)^m (\frac{1}{2})_m G_{0,(-1)^{m-1}}(t).$$

Finally, we obtain

$$F_{m-1/2}(z) = - \frac{4}{2m+1} \sum_{j=0}^{[m/2]} (-1)^j (m + \frac{3}{2} - 2j)_{2j} (2^{2j-1} - 1) \frac{\pi^{2j} B_{2j}}{(2j)!} z^{m+1/2-2j} +$$
$$+ 2(-1)^{m-1} \left(\frac{1}{2}\right)_m \sum_{k=1}^{\infty} \frac{(-1)^{k-1}}{k^{m+1/2}} G_{0,(-1)^{m-1}}\left(\sqrt{kz}\right).$$

The first term of the right-hand side coincides with Sommerfeld expansion of $F_{m-1/2}(z)$ (cf.,e.g., [3]). It suffices to find an evaluation method for the second term. It expresses itself by the two error functions:

$$G_{0,\pm 1}(t) = e^{-t^2} \mathrm{Erfi}(t) \pm e^{t^2} \mathrm{Erfc}(t).$$

For great z, taking into account asymptotic properties of the functions Erfi and Erfc, we introduce auxiliary functions

$$\mathrm{Bi}_q(y) := x e^{-x^2} \mathrm{Erfi}(x), \quad \mathrm{Bc}_q(y) := x e^{x^2} \mathrm{Erfc}(x) \qquad \text{where } x := q/y;$$

q is a fixed positive number.

Obviously, for each sequence of positive q_k's,

$$\sum_{k=1}^{\infty} \frac{(-1)^{k-1}}{k^{m+1/2}} G_{0,(-1)^{m-1}}\left(\sqrt{kz}\right)$$
$$= \sum_{k=1}^{\infty} \frac{(-1)^{k-1}}{k^{m+1/2}} \left[e^{-kz} \mathrm{Erfi}\left(\sqrt{kz}\right) + (-1)^{m-1} e^{kz} \mathrm{Erfc}\left(\sqrt{kz}\right) \right]$$
$$= \frac{1}{\sqrt{z}} \sum_{k=1}^{\infty} \frac{(-1)^{k-1}}{k^{m+1}} \left[\mathrm{Bi}_{q_k}\left(\frac{q_k}{\sqrt{kz}}\right) + (-1)^{m-1} \mathrm{Bc}_{q_k}\left(\frac{q_k}{\sqrt{kz}}\right) \right].$$

To simplify this expression we take $q_k := \sqrt{k}u$, where u is a fixed positive number. Then

$$\sum_{k=1}^{\infty} \frac{(-1)^{k-1}}{k^{m+1/2}} G_{0,(-1)^{m-1}}(\sqrt{kz})$$
$$= \frac{1}{\sqrt{z}} \sum_{k=1}^{\infty} \frac{(-1)^{k-1}}{k^{m+1}} \left[\mathrm{Bi}_{\sqrt{k}u}\left(\frac{u}{\sqrt{z}}\right) + (-1)^{m-1} \mathrm{Bc}_{\sqrt{k}u}\left(\frac{u}{\sqrt{z}}\right) \right]. \tag{4.1}$$

One can easily prove that Bi_q, Bc_q satisfy the differential equation

$$y^3 B'(y) + (y^2 + 2r)B(y) = r, \tag{4.2}$$

for $r := -q^2$ and $r := q^2$, respectively. In both cases this particular solution of (4.2) expands in $[-1, 1]$ into even Chebyshev series:

$$B(t) = {\sum_{j=0}^{\infty}}' b_j T_{2j}(t).$$

A standard method (cf., e.g., [5]), permits us to obtain from (4.2) a linear recurrence relation satisfied by the b_j's:

$$\begin{aligned} (2j+1)b_j + (6j+7+8r)b_{j+1} + (6j+11-8r)b_{j+2} + (2j+5)b_{j+3} & \\ = 8r \qquad (j = -1), & \qquad (4.3) \\ = 0 \qquad (j = 0, 1, \ldots). & \end{aligned}$$

Applying to (4.3) a backward recursion together with a Clenshaw suggestion (cf. [7], Section 4.7) one can evaluate the Chebyshev coefficients b_j.

In view of (4.1) two sequences of Chebyshev series are necessary. We expand, namely, the functions $\mathrm{Bi}_{\sqrt{k}u}$ and $\mathrm{Bc}_{\sqrt{k}u}$ for $k = 1, 2, \ldots$:

$$\mathrm{Bi}_{\sqrt{k}u}(x) = {\sum_{j=0}^{\infty}}' \mathrm{bi}_j(k) T_{2j}(x),$$
$$\mathrm{Bc}_{\sqrt{k}u}(x) = {\sum_{j=0}^{\infty}}' \mathrm{bc}_j(k) T_{2j}(x)$$

(the coefficients $\mathrm{bi}_j(k)$, $\mathrm{bc}_j(k)$ depend, of course, on u). Then

$$\sum_{k=1}^{\infty} \frac{(-1)^{k-1}}{k^{m+1/2}} G_{0,(-1)^{m-1}}(\sqrt{kz})$$
$$= \frac{1}{\sqrt{z}} {\sum_{j=0}^{\infty}}' \left[\sum_{k=1}^{\infty} \frac{(-1)^{k-1}}{k^{m+1}} \left(\mathrm{bi}_j(k) + (-1)^{m-1}\mathrm{bc}_j(k)\right) \right] T_{2j}\left(\frac{u}{\sqrt{z}}\right). \tag{4.4}$$

One can verify that

$$\mathrm{bi}_0(k) - 1 \sim c_0 k^{-1}, \quad \mathrm{bi}_j(k) \sim c_j k^{-j} \text{ for } j > 0 \qquad (k > 0;\ c_j \neq 0).$$

Analogous relations hold for the $\mathrm{bc}_j(k)$'s. Then the inner sums in the right-hand side of (4.4) can be calculated with the aid of a suitable acceleration method.

Table 4

j	a_j	j	a_j
0	0.843009021709026		
1	0.027424562083457	21	-0.000000001038832
2	0.006250637958580	22	0.000000001283979
3	-0.001283901704348	23	-0.000000000474767
4	-0.000720377017976	24	-0.000000000101217
5	0.000023305822313	25	0.000000000218766
6	0.000101707840697	26	-0.000000000114607
7	-0.000007102448003	27	0.000000000006820
8	-0.000016464450902	28	0.000000000033657
9	0.000003966790831	29	-0.000000000026621
10	0.000002403585618	30	0.000000000007995
11	-0.000001475532245	31	0.000000000003300
12	-0.000000076682581	32	-0.000000000005346
13	0.000000370753200	33	0.000000000002971
14	-0.000000124752673	34	-0.000000000000410
15	-0.000000041121734	35	-0.000000000000757
16	0.000000051138894	36	0.000000000000766
17	-0.000000013502508	37	-0.000000000000348
18	-0.000000007305423		
19	0.000000007727860		
20	-0.000000002175615		

The procedure described above gives, among others, the following expansion of $F_{1/2}$:

$$F_{1/2}(x) = \frac{2}{3}x\sqrt{x} + \frac{1}{\sqrt{x}} \sum_{j=0}^{\infty} a_j T_{2j}\left(\frac{2}{\sqrt{x}}\right) \qquad (x \geq 4).$$

The coefficients a_j are given in Table 4. They permit to calculate $F_{1/2}$ with a relative error not exceeding 10^{-11}.

References

[1] A.C.Beer, M.N.Chase, P.F.Choquard, Helv.Phys.Acta **28**(1955), 529-541.

[2] J.S.Blakemore, *Approximations for Fermi–Dirac integrals, especially the function $\mathcal{F}_{1/2}(\eta)$ used to describe electron density in a semiconductor*, Solid State Electron. **25**(1982), 1067-1076.

[3] W.J.Cody, H.C.Thacher, Jr., *Rational Chebyshev approximations for Fermi–Dirac integrals of orders $-\frac{1}{2}$, $\frac{1}{2}$ and $\frac{3}{2}$*, Math.Comput. **21**(1967), 30-40.

[4] R.B.Dingle, *The Fermi–Dirac integrals $F_p(n) = (p!)^{-1} \int_0^\infty \epsilon^p(e^{\epsilon-n}+1)^{-1}d\epsilon$*, Appl.Sci.Res.Ser.B **6**(1957), 225-239.

[5] L.Fox, I.B.Parker, *Chebyshev polynomials in numerical analysis*, London 1968.

[6] J.Mc.Dougall, E.C.Stoner, *The computation of Fermi–Dirac functions*, Philos.Trans.Roy.Soc. London Ser.A **237**(1939), 67-104.

[7] J.Wimp, *Computation with recurrence relations*, Boston 1984.

AN APPLICATION OF OPERATOR PADE APPROXIMANTS TO MULTIREGGEON PROCESSES

M. Pindor
Institute of Theoretical Physics
Warsaw University
00-681 Warsaw
ul Hoza 69
Poland

ABSTRACT. A series arising in a multireggeon description of scattering processes in Yang Mills theory, is considered. It is shown that after suitable transformation of integrands, it can be treated as a matrix element of a simple series of integral operators. This series resembles Neuman series and can be summed by [0/1] Pade Approximant. Direct solution of the integral equation leading to the series is however numerically difficult and unstable. The direct method is compared against the variational method of the calculation of matrix elements of OPA.

1. INTRODUCTION

The aim of this talk is to present an application of the operator Pade method to a problem arising in a theory of high energy elementary processes when they are described in a framework of the nonabelian gauge theory.

This approach has been studied in particular by L.N. Lipatov and coworkers [1] and they have arrived, using, so called, multireggeon formulation, at the following expression for the forward scattering amplitude continued to complex angular momenta in the crossed channel (apart from uninteresting factors):

$$F_g^T = \sum_{n=0}^{\infty} g^n \int K^T(q_1,q_2)\ldots K^T(q_n,q_{n+1}) \cdot \prod_{i=1}^{n+1} \frac{d^2q_i}{(q_i^2+m^2)^2[1+2ga(q_i^2)]} \tag{1}$$

(integration over d^2q_i runs through the whole plane)

A. Cuyt (ed.), Nonlinear Numerical Methods and Rational Approximation, 445–452.

$$K^T(q_1,q_2) = A_T + 2C_T\frac{(q_1^2+m^2)(q_2^2+m^2)}{(q_1-q_2)^2+m^2}$$

$$A_T=(-\frac{3}{2}C_T+1/2)m^2, \qquad C_T=2-T(T+1)/2 \qquad T=0,1,2 \tag{2}$$

$$a(q^2) = \frac{\pi(q^2+m^2)}{m^2(z^2-1)^{1/2}}\log[z+(z^2-1)^{1/2}]$$

$$z = 1+\frac{q^2}{2m^2} \tag{3}$$

Series (1) is not just power series in g, because g appears also in denominators of integrands. Howver, we can observe that all terms of the series have branch point for g=0 (because $a(q_1^2)$ increases to infinity when q_1^2 approaches infinity), so we can infer that also F_g^T has branch point for g=0. Therefore (1) should rather be considered as a formal series and to calculate a function defined by it we introduce:

$$W^T(q_1,q_2) = \frac{K^T(q_1,q_2)}{(q_1^2+m^2)(q_2^2+m^2)[1+2ga(q_1^2)]} \tag{4}$$

$$L(q) = \frac{1}{q^2+m^2}, \qquad L'(q) = \frac{1}{(q^2+m^2)[1+2ga(q^2)]} \tag{5}$$

and then we write the series (1) as:

$$F_g^T = \sum_{n=0}^{\infty}(g^nW^TL',L) \tag{6}$$

where we use the same symbol W^T for the integral operator in L^2 space and for its kernel, and obviously:

$$(f,g) = \int f(q)g(q)^* d^2q$$

Now we can sum the series (6) by [0/1] Pade approximant and we have:

$$F_g^T = (\frac{1}{I-gW^T}L', L) \tag{7}$$

and we shall compare below two methods of calculating this expression.

2. THE GAUSSIAN METHOD

We immediately see that:

$$F_g^T = \int L(q)G_g(q)d^2q \tag{8}$$

where $G_g(q)$ is the solution of the integral equation:

$$G_g(q) = L'(q) + g\int W^T(q,k)G_g(k)d^2k \tag{8a}$$

For T=1 we can guess the solution:

$$G_g(q) = \frac{1}{q^2+m^2} \cdot \frac{1}{1+\pi g} \tag{9}$$

and find F_g^1:

$$F_g^1 = \frac{1}{m^2(1+\pi g)} \tag{10}$$

For T=0,2 we also observe that G is a function of q^2 only, and after angular integration we find:

$$G_g(q^2) = \frac{1}{(q^2+m^2)[1+2ga(q^2)]} + \frac{1}{(q^2+m^2)[1+2ga(q^2)]} \cdot$$

$$\int [A^T + \frac{2C_T(q^2+m^2)(x+m^2)}{(x+q^2+m^2)^2-4q^2x}] \frac{G_g(x)}{x+m^2}dx \tag{11}$$

Replacing integral by the Gaussian quadrature rule at q^2's equal to the Gaussian nodes we can approximately calculate F_g^T. Unfortunately this process can be only very slowly convergent.

For example for T=1, g=-i we have the following numbers (the exact value is .2890+.9080i):

n	F_g^1
8	.1151+ .4432i
16	.2275+ .7979i
32	.3020+ .9211i
48	.4084+1.0198i
2*48	.2438+ .8155i
3*48	.2713+ .8770i

n*48 means that we divide (0,1) into n segments and apply 48-point Gaussian rule to each segment. Observe that 3*48 means inverting 144*144 complex, double precision matrix (occupying 331,776 bytes of storage), and therefore there is not much hope to achieve decent accuracy within reasonable amount of memory.

For T=0 we do not know an exact result but some approximate analysis of series (1) suggests that F_g^0 has a branch point (of the square root type) at $g_0=1/(16\pi)\log(2)\approx .0287$ [1]. Calculation of F^0 at g=.0284 in a way presented above gives the following numbers:

n	F^0
48	3.722
2*48	4.326
3*48	4.719
3*32	4.033
4*32	4.247
5*32	4.411

Numbers presented above justify a discussion of another method of calculating (7), what is the main point of the talk.

5. METHOD OF THE VARIATIONAL GRADIENT

The method is based on the concept of calculating the inverse of $I-gW^T$ in a finite dimensional subspace of L^2 and then changing this subspace so as to obtain the best approximation within the given dimension of the subspace [2,3].

In other words we replace (7) by:

$$F_g^T(P) = (P\frac{1}{P(I-gW^T)P}PL',L) \tag{12}$$

where P projects on our selected subspace, and then we change P while $\dim PL^2$=const, so as to minimize:

$$\left| F_g^T - F_g^T(P) \right|$$

For this method to be practical we must have a possibility to calculate easily matrix elements of K^T in the basis spanning PL^2: $f_n(x)$. This would, in most cases, need numerical integrations, making calculation of $[P(I-gW^T)P]^{-1}$ very slow. Therefore we use the trick common in theoretical physics. We shall take namely this basis to be of the form: $f_n(x)=\delta(x-x_n)$ - in such a case $(Wf_n,f_m)=W(x_n,x_m)$. Of course these f_n's do not belong to L^2, but we can consider this procedure as a limit $e \to 0$, while we use $f_n(x,e)=\delta_e(x-x_n)$ - model delta functions. We do not discuss this point any longer and simply assume, as is usually done in theoretical physics, that such limit exists.

However, now, a new difficulty arises: "matrix elements" of I in this "basis" do not exist. Therefore we transform (7) to:

$$F_g^T = (M\frac{1}{M-gW^TM}L',L) \tag{13}$$

For this relation to hold, it is sufficient that M is bounded and its inverse exists (even if unbounded), with $\frac{1}{I-gK}L'$ belonging to a domain of the inverse. We also assume that g is such that $\frac{1}{M-gKM}$ exists - if this was not the case, results would be numerically completely unstable with respect to $\dim PL^2$.

It was shown [2,3] that if we vary P, then one of the stationary points of $F_g^T(P)$ is just F_g^T. This stationary point is reached when both vectors:

$$h_r = [I-(M-gW^TM)P\frac{1}{P(M-gW^TM)P}P]L' \tag{14a}$$

and

$$h_l = [I-(M^+-g^*M^+W^{T+})P\frac{1}{P(M^+-g^*M^+W^{T+})P}P]M^+L \tag{14b}$$

(which could be called variational gradients of $F_g^T(P)$) vanish.

One can also easily find that:

$$\left|F_g^T - F_g^T(P)\right| = \left|(h_r,\frac{1}{(M-gW^TM)^+}M^+L)\right|$$

$$\|h_r\|\cdot\left\|\frac{1}{(M-gW^TM)^+}M^+L\right\|$$

$$= \left|(h_l,\frac{1}{M-gW^TM}L')\right|$$

$$\|h_l\|\cdot\left\|\frac{1}{M-gW^TM}L'\right\|$$

therefore, for $F_g^T(P) = F_g^T$ it is sufficient that any of the vectors h_r or h_l vanishes.

Our strategy, then, is to minimize $\|h_l\|\cdot\|h_r\|$ with respect to P, i.e. with respect to x_n's (we recall that our subspace is "spanned" by $\delta(x-x_n)$ n=1,...N). Though we cannot calculate neither $\|1/(M-gW^TM)L'\|$ nor $\|1/(M-gW^TM)^+ML\|$ we can have a "feeling" how far we are from $h_r=0$ or $h_l=0$ by comparing $\|h_r\|$ with $\|L'\|$ and $\|h_l\|$ with $\|M^+L\|$.

We used:

$$M(x,y) = \frac{1}{(x+y+2m^2)\,\mathrm{sqrt}(y+m^2)} \tag{15}$$

- such an operator M is bounded and allows us to calculate analytically $(W^T M)(x,y)$ what made calculations much faster.

For T=1 g=-i we obtained:

N	$\frac{\|h_r\|\cdot\|h_l\|}{\|M^+L\|\cdot\|L'\|}$	F^1
2	.40e-1	.2118+.7112i
3	.66e-2	.2637+.8344i
4	.12e-2	.3007+.9250i
5	.20e-3	.2942+.9162i
6	.56e-4	.2914+.9134i
7	.14e-4	.2903+.9101i

For T=0 g=.0284 we got:

N	$\frac{\|h_r\|\cdot\|h_l\|}{\|M^+L\|\cdot\|L'\|}$	F^0
2	.71e-2	4.372
3	.14e-2	4.817
4	.36e-3	5.005
5	.11e-3	5.116
6	.33e-4	5.229
7	.90e-5	5.226

4. CONCLUSIONS

We see that optimizing, via a nonlinear variational principle, a choice of the finite dimensional subspace in which $M-gW^TM$ acts we can achieve quite accurate values of $((M-gW^TM)^{-1}ML',L)$ with amazingly low dimension of such subspace.

Work supported in part by Polish Ministry of Higher Education and Technology Project CPBP 01.03

1. E.A. Kurayev, L.N. Lipatov, V.S. Fadin, "Pomeranchuk singularity in nonabelian gauge theories", (in Russian) JETP 22 (1977), 377-389
see also:
P. Rączka, R. Rączka, "On the rising cross section and the new approach to high energy hadron-hadron scattering", preprint ISAS 98/86/EP
2. J. Fleischer, M. Pindor, "Evaluation of operator Pade approximants for perturbative expansions in scattering theory", Physical Review 24 (1981), 1978-1986
3. M.Pindor, G. Turchetti, "Pade approximants and Variational Series for Operator Series", Il Nuovo Cimento 71A (1982), 171-186

INDEX

Q

R

S